Principles of Plasma Mechanics

Fourth Edition

Bishwanath Chakraborty
Formerly-Professor
Department of Mathematics
Jadavpur University, Kolkata

NEW AGE INTERNATIONAL (P) LIMITED, PUBLISHERS
New Delhi • Bangalore • Chennai • Guwahati
Hyderabad • Kolkata • Lucknow • Mumbai

Fourth Edition 2003

NEW AGE INTERNATIONAL (P) LIMITED, PUBLISHERS
4835/24, Ansari Road, Daryaganj,
New Delhi - 110 002

Offices at:
Bangalore, Chennai, Guwahati, Hyderabad, Kolkata,
Lucknow and Mumbai

ISBN : 81-224-1446-X

Published by K.K. Gupta for New Age International (P) Ltd., 4835/24, Ansari Road, Daryaganj, New Delhi-110 002 and printed in India at Taj Press, New Delhi-110 064.

To

The memory of my
father Gopal Chandra Chakraborty
and
mother Urmila Chakraborty

To

The memory of my

father Gopal Chandra Chakravorty

and

mother Urmila Chakraborty

PREFACE TO THE FOURTH EDITION

This edition is intended to be suitable for students (new learners) of physic science and engineering, knowing vector calculus and the basic laws of physic.

All those topics and items of the third edition which are not necessary for students have been dropped. Also, some new topics, useful for students have been included.

BISHWANATH CHAKRABORTY

CONTENTS

Chapter 1

BASIC ELECTRICITY AND MAGNETISM

1.1 INTRODUCTION

For convenience of study of plasma mechanics, here we develop the fundamental laws of electricity and magnetism and the basic equations of Maxwell, following from these laws in terms of charges, currents and intensities of electric field and magnetic field. Emphasis is given on those topics and on the mode of their presentation, which are most helpful for this purpose.

1.2 DIELECTRICS, CONDUCTORS, CHARGES, CURRENTS, DIPOLES, MAGNETIZATION

In electricity and magnetism the primary sources for fields are charges and currents, and secondary sources are polarizable and magnetizable matter. Charges in motion give rise to electric currents. Polarization is due to small displacement of charges of electric dipoles by fields. An electric dipole is formed by two charges of opposite sign, and bound together by Coulomb attraction. Magnetic poles cannot be isolated like charges. Magnetic dipoles are combinations of two equal magnetic monopoles of opposite polarity, and are formed by Coulomb like forces binding these poles. Magnetic dipoles and currents of electric charges cause magnetization in materials.

Motion of charges depends on their thermal state in a body, but their average motion is along the direction of the applied electric field. Electric current flows along the direction of motion of positive charges, though this electric current is due to electron motion.

Considering only the electric and magnetic properties, all material bodies are either dielectrics, or electrical conductors, or insulators. A dielectric material (glass, water, crystals) is an aggregate of electric dipoles. An applied electric field induces a dipole moment in molecules that have no dipole moment in the absence of the field, or reorients molecules that have permanent dipole moment. Since the electron mass is much smaller than the ionic mass, negative charges are displaced more by applied fields than ions. Matter may be polarized thus.

In conductors of electricity (copper, gold, aluminium, silver) charges are free to move under the action of applied electric field through distances large compared to atomic dimensions. This movement occurs even when the electric field is small. Electricity does not pass through insulators (Porcelain, Wood); so these are affected very little by applied fields.

Theory of electromagnetism, up to the times of Faraday and Maxwell (beginning of twentieth century), was purely phenomenological; an example is Ohm's law. The effects of electric and magnetic fields on materials were considered in terms of deformation of elastic media, which does not require the use of the concept of particle motion in presence of fields. This macroscopic theory, therefore gave birth to concepts of dielectrics and the material polarization effects. Concept of discrete charges of both sign, and the analytical microscopic (classical) theory of electrodynamics of Lorentz were very significant later developments.

The basic postulates for classical electromagnetics, based on a wide variety of experiments, are : (1) There exists two kinds of electric charges: a positive charge and a negative charge; (2) All charges are integral multiples of the electronic negative charge $-e$; (3) Electric charge is conserved in any isolated system or in the whole of universe; (4) An electric charge in motion may be acted on by a force independent of its velocity, and also by a force proportional to its velocity and directed at right angles to it; (5) Magnetic poles occur in pairs, so magnetic monopoles do not exist and the current of motion of magnetic monopoles vanishes.

Coulomb's law states that two charges of opposite signs attract and of same sign repel each other according to the law of inverse square of the distance. When charges are in motion their currents give rise to magnetic field. The smallest unit of positive charge is the charge of the hydrogen nucleus. Its mass is about 1837 times that of an electron, but the charge magnitude is the same. The electronic mass is 9.2×10^{-28} gm, and charge is 4.8×10^{-10} esu.

If e is the charge of a particle of mass m and $\mathbf{r}$ is its position vector at time t, and $\mathbf{v}\ (= \dot{\mathbf{r}})$ is its velocity, then the electric current of this charge is $e\mathbf{v}$. For N discrete charges, the electric current is the sum $\sum_{i=1}^{N} (e_i\ \dot{\mathbf{r}}_i)$, where e_i is the ith charge and $\mathbf{r}_i$ is its position vector at time t. If V is the volume containing these charges, then the density of charge, ρ^c, and the corresponding current, $\mathbf{j}$, are defined as $\left(\sum_{i=1}^{N} e_i\right)/V$ and $\left(\sum_{i=1}^{N} e_i\ \mathbf{v}_i\right)/V$, respectively. Using concepts of delta function $\delta(x)$ of Dirac (Appendix 1), these two quantities are defined as

$$\rho^c = \frac{1}{V} \sum_{i=1}^{N} [e_i\ \delta(\mathbf{r} - \mathbf{r}_i(t))], \tag{1.2.1}$$

$$\mathbf{j} = \frac{1}{V} \sum_{i=1}^{N} [e_i \, \mathbf{v}_i \, \delta (\mathbf{r} - \mathbf{r}_i (t))] \tag{1.2.2}$$

When a number of species of charged particles is present in a volume V, for the ith species, let $\mathbf{v}_i$, m_i and e_i be the velocity, mass and charge, respectively, per particle; and let N_i be the number density. Then for the ith species the density of charge, mass and current, respectively, are $(N_i \, e_i)$, $(N_i \, m_i)$, and $(N_i \, e_i \, \mathbf{v}_i)$; the average density of mass, charge and current in V being $\sum_{i=1}^{N} (N_i \, M_i)$, $\sum_{i=1}^{N} (N_i \, e_i)$ and $\sum_{i=1}^{N} (N_i \, e_i \, \mathbf{v}_i)$, respectively, where N is the number of species present in V. For a continuous distribution of all species of particles in V, these sums are replaced by the integrals

$$\sum_{i=1}^{N} \iiint_{v} (N_i \, m_i) \, dV, \sum_{i=1}^{N} \iiint_{v} (N_i \, e_i) \, dV, \sum_{i=1}^{N} \iiint_{v} (N_i \, e_i \, \mathbf{v}_i) \, dV$$

In a fully ionized hydrogen plasma, there are only free electrons, each of charge $-e$, and fully stripped ions, each of charge e, so $N = 2$.

The volume current I is defined as the time rate of flow of electric charge along a given direction in units of Coulomb per second or Ampere. Hence, if dq is the charge gain in an elementary volume dV, in time dt,

$$I = \frac{dq}{dt}, \; \mathbf{j} = \lim_{\delta \Sigma \to 0} \frac{\delta I}{\delta \Sigma},$$

where $\mathbf{j}$ is the surface current. Then

$$dI = (\mathbf{j} \, . \, d\,\boldsymbol{\Sigma}), \text{ so that } j_x = \frac{\partial I}{\partial \Sigma_x}, \text{ etc.}$$

where the surface element $\delta\Sigma_x$ is the projection of $\delta\boldsymbol{\Sigma}$ normal to the x-axis. The current passing through surface Σ is then given by

$$I = \iint_{\Sigma} (\mathbf{j} \, . \, d\boldsymbol{\Sigma}) = \iint_{\Sigma} j \cos (\hat{\mathbf{j}, \mathbf{n}}) \, d\,\Sigma = \iint_{\Sigma} j_n \, d\Sigma$$

where $(\hat{\mathbf{j}, \mathbf{n}})$ is the cosine of the angle between the surface current unit vector $\hat{\mathbf{j}}$ and the (unit vector $\mathbf{n}$ along normal to $\delta\boldsymbol{\Sigma}$; so $\delta\boldsymbol{\Sigma} = \mathbf{n}\,\delta\Sigma$. If l is the distance separating an electron from an ion, the electric dipole moment of this system is le. For a continuous distribution of charges of both sign, if the position vector of static ions is $\mathbf{R}$, and ρ^c is the density of electrons at $\mathbf{R} + \mathbf{r}$, the dipole moment vector per unit volume of the system, at $\mathbf{R}$, is $\rho^c \, \mathbf{r}$, and the total dipole moment in a volume V is the integral $\iiint_{v} \rho^c \, \mathbf{r} \; dV$.

Since, in dielectrics, applied electric fields cause a redistribution of the charges, without changing the total charge, we can write

$$\iiint_V \rho_i^c \, dV = 0$$

where the scalar quantity ρ_i^c is the average charge density of a body, and V is the dielectric volume. ρ_i^c is expressed as the divergence of a vector **P**; so

$$\rho_i^c = -\operatorname{div} \mathbf{P}, \iiint_V \rho_i^c \, dV = 0 = -\iiint_V \operatorname{div} \mathbf{P} \, dV \tag{1.2.3}$$

P is called the polarization vector. By Gauss's theorem (Appendix 2)

$$0 = -\iint_\Sigma P_N \, d\Sigma$$

where P_N is the normal component of **P** at a point of the surface Σ; this relation being valid for any surface, we put $P_N = 0$ for every point of Σ. Integrating the dipole density ρ_i^c **r**, over the whole volume V, we find that

$$\iiint_V \mathbf{r} \, \rho_i^c \, dV = -\iiint_V \mathbf{r} \operatorname{div} \mathbf{P} \, dV = -\iint_\Sigma \mathbf{r} \, (\mathbf{P} \, . \, d\mathbf{\Sigma}) + \iiint_V \mathbf{P} \, dV$$

Since $P_N = 0$ on Σ, we get $\iiint_V \mathbf{r} \, \rho_i^c \, dV = \iiint_V \mathbf{P} dV.$

Hence the polarization vector **P** is the dipole moment per unit volume. If, in addition, charges not belonging to a dielectric are brought from outside V, and their density is ρ_e^c, then the total electric charge density is $\rho^c = \rho_e^c + \rho_i^c$.

The angular momentum $\mathbf{M}_0$ and the magnetic moment **m** of a system are defined by

$$\mathbf{M}_0 = \sum (\mathbf{r} \times \mathbf{p}) \tag{1.2.4}$$

$$\mathbf{m} = \frac{1}{2c} \sum e \, (\mathbf{r} \times \mathbf{v}) \tag{1.2.5}$$

where **r** is the radius vector with respect to a given origin, **p** is the momentum of the particle, **v** is its velocity, e is its charge, and c is the vacuum speed of light, the summation notation Σ means summation over all the particles of the system.*

The mean value of the microscopic current density **j** $(= \rho^c \, \overline{\mathbf{v}})$ is in general, not zero either in dielectrics or conductors. However, in dielectrics, the surface integral of $\rho^c \, \overline{\mathbf{v}}$, taken over the area of any cross-section of the body, vanishes; so, following the preceding analysis, we can write

---

* The summation sign Σ should not be confused with the sign $\mathbf{\Sigma}$ for surfaces, and $d\mathbf{\Sigma}$ for surface elements.

$$\iint_{\Sigma} (\mathbf{j} \,.\, d\boldsymbol{\Sigma}) = 0 \tag{1.2.6}$$

For conductors, putting

$$\mathbf{j} = c \text{ curl } \mathbf{M}, \tag{1.2.7}$$

and integrating both sides over a cross-section of the body, bounded by a closed curve Γ, we find that

$$\iint_{\Sigma} (\mathbf{j} \,.\, d\boldsymbol{\Sigma} = c \iint_{\Sigma} (\text{curl } \mathbf{M} \,.\, d\boldsymbol{\Sigma}) = c \oint_{\Gamma} (\mathbf{M} \,.\, d\mathbf{l}) \tag{1.2.7 (a)}$$

The magnetization vector **M** is not uniquely defined by (1.2.7), because the gradient of any scalar quantity could be added to it there. Outside the body $\mathbf{M} = 0$. The total magnetic moment, due to all the charged particles moving in the body, following from (1.2.5), is given by

$$\frac{1}{2c} \iiint_{V} (\mathbf{r} \times \mathbf{j}) \, dV = \frac{1}{2} \iiint_{V} (\mathbf{r} \times \text{curl } \mathbf{M}) \, dV \tag{1.2.8}$$

Since $\mathbf{j} = 0$ outside V, integration is done over any volume which includes the body of the material.

Using the vector formula (A3.10) for $\mathbf{A} = \mathbf{r}$ and $\mathbf{B} = \mathbf{M}$, we find that

$$\nabla (\mathbf{r} \,.\, \mathbf{M}) = (\mathbf{M} \,.\, \nabla) \, \mathbf{r} + (\mathbf{r} \,.\, \nabla) \, \mathbf{M} + 0 + (\mathbf{r} \times \text{curl } \mathbf{M})$$

$$= \mathbf{M} + \frac{\partial}{\partial x} (x \, \mathbf{M}) + \frac{\partial}{\partial y} (y \, \mathbf{M}) + \frac{\partial}{\partial z} (z \, \mathbf{M}) - 3 \, \mathbf{M} + (\mathbf{r}x \text{ curl } \mathbf{M}), \tag{1.2.9}$$

Hence $$\iiint_{V} (\mathbf{r} \times (\nabla \times \mathbf{M})) \, dV = \iiint_{V} \nabla (\mathbf{r} \,.\, \mathbf{M}) \, dV - \iiint_{V} \left[\frac{\partial}{\partial x} (x \, \mathbf{M}) + \frac{\partial}{\partial y} (y \, \mathbf{M}) + \frac{\partial}{\partial z} (z\mathbf{M}) \right] dV + 2 \iiint_{v} \mathbf{M} \, dV$$

$$= \iint_{\Sigma} [(\mathbf{r} \,.\, \mathbf{M}) \, d\boldsymbol{\Sigma} - (\mathbf{r} \,.\, d\boldsymbol{\Sigma}) \, \mathbf{M}] + 2 \iiint_{V} \mathbf{M} \, dV \tag{1.2.10}$$

Since $\mathbf{M} = 0$ on Σ, we are left with the relation

$$\frac{1}{2c} \iiint_{V} (\mathbf{r} \times \rho \bar{\mathbf{v}}) \, dV = \iiint_{V} \mathbf{M} \, dV \tag{1.2.11}$$

So the magnetization **M** is the magnetic moment per unit volume.

(a) Point charge and field theory

A volume of plasma (classically defined and discussed in Chapter 3) has both electrical conductivity and dielectric polarization type properties. The conductivity is due to collisional exchange of momentum between the different plasma components. The polarization is due to the field induced reorientation of bound charges. The system exhibits characteristic oscillations due to collective effects of long range interaction, without disturbing, macroscopically, the charge neutrality.

The classical mechanics of point particles requires a well defined mass and a velocity. For macroscopic systems the position coordinates of particles are replaced by the centre of mass coordinates. The continuous systems are studied by the phenomenological *EM* theory of Maxwell, and the discrete systems by the microscopic analytical *EM* theory of Lorentz. The transformation from a system of discrete particles to a continuum, replacing the formalism of point mechanics by the corresponding field theory with the help of the concept of density quantities, like the momentum density and the energy density, is valid when the particles are very numerous. The ordinary particle derivative is then replaced by the substantial derivative or the variational derivative; and the variational procedure is generalised for continuous systems. Pressure gradients act on fluids but not on particles.

Since material bodies are continuous pieces in specific domains, the macroscopic mechanics of continuum should be regarded as a fundamental branch of mechanics, and particle mechanics as an idealization of it to treat very small bodies. However, at levels of subnuclear particles, collisions at relativistic speeds dominate. The collisions are manifestations of particle behaviour. So, the particle dynamics is more fundamental than continuum mechanics, and deserves to be treated first. Study of subnuclear physics also points to the strange conclusion that one should not enquire too closely as to what is meant by a particle, or by a collision. The two concepts, field and particles, are simultaneously utilized in some formalisms, in the development of the *EM* Lagrangian, for example.

Charge and associated mass are identified as riding on one and the same carrier, the particle. And though always the charge and the mass reside at the same point, the Newtonian concept of mass point emerges as charge point; and the charge and mass emerge as a particle. Relativistically, it is the mass velocity which cannot exceed C (the vacuum velocity of light). Since the *EM* laws do not directly involve mass, a charge, electromagnetically, is a primary thing. Mass is the energy manifestation of electric charge, and the associated *EM* fields. Since both charge and mass ride on the same carrier, the kinematics of charge is identified with the kinematics of mass, and both mass and charge are subjected to the relativistic restriction that the material velocity has the upper bound C.

(b) Magnetic moment

In the *CGS* units, the magnetic moment per unit volume, of a material, is defined as

$$\mu_1 = \frac{1}{2c}(\mathbf{r} \times \mathbf{j}), \text{ or as } \mu_2 = \frac{I\Sigma\mathbf{n}}{c}$$

where **r** is the charge displacement, **j** is the charge current per unit volume, I is the volume current, Σ is the area of the current loop, and **n** is the unit vector normal to the plane of Σ. Since $\boldsymbol{\mu}$, is perpendicular to the plane of the displacement **r** and current **j**, the two formulas for the magnetic moment **M** of a body of volume V, volume element dV, can be written as

$$\mathbf{M}_1 = \frac{1}{2c}\iiint_V (\mathbf{r}\times\mathbf{j})\, dV, \mathbf{M}_2 = \frac{1}{c}\iint_\Sigma I\, d\Sigma$$

where $d\boldsymbol{\Sigma} = (\mathbf{r} \times d\mathbf{l})/2$ and $d\mathbf{l}$ is the length element of the loop at the point P in the peripheri of Fig. 1.2.1. Evidently, then $\mathbf{j}\, dV = I\, d\mathbf{l}$, so that

$$d\mathbf{M}_1 = \frac{1}{2c}(\mathbf{r}\times\mathbf{j})\, dV = \frac{I}{2c}(\mathbf{r}\times d\mathbf{l}) = \frac{I}{c}d\boldsymbol{\Sigma} = d\mathbf{M}_2$$

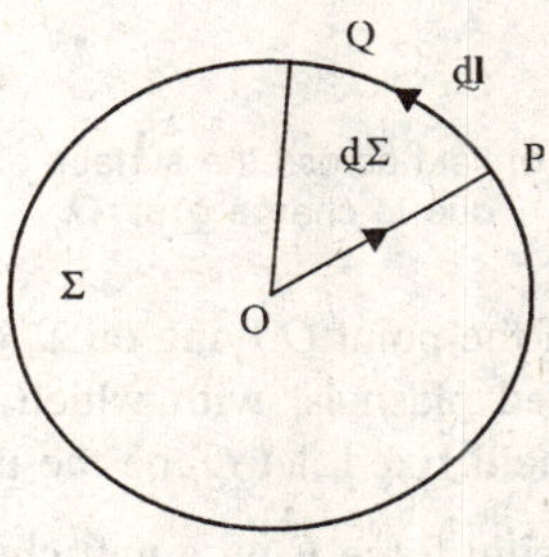

Fig. 1.2.1 The magnetic moment field geometry.

1.3 BASIC LAWS OF ELECTRODYNAMICS

(a) Divergence law of Coulomb for electric field

Coulomb's law states that two charges of opposite sign attract, and of same sign repel each other, according to the law of inverse square of the distance. We first consider the effect of an electric charge e at a point O on a unit charge at P lying on the surface element $d\boldsymbol{\Sigma}$. Here $\mathbf{i}$ (= $\mathbf{OP}/|OP|$), in Fig. 1.3.1, is the unit vector along **OP** and **n** is the unit vector drawn normal outward to the volume V. The electric field at P, therefore, is $\mathbf{E}$ (= $\mathbf{i}e/OP^2$), and

$$(\mathbf{E}\,.\, d\boldsymbol{\Sigma}) = \frac{e}{r^2}\, d\Sigma \cos\theta \tag{1.3.1}$$

where $\mathbf{r} = \mathbf{OP}$ and θ is the angle between **i** and **n**. If $d\Omega$ is the solid angle subtended at O by $d\boldsymbol{\Sigma}$, then $(\mathbf{E}\,.\,d\boldsymbol{\Sigma}) = ed\Omega$ because $r^2\, d\Omega = d\Sigma \cos\theta$. Accordingly,

$$\iint_\Sigma (\mathbf{E}\,.\,d\boldsymbol{\Sigma}) = e\iint_\Sigma d\Omega = 4\pi e \text{ if } O \text{ lies on the surface } \Sigma$$

$$= 2\pi e \quad \text{if } O \text{ lies on } \Sigma$$

$$= O \quad \text{if } O \text{ lies outisde } \Sigma$$

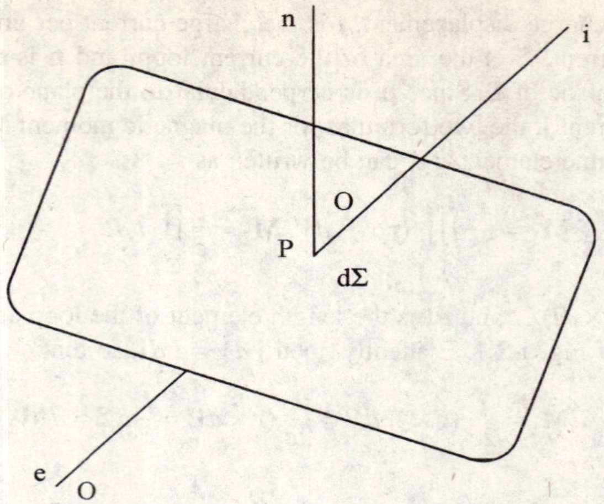

Fig. 1.3.1 Flux of field across the surface element $d\Sigma$ at P due to charge e at O.

We ignore the case of the point O lying on Σ, because then the results are not important for unbounded plasmas, with which we are mainly concerned. When several charges are held fixed, let O_i be the position of the ith charge e_i and let $O_i P = \mathbf{r}_i$. Then the sum force $\mathbf{E}$ on a unit charge at $P = \sum_i (e_i \mathbf{r}_i / r_i^3)$ and

$$\iint_\Sigma (\mathbf{E} \cdot d\mathbf{\Sigma}) = \iint_\Sigma \sum_i e_i (\mathbf{r}_i \cdot d\mathbf{\Sigma})/r_i^3$$

If θ_i is the angle between $\mathbf{r}_i$ and $d\mathbf{\Sigma}$, then

$$\iint_\Sigma (E \cdot d\mathbf{\Sigma}) = \sum_i \iint_\Sigma \frac{e_i}{r_i^2} \cos\theta_i \, d\Sigma = \sum_i \iint_\Sigma e_i \, d\Omega_i$$

where $d\Omega_i$ is the solid angle subtended by $d\Sigma$ at O_i. Using, in the summation, the indices p and q for charges inside and outside Σ, respectively, we can write

$$\iint_\Sigma (E \cdot d\mathbf{\Sigma}) = \sum_p e_p \iint_\Sigma d\Omega_p + \sum_q e_q \iint_\Sigma d\Omega_q$$

Since O_q, the seat of the charge e_q, is outside Σ, evidently $\iint_\Sigma d\Omega_q = 0$; and since O_p, the seat of e_p, is inside Σ, we have

$$\iint_\Sigma d\Omega_p = 4\pi; \text{ so } \iint_\Sigma (\mathbf{E} \cdot d\mathbf{\Sigma}) = 4\pi \sum_p e_p$$

If ρ^c is the charge density and dV is a volume element at O, then

$$\iint_\Sigma (\mathbf{E} \,.\, d\Sigma) = 4\pi \iiint_V \rho^c \, dV$$

where V is the volume within Σ. This is the integral form of Gauss's law for normal flux of the electic field. Then

$$\iint_\Sigma (\mathbf{E} \,.\, d\Sigma) - 4\pi \iiint_V \rho \, dV = 0 = \iiint_V (\text{div}\, \mathbf{E} - 4\pi\rho^c) \, dV$$

This macroscopic law shows that the total charge of a body is nothing but the surface integral of the normal component of the electric field over its entire surface. Since the integral relation holds good in any arbitrary volume, the integrand should vanish at each point, giving the differential equation

$$\text{div}\, \mathbf{E} = 4\pi\rho^c$$

for Gauss's law. We now write

$$\rho^c = \rho_e^c + \rho_i^c,$$

where ρ_e^c is the density of charges brought from outside, and ρ_i^c is that of a redistribution inside the closed surface Σ in presence of the electric field. Hence, putting $\rho_i^c = -\,\text{div}\ \mathbf{P}$ we get $\text{div}\ \mathbf{E} = 4\pi\,(\rho_e^c - \text{div}\ \mathbf{P})$, and the equation

$$\text{div}\, \mathbf{D} = 4\pi\rho_e^c \tag{1.3.2}$$

where

$$\mathbf{D} = \mathbf{E} + 4\pi\mathbf{P} \tag{1.3.3}$$

D is called the electric displacement vector. Dropping the subscript e from ρ_e^c, this law reads

$$\text{div}\, \mathbf{D} = 4\pi\rho^c \tag{1.3.4}$$

(b) The solenoidal law for magnetic induction

Since magnetic monopoles do not exist, only equal number of north and south poles occur together in any volume, small or large; so

$$\iint_\Sigma (\mathbf{H} \,.\, d\Sigma) = 0, \quad \iint_\Sigma (\mathbf{B} \,.\, d\Sigma) = 0 \tag{1.3.4 (a)}$$

for the magnetic field and $\mathbf{B} = \mathbf{H} + 4\pi\,\mathbf{M}$, **M** is the magnetization and the divergence law for magnetic induction reads

$$\text{div}\, \mathbf{H} = 0, \ \text{div}\, \mathbf{B} = 0, \ \text{div}\ \mathbf{M} = 0 \tag{1.3.5}$$

The vectors **M, B, H** are, therefore, solenoidal, *i.e.*, divergence free.

(c) Faraday's law of electromagnetic induction

When a closed circuit moves across a magnetic field, or if the magnetic field passing through the surface enclosed by the circuit, changes in time, then a current of electricity flows without batteries for current from outside sources. In either case, current lasts only as long as the circuit moves or the field changes. It makes no difference whether the magnetic field is due to a permanent magnetic field or an external circuit.

Consider a circuit Γ bounded by the surface Σ. The magnetic flux φ through this surface is given by

$$\varphi = \iint_{\Sigma} (\mathbf{B} \, . \, d\,\mathbf{\Sigma})$$

The electric circulation (or, the electromotive force) ε around the circuit is the line integral of the electric intensity, *E*. So,

$$\varepsilon = \oint_{\Gamma} (\mathbf{E} \, . \, d\mathbf{l})$$

where $d\mathbf{l}$ is the line element at *P* along the circuit Γ. The integral over the closed contour Γ is the electromotive force. It represents the work done by the electric field in moving a unit positive charge once around the contour. Since this integral depends only on the field and the contour Γ, the surface integral for the magnetic flux φ must also depend only on the field and the contour.

Experiments show that (*a*) if φ is changed in time in any way, then ε is proportional to $d\varphi/dt$, (*b*) the change in φ is opposed by ε; so $\varepsilon/(d\varphi/dt)$ is a negative quantity, and we can write

$$\varepsilon = -K \frac{d\varphi}{dt} \tag{1.3.6}$$

where *K* is the constant of proportionality. This is the integral form of Faraday's law of electromagnetic induction, because both ε and φ are given by integrals of fields. If the line integral of ε is transformed into a surface integral, by Stoke's theorem (Appendix 2), then (1.3.6) becomes

$$\iint_{\Sigma} (\text{curl } \mathbf{E} \, . \, d\,\mathbf{\Sigma}) = -K \frac{d}{dt} \iint_{\Sigma} (\mathbf{B} \, . \, d\,\mathbf{\Sigma})$$

When the surface Σ is fixed with respect to the observer, the total time derivative on the right hand side is replaced by the partial time derivative of **B** within the integral. And since Γ and Σ are arbitrary, we obtain the differential equation

$$\text{curl } \mathbf{E} = -K \frac{\partial \mathbf{B}}{\partial t},$$

for Faraday's law of electromagnetic induction. In the Gaussian (or mixed) system of units, the magnetic induction **B** is expressed in electromagnetic units and electric intensity **E** in electrostatic units, so *K* equals $1/C$ where $C\,(= 3 \times 10^{10} \text{ cm sec}^{-1})$ is the velocity of light in vacuum. Hence a standard form of the Faraday law is

$$\text{curl}\,\mathbf{E} = -\frac{1}{C}\frac{\partial \mathbf{B}}{\partial t} \tag{1.3.7}$$

(d) The Ampere Maxwell law of magnetoelectric induction

It is found that the magnetic circulation (or the magnetomotive force $(= \oint (\mathbf{H}\,.\,d\mathbf{l}))$ in the closed circuit Γ equals the sum of the conduction current and the displacement current passing through a given surface. So, we can write

$$\oint_{\Gamma} (\mathbf{H}\,.\,d\mathbf{l}) = \alpha I + \beta \frac{d}{dt}\iint_{\Sigma} (\mathbf{D}\,.\,d\boldsymbol{\Sigma})$$

where α and β are the constants of proportionality. Experiments have shown that if the volume current I is in stat amperes and D in stat volts cm^{-1} (that is, in electrostatic units), then $\alpha = 4\pi/C$, $\beta = 1/C$, and the integral form of the law is

$$\oint (\mathbf{H}\,.\,d\mathbf{l}) = \frac{4\pi}{C} I + \frac{1}{C}\iint_{\Sigma}\left(\frac{\partial \mathbf{D}}{\partial t}\,.\,d\boldsymbol{\Sigma}\right) \tag{1.3.8}$$

Again, applying Stoke's theorem (Appendix 2), on the integral in the left hand side, we obtain

$$\text{curl}\;\mathbf{H} = \frac{4\pi}{C}\mathbf{j} + \frac{1}{C}\frac{\partial \mathbf{D}}{\partial t} \tag{1.3.9}$$

where $I = \iint_{\Sigma} (\mathbf{j}\,.\,d\,\Sigma)$ and $\mathbf{j}$ is the current crossing unit area of Σ at the integration point. The quantity $\partial\mathbf{D}/\partial t$ is called the displacement current. It was first introduced by Maxwell. Motion of free charges gives rise to the conduction current $\mathbf{j}$. In dielectrics the conduction current is ignored; so (1.3.9) reduces to

$$\text{curl}\,\mathbf{H} = \frac{1}{C}\frac{\partial \mathbf{D}}{\partial t} \tag{1.3.10}$$

But, in electrical conductors the displacement current is neglizible; so (1.3.9) reduces to

$$\text{curl}\,\mathbf{H} = \frac{4\pi}{C}\,\mathbf{j} \tag{1.3.11}$$

In plasmas both conduction current and displacement current are effective; so the full equation (1.3.9), is necessary for plasma study.

1.4 FORCES

For two point charges e_1 and e_2, of masses m_1 and m_2, we denote by F_C the Coulomb force and by F_G the gravitational force. Then

$$F_C = \frac{e_1\, e_2}{r_{12}^2},\quad F_G = G\,\frac{m_1\, m_2}{r_{12}^2}$$

where r_{12} is the distance between the two particles and G is the gravitational constant ($= 6.67 \times 10^{-8}\ \text{cm}^2/\text{gm sec}^2$). For two electrons, since

$$F_C = 4 \times 10^{42} \times F_G,$$

the Coulomb force is far in excess of the gravitational force. These relations show that a kilogramme of mass is equivalent to about 8.6×10^{-11} Coulomb of electric charge. Hence the gravitational force is negligible in most cases of plasma study.

A particle P having charge e, moving with velocity $\mathbf{v}$, in presence of an electric field $\mathbf{E}$ and magnetic induction $\mathbf{B}$, experiences, on the basis of experiments the Lorentz force

$$e\mathbf{E} + e\,(\mathbf{v} \times \mathbf{H})/C$$

The Lorentz force density $\mathbf{F}$ is given by

$$\mathbf{F} = \rho^c\,\mathbf{E} + \frac{1}{C}\,(\mathbf{j} \times \mathbf{B}) \tag{1.4.1}$$

where ρ^c is the density of total charges and $\mathbf{j}$ is the current density. If $\mathbf{v}$ is the average velocity of motion of charges on unit volume at P, then there $\mathbf{j} = \rho^c\,\mathbf{v}$; so

$$\mathbf{F} = \rho^c\,\mathbf{E} + \frac{1}{C}\,\rho^c\,(\mathbf{v} \times \mathbf{B}) \tag{1.4.2}$$

The torque $\mathbf{N}$, per unit volume, is

$$\mathbf{N} = (\mathbf{r} \times \mathbf{F}), \tag{1.4.3}$$

where $\mathbf{r}$ is the position vector of P. For dipoles, it is

$$\mathbf{N} = (\boldsymbol{\mu}_e \times \mathbf{E}_d)$$

where $\boldsymbol{\mu}_e$ is the electric dipole moment and $\mathbf{E}_d$ is the electric field. For a system of n particles, if $\mathbf{p}_i$ is the momentum of the ith particle, then the angular momentum $\mathbf{M}_0$, given by (1.2.4), generates the torque

$$\mathbf{N} = \frac{d\mathbf{M}_0}{dt} \tag{1.4.4}$$

Microscopic systems have electric and magnetic moments. External fields interacting with these moments change the angular momentum and generate torques. If $\mathbf{m}$ is the magnetic dipole moment, then

$$\mathbf{N} = \mathbf{m} \times \mathbf{B} \tag{1.4.5}$$

This torque tends to direct the moment along the field and generates a force $\mathbf{F}$ given by

$$\mathbf{F} = (\mathbf{m} \,.\, \nabla)\,\mathbf{B} \tag{1.4.6}$$

which tries to drive the system to regions of strong field. This force vanishes if B is homogeneous; the only effect of it then is to produce a torque on the system. In the absence of magnetic field the angular momentum is constant.

The field **B** provided by a current in a loop C_1 acts on a small probing current loop C_2 forming a magnetic dipole. The dipole is given a rotational displacement such that the flux through the loop C_1 remains constant. The fall in energy for a given angular displacement $\delta\,\theta$ is greatest if the dipole is turned towards the direction of **B**. The torque acts so as to turn **m** (the dipole moment of C_2) in that direction. The corresponding angular displacement is given by

$$|\mathbf{N}|\,\delta\theta = -\,m\delta\,(\mathbf{B}\,.\,\mathbf{n}) = mB\sin\theta\,\delta\theta \tag{1.4.7}$$

where **N** is the torque on the dipole, θ is the angle between **m** and **B**, and **n** is the unit vector parallel to **m**.

The spin **S** of a system, defined by the relation (1.4.4), is its angular momentum; **S** is a pseudo vector*, whose rate of change in time equals the applied torque **N**. So

$$\mathbf{N} = \frac{d\mathbf{S}}{dt} \tag{1.4.8}$$

1.5 ANALYSIS OF THE MAXWELL EQUATIONS

The four Maxwell equations of electrodynamics can be expressed as a set of differential equations or integral relations.

Name of the law	*Integral relations*	*Differential equations*
Faraday's law of electromagnetic induction	$\oint_\Gamma (\mathbf{E}\,.\,d\mathbf{l}) = -\frac{1}{C}\frac{d}{dt}\iint_\Sigma (\mathbf{B}\,.\,d\mathbf{\Sigma})$	$\nabla\times\mathbf{E} = -\frac{1}{C}\frac{\partial\mathbf{B}}{\partial t}.$
Ampere-Maxwell's law of magneto-electric induction	$\oint_\Gamma (\mathbf{H}\,.\,d\mathbf{l}) = \frac{4\pi}{C}\iint_\Sigma (\mathbf{j}\,.\,d\mathbf{\Sigma}) + \frac{1}{C}\frac{d}{dt}\iint_\Sigma (\mathbf{D}\,.\,d\mathbf{\Sigma}),$	$\nabla\times\mathbf{H} = \frac{1}{C}\frac{\partial\mathbf{D}}{\partial t} + \frac{4\pi}{C}\mathbf{j}.$
Gauss's law for electric displacement	$\iint_\Sigma (\mathbf{D}\,.\,d\mathbf{\Sigma}) = 4\pi\iiint_v \rho^c\,d\tau,$	$\nabla\,.\,\mathbf{D} = 4\pi\rho^c$
Gauss's law for Magnetic induction	$\iint_\Sigma (\mathbf{B}\,.\,d\mathbf{\Sigma}) = 0$	$\nabla\,.\,\mathbf{B} = 0$

The integral relations are formulated on the basis of experimental findings. The set of differential equations follow from these for phenomena in stationary media. So, the integral set is more generally valid than the other set. Also, the

* The cross product $\mathbf{C} = \mathbf{A}\times\mathbf{B}$ of two vectors **A** and **B** is a pseudo vector, because under inversion **A** and **B** change sign but not vector **C**.

integral equations are valid over finite space domains at a time, whereas, the differential equations are valid over all space-time points in a region.

Some authors call the first and second equations of either column as the Maxwell equations, and the remaining two equations as the divergence equations. Since divergence of (1.3.7) gives $\partial\,(\text{div }\mathbf{B})/\partial t = 0$, the conclusion is that if $\nabla \cdot \mathbf{B} = 0$ at any moment, $(\nabla \cdot \mathbf{B}) = 0$ for all times. So, all the four equations are not linearly independent. Taking divergence of (1.3.9) and using (1.3.4), the condition for conservation of charges,

$$\nabla \cdot \mathbf{j} + \frac{\partial \rho^c}{\partial t} = 0 \tag{1.5.1}$$

is obtained. Together with the divergence of (1.3.9), it gives

$$\frac{\partial}{\partial t}(\nabla \cdot \mathbf{D} - 4\pi\rho^c) = 0$$

So, if the relation (1.3.4) is valid at any instant, it will remain valid always. Hence, the linearly independent set of Maxwell equations, depending on the field vectors **E, H, B, j, D** and the scalar field ρ^c, which are the unknown quantities, are (1.3.7), (1.3.9) and (1.3.4) or (1.5.1). The total number of field variables of this set is 16, and of the linearly independent Maxwell equations is only seven. A system is said to be closed when the number of independent equations of it equals the number of unknown quantities. To have a closed set of equations, for electrodynamics, therefore, nine other linearly independent connecting equations, depending on the nature of the material and on the intensity of the applied field, are necessary. A general form of these relations is

$$f_j(\mathbf{E}, \mathbf{H}, \mathbf{B}, \mathbf{D}, \mathbf{j}, \rho) = 0 \quad (i = 1, 2, 3, ..., 9)$$

For simple stationary media they simplify to the forms

$$\mathbf{D} = \mathbf{F}_1(\mathbf{E}), \ \mathbf{j} = \mathbf{F}_2(\mathbf{E}), \ \mathbf{B} = \mathbf{F}_3(\mathbf{H}) \tag{1.5.2}$$

The quantities $\mathbf{F}_1(\mathbf{E})$, $\mathbf{F}_2(\mathbf{E})$, $\mathbf{F}_3(\mathbf{H})$ indicate the functional dependence of **D, J, B** on **E** and **H**. These are the electrodynamical equations of state or, the phenomenological constitutive relations.

For weak electromagnetic fields in stationery, isotropic, unmagnetized, and non-collisional plasmas

$$\mathbf{D} = \varepsilon\mathbf{E}, \ \mathbf{j} = \sigma\mathbf{E}, \ \mathbf{B} = \mu\mathbf{H} \tag{1.5.3}$$

and, in stationary, isotropic crystals,

$$\mathbf{D} = \varepsilon\mathbf{E}, \ \mathbf{j} = 0, \ \mathbf{B} = \mu\mathbf{H}$$

In more complicated anisotropic crystals, the state equations are

$$D_i = \varepsilon_{ij}, \ E_j, \ \mathbf{j} = 0, \ B_i = \mu_{ij} H_j \tag{1.5.4}$$

where repeated index means summation over that index.

In crystals, like rubi, powerful beams of laser (light amplification by stimulated emission of radiation) develop bluish features due to doubling and trippling of frequencies, and several directions of ray. For these effects non-linear constitutive relations like

$$D_i = \varepsilon_{ij} E_j + \varepsilon_{ijk} E_j E_k + \varepsilon_{ijkl} E_j E_k E_l + \dots \tag{1.5.5}$$

are appropriate. Laser beams create potentials of the order of 10^5 volts/cm. Since charges inside a molecule are bound together by potentials of this magnitude, quantum mechanical effects are generated by such powerful beams.

Material constants like electrical conductivity, dielectric constant or magnetic permeability vary from material to material. The Maxwell equations, being fundamental laws, contain only the universal constants C and 4π, and are valid universally. These do not depend on the material properties, and the related constants. The constitutive relations depend on the material constants σ, μ, ε_{ij}, etc. Therefore, the essential difference between the fundamental laws and state relations is that the former are generally true and the latter are specifically true.

Also, to explain optical rotation of elliptically or circularly polarized waves, the dielectric tensor is related to a linear combination of **E** and its space derivatives, in the following manner :

$$D_i = \varepsilon_{ij} E_j + \gamma_{ijk} \frac{\partial E_j}{\partial x_k}$$

where γ_{ijk} is a third order tensor. For optically active, isotropic substances, we put $\gamma_{ijk} = \gamma\, \varepsilon_{ijk}$ (ε_{ijk} is the alternative unit tensor, Appendix 4) and obtain

$$D_i = \varepsilon E_i + \gamma \left(\frac{\partial E_k}{\partial x_j} - \frac{\partial E_j}{\partial x_k} \right)$$

The corresponding vector constitutive relation is

$$\mathbf{D} = \varepsilon \mathbf{E} + \gamma \nabla \times \mathbf{E}$$

This phenomenological theory for the constitutive relations is free from the Lorentz force $\rho \mathbf{E} + (\mathbf{j} \times \mathbf{B})/C$. Actually, the effect of the electromagnetic force is here indirectly included in Ohm's law

$$\mathbf{j} = \sigma \left\{ \mathbf{E} + \frac{1}{C} (\mathbf{v} \times \mathbf{B}) \right\}$$

which empirically accounts for the motion of charges in the presence of electric and magnetic fields and for dissipation of electromagnetic power into heat.

1.6 MACROSCOPIC AND MICROSCOPIC APPROACHES FOR ELECTRODYNAMICS

The Lorentz force is responsible for the momentum transfer by electric and magnetic fields. In the microscopic electron theory of Lorentz (1916) for electrodynamics, the Maxwell equations, together with the momentum transfer equations involving the Lorentz force, are employed. In the phenomenological macroscopic theory of Maxwell, the momentum transfer relations are replaced by some equations of state. In conductors charges are free to move, producing conduction currents whenever there is a potential difference between two points. In non-conductors charges are neither free nor wholly fixed in position. In presence of an electric field they produce displacement current. So in plasma there exists no conduction current; and currents are either of displacement or convection. There is no magnetism; magnetic fields are supposed to be produced only by motion of electricity. Space, at absolute rest, acts as the carrier of electric fields and charges move with respect to it.

In the Lorentz theory, motion of charges takes place in vacuum, so that $\varepsilon = 1$, $\mu = 1$, the vectors **B** and **D** are identified with **H** and **E** respectively; the displacement current is $\partial \mathbf{E}/C\,\partial t$, and the convection current is $\rho^c\,\mathbf{v}$ where ρ^c is the net charge density and **v** is the velocity. The Lorentz force per unit charge is $\mathbf{E} + (\mathbf{v} \times \mathbf{H})/C$. The condition for charge conservation, in the Maxwell theory, is

$$\frac{\partial \rho^c}{\partial t} + \operatorname{div}\,\mathbf{j} = 0$$

Integration over the volume V gives

$$\frac{\partial}{\partial t} \iiint_V \rho\, dV = - \iint_\Sigma j_n\, d\Sigma$$

where the surface Σ of V is large enough to include all matter in the universe. So the total charge in the universe is conserved. In the Lorentz theory, on the other hand, the charge conservation condition is

$$\frac{\partial \rho^c}{\partial t} + \operatorname{div}\,(\rho^c \mathbf{v}) = 0 \quad \text{or} \quad \frac{d\rho^c}{dt} = 0$$

where $\partial/\partial t$ is the partial derivative operator and d/dt is the substantial (or the total derivative) operator. So, ρ^c is conserved; then, charge is the primary concept in theory of Lorentz. This theory holds good throughout all space, including matter interior of an atom, molecule or electron. The equations of Lorentz are only those of Maxwell, extended to electrons; appropriate averaging of these should yield the macroscopic set of Maxwell equations. However, the electron field cannot be measured, because to do so we require a charge still smaller than the electron so that it does not influence the field, and this is not possible. For justification of the theory, Lorentz has shown that these equations explain the phenomenon of wave dispersion and normal Zeeman effect.

The macroscopic Maxwell equations (1.3.4), (1.3.5), (1.3.7) and (1.3.9) are

$$\operatorname{curl}\,\mathbf{E} = -\frac{1}{c}\frac{\partial \mathbf{H}}{\partial t}, \tag{1.6.1}$$

$$\operatorname{curl}\,\mathbf{H} = \frac{1}{c}\frac{\partial \mathbf{E}}{\partial t} + \frac{4\pi}{c}\left(\mathbf{j} + \frac{\partial \mathbf{P}}{\partial t}\right), \tag{1.6.2}$$

$$\text{div } \mathbf{E} = 4\pi (\rho^c + \text{div } \mathbf{P}), \tag{1.6.3}$$

$$\text{div } \mathbf{H} = 0 \tag{1.6.4}$$

In the microscopic theory the second and third of these relations are replaced by

$$\text{curl } \mathbf{H} = \frac{1}{c}\frac{\partial \mathbf{E}}{\partial t} + \frac{4\pi}{c}\mathbf{j}, \tag{1.6.5}$$

$$\text{div } \mathbf{E} = 4\pi\rho^c \tag{1.6.6}$$

Since these **J** and ρ^c of (1.6.6) are given by the delta function presentation of (1.2.1) and (1.2.2), these are not identical to the **j** and ρ^c of (1.6.2) and (1.6.3). The position vector $\mathbf{r}_i$ of (1.2.1) and (1.2.2), is determined by the momentum equation for the *i*th particle,

$$\frac{d}{dt}\left[\frac{m\mathbf{v}_i}{\sqrt{1 - v_i^2/c^2}}\right] = e_i\,\mathbf{E} + \frac{e_i}{c}(\mathbf{v}_i \times \mathbf{H}) + \mathbf{F}_i \tag{1.6.7}$$

where $\mathbf{F}_i$ represents the sum of the forces other than the Lorentz force. Since (1.2.1) and (1.2.2) satisfy a continuity condition, we find that (1.6.5) implies (1.6.6). Hence the system of equations (1.6.1), (1.6.5), (1.6.6) and (1.6.7) give $3N + 6$ linearly independent equations, because (1.6.6) is an extra equation. The field variables being **E**, **H**, $\mathbf{r}_i$, their total number is also $3N + 6$. The number of equations, therefore, in principle, is sufficient to solve problems. The delta function representations (1.2.1) and (1.2.2) are valid for ρ^c and **j**, due to a set of discrete particles. But, when the individuality of the particles is not important, the limiting process is adopted in which $e_i \to 0, m_i \to 0$ such that $N_i\,e_i \to$ constant, $N_i\,m_i \to$ constant, so, each particle is subdivided into smaller and smaller subparticles, gradually smearing out the delta functions and approaching limiting values which are in essence average values. One such limit is a distribution function in the space of position coordinates and momentum of all particles in a given volume (chapter 8). In a more reduced state this limit is a distribution in the six-dimensional phase space of position coordinates and momentum vector of only one particle (chapter 9). The still more reduced state is that of a continuous fluid, or of a mixture of several fluids (for example, a mixture of fluids of negative charges, positive charges and neutral atoms or molecules) (chapter 3).

If the distribution function for the *i*th species of particles in the six-dimensional phase space of position and velocity coordinates be $f_i\,(\mathbf{r}, \mathbf{v}, t)$, with $i = 1, 2, ..., N$ (chapters 9 and 10), then the field equations are

$$\frac{\partial f_1}{\partial t} + v_j \frac{\partial f_i}{\partial x_j} + F_j \frac{\partial f_i}{\partial x_j} = \left(\frac{\partial f_i}{\partial t}\right)_{\text{coll}}, \tag{1.6.8}$$

$$\text{curl } \mathbf{E} = -\frac{1}{c}\frac{\partial \mathbf{H}}{\partial t}, \tag{1.6.9}$$

$$\text{curl } \mathbf{H} = \frac{1}{c}\frac{\partial \mathbf{E}}{\partial t} + \frac{4\pi}{c}\sum_{i=1}^{N} e_i \iiint \mathbf{v}\, f_i\, d\mathbf{v}, \tag{1.6.10}$$

$$\text{div } \mathbf{E} = 4\pi \sum_i e_i \iiint f_i\, d\mathbf{v} \tag{1.6.11}$$

where $$\mathbf{F}_i = \frac{e_i}{m_i}\mathbf{E} + \frac{e_i}{m_i c}(\mathbf{v}\times\mathbf{H}) + \mathbf{F}_{0i},$$

F_j in (1.6.8) is the jth component of this force, and $\mathbf{F}_{0i}$ is the sum of the long range forces other than the Lorentz force, acting on the ith species. Since (1.6.11) obtains from (1.6.10), in all specific examples, the system of equations (1.6.8)-(1.6.11) consists of $6 + N$ equations for the $6 + N$ variables $\mathbf{E}, \mathbf{H}, f_i$.

When the medium is a mixture of N fluids, an element of the ith fluid at the point $\mathbf{r}$ at time t has the velocity $\mathbf{u}_i$, number density N_i and mass density ρ_i^m. Let m_i be the mass and e_i be the charge of each particle of this species. Then the field equations are (*i*) the momentum transfer condition for the ith particle:

$$\frac{d\mathbf{u}_i}{dt} = -\frac{1}{\rho_i^m}\nabla p_i + \frac{e_i}{m_i}\mathbf{E} + \frac{e_i}{m_i c}(\mathbf{u}_i\times\mathbf{H}) + \frac{1}{\rho_i^m}\mathbf{F}_i$$

where $\mathbf{F}_i$ contains the sum of the forces per unit volume not mentioned specifically; (*ii*) the mass conservation condition

$$\frac{\partial \rho_i^m}{\partial t} + \text{div}\,(\rho_i^m\,\mathbf{u}_i) = 0;$$

(*iii*) the state relations, for example, the adiabatic relation $dp_i = c_{is}^2\, d\rho_i^m$ where c_{is} is the velocity of sound in the ith fluid species (further discussion in section 3.4a); and (iv) the Maxwell equations

$$\text{curl } \mathbf{E} = -\frac{1}{c}\frac{\partial \mathbf{H}}{\partial t},\quad \text{curl } \mathbf{H} = \frac{1}{c}\frac{\partial \mathbf{H}}{\partial t} + \frac{4\pi}{c}\sum_{i=1}^{N}(N_i\, e_i\, \mathbf{u}_i)$$

These are a system of $3N + N + N + 3 + 3$ $(= 5N + 6)$ equations for the field variables $\mathbf{u}_i, \rho_i^m, p_i, \mathbf{E}, \mathbf{H}$ which are also $5N + 6$ in number.

Hence the microscopic equations contain only universal constants like charge and mass of a particle of each species and the vacuum velocity of light c. These do not contain constants which depend on the nature of the material used. But, the phenomenologically based equations contain material parameters, such as σ (electrical conductivity) and ε (dielectric constant), depending on the averaging of the moments of various orders of the mobile charges about their mean positions. In the macroscopic approach, electrodynamics is therefore combined with mechanics of deformation, which is suitable for elastic changes; and in the microscopic approaches, with analytical mechanics of particles in position space, or statistical mechanics of particles in phase space. The phenomenological theory is applicable to solids and dense fluids. When the mean interparticle distance as in a plasma is much greater than lengths of the order of the particle dimensions, microscopic equations are employed.

1.7 POYNTING THEOREM FOR ENERGY FLUX

The rate of work done by the Lorentz force density **F** of relation (1.4.1) is given by $(\mathbf{F}\,.\,\mathbf{v}) = \rho^c\,(\mathbf{E}\,.\,\mathbf{v}) + \{\mathbf{v}\,.\,(\mathbf{j}\times\mathbf{B})\}/c$. Since $\mathbf{j} = \rho^c\,\mathbf{v}$, this relation reduces to

$$(\mathbf{F}\,.\,\mathbf{v}) = (\mathbf{E}\,.\,\mathbf{j}) \tag{1.7.1}$$

To transform the right hand side of (1.7.1) into a function of **E**, **D**, **B** and **H** only, with the help of the Maxwell equations, we first put the value of **j** from (1.3.9) in one of these and get

$$(\mathbf{E}\,.\,\mathbf{j}) = \{c\,(\mathbf{E}\,.\,\text{curl }\mathbf{H})/4\pi\} - (\mathbf{E}\,.\,\partial\mathbf{D}/\partial t)/4\pi$$

Eliminating the first term on the right hand side, with the help of the vector identity (A 3.8) for **E** and **H**, and using equation (1.3.7) we obtain

$$(\mathbf{E}\,.\,\mathbf{j}) = \frac{c}{4\pi}(\mathbf{H}\,.\,\text{curl }\mathbf{E}) - \frac{c}{4\pi}\,\text{div }(\mathbf{E}\times\mathbf{H}) - \frac{1}{4\pi}\left(\mathbf{E}\cdot\frac{\partial D}{\partial t}\right)$$

$$= -\,\text{div }\mathbf{S}^t - \frac{1}{4\pi}\left\{\left(\mathbf{H}\cdot\frac{\partial\mathbf{B}}{\partial t}\right) + \left(\mathbf{E}\cdot\frac{\partial\mathbf{D}}{\partial t}\right)\right\} \tag{1.7.2}$$

where
$$\mathbf{S}^t = \frac{c}{4\pi}(\mathbf{E}\times\mathbf{H}) \tag{1.7.3}$$

$\mathbf{S}^t$ called the Poynting flux; it is the flux of energy of the electromagnetic fields. In linear isotropic media $\mathbf{D} = \varepsilon\,\mathbf{E}$, $\mathbf{B} = \mu\mathbf{H}$ and so

$$\left(\mathbf{E}\cdot\frac{\partial\mathbf{D}}{\partial t}\right) + \left(\mathbf{H}\cdot\frac{\partial\mathbf{B}}{\partial t}\right) = \frac{\partial}{\partial t}\left(\frac{\varepsilon E^2 + \mu H^2}{2}\right),\ (\mathbf{E}\,.\,\mathbf{j}) = -\,\text{div }\mathbf{S}^t - \frac{\partial}{\partial t}\left(\frac{\varepsilon E^2 + \mu H^2}{8\pi}\right) \tag{1.7.4}$$

The quantity $(\mathbf{E}\,.\,\mathbf{j})$ is an energy source or sink. Electrically conducting matter gains this amount of energy per unit volume per unit time in the presence of the field, and the field loses this energy. Since a part of the power density is the time rate of variation of $(\varepsilon E^2 + \mu H^2)/8\pi$, this quantity is the sum of energy densities in electric and magnetic fields. So we write

$$W_E = \frac{\varepsilon E^2}{8\pi},\quad W_H = \frac{\mu H^2}{8\pi} \tag{1.7.5}$$

where W_E is the electrical energy density and W_H is the magnetic energy density.

Integrating the second relation of (1.7.4) over a volume V, applying Green's theorem (Appendix 2) for transformation of the volume integral into a surface integral over the surface Σ of V, and using (1.7.5) we get

$$-\frac{\partial}{\partial t}\iiint\limits_V (W_E + W_H)\, dV = \iint\limits_\Sigma S_n^t\, d\Sigma + \iiint\limits_V (\mathbf{E}\,.\,\mathbf{j})\, dV \tag{1.7.6}$$

where $d\Sigma$ is a surface element and S_n^t is the normal component of S^t at a point on Σ. The expressions for the linear momentum and the angular momentum of the electromagnetic field, respectively, are

$$\iiint\limits_V \frac{1}{4\pi\, c}(\mathbf{E}\times\mathbf{H})\, dV \text{ and } \iiint\limits_V \frac{1}{4\pi c}[\mathbf{r}\times(\mathbf{E}\times\mathbf{H})]\, dV \tag{1.7.6 (a)}$$

Equation (1.7.6) shows that the time rate of loss of energy in electric and magnetic forms, inside V, equals the flow of electromagnetic energy through the surface Σ, plus the rate at which the field does work on matter within V. This condition of energy conservation is known as the Poynting theorem. When the entire electromagnetic energy is used to proudce joule heat, we put $\mathbf{j} = \sigma\mathbf{E}$ and get

$$(\mathbf{E}\,.\,\mathbf{j}) = \sigma E^2 \tag{1.7.7}$$

where σ is the electrical conductivity. The Poynting flux $\mathbf{S}^t$ is non-zero when an electrostatic field is crossed by a uniform magnetic field. But then div $\mathbf{S}^t$ is zero and $\mathbf{S}^t$ does not have any influence on the energy balance condition. A general energy conservation condition, based on a microscopic system of equations for plasma is considered in section 3.7.

1.8 MAXWELL STRESSES

Electromagnetic force and other forces generate stress and pressure on material bodies. To study these aspects consider the momentum transfer condition of continuous mechanics in the form

$$\frac{d\mathbf{G}}{dt} + \hat{\mathbf{X}}_j \frac{\partial T_{ij}}{\partial x_i} = \mathbf{F} \tag{1.8.1}$$

where $\mathbf{G}$ is the momentum of an element in a unit volume, T_{ij} is the ijth component of the pressure tensor, $\hat{\mathbf{X}}_j$ is the unit vector, x_j is the coordinate in the jth direction, and $\mathbf{F}$ is the force per unit volume, including the Lorentz force. We would find $\mathbf{G}$, T_{ij} and $\mathbf{F}$, in terms of the fields with the help of the Maxwell equations. Then the distribution of charges and currents, giving rise to the force $\mathbf{F}$, will be replaced by a continuous change of momentum and a pressure tensor, and the action-at-distance electrodynamics of this distribution will lead to the Faraday concept of a continuous distribution of stresses and a momentum density. For this purpose, the Lorentz force $\rho^c\,\mathbf{E} + (\mathbf{j}\times\mathbf{B})/c$, depending on the charges of density ρ^c and currents $\mathbf{j}$, would be expressed only in terms of the continuous field variables $\mathbf{E}, \mathbf{H}, \mathbf{D}, \mathbf{B}$. Replacing ρ^c by div $\mathbf{D}/4\pi$ and $\mathbf{j}$ by $(c \text{ curl } \mathbf{H}/4\pi) - (\partial\mathbf{D}/\partial t)/4\pi$ in the expression for the Lorentz force, we get

$$\mathbf{F}=\frac{1}{4\pi}\mathbf{E}\ \text{div}\ \mathbf{D}+\frac{\mu}{4\pi}(\text{curl}\ \mathbf{H}\times\mathbf{H})-\frac{1}{4\pi c}\left(\frac{\partial\mathbf{D}}{\partial t}\times\mathbf{B}\right) \tag{1.8.2}$$

The vector results (A 3.10) give

$$(\text{curl}\ \mathbf{E}\times\mathbf{E})_X=\frac{\partial}{\partial x}\left(E_X^2-\frac{1}{2}E^2\right)+\frac{\partial}{\partial y}(E_X E_Y)+\frac{\partial}{\partial z}(E_Z E_X)-E_X\ \text{div}\ \mathbf{E}, \tag{1.8.3}$$

$$(\text{curl}\ \mathbf{H}\times\mathbf{H})_X=\frac{\partial}{\partial x}\left(H_X^2-\frac{1}{2}H^2\right)+\frac{\partial}{\partial y}(H_X H_Y)+\frac{\partial}{\partial z}(H_Z H_X)$$

where we have also used the condition div $\mathbf{H}=0$. The simplest state relations $\mathbf{D}=\varepsilon\mathbf{E}$ and $\mathbf{B}=\mu\mathbf{H}$ show that

$$F_X=\frac{\partial}{\partial x}\left(\frac{\varepsilon E_X^2+\mu H_X^2}{4\pi}-\frac{\varepsilon E^2+\mu H^2}{8\pi}\right)+\frac{\partial}{\partial Y}\left(\frac{\varepsilon E_X E_Y+\mu H_X H_Y}{4\pi}\right)+\frac{\partial}{\partial z}\left(\frac{\varepsilon E_Z E_X+\mu H_Z H_X}{4\pi}\right)-\frac{e}{4\pi}(\text{curl}\ \mathbf{E}\times\mathbf{E})_X-\frac{1}{4\pi c}\left(\frac{\partial\mathbf{D}}{\partial t}\times\mathbf{B}\right)_X \tag{1.8.4}$$

Replacing curl $\mathbf{E}$ by $-\mu\partial\mathbf{H}/c\partial t$, we can write

$$F_X=\frac{\partial T_{XX}}{\partial x}+\frac{\partial T_{XY}}{\partial y}+\frac{\partial T_{ZX}}{\partial z}+\frac{\varepsilon\mu}{4\pi c}\left(\frac{\partial\mathbf{H}}{\partial t}\times\mathbf{E}\right)_X-\frac{\varepsilon\mu}{4\pi c}\left(\frac{\partial\mathbf{E}}{\partial t}\times\mathbf{H}\right)_X \tag{1.8.5}$$

$$\mathbf{g}=\frac{\varepsilon\mu}{c^2}\mathbf{S}^t,\quad \mathbf{S}^t=\frac{c}{4\pi}(\mathbf{E}\times\mathbf{H}), \tag{1.8.5 (a)}$$

$$F_i=\frac{\partial g_i}{\partial t}+\frac{\partial T_{ij}}{\partial x_j} \tag{1.8.6}$$

and obtain $F_i=\dfrac{\partial g_i}{\partial t}+\dfrac{\partial T_{ij}}{\partial x_j}$, (1.8.6)

where

$$\mathbf{T}=\begin{bmatrix}\dfrac{(\varepsilon E_X^2+\mu H_X^2)}{4\pi}-\dfrac{(\varepsilon E^2+\mu H^2)}{8\pi} & \dfrac{(\varepsilon E_X E_Y+\mu H_X H_Y)}{4\pi} & \dfrac{(\varepsilon E_X E_Z+\mu H_X H_Z)}{4\pi}\\ \dfrac{(\varepsilon E_X E_Y+\mu H_X H_Y)}{4\pi} & \dfrac{\varepsilon E_Y^2+\mu H_Y^2}{4\pi}-\dfrac{\varepsilon E^2+\mu H^2}{8\pi} & \dfrac{(\varepsilon E_Y E_Z+\mu H_Y H_Z)}{4\pi}\\ \dfrac{(\varepsilon E_X E_Z+\mu H_X H_Z)}{4\pi} & \dfrac{(\varepsilon E_Y E_Z+\mu H_Y H_Z)}{4\pi} & \dfrac{(\varepsilon E_Z^2+\mu H_Z^2)}{4\pi}-\dfrac{\varepsilon E^2+\mu H^2}{8\pi}\end{bmatrix}, \tag{1.8.6 (a)}$$

g is the electromagnetic momentum per unit volume, and T_{ij} is the ijth component of the Maxwell stress tensor. We can write

$$T_{ij} = \frac{1}{4\pi}(\varepsilon E_i E_j + \mu H_i H_j) - \frac{1}{8\pi}(\varepsilon E^2 + \mu H^2)\,\delta_{ij} \tag{1.8.7}$$

where the kronecker delta

$$\delta_{ij} = 1 \text{ if } i = j, \text{ and } \delta_{ij} = 0 \text{ if } i \neq j \tag{1.8.7 (a))}$$

Integrating $\hat{X}_i\,(\partial T_i/\partial x_j)$ in a volume V, and transforming the volume integral into a surface integral with the help of Gauss's theorem (Appendix 2), we obtain

$$\iiint_v \hat{\mathbf{X}}_i \frac{\partial T_{ij}}{\partial x_j}\,dV = \hat{\mathbf{X}}_i \iint_\Sigma T_{ij}\,d\Sigma_j \tag{1.8.7 (b))}$$

where $d\Sigma_j$ is the projection of the surface element $d\mathbf{\Sigma}$ on the plane whose normal is parallel to the jth direction. T_{ij} is the stress along the ith space direction, on a surface having normal along the jth direction. Hence charges and currents inside a material create electromagnetic stresses on its boundary surface.

To show this stress distribution, more clearly, consider the rectangular box PQLMNDEF of Fig. 1.8.1, having edges parallel to the coordinate axes. At the

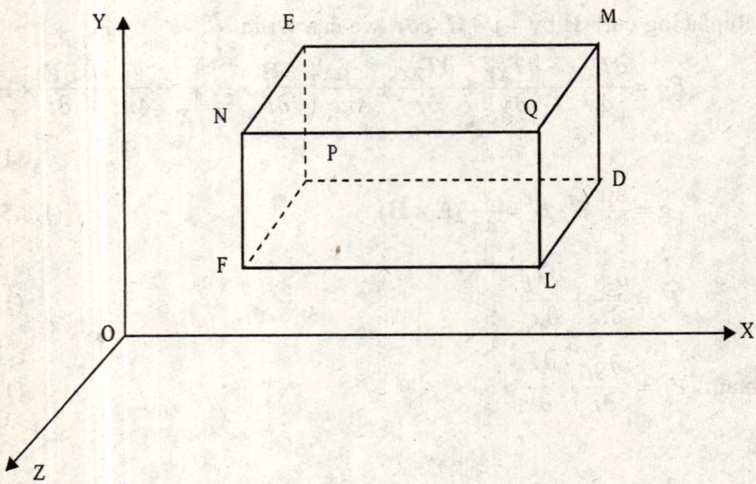

Fig. 1.8.1 The Maxwell stresses on the surfaces of a box.

point $P(x, y, z)$ dx is parallel to **OX**, dy to **OY** and dz to **OZ**. The stress component T_{XX} acts on the plane PENF parallel to **OX**; it is, therefore, a normal stress. T_{XY} is also parallel to **OX**; but it acts in the plane PDLF, so it is a tangential stress, T_{XZ} is, similarly, the tangential stress parallel to **OX**, acting in the plane PEMD. The other six componants of **T** are also either tangential or normal stresses, acting on the plane surfaces through P. The diagonal elements T_{XX}, T_{YY}, T_{ZZ} are normal stresses and the other elements are tangential stresses.

The pressure of electromagnetic radiation, generally, is a small quantity, and it is difficult to measure it experimentally. For example, in case of sun, it is 4.5×10^{-5} dynes/cm^2. For the whole earth, the force of this pressure is 5.7×10^7 k gm wt. Compared to the gravitational attraction, which is 3.6×10^{21} k gm wt. per unit volume, the earth pressure force is very small.

Since the radiation pressure is a surface force, and gravity is a volume force, their ratio becomes greater if the volume density of the body is smaller. For this reason, a comet is deflected by solar radiation, when it comes near the sun.

(a) A simple case of stress calculation

Consider the fields

$$\mathbf{E} = (0,\ 0,\ 0),\ \mathbf{H} = (H,\ 0,\ 0)$$

in the static problem $(\partial/\partial t = 0)$. Then, the stress tensor is

$$\mathbf{T} = \begin{bmatrix} \frac{\mu H^2}{4\pi} - \frac{\mu H^2}{8\pi} & 0 & 0 \\ 0 & -\frac{\mu H^2}{8\pi} & 0 \\ 0 & 0 & -\frac{\mu H^2}{8\pi} \end{bmatrix} = \begin{bmatrix} \frac{\mu H^2}{4\pi} & 0 & 0 \\ 0 & 0 & 0 \\ 0 & 0 & 0 \end{bmatrix} + \begin{bmatrix} -\frac{\mu H^2}{8\pi} & 0 & 0 \\ 0 & -\frac{\mu H^2}{8\pi} & 0 \\ 0 & 0 & -\frac{\mu H^2}{8\pi} \end{bmatrix} \tag{1.8.8}$$

and the components of $\mathbf{F}$ are

$$F_X = \frac{\partial}{\partial x}\left(\frac{\mu H^2}{4\pi}\right)^2 + \frac{\partial}{\partial x}\left(-\frac{\mu H^2}{8\pi}\right),\ F_y = \frac{\partial}{\partial y}\left(-\frac{\mu H^2}{8\pi}\right),\ F_Z = \frac{\partial}{\partial z}\left(-\frac{\mu H^2}{8\pi}\right) \tag{1.8.9}$$

These relations can be written as

$$p_0 = \frac{\mu H^2}{8\pi}, \tag{1.8.10}$$

$$F_X = \frac{\partial}{\partial x}\left(\frac{\mu H^2}{4\pi}\right) - \frac{\partial p_0}{\partial x},\ F_Y = \frac{\partial p_0}{\partial y},\ F_Z = -\frac{\partial p_0}{\partial z} \tag{1.8.11}$$

$$\mathbf{F} = \frac{1}{H^2}\mathbf{H}\,(\mathbf{H}\,.\,\nabla)\left(\frac{\mu H^2}{4\pi}\right) - \nabla\left(\frac{\mu H^2}{8\pi}\right)$$

The static magnetic field $\mathbf{H}$, therefore, provides an additional scalar pressure p_0 of amount $\mu H^2/8\pi$ dynes/cm^2, and a tension parallel to the direction of $\mathbf{H}$ increasing, of magnitude $\mu H^2/4\pi$, because F_x contains the additional term

$\partial(\mu H^2/4\pi)/\partial x$, which is positive in the direction of x increasing; so it acts like a tension in that direction. This increase of pressure means loss of mobility and transparency of matter when it is immersed in a magnetic field. So, a medium becomes stiff, (or, in other words, it is "frozen", or it gets ''fossilized'', or ''glued'') by the magnetic lines of force. Thus, (1) confinements is possible of charged particles for generation of thermonuclear fusion reaction at million degrees; (2) charged particles are held together, without any material vessel by these magnetic trays, and (3) fossilization of matter under pressure inside the earth is helped. A closed magnetic line of force is analogous to an endless elastic string (or spring) having tension $\mu H^2/4\pi$ and lateral pressure $\mu H^2/8\pi$. An elastic string collapses on release from a stretched state, and the magnetic lines do not collapse like that owing to lateral pressure. Sunspots, and other astrophysical phenomena are explained as consequences of the freezing effect of stellar magnetic fields. Numerically, the pressure of a field of intensity 5000 gauss $= (5000)^2/8\pi$ dynes/cm^2 $= 10^6$ dynes/cm^2 = one atmospheric pressure. Sunspots posses magnetic fields of the order of 2000 to 3000 gauss or even more. So, the sunspot regions appear to be dark when viewed from a very large distance.

(b) Alfven waves from analogy with strings

When an elastic string (or a spring) is physically disturbed, the disturbance propagates as transverse waves with phase velocity $(T/\rho^m)^{1/2}$ where T is the tension per unit length and ρ^m is the mass density of the medium. Since magnetic lines of force are like elastic strings, dispersive vibrations are generated perpendicular to these when MHD media are perturbed. Since $T = \mu H^2/4\pi$, the phase velocity of the wave propagations along the magnetic line of force, called the Alfven velocity C_A, is

$$\left(\frac{\mu H^2}{4\pi\rho}\right)^{1/2} = C_A \tag{1.8.12}$$

Alfven waves (or MHD waves), are elastic oscillations in the magnetic lines of force. In these the field provides the elasticity and the fluid continuum provides the inertia. Let $\mathbf{H}$ be perturbed to $\mathbf{H} + \mathbf{h}$, the current increment then is C curl $\mathbf{h}/4\pi$. But the assumption in (1.8.9) that $\mathbf{H}$ is a function of x only is not consistent with the system of field equations; this approximation is valid only for a strong applied magnetostatic field.

Exercise 1.8.1. Find the radiation pressure of a plane polarized *EM* wave which is incident normally from vacuum on a surface when (*a*) it absorbs, completely, the wave, (*b*) reflects, completely, the wave.

Answer. Let the electric field of the light wave be $\mathbf{E} = (E_0 \cos\theta, 0, 0)$ with $\theta = KZ - wt$. Then $H = (0, E_0 \cos\theta, 0)$,

$$T_{XX} = \frac{1}{4\pi} E_X^2 - \frac{1}{8\pi}(E_X^2 + H_Y^2) = \frac{1}{8\pi}(E_X^2 - H_Y^2) = 0,$$

$$T_{YY} = \frac{1}{4\pi} H_Y^2 - \frac{1}{8\pi}(E_X^2 + H_Y^2) = 0,$$

$$T_{ZZ} = 0 - \frac{1}{8\pi}(E_X^2 + H_Y^2) = -\frac{1}{4\pi} E_0^2 \cos^2\theta,\ T_{ij} = 0 \text{ when } i \neq j.$$

The negatgive sign in T_{ZZ} indicates that the surface force is a pressure; the average of T_{ZZ} over the time period $T\,(= 2\pi/W)$ is

$$\langle T_{ZZ} \rangle = \frac{1}{T}\int_0^T T_{ZZ}\ dt = \frac{E_0^2}{8\pi}$$

It is the energy density of the electric field. The energy ε, per unit area of the beam, is given by $\varepsilon = \langle E^2 + H^2 \rangle/8\pi = \langle E_X^2 + H_Y^2 \rangle/8\pi = E_0^2/8\pi$. Hence, the intensity of light (the average (over the time period T) amount of energy transported across unit area in unit time) $= \varepsilon c = cE_0^2/8\pi$.

(*a*) The radiation pressure on a perfectly absorbing surface equals the intensity of light divided by c (its velocity of propagation in vacuum).

(*b*) The opposite of a perfectly absorbing surface is a perfectly reflecting surface. For a perfectly reflecting surface the intensity of the radiation pressure is double, because the intensity of light is twice as great as in a perfectly absorbing surface for normal incidence.

Exericse 1.8.2. Find the radiation pressure for oblique incidence of light, from vacuum, on a perfectly absorbing surface.

Answer. For oblique incidence the force on the surface element $d\boldsymbol{\Sigma}$ is $pd\Sigma \cos\psi$ where p is the pressure on $d\Sigma$ and ψ is the angle between the direction of incidence and the direction normal to $d\boldsymbol{\Sigma}$. The component of this force along the normal to $d\boldsymbol{\Sigma}$ is $pd\Sigma\cos^2\psi$; so the pressure per unit area on $d\boldsymbol{\Sigma} = p\cos^2\psi$. For isotropic radiation*, the pressure on $d\Sigma$ is the average value of $p\cos^2\psi$ for all directions. On the curved surface of a sphere of unit radius, this average is

$$= \frac{1}{2\pi}\int_{\varphi=0}^{2\pi}\int_{\psi=0}^{\pi} p\cos^2\psi \sin\varphi\ d\psi\ d\varphi = \frac{1}{3}\frac{E_0^2}{4\pi} = \frac{2}{3}\varepsilon$$

because $p = E_0^2/8\pi$ and also $\varepsilon = \langle E^2 + H^2 \rangle/8\pi = E_0^2/8\pi$, ε being the average energy density.

---

* The radiation is isotropic when it is propagated in all directions such that the probability of the direction of propagation of any ray, selected at random, existing within a given solid angle, is proportional to the solid angle.

Exercise 1.8.3. Find the net force exerted by one hemispherical part of a uniformly charged sphere on the other hemisphere.

Answer. For a sphere of radius R and charge Q, the surface element $d\Sigma$ and the Coulomb field E are given by

$$d\Sigma = R^2 \sin\theta \; d\theta \; d\phi \tag{1}$$

$$\mathbf{E} = \frac{Q}{R^3}\mathbf{R}$$

See Fig. 1.8.2. The direction cosines of **OP** are $\sin\theta \cos\phi$, $\sin\theta \sin\phi$ $\cos\theta$. So $E_X = E\sin\theta \cos\phi$, $E_Y = E\sin\theta \sin\phi$, $E_Z = E\cos\theta$. Only the stress

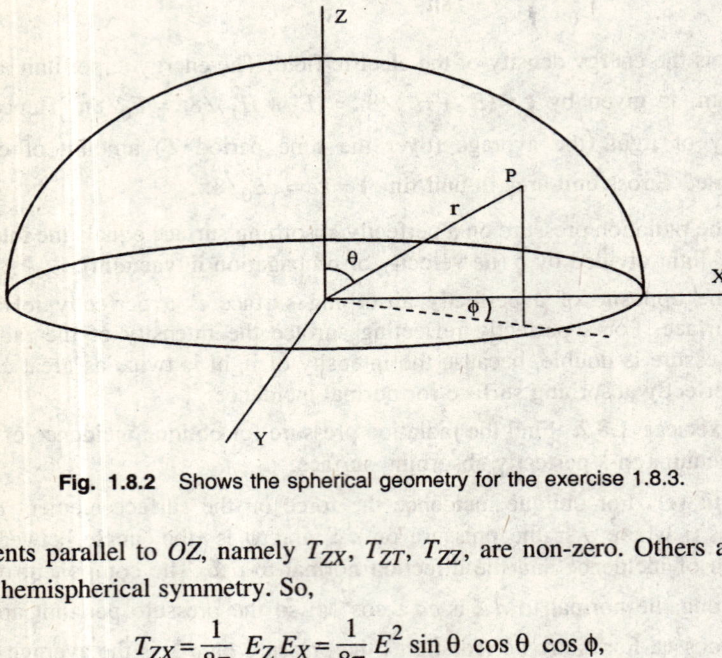

Fig. 1.8.2 Shows the spherical geometry for the exercise 1.8.3.

elements parallel to OZ, namely T_{ZX}, T_{ZT}, T_{ZZ}, are non-zero. Others are zero from hemispherical symmetry. So,

$$T_{ZX} = \frac{1}{8\pi} E_Z E_X = \frac{1}{8\pi} E^2 \sin\theta \cos\theta \cos\phi,$$

$$T_{ZY} = \frac{1}{8\pi} E_Z E_X = \frac{E^2}{8\pi} \sin\theta \cos\theta \sin\phi;$$

$$T_{ZZ} = \frac{1}{8\pi}(E_Z^2 - E_X^2 - E_Y^2) = \frac{E^2}{8\pi}(\cos^2\theta - \sin^2\theta)$$

$$\text{Hence } (\mathbf{T}\,.\,d\mathbf{\Sigma}) = T_{ZX}\,d\Sigma_X + T_{ZY}\,d\Sigma_Y + T_{ZZ}\,d\Sigma_Z = \frac{Q^2}{8\pi R^2}\sin\theta \cos\theta \; d\theta \; d\phi$$

$$= \frac{Q^2}{8\pi R^2}\sin^2\theta \cos\theta \; d\theta \; d\phi$$

The force on the upper hemisphere,

$$F_U = \frac{Q^2}{8\pi R^2} \int_{\theta=0}^{\pi/2} \int_{Q=0}^{2\pi} \sin\theta \, d\theta \, d\phi = \frac{Q^2}{8R^2}$$

The circular disc of the upper bowl is in the plane $z = 0$. The normal which is outward with respect to the bowl for the disc, is along the negative direction of the z-axis. So

$$d\mathbf{\Sigma} = (r\, d\theta\, dr\, (-\hat{\mathbf{k}}))$$

where $\hat{\mathbf{k}}$ is the unit vector along OZ. The electric field in the plane of the disc is

$$\mathbf{E} = (E \cos\phi,\ E \sin\phi,\ 0) = \frac{Q}{R^3} r\, \hat{\mathbf{r}}$$

where $\hat{\mathbf{r}} = (\cos\phi,\ \sin\phi,\ 0)$. So, $\hat{\mathbf{r}}$ is the unit vector through the centre at a point P in the plane of the disc. Here only T_{ZZ} is not zero;

$$T_{ZZ} = \frac{1}{8\pi}\,(E_Z^2 - E_X^2 - E_Y^2) = -\frac{Q^2}{8\pi}\,\frac{r^2}{r^6};\ T_{ZZ}\, d\Sigma_Z = -\frac{Q^2 r^2}{8\pi R^2}\,(-r\, d\theta\ dr)$$

Integration gives

$$F_D = \frac{Q^2}{8\pi R^2} \int_{r=0}^{R} \int_{\theta=0}^{2\pi} r^3\, d\theta\, dr = \left(\frac{Q}{4R}\right)^2$$

for the force on the disc. Then the net force on the upper bowl

$$= F = F_U + F_D = \frac{3Q^2}{16R^Z}.$$

1.9 VECTOR POTENTIAL AND SCALAR POTENTIAL

The vector **A** is defined by the relation

$$\mathbf{B} = \text{curl } \mathbf{A} \tag{1.9.1}$$

where **B** is the magnetic induction vector, so div $\mathbf{B} = 0$. Then using equation (1.3.7) we find that

$$\text{curl}\left(\mathbf{E} + \frac{1}{c}\,\frac{\partial \mathbf{A}}{\partial t}\right) = 0, \tag{1.9.2}$$

$$\mathbf{E} + \frac{1}{c}\,\frac{\partial \mathbf{A}}{\partial t} = \nabla\varphi \tag{1.9.3}$$

where φ is a scalar function of space coordinates and time. For electrostatic fields we then write $\mathbf{E} = \nabla\varphi$. Putting the values of **E** and **H** from (1.9.1) and (1.9.3) in the Maxwell equation (1.3.9) we obtain

$$\text{curl curl } \mathbf{A} + \frac{\varepsilon\,\mu}{c^2}\frac{\partial^2 \mathbf{A}}{\partial t^2} - \frac{\varepsilon\,\mu}{c}\text{ grad}\left(\frac{\partial \varphi}{\partial t}\right) = -\frac{4\pi\,\mu}{c}\mathbf{j}; \tag{1.9.4}$$

and the vector results (A 3.11) reduce this equation to

$$\nabla^2 \mathbf{A} - \frac{1}{c_1^2}\frac{\partial^2 \mathbf{A}}{\partial t^2} - \text{grad}\left(\text{div } \mathbf{A} - \frac{\varepsilon\mu}{c}\frac{\partial \varphi}{\partial t}\right) = -\frac{4\pi\,\mu}{c}\mathbf{j}$$

where $c_1^2 = c^2/\varepsilon\,\mu$. The vector **A** is defined uniquely by the Lorentz gauge condition

$$\text{div } \mathbf{A} - \frac{\varepsilon\,\mu}{c}\frac{\partial \rho^c}{\partial t} = 0 \tag{1.9.5}$$

This **A** is the vector potential and φ is the scalar potential of the Lorentz gauge. Putting (1.9.5) in (1.9.4) we obtain

$$\nabla^2 \mathbf{A} - \frac{1}{c_1^2}\frac{\partial^2 \mathbf{A}}{\partial t^2} = -\frac{4\pi}{c}\mathbf{j} \tag{1.9.6}$$

Taking divergence of (1.9.3), and using (1.9.5) and (1.3.4), the equation for ϕ is obtained :

$$\nabla^2 \varphi - \frac{1}{c_1^2}\frac{\partial^2 \varphi}{\partial t^2} = \frac{4\pi}{\varepsilon}\rho \tag{1.9.7}$$

Solution of these inhomogeneous wave equations for **A** and ϕ depends on the source fields, which are the current **J** and charge density ρ^c. Besides, characteristic source-free wave solutions (one going forward and one backward) are possible for $\mathbf{j} = 0$ in **A**, and $\rho^c = 0$ in ϕ. The source-dependent solutions for **A** and ϕ are the radiation solutions, section 1.10.

With the help of (1.9.1), (1.9.3) and (A.3.10) the Lorentz force of (1.4.2) reads

$$\mathbf{F} = \rho^c\left\{\nabla\,\varphi - \frac{1}{c}\frac{\partial \mathbf{A}}{\partial t} + \frac{1}{c}(\mathbf{v}\times(\nabla\times\mathbf{A}))\right\} \tag{1.9.8}$$

$$= \rho^c\left\{\nabla\varphi - \frac{1}{c}\frac{\partial \mathbf{A}}{\partial t} + \frac{1}{c}\nabla(\mathbf{v}\,.\,\mathbf{A}) - \frac{1}{c}(\mathbf{v}\,.\,\nabla)\,\mathbf{A} - \frac{1}{c}(\mathbf{A}\,.\,\nabla)\,\mathbf{v} - \frac{1}{c}(\mathbf{A}\times(\nabla\times\mathbf{v}))\right\} \tag{1.9.9}$$

1.10 RADIATION SOLUTION FOR THE POTENTIALS

The solutions of the wave equations (1.9.6) and (1.9.7), Appendix 4, are

$$\mathbf{A}(\mathbf{r}, t) = \frac{4\pi\mu}{c}\iiiint G(\mathbf{r}', t'; \mathbf{r}, t)\,\mathbf{j}(\mathbf{r}', t')\,dr'dt' = \frac{\mu}{c}\iiint_\tau \frac{1}{R}\mathbf{j}(\mathbf{r}', t')\,d\tau', \tag{1.10.1}$$

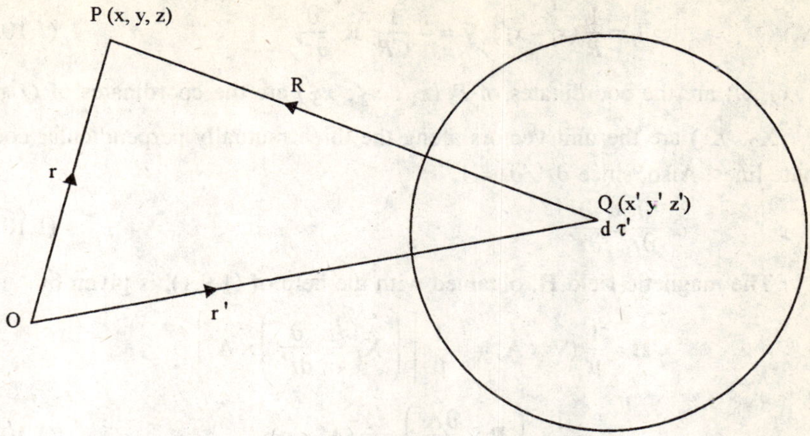

Fig. 1.10.1 The source volume V, for charges and currents, origin O, observation point P and integration point Q, are shown.

$$\varphi(\mathbf{r}, t) = -\frac{4\pi}{c} \iiiint G(\mathbf{r}', t' \, \mathbf{r}, t)\, d\mathbf{r}'\, dt' = -\frac{1}{c} \iiint_V \frac{1}{R} \rho(\mathbf{r}', t')\, dv' \quad (1.10.2)$$

where G is Green's function, $\mathbf{r}'$ is the position vector of the integration point Q, $\mathbf{r}$ is the position vector of the observation point P, $\mathbf{R} = \mathbf{r} - \mathbf{r}'$, dv' is the volume element at the integration point $\mathbf{r}'$ and

$$t' = t - \frac{1}{c}|\mathbf{r} - \mathbf{r}'| = t - \frac{1}{c} R(t'); \quad (1.10.3)$$

t' is the retarded time at which the effect occurring at Q is observed at P at time t. Integration has therefore to be performed over those regions where $\mathbf{j}$ and ρ^c are non-zero at the retarded time. The other solution at the advanced time $t + PQ/C$, called the advanced solution, is non-physical. Since $\mathbf{A}$ and φ decrease inversely as distance from the source, there is no source at infinity. The components of $\mathbf{r}$ are either (x, y, z) or (x_1, x_2, x_3); the components of $\mathbf{r}'$ also are either (x', y', z') or (x_1', x_2', x_3').

The expanded form of the gradient operator ∇ is

$$\nabla = \hat{\mathbf{X}}_i \frac{\partial}{\partial x_i} = \hat{\mathbf{X}}_i \frac{\partial t'}{\partial x_i} \frac{\partial}{\partial t'}$$

$$= \hat{\mathbf{X}}_i \frac{\partial}{\partial x_i} \left[t - \frac{1}{c} \{(x_1 - x_1')^2 + (x_2 - x_2')^2 + (x_3 - x_3')^2\}^{1/2} \right] \frac{\partial}{\partial t'}$$

$$= -\hat{\mathbf{X}}_i \frac{l_i}{c} \frac{\partial}{\partial t'} \quad (1.10.4)$$

where repeated index means summation over that (Einstein convention),

$$l_i = \frac{1}{R}(x_i - x_i'), \nabla = -\frac{1}{CR}\ \mathbf{R}\ \frac{\partial}{\partial t'} \tag{1.10.5}$$

(x_1, x_2, x_3) are the coordinates of P, (x_1', x_2', x_3') are the coordinates of Q and $(\hat{\mathbf{X}}_1, \hat{\mathbf{X}}_2, \hat{\mathbf{X}}_3)$ are the unit vectors along the three mutually perpendicular coordinate lines. Also, since $\partial t'/\partial t = 1$,

$$\frac{\partial}{\partial t} = \frac{\partial}{\partial t'} \tag{1.10.6}$$

The magnetic field $\mathbf{H}$, obtained with the help of (1.9.1), is given by

$$\mathbf{H} = \frac{1}{\mu}(\nabla \times \mathbf{A}) = -\frac{1}{\mu}\left[\left(\hat{\mathbf{X}}_i\ \frac{l_i}{c}\ \frac{\partial}{\partial t'}\right) \times \mathbf{A}\right]$$

$$= -\frac{1}{\mu c R}\left(\mathbf{R} \times \frac{\partial \mathbf{A}}{\partial t'}\right) = \frac{1}{c}(\mathbf{A}' \times \mathbf{n}) \tag{1.10.7}$$

where $$\mathbf{A}' = \frac{\partial}{\partial t'}\mathbf{A}(t'), \mathbf{n} = \frac{\mathbf{R}}{R} \tag{1.10.8}$$

Hence (1.9.3) and (1.10.4) give

$$\mathbf{E} = -\frac{1}{c}\mathbf{A}' - \frac{1}{c}\varphi'\,\mathbf{n} \tag{1.10.9}$$

where $$\varphi' = \frac{\partial}{\partial t'}\varphi(t') \tag{1.10.10}$$

Neglecting φ and ρ, the electromagnetic radiation from dipoles is evaluated from the vector potential solution. But, ρ^c and φ are also important in plasmas, whether magnetized or not. Actually, in plasmas, where the microscopic theory of Lorentz for electrodynamics is employed, $\mathbf{A}$, φ, ρ^c and $\mathbf{J}$ are mutually connected and $\varepsilon = 1$ and $\mu = 1$.

1.11 FAR FIELD APPROXIMATION

For the far field radiation solution, the largest length of the source volume V is small compared to its distances from the observation point P, in the radiation zone of Fig. 1.11.1. If the origin O is taken within V, then $OP \approx OQ$,

$$\mathbf{n} = \frac{1}{r}\mathbf{r}, \frac{1}{R} \approx \frac{1}{r}, R \approx |\mathbf{r} - \mathbf{r}'| \approx r - (\mathbf{n} \cdot \mathbf{r}'), \tag{1.11.1}$$

$$\mathbf{A} = \frac{\mu}{cr}\iiint_V \mathbf{j}\left(t - \frac{r}{c} + \frac{1}{c}(\mathbf{n} \cdot \mathbf{r}')\right) dV' \tag{1.11.2}$$

$$\approx \frac{\mu}{cr}\iiint_V \mathbf{j}\left(t' + \frac{1}{c}(\mathbf{n} \cdot \mathbf{r}')\right) dV' \tag{1.11.3}$$

$$\varphi = -\frac{1}{\varepsilon r}\iiint_V \rho\left(t - \frac{r}{c} + \frac{(\mathbf{n} \cdot \mathbf{r}')}{c}\right) dV' \tag{1.11.4}$$

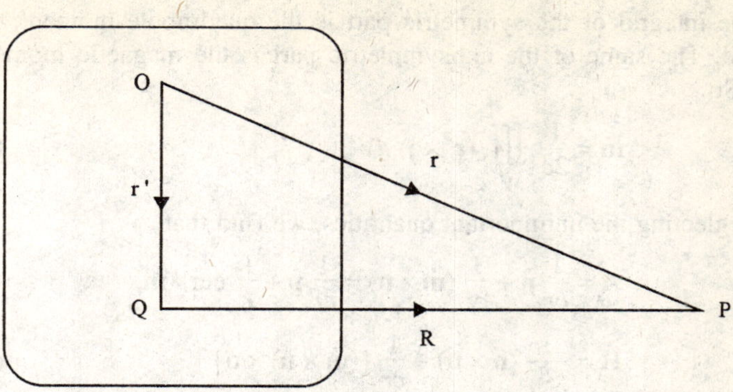

Fig. 1.11.1 The far-field approximation in the radiation problem.

$$\approx -\frac{1}{\varepsilon r} \iiint_V \rho \left(t' + \frac{1}{c} (\mathbf{n} \cdot \mathbf{r}') \right) dV' \tag{1.11.5}$$

where $t' = t - r/c$, t' is a retarded time. Retaining the first and second terms of the expansion of the integrands in powers of $(\mathbf{n} \cdot \mathbf{r}')/c$, we obtain

$$\mathbf{A}(\mathbf{r}, t) = \frac{\mu}{cr} \iiint_V \mathbf{j}(t')\, dV' + \frac{\mu}{c^2 r} \frac{\partial}{\partial t'} \iiint_V \mathbf{j}(t')\, (\mathbf{n} \cdot \mathbf{r}')\, dV', \tag{1.11.6}$$

$$\varphi(\mathbf{r}, t) = -\frac{1}{r} \iiint_V \rho^c(t')\, dV' - \frac{1}{cr} \frac{\partial}{\partial t'} \iiint_V \rho(t')\, (\mathbf{n} \cdot \mathbf{r}')\, dV' \tag{1.11.7}$$

The dipole moment $\mathbf{p}$ of the radiating volume V is defined as

$$\mathbf{p}(t') = \iiint_v \mathbf{r}' \rho^c(\mathbf{r}', t')\, dV' \tag{1.11.8}$$

Then,

$$\dot{\mathbf{p}}(t') = \frac{\partial \mathbf{p}}{\partial t'} = \iiint_V \mathbf{j}\, dV' \tag{1.11.9}$$

$$\mathbf{A}(\mathbf{r}, t) = \frac{1}{cr} \dot{\mathbf{p}} + \frac{1}{c^2 r} \frac{\partial}{\partial t'} \iiint_V \mathbf{j}(t')\, (\mathbf{n} \cdot \mathbf{r}')\, dV' \tag{1.11.10}$$

where a dot denotes differentiation with respect to the retarded time t', at which radiation occurs in V, and not with respect to the time t at the distant observation point P. Resolution of the integrand in the second term into the sum of a symmetric and an antisymmetric parts, gives

$$\frac{1}{c} (\mathbf{n} \cdot \mathbf{r}')\, \mathbf{j}(t') = \frac{1}{2c} \{(\mathbf{n} \cdot \mathbf{r}')\, \mathbf{j} + (\mathbf{n} \cdot \mathbf{j})\, \mathbf{r}'\} + \frac{1}{2c} (\mathbf{r}' \times \mathbf{j}) \times \mathbf{n} \tag{1.11.11}$$

The integral of the symmetric part is the quadrupole moment, which is neglected. The same of the antisymmetric part is the magnetic moment **m** of (1.2.5). So

$$\mathbf{m} = \frac{1}{2c} \iiint_V (\mathbf{r}' \times \mathbf{j})\, dV', \tag{1.11.12}$$

Neglecting the unimportant quantities, we find that

$$\mathbf{A} = \frac{1}{cr}\dot{\mathbf{p}} + \frac{1}{cr}(\dot{\mathbf{m}} \times \mathbf{n}) = \frac{1}{cr}\dot{\mathbf{p}} + \frac{1}{r}\ \text{curl}\ \mathbf{m},$$

$$\mathbf{H} = \frac{1}{c^2 r}(\ddot{\mathbf{p}} \times \mathbf{n}) + \frac{1}{cr}\{(\ddot{\mathbf{m}} \times \mathbf{n}) \times \mathbf{n}\} \tag{1.11.13}$$

The intensity dP of radiation in the elementary solid angle $d\Omega$ determines the energy flowing in unit time through the elementary surface $d\Sigma\ (= r^2\, d\Omega)$ of the sphere having centre at the origin O of the coordinate system in the volume V and the radial distance r. Hence, if **S** is the Poynting energy flux, $dP = |\mathbf{S}|\, r^2\, d\Omega$, and equations (1.10.7) and (1.10.8) show that

$$(\mathbf{n}\,.\,\mathbf{H}) = 0,\ \ \mathbf{S} = \frac{cH^2}{4\pi}\mathbf{n}$$

Moreover, in the radiation zone, since $\mathbf{E} = (\mathbf{H} \times \mathbf{n})$, putting $\mathbf{m} = 0$ we obtain

$$dP = \frac{1}{4\pi c^3} |(\ddot{\mathbf{p}} \times \mathbf{n})|^2\, d\Omega \tag{1.11.14}$$

$$\approx \frac{1}{4\pi c^3} |\ddot{p}|^2 \sin^2\theta\ 2\pi \sin\theta\, d\theta \tag{1.11.15}$$

where θ is the angle between **p** and **n**, and $d\Omega = 2\pi \sin\theta\, d\theta$. Integrating now with respect to θ, from 0 to π, gives

$$P = \frac{|\ddot{\mathbf{p}}|^2}{4\pi c^3} \int_0^{\pi} 2\pi \sin^3\theta\, d\theta = \frac{2}{3c^3} |\ddot{\mathbf{p}}|^2 \tag{1.11.16}$$

for the flow of energy of radiation. Its value depends on the acceleration of the charges inside the source volume V.

1.12 LIENARD-WIECHERT POTENTIALS

We now consider the field of a single electron of charge e at the point $\mathbf{r}'(t')$ moving in vacuum with velocity $\mathbf{v}\ (= d\mathbf{r}'/dt')$ along the trajectory $\mathbf{r} = \mathbf{r}'(t')$. The retarded radiative process is described by two-body Coulomb interaction in the absence of any external static magnetic field, and so in the absence of any anisotropy in the dielectric behaviour. The results are valid for long range interactions in dilute gases, so that at any time only a negligible fraction of the particles is interacting, and there is no correlation in their motion;

collective processes involving many bodies are unimportant in these radiation processes. The delta function representations of charge and current for a single charge are

$$\rho^c(\mathbf{r}, t) = e\,\delta(\mathbf{r} - \mathbf{r}'(t')), \tag{1.12.1}$$

$$\mathbf{j}(\mathbf{r}, t) = e\,\dot{\mathbf{r}}'(t')\,\delta(\mathbf{r} - \mathbf{r}'(t')) \tag{1.12.1}$$

Then

$$\mathbf{A}(\mathbf{r}, t) = \frac{e}{c}\iiiint \frac{1}{R}\,\mathbf{r}'(t')\,\delta\left(T - \frac{R}{c}\right)\delta(\mathbf{r} - \mathbf{r}'(t'))\,dv'\,dt' \tag{1.12.3}$$

where $T = t - t'$, $\mathbf{R}(t') = \mathbf{r}(t) - \mathbf{r}'(t') = R\,\mathbf{n}(t')$, $\mathbf{n} = \mathbf{R}/R$. Integrating over $\mathbf{r}'$ we obtain

$$\mathbf{A}(\mathbf{r}, t) = \frac{e}{c}\int \frac{1}{R(t')}\,r'(t')\,\delta\left(t' - t + \frac{1}{c}R(t')\right)dt' \tag{1.12.4}$$

We use the delta function property 1 of Appendix 1,

$$\int_{-\infty}^{\infty} \delta\{\psi(x)\}\,f(x)\,dx = \sum_j \frac{f(x_i)}{\left|\dfrac{d\psi}{dx}\right|_{x = x_j}} \tag{1.12.5}$$

where the $x_j\,s$ are the roots of the equation $\psi(x) = 0$, to find $\mathbf{A}$ from equation (1.12.4), and obtain

$$\mathbf{A}(\mathbf{r}, t) = \frac{e}{c}\left[\frac{\dot{\mathbf{r}}'(t')}{R(t')\left(1 + \dfrac{dR}{cdt'}\right)}\right]_{t' = t - \frac{1}{c}R(t')} \tag{1.12.6}$$

Since $R^2 = c^2(t - t')^2 = (x - x')^2 + (y - y')^2 + (z - z')^2$, (1.12.7)

we find that

$$\frac{dR}{dt'} = \left[\frac{\partial R}{\partial x'}\frac{dx'}{dt'} + \frac{\partial R}{\partial y'}\frac{dy'}{dt'} + \frac{\partial R}{\partial z'}\frac{dz'}{dt'}\right]$$

$$= -\left[(x - x')\frac{dx'}{dt'} + (y - y')\frac{dy'}{dt'} + (z - z')\frac{dz'}{dt'}\right]$$

$$= -\frac{1}{R}(\mathbf{R}\,.\,\mathbf{v}) = -(\mathbf{n}\,.\,\mathbf{v}) = -v_R \tag{1.12.8}$$

Hence (1.12.6) reduces to

$$\mathbf{A}(\mathbf{r}, t) = \left[\frac{e\mathbf{v}}{cR - (\mathbf{v}\,.\,\mathbf{R})}\right]_{t'} = \left[\frac{e\boldsymbol{\beta}}{s}\right]_{t'} \tag{1.12.9}$$

where $$s = gR = R - \frac{1}{c} R v_R = R - \frac{1}{c} (\mathbf{R} \cdot \mathbf{v}) = R\,(1 - (\mathbf{n} \cdot \boldsymbol{\beta})) \quad (1.12.10)$$

$$\boldsymbol{\beta} = \mathbf{v}/c, \; g = 1 - v_R/c = 1 - (\mathbf{n} \cdot \boldsymbol{\beta}) \quad (1.12.10\,(a))$$

Then, similarly proceeding, we obtain

$$\varphi\,(\mathbf{r},\, t) = -\left[\frac{e}{R - \frac{1}{c} (\mathbf{v} \cdot \mathbf{R})} \right]_{t'} = -\left[\frac{e}{s} \right]_{t'} \quad (1.12.11)$$

The formulas (1.12.9) and (1.12.11) are the Lienard-Weichert potential solutions for radiation of an accelerated charge. The term $(\mathbf{n} \cdot \boldsymbol{\beta})$ in s is due to relativistic motion. So the LW potentials are relativistically correct. The corresponding non-relativistic relations are

$$\mathbf{A}\,(\mathbf{r},\, t) = \left[\frac{e\boldsymbol{\beta}}{R} \right]_{t'}, \quad \varphi\,(\mathbf{r},\, t) = -\left[\frac{e}{R} \right]_{t'} \quad (1.12.12)$$

(a) The LW potential solution

Regard the electron charge e to be continuously distributed in a small volume V and not concentrated at a given point. To include the retarded effects of radiation from all points of V, at times differing slightly from each other, and reaching the observation point P at time t, Fig. 1.10.1, take an elementary volume dV in V, containing the charge element de, at the point Q. Actually, the charges in V are those found in the respective volume elements at the corresponding retarded times $t - PQ/C$; so the charges and their currents are not their normal values associated with a space point, and due consideration should be given to these retarded times in the expressions for $\mathbf{A}$ and ϕ at P at time t.

We now consider a sphere of radius R with centre at Q. The radiation from the surface element $d\Sigma$ of the sphere, at time t', reaches P at time t. At time $t' + dt'$, this radiation is on the surface of radius $R - c\,dt'$. If the electron is at rest, the amount of charge that the sphere of radius R will cross during the time dt' as it shrinks by dR, is $\rho^c\, d\Sigma\, dR$. But if $\mathbf{v}$ is the electron velocity, a charge which is less than $\rho^c\, d\Sigma\, dR$ by the amount $\rho^c\, d\Sigma\, v_R\, dt'$, will be crossed by the sphere. Since $cdt' = dR$,

$$de = \rho^c\, d\Sigma\, dR - \rho^c\, d\Sigma\, v_R\, dt' \quad (1.12.13)$$

$$= \rho^c\, dv - \rho^c\, dv\, \frac{v_R}{c} = \rho^c\, dv \left[1 - \frac{1}{RC} (\mathbf{R} \cdot \mathbf{v}) \right], \quad (1.12.14)$$

$$\rho^c\, dv = \frac{de}{1 - \beta \cos\theta}, \quad (1.12.15)$$

$$\mathbf{A} = \frac{1}{c} \iiint \left[\frac{\mathbf{v}\, de}{R\,(1 - \beta \cos\theta)} \right]_{t'}, \quad (1.12.16)$$

$$\varphi = -\iiint \left[\frac{de}{R\,(1-\beta\cos\theta)} \right]_{t'} \tag{1.12.17}$$

because *de* is the charge observed thus. These far-field solutions recover (1.12.9) and (1.12.11) for the point-like electron charge.

A three-dimensional consideration is possible of the volume distortion at a particular instant, in the charge-carrying volume element due to the retarded effects at different points in it. For this retarded solution, in addition to Q at the retarded time t' with coordinates x', y', z', consider another point Q_1 in its neighbourhood with coordinates $(x'+x_1', y'+y_1', z'+z_1')$ also at the same retarded time t'. Since, in general, $PQ_1 \neq PQ$, radiation from Q at the time t' cannot reach P at time t. But the motion of the electron being known, we find the position of Q_1^e at another time, from where the radiation reached P at time t. Let the coordinates of Q_1^e be $(x'+x_e', y'+y_e', z'+z_e')$. Since the electron volume is small, the difference in time $\delta t\ (= t_e' - t')$ is small and

$$Q_1^e P = c\,(t - t_e') = R - c\,\delta t, \tag{1.12.18}$$

$$x_e' = x_1' + v_x\,\delta t,\ y_e' = y_1' + v_y\,\delta t,\ z_e' = z_1' + v_z\,\delta t, \tag{1.12.19}$$

$$(Q_1^e P)^2 = (x - x' - x_e')^2 + (y - y' - y_e')^2 + (z - z' - z_e')^2$$
$$= (x - x' - x_1' - v_x\,\delta t)^2 + (y - y' - y_1' - v_y\,\delta t)^2 + (z - z' - z_1' - v_z\,\delta t)^2 \tag{1.12.20}$$

Since both QQ_1 and QQ_1^e are small,

$$(Q_1^e P)^2 \approx \Sigma\,(x - x')^2 - 2\Sigma\,(x - x')\,(x_1' + v_x\,\delta t)$$
$$\approx R^2 - 2\,(\mathbf{R}\,.\,(\delta\mathbf{R}' + \mathbf{v}\delta t)) \tag{1.12.21}$$

where $\delta\mathbf{R}' \equiv (x_1', y_1', z_1')$. Also, squaring (1.12.19) we obtain

$$(Q_1^e P)^2 \approx R^2 - 2RC\,\delta t \tag{1.12.22}$$

Equating these two values of $(Q_1^e P)^2$, we obtain

$$\delta t = \frac{(\mathbf{R}\,.\,\delta\mathbf{R}')}{CR\,(1 - (\boldsymbol{\beta}\,.\,\mathbf{n}))} = \frac{(\mathbf{R}\,.\,\delta\mathbf{R}')}{Cs} \tag{1.12.23}$$

Differentiating this relation with respect to x_1', we have

$$\frac{\partial\,\delta t}{\partial x_1'} = \frac{x - x'}{cs} \tag{1.12.24}$$

Let the volume elements at Q_1 and Q_1^e be dV and dV_e; respectively. Then $dV_e = j\,dV$ where the Jacobian of the volume transformation of the relations of (1.12.19) is

$$J = \begin{vmatrix} 1 + v_x \dfrac{\partial \delta t}{\partial x_1'} & v_x \dfrac{\partial \delta t}{\partial y_1'} & v_x \dfrac{\partial \delta t}{\partial z_1'} \\ v_y \dfrac{\partial \delta t}{\partial x_1'} & 1 + v_y \dfrac{\partial \delta t}{\partial y_1'} & v_y \dfrac{\partial \delta t}{\partial z_1'} \\ v_z \dfrac{\partial \delta t}{\partial x_1'} & v_z \dfrac{\partial \delta t}{\partial y_1'} & 1 + v_z \dfrac{\partial \delta t}{\partial z_1'} \end{vmatrix}$$

$$= 1 + v_x \frac{\partial \delta t}{\partial x_1'} + v_y \frac{\partial \delta t}{\partial y_1'} + v_z \frac{\partial \delta t}{\partial z_1'} + \ldots \tag{1.12.25}$$

$$= 1 + \frac{1}{cs} \{ v_x (x - x') + v_y (y - y') + v_z (z - z') \}$$

$$= 1 + \frac{1}{cs} (\mathbf{R} \, . \, \mathbf{v}) \tag{1.12.26}$$

Hence

$$dV_e = \left\{ 1 + \frac{1}{cs} (\mathbf{R} \, . \, \mathbf{v}) \right\} dv = \frac{1}{s} R dV \tag{1.12.27}$$

The volume element dv_e in (1.12.3) is actually the effective volume dv' of (1.12.27). So

$$\mathbf{A} = \iiint_v \left[\frac{1}{R} \rho^c \boldsymbol{\beta} \, dV_e \right]_{t'}, \quad \varphi = - \iiint_v \left[\frac{1}{R} \rho^c \, dV_e \right]_{t'} \tag{1.12.28}$$

(b) The four-vector method of solution of the LW potential

In the coordinate system moving with the electron, since $\mathbf{v} = 0$, we obtain

$$\mathbf{j} = 0, \; \mathbf{A} = 0, \; \varphi = - \frac{e}{R(t')} \tag{1.12.29}$$

Hence, in the laboratory frame, the 4-potential is

$$A_\mu \equiv (\mathbf{A}, \, -i\varphi) = \frac{e u_\mu}{R_\lambda \, u_\lambda} \tag{1.12.30}$$

in terms of the 4-velocity $u_\mu \equiv (\mathbf{v}, ic)$ and the 4-coordinates $R_\lambda = (\mathbf{R}, \; ic\,(t - t'))$. Since

$$t' = t - \frac{1}{c} R(t'), \tag{1.12.31}$$

we find that

$$R_\mu R_\mu = 0, \tag{1.12.32}$$

$$R_\mu u_\mu = \gamma \, [(\mathbf{v} \, . \, \mathbf{R}) - cR], \tag{1.12.33}$$

and so

$$\mathbf{A} = \left[\frac{e\mathbf{v}}{sc}\right]_{t'},\ \varphi = \left[-\frac{e}{s}\right]_{t'},\ A_\mu = \left[\frac{e\mathbf{v}}{cS}, \frac{ie}{S}\right] \tag{1.12.34}$$

1.13 RADIATION FROM AN ACCELERATED ELECTRON

Using (1.12.9) and (1.12.11), for **A** and φ, in (1.9.3) and (1.9.1) for **E** and **B**, we find that

$$\mathbf{E} = -\frac{e}{c^2}\frac{\partial}{\partial t}\left(\frac{\mathbf{v}}{\mathbf{s}}\right) - e\,\nabla\left(\frac{1}{s}\right)$$

$$= -\frac{e}{c^2 s}\frac{\partial \mathbf{v}}{\partial t} + \frac{e\mathbf{v}}{c^2 s^2}\frac{\partial s}{\partial t} + \frac{e}{s^2}\nabla s, \tag{1.13.1}$$

$$\mathbf{B} = \nabla\times\mathbf{A} = \frac{e}{c}\nabla\times\left(\frac{\mathbf{v}}{\mathbf{s}}\right)$$

$$= \frac{e}{cs}(\nabla\times\mathbf{v}) + \frac{e}{c}\left(\left(\nabla\frac{1}{s}\right)\times\mathbf{v}\right)$$

$$= \frac{e}{cs}(\nabla\times\mathbf{v}) + \frac{e}{cs^2}(\mathbf{v}\times\nabla s) \tag{1.13.2}$$

where, because of involvement of two times t and t', we ignore the variation of t' while differentiating with respect to t, and then replace t by t'. To consider the influence of t' in these results, we write

$$\frac{\partial \mathbf{v}(t')}{\partial t} = \dot{\mathbf{v}}\frac{dt'}{dt} \tag{1.13.3}$$

where dot denotes derivative with respect to t'. Thus

$$\frac{ds}{dt} = \dot{s}\frac{\partial t'}{\partial t} = \frac{dR}{dt} - \frac{1}{c}(\mathbf{R}\,.\,\dot{\mathbf{v}}) + \frac{v^2}{c}, \tag{1.13.4}$$

where we have used (1.12.10), and the fact that

$$\mathbf{v} = -\frac{d\mathbf{R}}{dt'} \tag{1.13.5}$$

Since

$$[\text{curl } \mathbf{v}]_x = \frac{\partial v_z}{\partial y} - \frac{\partial v_y}{dz} = \dot{v}_z\frac{\partial t'}{dy} - \dot{v}_y\frac{dt'}{dz} = (\text{grad } t'\times\dot{\mathbf{v}})_x,$$

we find that

$$\text{curl } \mathbf{v} = (\text{grad } t'\times\dot{\mathbf{v}}) \tag{1.13.6}$$

Also, since $\mathbf{v} = -d\mathbf{R}/dt'$, we see that

$$(\mathbf{R}\,.\,\mathbf{v}) = -\left(\mathbf{R}\cdot\frac{d\mathbf{R}}{dt'}\right) = -\frac{1}{2}\frac{dR^2}{dt'} = -R\frac{dR}{dt'}, \tag{1.13.7}$$

$$\frac{dR}{dt'} = \frac{1}{R}(\mathbf{R}\,.\,\mathbf{v}) = -v\cos\theta = -v_R = -\frac{C}{R}(R-s), \tag{1.13.8}$$

$$\mathbf{v} = -\frac{dR}{dt'}\mathbf{n} = \mathbf{n}v\cos\theta, \tag{1.13.9}$$

$$\text{grad}\, s\,(t') = \text{grad}\, s\,(R, t) + \frac{ds}{dt'}\,\text{grad}\, t', \tag{1.13.10}$$

$$\text{grad}\, s\,(t') = \mathbf{n} - \boldsymbol{\beta} + \frac{ds}{dt'}\,\text{grad}\, t', \tag{1.13.11}$$

where θ is the angle between **R** and **v**, and in the last result the relation (1.12.10) has been used. Using (1.13.8) in the first term of (1.13.4), and taking gradient of t', we obtain

$$\frac{ds}{dt'} = -v_R - \frac{1}{c}(\mathbf{R}\,.\,\dot{\mathbf{v}}) + \frac{1}{c}v^2, \tag{1.13.12}$$

$$\text{grad}\, t' = \text{grad}\left(t - \frac{1}{c}R\,(t')\right) = 0 - \frac{1}{c}\,\text{grad}\, R\,(t')$$

$$= -\frac{1}{c}\,\text{grad}\, R - \frac{1}{c}(\text{grad}\, t')\frac{dR}{dt'}$$

$$= -\frac{\mathbf{n}}{c} + \frac{1}{c}v_R\,\text{grad}\, t' \tag{1.13.13}$$

Solving for grad t' from this relation we obtain

$$\text{grad}\, t' = -\frac{1}{cs}\mathbf{R} \tag{1.13.14}$$

Differentiation of (1.12.31) with respect to t gives

$$\frac{dR}{dt} = c - c\frac{dt'}{dt} = \frac{dR}{dt'}\frac{dt'}{dt} \tag{1.13.15}$$

Hence, solving for dt'/dt gives

$$\frac{dt'}{dt} = \frac{R}{s}; \tag{1.13.16}$$

so (1.13.12) reduces to

$$\frac{ds}{dt'} = \frac{R}{s}\left\{-v_R - \frac{1}{c}(\mathbf{R}\,.\,\dot{\mathbf{v}}) + \frac{1}{c}v^2\right\} \tag{1.13.17}$$

Then (1.13.3) and (1.13.11) transform respectively to

$$\frac{d\mathbf{v}}{dt} = \frac{R}{s}\dot{\mathbf{v}}, \tag{1.13.18}$$

and

$$\text{grad}\, s = \mathbf{n} - \boldsymbol{\beta} - \frac{1}{cs}\mathbf{R}\left\{-v_R - \frac{1}{c}(\mathbf{R}\,.\,\dot{\mathbf{v}}) + \frac{v^2}{c}\right\} \tag{1.13.19}$$

The changes in the time scale of the three terms of (1.13.1) for **E** and the two terms of (1.13.2) for **B** are now complete. We find that

$$\mathbf{E} = -\frac{eR}{c^2 s^2}\dot{\mathbf{v}} + \frac{eR\mathbf{v}}{c^2 s^3}\left\{-v_R - \frac{1}{c}(\mathbf{R}\,.\,\dot{\mathbf{v}}) + \frac{1}{c}v^2\right\} +$$

$$+\frac{e}{s^2}\left[\mathbf{n} - \boldsymbol{\beta} - \frac{1}{cs}\mathbf{R}\left\{-v_R - \frac{1}{c}(\mathbf{R}\,.\,\dot{\mathbf{v}}) + \frac{1}{c}v^2\right\}\right] \quad (1.13.20)$$

$$= -\frac{eR}{cs^2}\dot{\boldsymbol{\beta}} + \frac{eR}{s^3}\boldsymbol{\beta}\left\{-(\mathbf{n}\,.\,\boldsymbol{\beta}) - \frac{R}{c}(\mathbf{n}\,.\,\dot{\boldsymbol{\beta}}) + \beta^2\right\} +$$

$$+\frac{e}{s^2}\left[-\boldsymbol{\beta} + \mathbf{n} - \frac{R}{s}\mathbf{n}\left\{-(\mathbf{n}\,.\,\boldsymbol{\beta}) - \frac{1}{c}R(\mathbf{n}\,.\,\dot{\boldsymbol{\beta}}) + \beta^2\right\}\right], \quad (1.13.21)$$

$$\mathbf{B} = \left[\frac{e}{c^2 s^2}(\dot{\mathbf{v}} \times \mathbf{R}) + \frac{e}{cs^2}(\mathbf{v} \times \nabla S)\right]_{t'} \quad (1.13.22)$$

Collecting the terms containing $\dot{\boldsymbol{\beta}}$, **n** and $\boldsymbol{\beta}$ together we obtain

$$\mathbf{E} = \frac{eR^2}{cs^3}\left[(\mathbf{n} - \boldsymbol{\beta})(\mathbf{n}\,.\,\dot{\boldsymbol{\beta}}) - \frac{s}{R}\dot{\boldsymbol{\beta}}\right] + \frac{eR}{s^3}\boldsymbol{\beta}\left[\beta^2 - (\mathbf{n}\,.\,\boldsymbol{\beta}) - \frac{s}{R}\right] +$$

$$+\frac{eR}{s^3}\mathbf{n}\,[1 - (\mathbf{n}\,.\,\boldsymbol{\beta}) + (\mathbf{n}\,.\,\boldsymbol{\beta}) - \beta^2]$$

Since $1 - (\mathbf{n}\,.\,\boldsymbol{\beta}) = (\mathbf{n}\,.\,(\mathbf{n} - \boldsymbol{\beta}))$, this relation becomes

$$\mathbf{E} = \frac{eR^2}{cs^3}\mathbf{n} \times ((\mathbf{n} - \boldsymbol{\beta}) \times \dot{\boldsymbol{\beta}}) + \frac{eR}{s^3}(\mathbf{n} - \boldsymbol{\beta})\ (1 - \beta^2) \quad (1.13.23)$$

$$= \left[\frac{e}{R^2 g^3}(\mathbf{n} - \boldsymbol{\beta})\ (1 - \beta^2) + \frac{e}{cRg^2}\mathbf{n} \times \{(\mathbf{n} - \boldsymbol{\beta}) \times \dot{\boldsymbol{\beta}}\}\right]_{t'} \quad (1.13.23\ (a))$$

Taking cross product of **E** of (1.13.20) with **n** we obtain

$$(\mathbf{n} \times \mathbf{E}) = -\frac{eR}{cs^2}(\mathbf{n} \times \dot{\boldsymbol{\beta}}) + \frac{eR}{s^3}(\mathbf{n} \times \boldsymbol{\beta})\,\{-(\mathbf{n}\,.\,\boldsymbol{\beta}) - \frac{R}{c}(\mathbf{n}\,.\,\dot{\boldsymbol{\beta}}) + \beta^2\Big\} - \frac{e}{s^2}(\mathbf{n} \times \boldsymbol{\beta})$$

$$= -\frac{eR}{cs^2}(\mathbf{n} \times \dot{\boldsymbol{\beta}}) - \frac{eR}{s^3}(\mathbf{n} \times \boldsymbol{\beta})\left\{1 - \beta^2 + \frac{R}{c}(\mathbf{n}\,.\,\dot{\boldsymbol{\beta}})\right\} \quad (1.13.24)$$

And since

$$(\boldsymbol{\beta} \times \nabla s) = -(\mathbf{n} \times \boldsymbol{\beta}) + \frac{R}{cs}(\mathbf{n} \times \boldsymbol{\beta})\ \{-(\mathbf{n}\,.\,\mathbf{v}) - (\mathbf{R}\,.\,\dot{\boldsymbol{\beta}}) + c\beta^2\}$$

$$= -\frac{R}{s}(\mathbf{n} \times \boldsymbol{\beta})\left\{1 - \beta^2 + \frac{R}{c}(\mathbf{n}\,.\,\dot{\boldsymbol{\beta}})\right\}, \quad (1.13.25)$$

we can write (1.13.24) as

$$(\mathbf{n} \times \mathbf{E}) = -\frac{eR}{cs^2}(n \times \dot{\boldsymbol{\beta}}) + \frac{e}{s^2}(\boldsymbol{\beta} \times \nabla s) \tag{1.13.26}$$

Again, using (1.13.21),

$$\mathbf{B} = -\frac{eR}{cs^2}(\mathbf{n} \times \dot{\boldsymbol{\beta}}) + \frac{e}{s^2}(\boldsymbol{\beta} \times \nabla s) \tag{1.13.27}$$

it follows that

$$(\mathbf{n} \times \mathbf{E}) = \mathbf{B} \tag{1.13.28}$$

Hence the electric field, the magnetic field and the direction of radiation at a point are three mutually perpendicular vectors. With the help of the relation (1.13.19) for ∇s, the formula (1.13.27) can be written as

$$\mathbf{B} = \frac{eR}{cs^2}(\dot{\boldsymbol{\beta}} \times \mathbf{n}) + \frac{e}{s^2}\left[1 - \frac{R}{s}\left\{(\mathbf{n} \,.\, \boldsymbol{\beta}) + \frac{R}{C}(\mathbf{n} \,.\, \dot{\boldsymbol{\beta}}) - \beta^2\right\}(\boldsymbol{\beta} \times \mathbf{n})\right]$$

$$= \frac{eR}{cs^2}(\dot{\boldsymbol{\beta}} \times \mathbf{n}) + \frac{eR\rho}{s^3}(\boldsymbol{\beta} \times \mathbf{n}) \tag{1.13.28 (a)}$$

where
$$\rho = 1 + \frac{R}{c}(\mathbf{n} \,.\, \dot{\boldsymbol{\beta}}) - \beta^2 \tag{1.13.28 (b)}$$

(a) Deduction of the results by another method

Putting the relations (1.11.2), (1.11.3), (1.12.1) and (1.12.2) in (1.10.9) we find that

$$\mathbf{E}(\mathbf{r}, t) = -\frac{e}{c^2}\frac{\partial}{\partial t}\int \frac{1}{R(t')}\mathbf{v}(t')\,\delta\left(t' - t + \frac{R(t')}{c}\right)dt'$$

$$- e\mathbf{n}\frac{\partial}{\partial R}\int \frac{1}{R(t')}\,\delta\left(t' - t + \frac{R(t')}{c}\right)dt' \tag{1.13.29}$$

$$= -\frac{e}{c^2}\frac{\partial}{\partial t}\int \frac{1}{R(t')}\,\mathbf{v}(t')\,\delta(t' - t + R(t')/C)\,dt'$$

$$+ e\mathbf{n}\left[\int \frac{1}{R^2}\,\delta\left(t' - t + \frac{R}{c}\right)dt' - \int \frac{1}{R}\,\delta'\left(t' - t + \frac{R}{c}\right)dt'\right] \tag{1.13.30}$$

where the second term has been differentiated with respect to R, and the primed delta function stands for its derivative with respect to a new argument T, defined by

$$T = t' - t + \frac{1}{c}R(t') \tag{1.13.31}$$

In the first term of (1.13.29) differentiation is done with respect to t and not t', so that this operation does not affect $\mathbf{v}(t')$ and $R(t')$ in the integrand. So,

$$\mathbf{E}(\mathbf{r},\ t) = e \int \left[\frac{1}{R^2} \mathbf{n}\, \delta T + \frac{1}{c_R} (\boldsymbol{\beta} - \mathbf{n})\, \delta^1\, T \right] \frac{dT}{g} \tag{1.13.32}$$

where from (1.12.10 (*a*))

$$g = 1 - \frac{1}{c} v_R = 1 - \frac{1}{RC} (\mathbf{R} \,.\, \mathbf{v}) = 1 - (\mathbf{n} \,.\, \boldsymbol{\beta}) \tag{1.13.33}$$

The second integral of **E**

$$= \int \frac{1}{gCR} (\boldsymbol{\beta} - \mathbf{n})\, \delta'\, T\, dT \tag{1.13.34}$$

$$= \int \left[\frac{d}{dT} \left\{ \frac{1}{cRg} (\boldsymbol{\beta} - \mathbf{n})\, \delta T \right\} - \delta T \frac{d}{dT} \left(\frac{1}{cRg} (\boldsymbol{\beta} - \mathbf{n}) \right) \right] dT$$

$$= 0 - \int \left\{ \frac{d}{dT} \left(\frac{1}{cRg} (\boldsymbol{\beta} - \mathbf{n}) \right) \right\} \delta T\, dT$$

$$= - \int \delta \left(t' - t + \frac{1}{c} R(t') \right) \left\{ \frac{d}{dt'} \left(\frac{1}{cRg} (\boldsymbol{\beta} - \mathbf{n}) \right) \right\} dt'$$

$$= - \left[\frac{1}{cg} \frac{d}{dt'} \left(\frac{1}{Rg} (\boldsymbol{\beta} - \mathbf{n}) \right) \right]_{t'}, \tag{1.13.35}$$

where the delta function relation (1.12.5) has been used. Hence, the relation (1.13.32) becomes

$$\mathbf{E}(\mathbf{r},\ t) = e \left[\frac{1}{R^2 g} \mathbf{n} + \frac{1}{cg} \frac{d}{dt'} \left(\frac{1}{Rg} (\mathbf{n} - \boldsymbol{\beta}) \right) \right]_{t'} \tag{1.13.36}$$

We now differentiate $R\mathbf{n}$ with respect to t', and using (1.12.8) for dR/dt', obtain

$$\frac{d}{dt'} (R\mathbf{n}) = R \frac{d\mathbf{n}}{dt'} + \mathbf{n} \frac{dR}{dt'} = R \frac{d\mathbf{n}}{dt'} - \mathbf{n}\, (\mathbf{n} \,.\, \mathbf{v})$$

$$= \frac{d}{dt'} (\mathbf{r}(t) - \mathbf{r}(t')), \tag{1.13.37}$$

$$\frac{1}{c} \frac{d\mathbf{n}}{dt'} = \frac{1}{Rc} \left[\frac{d}{dt'} (\mathbf{r}(t) - \mathbf{r}(t')) + \mathbf{n}\, (\mathbf{n} \,.\, \mathbf{v}) \right]$$

$$= \frac{1}{Rc} [-\mathbf{v} + \mathbf{n}\, (\mathbf{n} \,.\, \mathbf{v})] = \frac{1}{R} [-\boldsymbol{\beta} + \mathbf{n}\, (\mathbf{n} \,.\, \boldsymbol{\beta})]$$

$$= \frac{1}{R} (\mathbf{n} \times (\mathbf{n} \times \boldsymbol{\beta})), \tag{1.13.38}$$

$$\frac{\boldsymbol{\beta}}{R} = \frac{\mathbf{n}}{R} (\mathbf{n} \,.\, \boldsymbol{\beta}) - \frac{1}{c} \frac{d\mathbf{n}}{dt'}, \tag{1.13.39}$$

$$\frac{1}{c} \frac{d}{dt} \left(\frac{1}{R} \mathbf{n} \right) = \frac{R}{c} \left[\frac{d}{dt} \left(\frac{1}{R^2} \mathbf{n} \right) - \frac{1}{R} \mathbf{n} \frac{d}{dt} \left(\frac{1}{R} \right) \right]$$

$$= \frac{R}{c}\frac{d}{dt}\left(\frac{1}{n^2}\mathbf{R}\right) - \frac{\mathbf{n}}{\mathbf{c}}\frac{d}{dt}\left(\frac{1}{R}\right)\frac{dt'}{dt}$$

$$= \frac{R}{c}\frac{d}{dt}\left(\frac{1}{R^2}\mathbf{n}\right) - \frac{1}{gR^2}\mathbf{n}\,(\mathbf{n}\,.\,\boldsymbol{\beta}) \tag{1.13.40}$$

For finding (1.13.40) we have used (1.13.8) and (1.13.16). Next, taking $\boldsymbol{\beta}$ from (1.13.39) in (1.13.36), and transforming equation (1.13.36) to that in terms of t, we obtain

$$\mathbf{E}\,(\mathbf{r},\ t) = \frac{1}{gR^2}\,e\mathbf{n} + \frac{e}{c}\frac{d}{dt}\left\{\frac{1}{Rg}\mathbf{n} - \frac{1}{Rg}(\mathbf{n}\,.\,\boldsymbol{\beta})\,\mathbf{n} + \frac{1}{cg}\frac{d\mathbf{n}}{dt'}\right\}$$

$$= \frac{1}{R^2 g}\,e\mathbf{n} + \frac{e}{c}\frac{d}{dt}\left\{\frac{1}{R}\mathbf{n} + \frac{1}{c}\frac{d\mathbf{n}}{dt}\right\} \tag{1.13.41}$$

$$= \frac{1}{R^2 g}\,e\mathbf{n} + \frac{eR}{c}\frac{d}{dt}\left(\frac{1}{R^2}\mathbf{n}\right) - \frac{e}{R^2 g}(\mathbf{n}\,.\,\boldsymbol{\beta})\,\mathbf{n} + \frac{e}{c^2}\frac{d^2\mathbf{n}}{dt^2} \tag{1.13.42}$$

where g is given by the relation (1.13.33) and

$$\frac{d}{dt} = \frac{1}{g}\frac{d}{dt'} \tag{11.13.43}$$

To get the results in terms of the retarded time t' we carry out the differentiation in the second term of (1.13.36); denoting differentiation with respect to t' by dot, with the help of (1.12.8) we obtain,

$$\mathbf{E}\,(\mathbf{r},\ t) = \left[\frac{e\mathbf{n}}{R^2 g}\right]_{t'} + \frac{e}{c}\left[\frac{1}{g}(\mathbf{n}-\boldsymbol{\beta})\left\{\frac{1}{Rg^2}((\dot{\mathbf{n}}\,.\,\boldsymbol{\beta}) + (\mathbf{n}\,.\,\dot{\boldsymbol{\beta}})) + \right.\right.$$

$$\left.\left. + \frac{c}{R^2 g}(\mathbf{n}\,.\,\boldsymbol{\beta})\right\} + \frac{1}{Rg}(\dot{\mathbf{n}} - \dot{\boldsymbol{\beta}})\right]_{t'} \tag{1.13.44}$$

Collecting next all the terms containing $\dot{\boldsymbol{\beta}}$ in one side we find that

$$\mathbf{E}\,(\mathbf{r},\ t) = \left[\frac{e}{cg}\left\{\frac{1}{Rg^2}(\mathbf{n}-\boldsymbol{\beta})\ (\mathbf{n}\,.\,\dot{\boldsymbol{\beta}}) - \frac{1}{Rg}\dot{\boldsymbol{\beta}}\right\} + \frac{e\dot{\mathbf{n}}}{R^2 g}\right.$$

$$\left. + \frac{e}{Rcg^3}(\dot{\mathbf{n}}\,.\,\boldsymbol{\beta})\ (\mathbf{n}-\boldsymbol{\beta}) + \frac{e}{R^2 g^2}(\mathbf{n}\,.\,\boldsymbol{\beta})\ (\mathbf{n}-\boldsymbol{\beta}) + \frac{e\dot{\mathbf{n}}}{Rcg^2}\right]_{t'}$$

$$= \left[\frac{e}{cRg^3}\,\mathbf{n}\times\{(\mathbf{n}-\boldsymbol{\beta})\times\dot{\boldsymbol{\beta}}\}\right]_{t'} + \left[\frac{e\mathbf{n}}{R^2 g} + \frac{e}{R^2 g^3}\ (\mathbf{n}-\boldsymbol{\beta})\ \{(\mathbf{n} + (\mathbf{n}+\boldsymbol{\beta}))\,.\,\boldsymbol{\beta}\}\right.$$

$$\left. + \frac{e}{R^2 g^2}(\mathbf{n}-\boldsymbol{\beta})\ (\mathbf{n}\,.\,\boldsymbol{\beta}) + \frac{e}{R^2 g^2}(\mathbf{n}\times(\mathbf{n}\times\boldsymbol{\beta}))\right]_{t'} \tag{1.13.45}$$

where we have used (1.12.8) in the first term and (1.13.40) for $\dot{\mathbf{n}}$. Hence the second term

$$= \left[\frac{e\mathbf{n}}{R^2 g} - \frac{e\mathbf{n}}{R^2 g^3} (\mathbf{n} \times \boldsymbol{\beta})^2 (\mathbf{n} - \boldsymbol{\beta}) + \frac{e}{R^2 g^2} \{ \mathbf{n} (\mathbf{n} \,.\, \boldsymbol{\beta}) - \boldsymbol{\beta} (\mathbf{n} \,.\, \boldsymbol{\beta}) + (\mathbf{n} \,.\, \boldsymbol{\beta}) \, \mathbf{n} - \boldsymbol{\beta} \} \right]_{t'}$$

$$= \left[\frac{e}{R^2} \left(\frac{\mathbf{n}}{g} - \frac{1}{g^3} (\mathbf{n} \times \boldsymbol{\beta})^2 (\mathbf{n} - \boldsymbol{\beta}) + \frac{2}{g^2} (\mathbf{n} \,.\, \boldsymbol{\beta}) \, \mathbf{n} - \frac{1}{g^2} \boldsymbol{\beta} \{ 1 + (\mathbf{n} \,.\, \boldsymbol{\beta}) \} \right) \right]_{t'}$$

$$= \left[\frac{e}{R^2} \left(\frac{\mathbf{n}}{g} (1 + (\mathbf{n} \,.\, \boldsymbol{\beta})) - \frac{1}{g^3} (\mathbf{n} \times \boldsymbol{\beta})^2 (\mathbf{n} - \boldsymbol{\beta}) - \frac{\boldsymbol{\beta}}{g^2} \{ 1 + (\mathbf{n} \,.\, \boldsymbol{\beta}) \} \right) \right]_{t'}$$

$$= \left[\frac{e}{R^2} \left\{ \frac{1}{g^2} (\mathbf{n} - \boldsymbol{\beta}) \; (1 + (\mathbf{n} \,.\, \boldsymbol{\beta})) - \frac{1}{g^3} (\mathbf{n} \times \boldsymbol{\beta})^2 (\mathbf{n} - \boldsymbol{\beta}) \right\} \right]_{t'}$$

$$= \left[\frac{e \, (\mathbf{n} - \boldsymbol{\beta})}{R^2 g^3} \{ 1 - (\mathbf{n} \,.\, \boldsymbol{\beta})^2 - (\mathbf{n} \times \boldsymbol{\beta})^2 \} \right]_{t'}$$

$$= \left[\frac{e \, (\mathbf{n} - \boldsymbol{\beta})}{R^2 g^3} (1 - |\mathbf{n}|^2 \; |\boldsymbol{\beta}|^2) \right]_{t'}$$

$$= \left[\frac{e}{R^2 g^3} (\mathbf{n} - \boldsymbol{\beta}) \; (1 - \beta^2) \right]_{t'} \tag{1.13.46}$$

and the relativistically correct expression for the radiation electric field in terms of t', is given by

$$\mathbf{E} = \left[\frac{e}{R^2 g^3} (\mathbf{n} - \boldsymbol{\beta}) \; (1 - \beta^2) \right]_{t'} + \left[\frac{e}{cRg^3} \mathbf{n} \times \{ (\mathbf{n} - \boldsymbol{\beta}) \times \dot{\boldsymbol{\beta}} \} \right]_{t'} \tag{1.13.47}$$

1.14 RADIATION POWER

The first term of (1.13.47) is independent of the acceleration $\dot{\mathbf{v}}$ and is inversely proportional to R^2, the second term depends on $\dot{\mathbf{v}}$ and is inversely proportional to R. So the first term gives the static part of the field of the moving charge, and, at large distances, vanishes faster than the second term. Neglecting that term we have

$$\mathbf{E} \, (\mathbf{r}, \, t) = \left[\frac{e}{cRg^3} \mathbf{n} \times \{ (\mathbf{n} - \boldsymbol{\beta}) \times \dot{\boldsymbol{\beta}} \} \right]_{t'} \tag{1.14.1}$$

The Poynting flux of energy, using (1.13.28), is

$$\mathbf{S} = \frac{c}{4\pi} (\mathbf{E} \times \mathbf{H}) = \frac{c}{4\pi} (\mathbf{E} \times (\mathbf{n} \times \mathbf{E})) = \frac{cE^2}{4\pi} \mathbf{n} \tag{1.14.2}$$

Hence, in the far field zone, the Poynting energy flows out radially and the field has the character of spherical waves spreading out radially from the accelerated electron. The total radiation passing through the surface Σ of a sphere of radius R, is

$$\iint_{\Sigma} (\mathbf{S} \,.\, d\boldsymbol{\Sigma}) = \frac{c}{4\pi} \iint_{\Sigma} R^2 E^2 \, d\Omega \tag{1.14.3}$$

$$= \frac{c}{4\pi} \iint_{\Sigma} R^2 E^2 \sin\theta \, d\theta \, d\varphi \tag{1.14.4}$$

where $d\Omega \; (= d\Sigma/R^2 = \sin\theta \, d\theta \, d\varphi)$ is the differential element of the solid angle subtended at the seat Q of Fig. 1.11.1 of the charge and the integral of $\mathbf{S}$ is taken over Σ. If $dP\,(t)/d\Omega$ is the energy flux per unit time, measured by the observer at time t, due to emission by the charge at time t', then

$$\frac{dP\,(t)}{d\Omega} = R^2 \, (\mathbf{S} \,.\, \mathbf{n}) \tag{1.14.5}$$

Further, if $dP\,(t')$ is the energy flowing out per unit time at time t' at the charge, then

$$\frac{dP\,(t')}{d\Omega} = (\mathbf{S} \,.\, \mathbf{n}) \, R^2 \frac{dt}{dt'} = R^2 g \, (\mathbf{S} \,.\, \mathbf{n}) \tag{1.14.6}$$

where g is given by (1.13.33) and (1.13.16) has been used. Hence

$$\frac{dP\,(t')}{d\Omega} = \frac{c^2}{4\pi c g^5} \left| \mathbf{n} \times \{(\mathbf{n} - \boldsymbol{\beta}) \times \dot{\boldsymbol{\beta}}\} \right|^2 \tag{1.14.7}$$

When the electron is momentarily at rest, $\mathbf{v} = 0$ though $\dot{\mathbf{v}} \neq 0$. The analogous conditions $\boldsymbol{\beta} = 0$, $\dot{\boldsymbol{\beta}} \neq 0$, valid approximately when $v << c$, give

$$\frac{dt'}{dt} = 1, \;\; \mathbf{E} = \frac{e}{cR} \{(\mathbf{n} \,.\, \dot{\boldsymbol{\beta}}) \, \mathbf{n} - \dot{\boldsymbol{\beta}}\},$$

$$E^2 = \frac{e^2}{c^2 R^2} \{\dot{\boldsymbol{\beta}}^2 - (\mathbf{n} \,.\, \dot{\boldsymbol{\beta}})^2\} = \frac{e^2}{c^2 \beta^2} (\mathbf{n} \times \dot{\boldsymbol{\beta}})^2, \tag{1.14.8}$$

$$\frac{dP}{d\Omega} = \frac{e^2 \dot{v}^2}{4\pi c^3} \sin^2\theta$$

where θ is the angle made by $\mathbf{n}$ about $\dot{\boldsymbol{\beta}}$ (polar axis), $dP/d\Omega$ is then the instantaneous power radiated per unit solid angle. Hence, power radiated in all directions (the Larmor formula), found by integrating $dP/d\Omega$ over all solid angles, is

$$P = \frac{2e^2}{3c^3} \dot{v}^2 \tag{1.14.9}$$

This relation can be written as

$$P = \frac{2e^2}{3m^2 c^3}\left(\frac{d\mathbf{p}}{dt} \cdot \frac{d\mathbf{p}}{dt}\right) \tag{1.14.10}$$

where $\mathbf{p}$ is the particle momentum $m\mathbf{v}$. The Lorentz invariant generalization of the formula (1.14.10), in four-space, for deriving the relativistically correct formula for P, is

$$P = \frac{2e^2}{3m^2 c^3}\left(\frac{dp_\mu}{dT'} \frac{dp_\mu}{dT'}\right) \tag{1.14.11}$$

where dT' is the proper time element at the retarded time in the source, p_μ is the energy momentum four-vector, Einstein summation convention for repeated index is used, $dT' = dt'/\gamma$ where γ is the relativistic factor $(1-\beta^2)^{-1/2}$. Since $\mathbf{p} = m\gamma\,\mathbf{v}$, m is the rest mass, $mc\gamma = \sqrt{m^2 c^2 + p^2}$, we find that

$$\frac{dp_\mu}{dT'}\frac{dp_\mu}{dT'} = \left(\frac{dp}{dT'}\right)^2 - \frac{1}{c^2}\left(\frac{d}{dT'}(\gamma mc^2)\right)^2 \tag{1.14.12}$$

$$= \left(\frac{d\mathbf{p}}{dT'}\right)^2 - \left(\boldsymbol{\beta} \cdot \frac{d\mathbf{p}}{dT'}\right)^2, \tag{1.14.13}$$

$$\left\{\frac{d}{dT'}(mc\gamma)\right\}^2 = \left\{\frac{1}{mc\gamma}\left(\mathbf{p} \cdot \frac{d\mathbf{p}}{dT'}\right)\right\}^2 = \left(\boldsymbol{\beta} \cdot \frac{d\mathbf{p}}{dT'}\right)^2, \tag{1.14.14}$$

$$\frac{d\mathbf{p}}{dT'} = \gamma\frac{d\mathbf{p}}{dt'} = mc\gamma^2\,\{\dot{\boldsymbol{\beta}} + \gamma^2(\boldsymbol{\beta} . \dot{\boldsymbol{\beta}})\,\boldsymbol{\beta}\} \tag{1.14.15}$$

where the derivative with respect to the retarded time t' is denoted by dot, as in some preceding sections. Hence

$$\begin{aligned}\left(\frac{d\mathbf{p}}{dT'}\right)^2 - \left(\boldsymbol{\beta} \cdot \frac{d\mathbf{p}}{dT'}\right)^2 &= m^2 c^2 \gamma^4\,\{\dot{\beta}^2 + 2\gamma^2(\boldsymbol{\beta} . \dot{\boldsymbol{\beta}})^2 + \gamma^4(\boldsymbol{\beta} . \dot{\boldsymbol{\beta}})^2 \beta^2\}\\ &\quad - m^2 c^2 \gamma^4\,\{(\boldsymbol{\beta} . \dot{\boldsymbol{\beta}}) + \beta^2\gamma^2(\boldsymbol{\beta} . \dot{\boldsymbol{\beta}})\}^2\\ &= m^2 c^2 \gamma^4\,\{\dot{\beta}^2 + (\boldsymbol{\beta} . \dot{\boldsymbol{\beta}})^2(2\gamma^2 + \gamma^4\beta^2 - \gamma^4)\}\\ &= m^2 c^2 \gamma^4\,\{\dot{\beta}^2 + \gamma^2(\boldsymbol{\beta} . \dot{\boldsymbol{\beta}})^2\}\\ &= m^2 c^2 \gamma^6\,(\dot{\beta}^2 - \beta^2\dot{\beta}^2 \sin^2\psi)\\ &= m^2 c^2 \gamma^6\,\{\dot{\beta}^2 - (\boldsymbol{\beta} \times \dot{\boldsymbol{\beta}})^2\}\end{aligned} \tag{1.14.16}$$

where ψ is the angle between $\boldsymbol{\beta}$ and $\dot{\boldsymbol{\beta}}$. Putting this result in (1.14.11) we find that

$$P = \frac{2e^2}{3c}\gamma^6\,\{\dot{\beta}^2 - (\boldsymbol{\beta} \times \dot{\boldsymbol{\beta}})^2\} = \frac{2e^2}{3c}\gamma^4\,\{\dot{\beta}^2 + \gamma^2(\boldsymbol{\beta} . \dot{\boldsymbol{\beta}})^2\} \tag{1.14.17}$$

1.15 DAMPING FORCE OF RADIATION

Radiation from an accelerated electron has energy, momentum and angular momentum. This being a loss of energy to the electron, the electron motion is damped. The reaction generates a self-force of damping, $\mathbf{F}_S$, which is evaluated non-relativistically, here, from the condition of conservation of energy.

Assume that the motion is the same at the two prescribed times t_1 and t_2, then the work done by the force $\mathbf{F}_s$, on the electron, must equal the negative of the energy radiated in the time interval (t_1, t_2), so

$$\int_{t_1}^{t_2} (\mathbf{F}_s \cdot \mathbf{v})\, dt = -\frac{2e^2}{3c^3} \int_{t_1}^{t_2} \dot{v}^2\, dt \tag{1.15.1}$$

Now assume that $\dot{\mathbf{v}}$ is the same at the two times t_1 and t_2, and write

$$\int_{t_1}^{t_2} (\dot{\mathbf{v}} \cdot \dot{\mathbf{v}})\, dt = \int_{t_1}^{t_2} \left\{ \frac{d}{dt} (\dot{\mathbf{v}} \cdot \mathbf{v}) - (\ddot{\mathbf{v}} \cdot \mathbf{v}) \right\} dt$$

$$= 0 - \int_{t_1}^{t_2} (\ddot{\mathbf{v}} \cdot \mathbf{v})\, dt \tag{1.15.2}$$

Therefore (1.15.1) becomes

$$\int_{t_1}^{t_2} (\mathbf{F}_s \cdot \mathbf{v})\, dt = \frac{2e^2}{3c^3} \int_{t_1}^{t_2} (\ddot{\mathbf{v}} \cdot \mathbf{v})\, dt, \tag{1.15.3}$$

giving, for the self-force of radiation reaction, at any instant, the relation

$$\mathbf{F}_s = \frac{2e^2}{3c^3} \ddot{\mathbf{v}}, \tag{1.15.4}$$

Including this additional force, the equation of motion of a particle of charge $-e$ and mass m, reads

$$m\dot{\mathbf{v}} = -e\mathbf{E} - \frac{e}{mc} (\mathbf{v} \times \mathbf{B}) + \frac{2e^2}{3c^3} \ddot{\mathbf{v}} \tag{1.15.5}$$

These formulas can be also derived by other methods, Dutta (1977, sec. 10.2). The corresponding relativistically correct equation of motion, is

$$m\ddot{x}_\mu = -\frac{e}{c} \ddot{x}_\nu F_{\mu\nu} + \frac{2e^2}{3c^3} \left(\dddot{x}_\mu - \frac{1}{c^2} \dot{x}_\mu \dot{x}_\nu \ddot{x}_\nu \right) \tag{1.15.6}$$

where m is the rest mass, x_μ $(= \mathbf{r}, ict)$ is the four-vector of coordinates, the electromagnetic field tensor

$$F_{\mu\nu} = \underset{\mu}{\downarrow}\overset{\rightarrow \nu}{\begin{bmatrix} 0 & H_3 & -H_2 & -iE_1 \\ -H_3 & 0 & H_1 & -iE_2 \\ H_2 & -H_1 & 0 & -iE_3 \\ iE_1 & iE_2 & iE_3 & 0 \end{bmatrix}}, \tag{1.15.7}$$

and differentiation is done with respect to proper time of equation (1.14.11).

Some conceptual difficulties exist in the theory of radiation damping of an accelerated electron. The relativistically correct force of radiation damping in three-space, is given by

$$\mathbf{F}_S = \frac{2e^2}{3c^2}\gamma^2\,[\ddot{\boldsymbol{\beta}} + \gamma^2\,(\boldsymbol{\beta}\,.\,\ddot{\boldsymbol{\beta}})\,\boldsymbol{\beta} + 3\gamma^2\,(\boldsymbol{\beta}\,.\,\dot{\boldsymbol{\beta}})\,\dot{\boldsymbol{\beta}} + 3\gamma^4\,(\boldsymbol{\beta}\,.\,\dot{\boldsymbol{\beta}})^2\,\boldsymbol{\beta}] \tag{1.15.8}$$

where $\gamma^2 = (1-\beta^2)^{-1}$

In the classical model of a bound electron, executing a simple harmonic oscillation of angular frequency ω_0, and under the influence of an applied electric field $\mathbf{E}$, in place of (1.15.5), the equation of motion is

$$m\ddot{\mathbf{r}} = -m\omega_0^2\,\mathbf{r} + \frac{2e^2}{3c^3}\dddot{\mathbf{r}} - e\mathbf{E} - \frac{e}{c}(\dot{\mathbf{r}} \times \mathbf{H}) \tag{1.15.9}$$

where $\mathbf{r}$ is the displacement of the electron from its nucleus. This equation is used to find the Rayleigh scattering cross-section in plasmas and other gaseous media, and for modification in Rutherford scattering (section 6.2).

1.16 BASICALLY RELATIVISTIC NATURE OF FLUCTUATING MAGNETIC FIELDS

We consider a collection of n point charges in a volume V, e_i is the charge of the ith particle situated at $\mathbf{r}_i\,(t)$ at time t, $\dot{\mathbf{r}}_i\,(t)$ is its velocity. Then the Lorentz equations of electrodynamics are

$$\text{div }\mathbf{E} = \frac{4\pi}{v}\sum_{i=1}^{n} e_i\,\delta\,(\mathbf{r}_i - \mathbf{r}), \tag{1.16.1}$$

$$\text{curl }\mathbf{H} = \frac{1}{c}\frac{\partial \mathbf{E}}{\partial t} + \frac{4\pi}{cv}\sum_{i=1}^{n} e_i\,\dot{\mathbf{r}}_i\,\delta\,(\mathbf{r}_i - \mathbf{r}), \tag{1.16.2}$$

$$\text{div }\mathbf{B} = 0,\ \text{curl }\mathbf{E} = -\frac{1}{c}\frac{\partial \mathbf{B}}{\partial t} \tag{1.16.3}$$

where $\mathbf{r}$ is the position vector, and the sources of charges and currents are three-dimensional delta functions of position coordinates. Omitting the terms in $1/C^2$, the equations for the potentials $\varphi\,(\mathbf{r}, t)$ and $\mathbf{A}\,(\mathbf{r},\ t)$ (section 1.9), are

$$\nabla^2 \varphi - \frac{1}{c}\frac{\partial}{\partial t} \text{ div } \mathbf{A} = \frac{4\pi}{v} \sum_i e_i \delta(\mathbf{r}_i - \mathbf{r}), \tag{1.16.4}$$

$$\nabla^2 \mathbf{A} - \text{grad}\left(\text{div } \mathbf{A} - \frac{1}{c}\frac{\partial \varphi}{\partial t}\right) = -\frac{4\pi}{cv} \sum_i e_i \dot{\mathbf{r}}_i \delta(\mathbf{r}_i - \mathbf{r}) \tag{1.16.5}$$

Introducing the Lorentz gauge condition (1.9.5), these equations reduce to

$$\nabla^2 \varphi = \frac{4\pi}{\tau} \sum_i e_i \delta(\mathbf{r}_i - \mathbf{r}),\ \nabla^2 \mathbf{A} = -\frac{4\pi}{cv} \sum_i e_i \dot{\mathbf{r}}_i \delta(\mathbf{r}_i - \mathbf{r}) \tag{1.16.6}$$

where also terms proportional to $1/c^2$ have been rejected. The delta function property 12 of Appendix 1 leads to the following solutions of equations of (1.16.6),

$$\varphi = \frac{1}{v} \sum_i \frac{e_i}{|\mathbf{r}_i - \mathbf{r}|},\ \mathbf{A} = \frac{1}{cv} \sum_i \frac{e_i \dot{\mathbf{r}}_i}{|\mathbf{r}_i - \mathbf{r}|} \tag{1.16.7}$$

Then since $\mathbf{B}$ = curl $\mathbf{A}$ and $\mathbf{E}$ = grad φ, we obtain

$$\mathbf{E} = \frac{1}{v} \sum_i \nabla \left(\frac{e_i}{|\mathbf{r}_i - \mathbf{r}|} \right),\ \mathbf{B} = \sum_i \left\{ \frac{1}{cv} \nabla \times \left(\frac{e_i \dot{\mathbf{r}}_i}{|\mathbf{r}_i - \mathbf{r}|} \right) \right\} \tag{1.16.8}$$

The magnetic field is therefore of the order of $1/c$ times $\mathbf{E}$. This $\mathbf{E}$ is irrotational and is in agreement with the second equation of (1.16.3) because $\partial \mathbf{B}/c\partial t$, being of the order of $1/c^2$, is negligible in the non-relativistic theory. So, strictly speaking, the Faraday law of induction (the second equation in (1.16.3)) is a relativistic equation; non-relativistically it should be replaced by $\nabla \times \mathbf{E} = 0$ because the temporally fluctuating magnetic field is basically a quantity of relativistic order of magnitude.

Chapter 2

DYNAMICS OF CHARGED PARTICLES

2.1 INTRODUCTION

In this chapter we consider classical dynamics of charged particles in presence of constant and slowly varying magnetic field, electric field and gravitational and other fields of non-electromagnetic origin. These fields generate acceleration, polarization current and other effects on charged particles, when the mean free path (average distance between two successive collisons) is large. A Taylor series expansion of the magnetic field over fractions of a gyration period of electrons is not useful to follow plasma behaviour over many such periods. So, gyration and motion parallel to a field line are introduced explicitly into the Taylor series expansion and deviations from helical motion as perturbations.

2.2 EQUATION OF MOTION

Formal solution of the full set of Maxwell equations is not possible because of non-linearity, unknown boundary conditions and other mathematical difficulties. But the momentum transfer equation for motion of a charged particle can be solved for given **E**, **H**, **F** in some cases, ignoring influence of induced field. So, these results are valid for short duration of motion due to strong applied fields.

In the equation of motion of a charged particle,

$$m\frac{d\mathbf{v}}{dt} = e\mathbf{E} + \frac{e}{c}(\mathbf{v} \times \mathbf{H}) + \mathbf{F} \tag{2.2.1}$$

the Lorentz force acts in opposite directions on electrons and ions; so it splits their motion. But the forces of non-electrical origin act in the same direction on all particles; so these drag all charged and uncharged particles in their directions.

Equation (2.2.1) can be written as

$$m\frac{d\mathbf{v}}{dt} = e\mathbf{E}' + \frac{e}{c}(\mathbf{v} \times \mathbf{H}) \tag{2.2.2}$$

where $$\mathbf{E}' = \mathbf{E} + \frac{\mathbf{F}}{e} \tag{2.2.3}$$

So **F** is equivalent to an electric force $\mathbf{F}/e$ per unit charge. Taking scalar product of all the terms of (2.2.2) with **v** we get

$$\left(\mathbf{v} \cdot \frac{d\mathbf{v}}{dt}\right) = \frac{e}{m}(\mathbf{E}' . \mathbf{v}) \tag{2.2.4}$$

Integration gives

$$mv^2 - mv_0^2 = 2e \int_{\mathbf{r}_0}^{\mathbf{r}} (\mathbf{E}' \,.\, d\mathbf{v}) \tag{2.2.5}$$

where $\mathbf{v}_0$ is the velocity at $\mathbf{r}_0$. If both **E** and **F** are conservative fields, then we can write $\mathbf{E}' = \nabla\varphi$ and obtain

$$mv^2 - mv_0^2 = 2e\,(\varphi - \varphi_0) \tag{2.2.6}$$

where φ is the potential of $\mathbf{E}'$ at **r** and $\varphi = \varphi_0$ at $\mathbf{r} = \mathbf{r}_0$.

2.3 PARTICLE MOTION IN PRESENCE OF MAGNETIC FIELD

Putting $\mathbf{F} = 0$ and $\mathbf{E} = 0$ and assuming **H** to be constant in time and space, and parallel to the Z-axis, equation (2.2.1) reduces to

$$\ddot{x} = \Omega\,\dot{y},\; \ddot{y} = -\Omega\,\dot{x},\; \ddot{z} = 0 \tag{2.3.1}$$

where $\Omega = eH/mc$ and dot means the time derivative. These equations show that particle motion cannot be started from rest by sudden application of a constant magnetic field. Hence the transverse kinetic energy

$$W = \frac{1}{2} mv_\perp^2 = \frac{1}{2} m\,(\dot{x}^2 + \dot{y}^2) = \text{constant}, \tag{2.3.2}$$

and also the longitudinal kinetic energy

$$W_\parallel = \frac{1}{2} m\dot{z}^2 = \text{constant} \tag{2.3.3}$$

where $\mathbf{v}_\perp$, in the xy plane, is the particle velocity perpendicular to the magnetic field. The force $e\,(\mathbf{v} \times \mathbf{H})/c$ can only change the direction of motion in planes perpendicular to **H**, and cannot change the magnitude of the velocity

The first and second equations of (2.3.1), on integration, give

$$\dot{x} = \Omega\,(y - y_0) + u_1,\; \ddot{y} = -\Omega\,(x - x_0) + u_2 \tag{2.3.4}$$

So, the differential equations of motion, parallel to the x and y-axes, are

$$\ddot{x} = -\Omega^2 \left(x - x_0 - \frac{u_2}{\Omega} \right),\; \ddot{y} = -\Omega^2 \left(y - y_0 + \frac{u_1}{\Omega} \right) \tag{2.3.5}$$

where the initial coordinates and velocity components are (x_0, y_0, z_0) and (u_1, u_2, u_3) respectively. Ω is called the Larmor frequency or cyclotron gyration frequency, or simply the gyrofrequency. The solutions of equations of (2.3.5) are

$$x - x_0 - \frac{u_2}{\Omega} = -\frac{u_2}{\Omega} \cos \Omega t + \frac{u_1}{\Omega} \sin \Omega t, \tag{2.3.6}$$

$$y - y_0 + \frac{u_1}{\Omega} = \frac{u_1}{\Omega} \cos \Omega t + \frac{u_2}{\Omega} \sin \Omega t \tag{2.3.7}$$

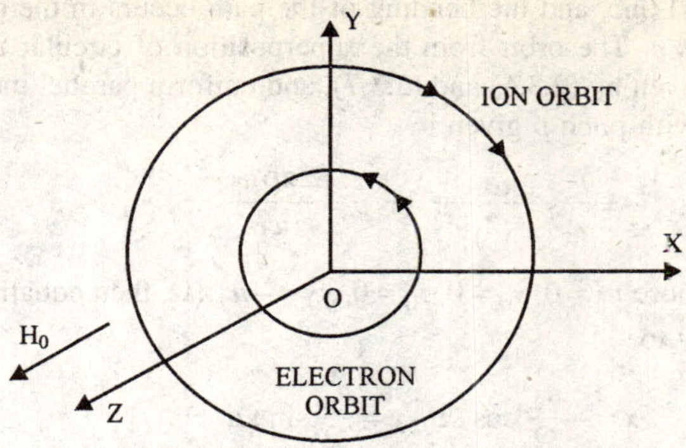

Fig. 2.3.1 Concentric orbits of an electron and an ion to an observer looking antiparallel to H_0.

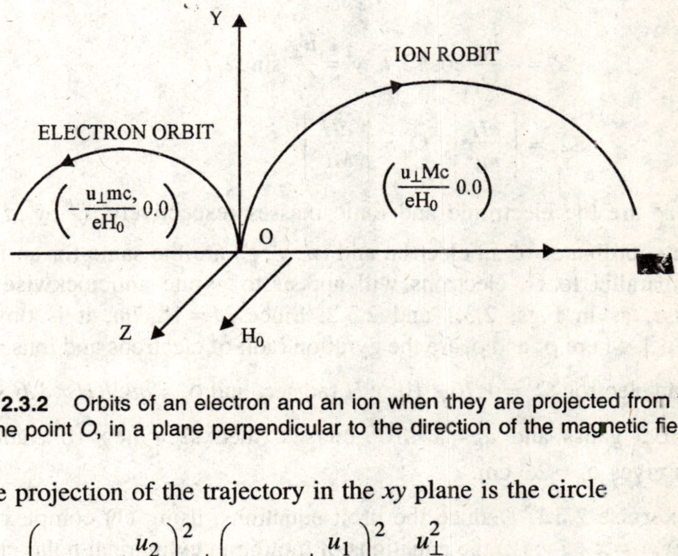

Fig. 2.3.2 Orbits of an electron and an ion when they are projected from the same point *O*, in a plane perpendicular to the direction of the magnetic field.

The projection of the trajectory in the *xy* plane is the circle

$$\left(x - x_0 - \frac{u_2}{\Omega}\right)^2 + \left(y - y_0 + \frac{u_1}{\Omega}\right)^2 = \frac{u_\perp^2}{\Omega^2} \tag{2.3.8}$$

where $$u_\perp^2 = u_1^2 + u_2^2 \tag{2.3.9}$$

If ρ is the radius of this circle, then

$$\rho = \frac{u_\perp}{\Omega} \tag{2.3.10}$$

So the radius of gyration

$$\rho = \frac{(\mathbf{v} \times \mathbf{H})}{\Omega H} = \frac{(\mathbf{v} \times \mathbf{\Omega})}{\Omega^2} \tag{2.3.11}$$

because $\mathbf{\Omega} = e\mathbf{H}/mc$, and the bending of the path occurs in the direction of the force $e\,(\mathbf{v} \times \mathbf{H})/c$. The orbit from the superposition of circular motion perpendicular to $\mathbf{H}$, given by (2.3.6) and (2.3.7), and uniform parallel motion, is a right circular helix with pitch p given by

$$\frac{p}{\dot{z}} = \frac{2\pi\rho}{(\dot{x}^2 + \dot{y}^2)^{1/2}} \quad \text{or} \quad p = \frac{2\pi\rho\, u_{\parallel}}{u_{\perp}} \tag{2.3.12}$$

If we choose $u_{\parallel} = 0$, $y_0 = 0$, $z_0 = 0$, $x_0 = -u_{\perp}/\Omega$, then equations (2.3.6) and (2.3.7) simplify to

$$x = -\frac{u_{\perp}}{\Omega}\cos \Omega t, \quad y = \frac{u_{\perp}}{\Omega}\sin \Omega t \tag{2.3.13}$$

For an electron and an ion we can therefore write

$$x^e = \frac{u_{\perp}^e}{\Omega_e}\cos \Omega_e\, t, \quad y^e = \frac{u_{\perp}^e}{\Omega_e}\sin \Omega_e\, t; \tag{2.3.14}$$

$$x^i = -\frac{u_{\perp}^i}{\Omega_i}\cos \Omega_i\, t, \quad y^i = \frac{u_{\perp}^i}{\Omega_i}\sin \Omega_i\, t \tag{2.3.15}$$

where
$$\Omega_e = \left|\frac{eH}{mc}\right|, \quad \Omega_i = \left|\frac{eH}{Mc}\right| \tag{2.3.16}$$

m and M are the electronic and ionic masses respectively, (x^e, y^e, z^e) are the position coordinates of an electron and (x^i, y^i, z^i) are the same for an ion. Looking antiparallel to $\mathbf{H}$, electrons will appear to gyrate anticlockwise and ions clockwise, as in Figs. 2.3.1 and 2.3.2. Since $M = 1837m$, it is obvious that $|\rho_i| >> |\rho_e|$, where ρ_e and ρ_i are the gyration radii of electrons and ions respecively. For an electron $\Omega_e = 1.76 \times 10^7 \times H$ rad/sec, and $\rho_e = au_{\perp}/(H \times 1.6 \times 10^8)$ cm. If $H_0 = 0.5$ gauss and $u_e = 2 \times 10^7$ cm/sec, then $\Omega_e = 8.8 \times 10^6$ rad/sec., and (2.3.10) gives $\rho_e = 2.3$ cm.

Exercise 2.3.1. Deduce the orbit equations, using (1) complex variables $r = x + iy$, $v = \dot{x} + i\,\dot{y}$; (2) the equations of motion in cylindrical polar coordinates (r, θ, z).

Exercise 2.3.2. Determine the radius and frequency Ω of gyration for a 10 Kev electron in the earth's magnetic field 0.5 gauss.

Answer. The Kinetic energy of the electron is

$$\varepsilon = 10^3 \times 10 = 10^4 \times 1.602 \times 10^{-12} \text{ erg} = mv_{\perp}^2/2,$$

where $m = 9.1 \times 10^{-28}$ gm. Hence

$$v_{\perp} = \left[\frac{2 \times 10^4 \times 1.6 \times 10^{-12}}{9.1 \times 10^{-28}}\right]^{1/2} \text{cm/sec.} = 6 \times 10^9 \text{ cm/sec.}$$

$$\Omega = \frac{eH}{mc} = \frac{4.8 \times 10^{-10} \times 0.5}{9.1 \times 10^{-28} \times 3 \times 10^{10}} \approx 8.8 \times 10^{6} \text{ rad/sec., } \rho = \frac{v_\perp}{\Omega} = 6.7 \times 10^{2} \text{ cm.}$$

Exercise 2.3.3. For an ion in a solar wind, blowing at 300 Km/sec, find ρ for $B = 5 \times 10^{-9}$ tesla.

Answer. Since ion mass $M_i = 1837\, m = 1.7 \times 10^{-24}$ gm, and 1 tesla = 10^4 gauss, we find that $\rho = 6.8 \times 10^7$ cm.

Exercise 2.3.4. Near the sun spot find $v_\perp$ and ρ for a 1 Kev H_e^+ ion if $B = 5 \times 10^{-2}$ tesla.

Answer. For a H_e^+ ion the mass is $M = 4 \times M_i = 6.8 \times 10^{-24}$ gm. Hence

$$v_\perp = 2.2 \times 10^7 \text{ cm/sec and } \rho = 18.3 \text{ cm.}$$

For estimating the time interval T of slow variation of the magnetic field H, the Taylor series expression in terms of **H** at t_0, is required:

$$\mathbf{H}(t_1) = \mathbf{H}(t_0) + (t_1 - t_0)\left(\frac{\partial \mathbf{H}}{\partial t}\right)_{t_0} + \ldots \tag{2.3.17}$$

Then, for the ith component of **H**, choose t_1 relative to t_0 such that

$$T = |H_i| / \left|\frac{\partial H_i}{\partial t}\right| \tag{2.3.18}$$

where $T = t_1 - t_0$. Similarly, the characteristic length R_{ij} of the variation of H_i parallel to the jth direction is given by

$$R_{ij} = |H_i| / \left|\frac{\partial H_i}{\partial x_j}\right| \tag{2.3.19}$$

The conditions of slow variation in time and position space, respectively, valid outside of all boundaries, are

$$T >> \frac{2\pi}{\Omega}, \tag{2.3.19 (a)}$$

$$R_{ij} >> \rho \tag{2.3.19 (b)}$$

The deviation of the path of a particle due to collisions with others (Chapter 6) depends on the mean free path length l and collision frequency ν. The relation between l and ν is

$$\frac{v_\perp}{l} = \nu \tag{2.3.20}$$

where $v_\perp$ is the velocity of the particle in the plane perpendicular to the magnetic field. Since the mean time T between collisions is $1/\nu$, we find that

$$\frac{v_\perp}{\rho} = \Omega, \Omega T = \frac{l}{\rho} \tag{2.3.21}$$

Therefore, if $l >> \rho$ (or $\Omega T >> 1$), the particle will complete several orbits before a collision occurs; but if $l << \rho$ (or $\Omega T << 1$), the path will change by collisions before a circular orbit is completed. Collision frequency ν is high when the gas pressure is above the atmospheric pressure; then $\Omega T < 1$ for about 40,000 gauss. At subatmospheric pressures $\Omega T > 1$ for B greater than 60,000 gauss.

Moving charges gyrate about magnetic lines of forces. So these lines of force act like material trays for holding the moving charges, and for providing their thermal insulations. Magnetic trays and magnetic thermal insulation eliminate the problem of material container evaporation at high temperature. Magnetic thermal insulation is used in fusion reactors to keep charges away from the material walls. Leakage parallel to the magnetic field is checked by the mirror effect (section 2.12 (*b*)) through increase of the field strength at the two ends. Even then, particles escape because mirror confinement arrangements are never full proof. Bending of the magnetic field lines to form closed curves by coils carrying current and wound round the surface of a torus also prevents particle loss along field lines. Thus toroidal magnetic fields may be used for confinement.

For confinement perpendicular to the field lines the plasma pressure $p_\perp$, several hundred atmosphere, must balance the magnetic pressure $\mu H^2/8\pi$; for instance, at temperatures above 10 KeV and a density of $10^{14}/\text{cm}^3$, since $p = NK_B T$ where N is the number density, T is the temperature in degrees Kelvin and K_B is the Boltzmann constant, 50 to 100 Killogauss magnetic field, which means a pressure of 400 atmosphere, can do the job.

The confinement condition $NK_B T = H^2/8\pi$ at the boundary of the plasma shows that the electrons arriving there are curved back into the plasma with a radius of curvature ρ given by

$$\rho = \left(\frac{mc}{eH}\right)\left(\frac{K_B T}{m}\right)^{1/2} = \left(\frac{mc^2}{8\pi Ne^2}\right)^{1/2} \approx \frac{c}{\omega_p}$$

This is roughly the thickness of the boundary layer. It is small compared to the wavelength $c/(\omega^2 - \omega_p^2)^{1/2}$ for waves of low frequency. Since collisions can remove the centres of curvature of particles from one line of force to another, some particles can be driven away from the region of confinement.

Since the radius of curvature ρ for gyrating particles is small, the displacement away from the guiding centre is not much. So the actual complicated trajectories due to the fields can be replaced by the simpler trajectories of the guiding centres for all practical purposes.

Another procedure, discussed in sections 2.7 and 2.8, is that if the motion of an observer, say A, can be found out, to whom the particle motion appears very simple, then the orbit seen by the laboratory observer, B say, is the superposition of the motion of A and the trajectory seen by A.

2.4 ORBITS IN PRESENCE OF CONSTANT MAGNETIC FIELD AND ELECTRIC FIELDS

The equation of motion,

$$\dot{\mathbf{v}} = \frac{e}{m}\mathbf{E} + (\mathbf{v} \times \mathbf{\Omega}) \tag{2.4.1}$$

can be resolved into the following two independent equations

$$\dot{\mathbf{v}}_{\parallel} = \frac{e}{m} E_{\parallel} \tag{2.4.2}$$

$$\dot{\mathbf{v}}_{\perp} = \frac{e}{m}\mathbf{E}_{\perp} + (\mathbf{v}_{\perp} \times \mathbf{\Omega}) \tag{2.4.3}$$

Putting $$\mathbf{v}_{\perp} = \mathbf{v}_i + \mathbf{v}_E \tag{2.4.4}$$

with $$\mathbf{v}_E = \frac{c}{H^2}(\mathbf{E}_{\perp} \times \mathbf{H}) \tag{2.4.5}$$

in (2.4.3), we get $$\dot{\mathbf{v}}_i = (\mathbf{v}_1 \times \mathbf{\Omega}). \tag{2.4.6}$$

Here v_E is constant, its relativistic limit being c, we must have $|E_{\perp}| < |H|$ The observer, for whom (2.4.6) is valid, finds that the particle motion is a right circular helix with axis parallel to the direction of the magnetic field. This observer moves with the drift velocity* $\mathbf{v}_E$ which is perpendicular to the plane of **E** and **H**. $\mathbf{v}_E$ is independent of charge and mass, and so is same for all species of charges. The orbit seen by the laboratory observer is the superposition of the electric drift $\mathbf{v}_E$ and the motion seen by the observer moving with this velocity. The property that both electrons and ions move with the same drift velocity can be utilized to drive a jet of neutral plasma from a plasma engine in the direction of $(\mathbf{E} \times \mathbf{H})$.

Take OZ parallel to the direction of the magnetic field, and the X-axis in the plane of **E** and **H**. If the components of **E** are $(E, O, E_{\parallel})$, the equations of motion, obtained from (2.4.1), become

$$\ddot{x} = \frac{e}{m}E + \Omega\dot{y},\ \ \ddot{y} = -\Omega\dot{x},\ \ \ddot{z} = \frac{e}{m}E \tag{2.4.7}$$

---

* The drift velocity is defined as the velocity of the particles in a small volume, and strictly speaking, is the limit of this velocity when the volume becomes very small, but not smaller than the sphere of radius equal to the gyration radius.

Solution of these equations can be written as

$$x - x_0 - \frac{1}{\Omega} u_2 - \frac{eE}{m\Omega^2} = \left(-\frac{u_2}{\Omega} - \frac{eE}{m\Omega^2} \right) \cos \Omega t + \frac{u_1}{\Omega} \sin \Omega t, \quad (2.4.8)$$

$$y - y_0 + \frac{1}{\Omega} u_1 = -\frac{eEt}{m\Omega} + \frac{u_1}{\Omega} \cos \Omega t + \frac{eE}{m\Omega^2} \sin \Omega t \quad (2.4.9)$$

$$z = z_0 + ut + \frac{eE_0}{2m} t^2 \quad (2.4.10)$$

where the initial position coordinates and velocity components are (x_0, y_0, z_0) and $(u_1, u_2, u_{\|})$ respectively. If, initially, the particle is at rest at the origin, these equations reduce to

$$x = \frac{eE}{m\Omega^2}(1 - \cos \Omega t), \; y = -\frac{eE}{m\Omega^2}(\Omega t - \sin \Omega t), \; z = \frac{eE_{\|}}{2m} t^2 \quad (2.4.11)$$

The projection of this trajectory in the *xy* plane is a cycloid. When $E_{\|} = 0$, the only force which can act is eE along OX, the gain in velocity due to this acceleration, say u, is parallel to OX; the force due to this velocity, being $e(\mathbf{u} \times \mathbf{H})/c$, is parallel to the Y-axis. So the bending begins towards the Y-axis. The particle motion again becomes parallel to **E** when it touches the Y-axis; the motion is thereafter repeated. The ions, being very heavy relative to the electrons, can be regarded as starting from rest and describing approximately the cycloid generated by the first two equations of (2.4.11). When $E_{\|} \neq 0$ the particle is also accelerated parallel to the magnetic field; so the trajectory is the superposition of this acceleration and the trajectory given by (2.4.8) and (2.4.9).

Exercise 2.4.1. An ion engine has the magnetic field 10^4 gauss. A hydrogen plasma is placed at a $\mathbf{v}_E$ drift of 1000 km/sec. Find the electric field present in the plasma.

Answer. Since the electric drift $\mathbf{v}_E = c\mathbf{E}_{\perp}/H$, we find that

$$E_{\perp} = \frac{H}{C} \mathbf{v}_E = \frac{10^4 \times 10^3 \times 10^5}{3 \times 10^{10}} = 3.3 \times 10^2 \text{ esu.}$$

Exercise 2.4.2. In a cylindrically symmetric plasma column let the electron number density $n(r)$ vary with the scale length λ (so that $\partial n/\partial r = -n/\lambda$), and let the electron concentration obey the Boltzmann law $n = n_0 \exp(-q\varphi/K_B T)$, where $q = -e$. Then (*a*) since $\mathbf{E} = -\nabla \phi$, find the radial electric field E_r depending on λ; (*b*) For electrons show that the finite Larmor radius effects are large if v_E is as large as v_{th}. Specifically show that $\rho = 2\lambda$ if $v_E = v_{th}$; (*c*) For ions also find the gyration radius if $v_E = v_{thi}$ and $T_e = T_i$.

Answer. (*a*) Since $\phi = \dfrac{K_B T}{e} \log \dfrac{n}{n_0}$, it is clear that

$$E_r = -\frac{\partial\varphi}{\partial r} = -\frac{K_B T}{e}\frac{1}{n}\frac{\partial n}{\partial r} = \frac{K_B T}{e\lambda},\ E_\theta = 0,\ E_z = 0$$

(*b*) Since $\mathbf{H} = (0, 0, H)$, we find that $\mathbf{v}_E = \frac{c}{H^2}(\mathbf{E}\times\mathbf{H}) = \left(0,\ -\frac{CK_B T}{e\lambda H},\ 0\right)$.

And since $v_{th} = (2K_B T/m)^{1/2}$, for electrons

$$|v_E| = CK_B T_e/e\lambda H = \frac{cm\, v_{th}^2}{2e\lambda H} = v_{th}^2/2\lambda\Omega.$$

In the formula $\rho = v_\perp/\Omega$, we take the average $\langle v_\perp\rangle$ for a distribution of velocities. So $\langle mv_\perp^2/2\rangle = m\langle(v_x^2 + v_y^2/2)\rangle = 2K_B T/2 = K_B T$. Hence $\langle v_\perp^2\rangle = 2K_B T/m$; $v_\perp = (2K_B T/m)^{1/2} = v_{th}$, and the relation $\rho = v_{th}/\Omega$ gives $|v_E| = (\rho v_{th}^2/2\lambda)\, v_{th}$. This shows that the relation $|v_E| = v_{th}$ implies $\rho = 2\lambda$.

(*c*) For ions $v_{thi} = (2K_B T_i/M)^{1/2} = v_{\perp i}$. So $\rho_i = v_{\perp i}/\Omega_i = v_{thi}/\Omega_i$, and we can write $|v_E| = CK_B T_e/\lambda eH = 2K_B T_e \,.\, M_C/2\lambda\, MeH$

$$= 2K_B T_i T_e/2\lambda T_i M\Omega_i = T_e v_{\text{th}}^2 \rho_i/2\lambda T_i v_{\perp i} = (T_e/2\lambda T_i)\,\rho_i\, v_{\text{thi}}$$

Hence if $|v_E| = v_{thi}$, we obtain

$$T_e \rho_i = 2\lambda T_i,\ \text{so that}\ \rho_i = 2\lambda\ \text{if}\ T_e = T_i.$$

Exercise 2.4.3. In a *Q* machine containing a cylindrical plasma, with equal electron and ion temperatures, each equal to 0.2 eV, assume that $n(r) = n_0 \exp[\exp(-r^2/a^2) - 1]$ and that the electron density obeys the Boltzmann relation $n(r) = n_0 \exp(e\phi/K_B T)$. Then for $B = 2\times 10^3$ gauss find max v_E if $a = 1$ cm. Compare this v_E with Mg due to earth's gravitational field.

Answer. We have $K_B T_e = K_B T_i = 0.2$ eV, $\frac{e\phi}{K_B T} = e^{-r^2/a^2} - 1$,

$$E_r = -\frac{\partial\phi}{\partial r} = -\frac{K_B T}{e}\cdot\frac{2r}{a^2}\, e^{-r^2/a^2},\ E_\theta = 0,\ E_r = 0.50.$$

So, evidently, $\frac{E_r}{r} = \frac{2K_B T}{ea^2}\left(1 - \frac{2r^2}{a^2}\right) e^{-r^2/a^2}$. For a maximum of E_r, since $\partial E_r/\partial r = 0$, we obtain $r = a/\sqrt{2}$. Hence $(E_r)_{\max} = \frac{K_B T}{e}\frac{2}{a^2}\frac{a}{\sqrt{2}} e^{-1/2}$. Since $K_B T_e = 0.2\ \text{eV} = \frac{2}{10}\times 1.6\times 10^{-12}$ erg and since $a = 1$ cm we find that

$$(E_r)_{\max} = \frac{2\times 1.6\times 10^{-13}}{4.8\times 10^{-10}\times 1}\cdot\frac{2}{\sqrt{2}} e^{-1/2}\ \text{ESU} = \frac{2\times 10^{-3}}{4.3}\ \text{ESU}$$

$$= \frac{2 \times 10^{-3}}{4.2} \times 3 \times 10^4 \text{ Volt/m} = 15 \text{ v/m}.$$

Hence

$$|v_E|_{max} = \frac{c}{H}(E_r)_{max} = \frac{c \times 2 \times 10^{-3}}{4.2 \times 2 \times 10^3} = \frac{6 \times 10^3}{0.84} = 8 \times 10^3 \text{ cm/sec}.$$

The gravitational field of an ion =

$$Mg = 1.7 \times 10^{-24} \times 981 \text{ dynes/cm}^2 = 1.6 \times 10^{-21} \text{ dynes/cm}^2.$$

Hence $e\,(E_r)_{max} = 4.8 \times 10^{-10} \times \frac{2 \times 10^{-3}}{4.2} = 2.3\ 10^{-13}$ dynes/cm^2

$$= 1.5 \times 10 \text{ Mg dynes/cm}^2.$$

2.5 ADIABATIC INVARIANCE OF MAGNETIC MOMENT

The magnetic moment μ is an adiabatic invariant for slowly varying magnetic field, both in space and time. The effect of slow temporal variation is discussed in section 2.14, the weak longitudinal variation in section 2.12, and here briefly the longitudinal gradient effect on motion of a charged particle. The property of adiabatic invaraince does simplify the mathematics in many cases of complicated motion.

In a cylindrical polar coordinate system the solenoidal condition $\nabla \cdot \mathbf{H} = 0$ becomes

$$\frac{1}{r}\frac{\partial}{\partial r}(rH_r) + \frac{\partial H_z}{\partial z} = 0 \tag{2.5.1}$$

where for cylindrical symmetry, the θ variation is absent. Since the magnetic field is approximately constant in distances of the order of the gyration radius, integration gives

$$H_r = -\frac{r}{2}\frac{\partial H_z}{\partial z} \tag{2.5.2}$$

so that using (2.3.11) we can write

$$H_r = -\frac{\rho}{2}\frac{\partial H_z}{\partial z} \tag{2.5.3}$$

Also, for weak longitudinal gradients,

$$\frac{\partial H_z}{\partial z} = \frac{\partial H}{\partial z} \tag{2.5.4}$$

Hence equation (2.5.3) becomes

$$H_r = -\frac{\rho}{2}\frac{\partial H}{\partial z} \tag{2.5.5}$$

Approximately the equation of motion is

$$m\frac{dv_{\parallel}}{dt}=\frac{|e|}{c}v_{\perp}H_r=-\frac{|e|}{2c}v_{\perp}\rho\frac{\partial H}{\partial z} \tag{2.5.6}$$

The magnetic moment μ, per particle of charge e, is defined as

$$\mu=\frac{|e|}{2c}v_{\perp}\rho; \tag{2.5.7}$$

so (2.5.6) reads

$$m\frac{dv_{\parallel}}{dt}=-\mu\frac{\partial H}{\partial z} \tag{2.5.8}$$

This equation can be written as

$$\frac{dW_{\parallel}}{dt}=-\mu v_{\parallel}\frac{\partial H}{\partial z} \tag{2.5.9}$$

or as

$$\frac{dW_{\parallel}}{dt}=-\frac{dW_{\perp}}{dt}=-\mu\frac{dH}{dt} \tag{2.5.10}$$

because $$W=W_{\perp}+W_{\parallel}=\text{constant}, \tag{2.5.11}$$

as may be seen from (2.3.2) and (2.3.3) or directly from (2.2.10) by putting $\phi=\phi_0$. Again, from (2.3.11), since $v_{\perp}=\rho\Omega$, equation (2.5.7) takes the vector form

$$\boldsymbol{\mu}=\frac{W_{\perp}}{H}\frac{\mathbf{H}}{H} \tag{2.5.12}$$

because $\boldsymbol{\mu}$ is directed along the line of free through the guiding centre. Then

$$\frac{d}{dt}(\mu H)=\frac{dW_{\perp}}{dt}=-\frac{dW_{\parallel}}{dt}=\mu\frac{dH}{dt} \tag{2.5.13}$$

Hence

$$\frac{d\mu}{dt}=0 \tag{2.5.14}$$

and the adiabatic invariance of the magnetic moment of a charged particle for weak longitudinal inhomogenities of the magnetic field is proved. The fields being quasistatic, equation (2.5.8) with the help of (2.5.11) and (2.5.12) gives

$$mv_{\parallel}\frac{dv_{\parallel}}{dz}=\frac{dW_{\parallel}}{dz}=-\frac{dW_{\perp}}{dz}=-\frac{W_{\perp}}{H}\frac{dH}{dz} \tag{2.5.15}$$

If, therefore, the quasistatic force $F_{\parallel}$ is defined in the manner

$$F_{\parallel}=mv_{\parallel}\frac{dv_{\parallel}}{dz}=-\mu\frac{\partial H}{\partial z}$$

$$= -\frac{dW_\perp}{dz} = -\frac{d}{dz}(\mu H) \tag{2.5.16}$$

by equation (2.5.12), then we have

$$F_\| = -\frac{d}{dz}(\mu H) \tag{2.5.17}$$

So, $F_\|$ is the conservative force with this potential.

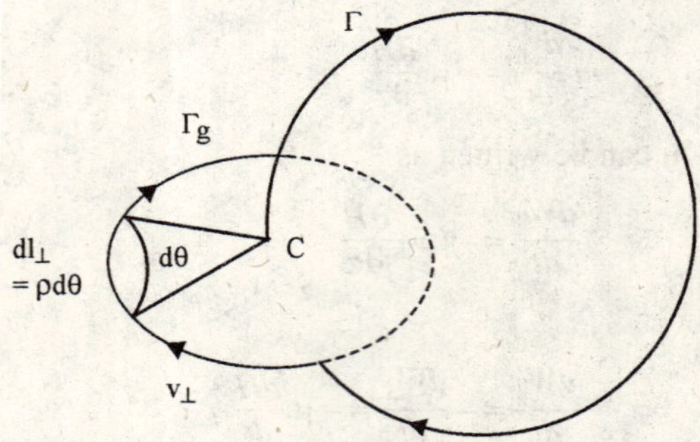

Fig. 2.5.1

Fig. 2.5.1, shows two of the three periodic motions for magnetic confinement (see section 2.4). These are (*i*) The gyromotion curve Γ_g, and (*ii*) The closed magnetic line Γ through the centre of gyration. Here ρ is the radius of gyration and $d\theta$ is the angle increment subtended by arc increment $dl_\perp$ about the guiding centre.

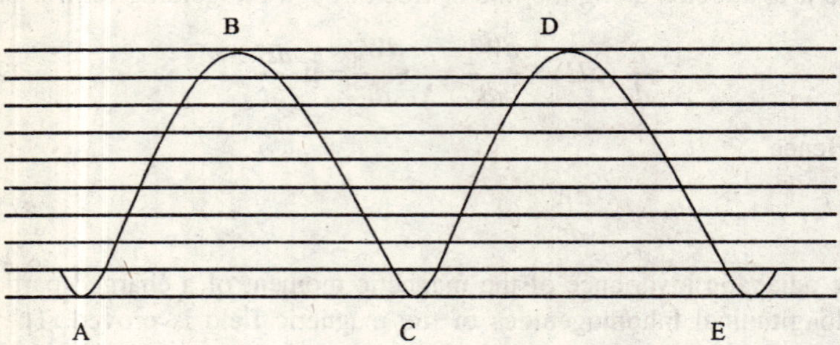

Fig. 2.5.2

Fig. 2.5.2, shows the reflection between two regions of strong magnetic fields.

2.6 SECOND AND OTHER ADIABATIC INVARIANTS

Let the circuit of gyromotion of a charged particle be Γ_g of Fig. 2.5.1, about the guiding centre C which lies on a closed line of force Γ. The separation of Γ and Γ_g is possible when the magnetic moment μ is an adiabatic invariant. Moreover, the particle is trapped between two regions of relatively strong fields (the magnetic mirror effect) of Fig. 2.5.2 and section 2.12 (*b*) which is due to a second adiabatic invariant.

Consider the trapping of a charged particle between two magnetic mirror fields (see section 2.12 (*b*)), at points A and B of Fig. 2.5.2. Let l be the distance between the magnetic lines of force through A and B. The particle motion is defined by the integral

$$j = \oint_{ABCDE} (\mathbf{v} \,.\, \mathbf{d}l) = \oint_{\Gamma} u_n \, dl \tag{2.6.1}$$

taken over one period of oscillation of the particles back and forth between the mirrors at A and B covering the path $ABCDE$. It will be shown that the motion of the particle has i of equation (2.6.1) as an adiabatic invariant (also called the longitudinal adiabatic invariant) where the magnetic moment μ is an adiabatic invariant. So, the variation exists of the component $v_{\|}$ of the velocity v parallel to the curve at a point, from motion governed by the static potential

$$\varphi(l) = \mu H \tag{2.6.2}$$

of the conservative force field $\nabla(\mu H)$.

For a simple point, assume that the magnetic field H is uniform in space except at the levels A and B where the field is large enough for the bounce due to the mirror effect. The mirror at A approaches the mirror B with the velocity

$$v_0 = -\frac{dl}{dt} \tag{2.6.3}$$

The negative sign is due to the fact that l decreases with time t. It is also assumed that

$$|v_0| << |v_{\|}| \tag{2.6.3 (a)}$$

Since H is uniform between the levels A and B, except there, we replaced (2.6.1) by

$$j = \int_0^{2l} v_{\|} \, dl = 2 v_{\|} \, l \tag{2.6.4}$$

Hence the time derivative of j is given by

$$\begin{aligned} \frac{dj}{dt} &= 2v_{\|} \frac{dl}{dt} + 2l \frac{dv_{\|}}{dt} \\ &= -2v_{\|} \, v_0 + 2l \frac{dv_{\|}}{dt} \end{aligned} \tag{2.6.5}$$

Moreover, we have

$$\frac{dv_{\parallel}}{dt} = \frac{\delta v_{\parallel}}{2l/v_{\parallel}} \tag{2.6.6}$$

where $\delta v_{\parallel}$ represents the change in particle velocity $v_{\parallel}$ on reflection by the moving mirrors. So

$$\delta v_{\parallel} = v_0 - (-v_0) = 2v_0 \tag{2.6.7}$$

and (2.6.6) becomes

$$\frac{dv_{\parallel}}{dt} = \frac{2v_0}{2l/v_{\parallel}} = \frac{v_0 v_{\parallel}}{l} \tag{2.6.8}$$

Then (2.6.4) and (2.6.5) give

$$\frac{dj}{dt} = \frac{d}{dt}(2v_{\parallel} l) = 2l\frac{dv_{\parallel}}{dt} + 2v_{|}\frac{dl}{dt}$$

$$= 2l\frac{v_0 v_{\parallel}}{2} - 2v_{\parallel} v_0 = 0 \tag{2.6.9}$$

Hence j is an adiabatic invariant in the particle motion, where the magnetic moment μ is an adiabatic invariant. For a charged particle in motion there are three adiabatic invariants, namely, the magnetic moment μ, the longitudinal or the second adiabatic invariant j and the flux invariant φ.

The magnetic moment μ of (2.5.12) is an adiabatic invariant in the guiding centre approximation. When the field can, moreover, trap the particle for oscillation between mirror fields in addition to the periodicity due to μ, the particle motion will be further controlled by the longitudinal invariance of j of relation (2.6.1). And, if, in addition to adiabatic invariance of μ and j, the drift from one line of force to another carries the particle repeatedly around a closed surface, then it will have a third periodicity in the motion due to the existence of a third adiabatic invariant φ.

The number of adiabatic invariants is determined by the number of periodicities in the motion. This number does not exceed the number of degrees of freedom in the motion of constituents of a system. A moving charged particle has three degrees of freedom. Depending on the field geometry it can have one, two or three adiabatic invariants. In the alsence of fields providing mirror effects for confinement, there is only gyrating motion of the guiding centre approximation due to the adiabatic invariance of the magnetic moment μ. So, this motion has only one adiabatic invariant though it has three degrees of freedom. In the Van Allen radiation belt, the periodicity of gyration about geomagnetic lines of force provides the adiabatic invariance of μ; the second periodicity is that of the north-south oscillation. Periodicity is that of the precessional (adjective of the word PRECESSION) rotation about the earth. So, this motion is governed by three adiabatic invariants. The guiding centre motion remains at all times on a surface with same electric current j, μ, and φ. The number of degrees of freedom is reduced by one for each adiabatic invariant.

2.7 ORBIT IN HOMOGENEOUS MAGNETIC FIELD AND A FORCE OF NON-ELECTROMAGNETIC NATURE

In this case the equation of motion is

$$\dot{\mathbf{v}} = \frac{e}{m}\mathbf{E}' + (\mathbf{v} \times \Omega) \tag{2.7.1}$$

where $\mathbf{E}' = \mathbf{F}/e$. So, all results of the preceding section hold good if $\mathbf{E}$ is replaced by $\mathbf{F}/e$, the drift, corresponding to v_E, is now denoted by v_F,

$$v_F = \frac{c}{eH^2}(\mathbf{F} \times \mathbf{H}) \tag{2.7.2}$$

It depends on the sign of charge. The gravitational drift $\mathbf{v}_g$ is obtained by replacing $\mathbf{F}$ by $m\mathbf{g}$ where m is the particle mass and $\mathbf{g}$ is the gravitational acceleration.

$$v_g = \frac{mcg}{eH} \tag{2.7.3}$$

If $\mathbf{H}$ is directed northward, $\mathbf{g}$ being downward, the gravitational drift for the ions is eastward and for the electrons westward. For an electron

$$v_g = \frac{1}{H} \times 5.5 \times 10^{-5} \text{ cm/sec} \tag{2.7.4}$$

The drift v_F, due to charge separation gives rise to a current density $\mathbf{j}_F$, given by

$$\mathbf{j}_F = \sum_r (N_r e_r v_{F_r}) \tag{2.7.5}$$

The gravitational current, therefore, is

$$\mathbf{j}_g = \frac{c}{H^2}(\mathbf{g} \times \mathbf{H})\ (\Sigma \rho_r^m) \tag{2.7.6}$$

where ρ_r is the mass density of the rth species. In the ionosphere this drift is a circulation of currents of electrons and ions in opposite senses in planes normal to the north-south axis of earth. It gives rise to an induced magnetization parallel to this axis.

Exercise 2.7.1. Calculate v_g when H is about one gauss. Find, at 300 kilometers above earth's surface, the time taken for an electron in the equitorial plane to complete one circle at that height (Radius of earth = 6370 km). Find the same for an ion also and compare.

Answer. $v_g = 5.5 \times 10^{-5}$ cm/sec. Time for an electron and an ion are 7.6×10^{13} sec, and 4.2×10^{10} sec, respectively.

Exercise 2.7.2. Find an expression for the magnetic field of the gravitational drift of electrons and ions of ionosphere.

Answer. In the ionosphere the gravitational drifts and the corresponding current are circular round the earth. The planes of these are vertical and through the east-west directions. This circular current of both sign gives rise to an induced magnetization $\mathbf{B}_i$ along the north-south line, which follows from (1.2.11).

$$\mathbf{B}_i = 4\pi \iiint_v \mathbf{M}\, dv = \frac{4\pi}{2c} \iiint_v (\mathbf{r} \times \mathbf{j}_g)\, dV$$

$$= \frac{4\pi}{2c} \iiint_v \frac{g\rho_m}{H} (\mathbf{r} \times \hat{\mathbf{E}})\, dV \tag{1}$$

where $\mathbf{r}$, (the position vector), is vertically upward, $\hat{\mathbf{E}}$, $\hat{\mathbf{N}}$, $\hat{\mathbf{V}}$ are the unit vectors along east, north and upward vertical respectively, $\mathbf{r} = (R+h)\,\hat{\mathbf{V}}$, R is the radius of earth and h is the height of the volume element dV above earth's surface. Hence

$$\mathbf{B}_i = \frac{4\pi}{2} \hat{\mathbf{N}} \int_{h_1}^{h_2} \int_{\theta=0}^{\pi} \int_{\varphi=0}^{2\pi} \frac{1}{H} (R+h)^3\, g\rho_m \sin\theta\, dh\, d\theta\, d\varphi \tag{2}$$

where h_1 and h_2 are respectively the lower and upper values of ionospheric height. Assuming g, ρ_m and H to be constant in this region, we find that

$$B_i = \frac{4\pi^2\, g\rho_m}{H} R^2\, \delta h \text{ gauss}$$

where $\delta h = h_2 - h_1$.

2.8 ORBIT IN PRESENCE OF CONSTANT MAGNETIC FIELD AND ELECTRIC FIELD SLOWLY VARYING WITH TIME

For slow variation of the electric field $\mathbf{E}$, its first derivative, partially with respect to time, is constant; so all derivatives of higher order vanish. Therefore, writing $\boldsymbol{v}_\perp = \boldsymbol{v}_1 + \boldsymbol{v}_E$ in (2.4.3) we get

$$\dot{\boldsymbol{v}}_1 = -\dot{\boldsymbol{v}}_E + (\boldsymbol{v}_1 \times \boldsymbol{\Omega}) \tag{2.8.1}$$

where $\boldsymbol{v}_E$ is given by (2.4.5). The effect of slowly varying $\mathbf{E}$ is therefore equivalent to a non-electric force per unit mass equal to $-\dot{\boldsymbol{v}}_E$. Regarding $\dot{\boldsymbol{v}}_E$ as constant we put

$$\boldsymbol{v}_1 = \boldsymbol{v}_2 + \boldsymbol{v}_P, \tag{2.8.2}$$

with

$$\boldsymbol{v}_P = -\frac{mc}{eH^2} (\dot{\boldsymbol{v}}_E \times \mathbf{H}), \tag{2.8.3}$$

and obtain

$$\dot{\boldsymbol{v}}_2 = (\boldsymbol{v}_2 \times \boldsymbol{\Omega}) \tag{2.8.4}$$

Therefore, for crossed fields,

$$v_P = \frac{mc^2}{eH^2} \dot{\mathbf{E}}_\perp \tag{2.8.5}$$

v_P is called the polarization drift. Hence, for an observer moving with the velocity $v_E + v_P$, the orbit is the right circular helix, section 2.3. In the laboratory frame the orbit is obtained by superposing the velocity $v_E + v_P$ on this helix.

Charges of both signs are therefore separated by the v_P drift, and the medium is polarized; the polarization current density $\mathbf{j}_P$ is given by

$$\mathbf{j}_P = \sum_r (N_r e_r v_{Pr}) = \frac{c^2}{H^2} \frac{\partial \mathbf{E}}{\partial t} (\sum_r N_r m_r) \tag{2.8.6}$$

where ρ is the total mass density ($= \sum_r N_r m_r$). If ρ_ρ is the corresponding space charge density, then

$$\operatorname{div} \mathbf{j}_\rho + \frac{\partial \rho_P}{\partial t} = 0 \tag{2.8.7}$$

Equation (2.8.6), on integration, gives

$$\rho_P = -\frac{\rho_0 c^2}{H^2} \operatorname{div} \mathbf{E} \tag{2.8.8}$$

Writing the equation for curl **H** as

$$\operatorname{curl} \mathbf{H} = \frac{\epsilon}{c} \frac{\partial \mathbf{E}}{\partial t} = \frac{4\pi}{c} \mathbf{j}_\rho + \frac{1}{c} \frac{\partial \mathbf{E}}{\partial t} = \frac{1}{c} \left(1 + \frac{4\pi\rho_0}{H^2} c^2 \right) \frac{\partial \mathbf{E}}{\partial t}, \tag{2.8.9}$$

and comparing the two values of $\partial \mathbf{E}/\partial t$, thus obtained, we find the formula

$$\epsilon = 1 + \frac{4\pi \rho_0}{H^2} c^2 \tag{2.8.10}$$

for the dielectric constant ϵ. Since this ϵ has a large value (for instance, when $H = 10^4$ gauss, $\rho_0 = 2 \times 10^{-9}$ gm/cc and $N = 10^{15}$/cc, we find that $\epsilon = 1 + 2 \times 10^5$), we can write approximately

$$\epsilon = \frac{4\pi \rho_0}{H^2} c^2 \tag{2.8.11}$$

The phase velocity of the wave, therefore, is given by

$$v_{Ph} = \frac{c}{\sqrt{\epsilon}} = \left(\frac{H^2}{4\pi\rho_0} \right)^{1/2} \tag{2.8.12}$$

These are for Alfven waves; (see sections 1.8 and 4.10).

If W_E and W_{KE} are respectively the densities of electric and kinetic energies due to the v_E drift, then

$$W_E + W_{KE} = \frac{E^2}{8\pi} + \frac{\rho_0 C^2}{2H^2} E^2 = \frac{E^2}{8\pi}\left(1 + \frac{4\pi\rho_0}{H^2} c^2\right) \tag{2.8.13}$$

Since $\in = 1 + 4\pi \rho_0 c^2/H^2$, writing $\mathbf{D} = \in \mathbf{E}$, this relation becomes

$$W_E + W_{KE} = \frac{1}{8\pi} (\mathbf{E} \,.\, \mathbf{D}) \tag{2.8.14}$$

If $\delta\mathbf{r}$ is the displacement due to v_ρ in time δt, then

$$\delta\mathbf{r} = v_\rho \, \delta t = \left(\frac{mc^2}{eH^2}\right) \delta\mathbf{E} \tag{2.8.15}$$

and the work done on unit volume

$$= Ne\,(\mathbf{E} \,.\, \delta\mathbf{r}) = \delta\left(\frac{1}{2} mN\, v_E^2\right) \tag{2.8.16}$$

It is the potential energy stored in the medium, and is responsible for the polarization drift. Hence the energy of the v_E drift is generated by the work done on the particle by the polarizing field. The condition of slow variation is $|\dot{\mathbf{v}}_p| << |\dot{\mathbf{v}}_2|$. So

$$\frac{|\dot{v}_P|}{|\dot{v}_2|} = \frac{|mc^2\, \ddot{\mathbf{E}}/eH^2|}{|\mathbf{v}_2\, \Omega|} = \frac{m^2 c^3\, \ddot{E}}{(e^2 H^3)|\, v_2|} << 1$$

Hence, if ω is the frequency of this $\mathbf{E}$ field, in view of (2.6.5) we can write

$$\frac{|\dot{v}_P|}{|\dot{v}_2|} = \left(\frac{\omega}{\Omega}\right)^2 \frac{|v_E|}{|v_2|} \tag{2.8.17}$$

2.9 SPATIALLY VARYING MAGNETIC FIELD AND THE GUIDING CENTRE APPROXIMATION

A weak inhomogeneity in the static magnetic field $\mathbf{H}$ is represented by the first order space derivatives, nine in number, of it. These are the elements of the matrix

$$\frac{\partial H_i}{\partial x_j} = \begin{bmatrix} \dfrac{\partial H_x}{\partial x} & \dfrac{\partial H_x}{\partial y} & \dfrac{\partial H_x}{\partial z} \\ \dfrac{\partial H_y}{\partial x} & \dfrac{\partial H_y}{\partial y} & \dfrac{\partial H_y}{\partial z} \\ \dfrac{\partial H_z}{\partial x} & \dfrac{\partial H_z}{\partial y} & \dfrac{\partial H_z}{\partial z} \end{bmatrix} \tag{2.9.1}$$

Only two of the three diagonal elements are independent, because the third one is determined from the solenoidal condition

$$\frac{\partial H_x}{\partial x}+\frac{\partial H_y}{\partial y}+\frac{\partial H_z}{\partial z}=0 \qquad (2.9.2)$$

Moreover, in the static case, the volume currents are negligible; so the Maxwell equation curl $\mathbf{H} = 0$ is valid; it gives

$$\frac{\partial H_z}{\partial y}=\frac{\partial H_y}{\partial z},\ \frac{\partial H_z}{\partial x}=\frac{\partial H_x}{\partial z},\ \frac{\partial H_x}{\partial y}=\frac{\partial H_y}{\partial x} \qquad (2.9.3)$$

These determine three of the six non-diagonal elements in terms of the remaining three. Hence the matrix of (2.9.1) possess only five independent elements; so

$$\frac{\partial H_i}{\partial x_j}=\begin{bmatrix} \dfrac{\partial H_x}{\partial x} & \dfrac{\partial H_y}{\partial x} & \dfrac{\partial H_z}{\partial x} \\ \dfrac{\partial H_y}{\partial x} & \dfrac{\partial H_y}{\partial y} & \dfrac{\partial H_y}{\partial z} \\ \dfrac{\partial H_z}{\partial x} & \dfrac{\partial H_y}{\partial z} & -\dfrac{\partial H_x}{\partial x}-\dfrac{\partial H_z}{\partial z} \end{bmatrix} \qquad (2.9.4)$$

Of these the divergence sources are $\partial H_x/\partial x$, $\partial H_y/\partial y$, $\partial H_z/\partial z$, the longitudinal gradient source is $\{(\mathbf{H}\,.\,\nabla)\ \mathbf{H}\}/|\mathbf{H}|$, the transverse gradient source is $\{(\mathbf{H}\times\nabla)\ \mathbf{H}\}/|\mathbf{H}|$, the curvature sources are $\partial H_x/\partial z$ and $\partial H_y/\partial z$ and the shear sources are $\partial H_x/\partial y$ and $\partial H_y/\partial x$. The shear sources do not produce any of the first order drifts because they twist the lines of force about each other and act in the direction of the magnetic field. So these will be neglected in our study.

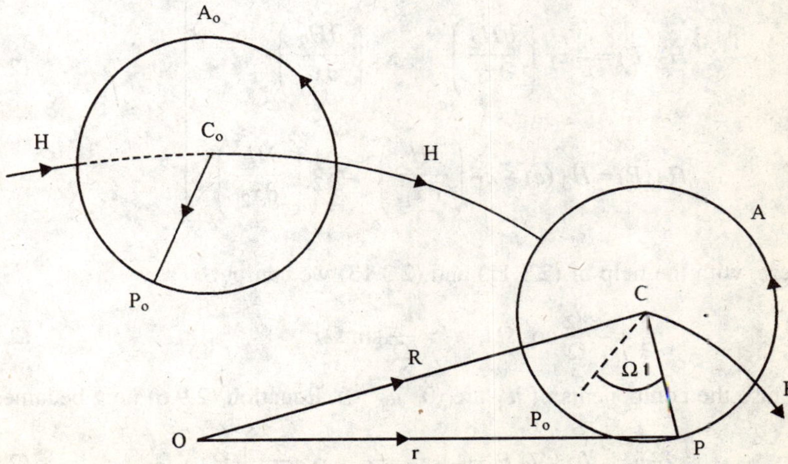

Fig. 2.9.1 Guiding centre motion.

The instantaneous position of the plane containing the tangent and the principal normal to the particle trajectory is called the guiding plane; it moves in the closed line of force through the slowly drifting guiding centre, which is the centre of curvature of the particle path, as in Fig. 2.9.1. As both gyration radius and time period decrease compared respectively to the system size and the variation time of the fields, the picture of a slowly drifting circle fits better with the real situation.

At time t the particle is at the point P, the guiding centre of its orbit is at C, the position vectors of C and P are $\mathbf{R}$ and $\mathbf{r}$ respectively; the vectorial radius of gyration ρ is directed towards C. The magnetic field at P is expanded in Taylor's series of the field at C.

$$\mathbf{H}(\mathbf{r}) = \mathbf{H}(\mathbf{R}-\mathbf{P}) = \mathbf{H}(\mathbf{R}) + \sum_n \frac{1}{n!}(-\boldsymbol{\rho}\,.\,\nabla)^n\,\mathbf{H} = \mathbf{H} + \delta\mathbf{H} \tag{2.9.5}$$

where the derivatives within the summation sign are taken at the guiding centre C, and $\delta\mathbf{H}$ (evaluated at C) is the difference between the values of $\mathbf{H}$ at P and C. The effects of spatial inhomogeneity are obtained from the first term of the series for $\delta\mathbf{H}$; so

$$\delta\mathbf{H} = -(\boldsymbol{\rho}\,.\,\nabla)\,\mathbf{H}(\mathbf{R}) \tag{2.9.6}$$

OX_3 is the tangent to the curve at C of the guiding centre; the principal normal CP coincides with OX_1 and OX_2 is along the binormal. The components of $\boldsymbol{\rho}$ along these directions may be written as $(x_1, x_2, 0)$. Hence, if the components of $\mathbf{H}$ at P, along these directions, are H_1, H_2, H_3, then up to the first order of correctness,

$$\left.\begin{aligned} H_1(P) &= -x_1\left(\frac{\partial H_i}{\partial x_1}\right)_c - x_2\left(\frac{\partial H_1}{\partial x_2}\right)_c, \\ H_2(P) &= -x_1\left(\frac{\partial H_2}{\partial x_1}\right)_c - x_2\left(\frac{\partial H_2}{\partial x_2}\right)_c, \\ H_3(P) &= H_3(c) - x_1\left(\frac{\partial H_3}{\partial x_1}\right)_c - x_2\left(\frac{\partial H_3}{\partial x_2}\right)_c \end{aligned}\right\} \tag{2.9.7}$$

where, with the help of (2.3.11) and (2.3.13) we can put

$$x_1 = -\frac{u_2}{\Omega}\cos\Omega t,\quad x_2 = \frac{u_2}{\Omega}\sin\Omega t \tag{2.9.8}$$

because the components of $u_\perp$ are $(0, u_2, 0)$. Equation (2.9.6) now becomes

$$\delta\mathbf{H} = -(\boldsymbol{\rho}\,.\,\nabla)\,\mathbf{H} = -\left(x_1\frac{\partial}{\partial x_1} + x_2\frac{\partial}{\partial x_2}\right)\mathbf{H} \tag{2.9.9}$$

because $$(\boldsymbol{\rho}\,.\,\nabla) = x_1 \frac{\partial}{\partial x_1} + x_2 \frac{\partial}{\partial x_2} \tag{2.9.10}$$

and the equation of motion of the particle is

$$\ddot{\mathbf{R}} - \ddot{\mathbf{p}} = (\dot{\mathbf{R}} \times \boldsymbol{\Omega}(\mathbf{R})) + (\dot{\mathbf{R}} \times \delta\,\boldsymbol{\Omega}) - (\dot{\boldsymbol{\rho}} \times \boldsymbol{\Omega}(\mathbf{R})) - (\dot{\boldsymbol{\rho}} \times \delta\,\boldsymbol{\Omega}) \tag{2.9.11}$$

Since both ρ and $\delta\,\mathbf{H}$ are linear combinations of $\cos \Omega t$ and $\sin \Omega t$, the terms of $\dot{\boldsymbol{\rho}} \times \delta\,\boldsymbol{\Omega}$ in (2.9.11) are linear combinations of $\cos^2 \Omega t$, $\sin^2 \Omega t$ and $\sin \Omega t \cos \Omega t$. Denoting the average of a quantity x over the time period $2\pi/\Omega$ by $\langle x \rangle$, we find that

$$\langle \ddot{\rho} \rangle = 0,\ \langle (\dot{\boldsymbol{\rho}} \times \mathbf{H}(\mathbf{R})) \rangle = 0,\ \langle \delta\mathbf{H} \rangle = 0,\ \langle (\dot{\boldsymbol{\rho}} \times \delta\mathbf{H}) \rangle \neq 0$$

Averaging all terms of (2.9.11) we obtain

$$\langle \ddot{\mathbf{R}} \rangle = \langle (\dot{\mathbf{R}} \times \boldsymbol{\Omega}(\mathbf{R})) \rangle - \langle (\dot{\boldsymbol{\rho}} \times \delta\,\boldsymbol{\Omega}) \rangle \tag{2.9.12}$$

This equation determines the static drift of the guiding centre in units of the Larmor gyration time period.

2.10 PARTICLE MOTION IN PRESENCE OF SPATIALLY INHOMOGENEOUS MAGNETIC FIELD

For study of effects of transverse gradients of inhomogeneous magnetic field, assume that

$$\mathbf{H} \equiv [0,\ 0,\ H(y)] \tag{2.10.1}$$

and, for the curvature affect,

$$\mathbf{H} \equiv [0,\ H_y(z),\ H] \tag{2.10.2}$$

The magnetic field H of (2.10.2), parallel to the z direction, is a constant; $H(y)$ of (2.10.1) is a slowly varying function of y and $H_y(z)$ of (2.10.2) is a slowly varying function of z. So, obviously $H_y(z)$, $\partial H_y(z)/\partial z$ and $\partial H(y)/\partial y$ are small quantities, and $|\delta\mathbf{H}| \ll |\mathbf{H}|$ where $\delta\,\mathbf{H} = [0,\ H_y(z),\ 0]$. This inequality is also equivalent to the condition $|\rho| \ll L$ where ρ is the radius of curvature of (2.3.11) and L is a characteristic distance, over which the magnetic field changes significantly (see section 2.3). Both transverse gradient and curvature effects can be studied simultaneously by assuming that

$$\mathbf{H} \equiv [0,\ H_y(z),\ H(y)] \tag{2.10.3}$$

As in Fig. 2.9.1, let $\mathbf{R}$ be the position vector of the centre of curvature and let $\mathbf{r}$ be the position vector of the particle itself, of charge e and mass m. Then the Taylor expansion of the field components about $\mathbf{r}$ gives

$$\begin{aligned} H(y) &= H(y_1) + (y - y_1)\,\frac{dH}{dy} + \ldots \\ &= H_0 + (y - y_1)\,H' + \varepsilon \end{aligned} \tag{2.10.4}$$

where the derivative with respect to y is denoted by a prime, ε is of the order of $(\rho/L)^2$ and $H' = dH/dy$; L being a distance over which the field $H(y)$ changes significantly; the components of $\mathbf{R}$, section 2.9, along the coordinate axes are taken to be (x_1, y_1, z_1). So, using (2.3.8), the coordinates of the centre of gyration are given by

$$x_1 = x_0 + \frac{u_2}{\Omega},\quad y_1 = y_0 - \frac{u_1}{\Omega} \tag{2.10.5}$$

(a) Transverse gradient effects

In this case, using (2.10.1) in (2.3.1), we find that

$$\ddot{x} = \Omega(y)\,\dot{y}, \tag{2.10.6}$$

$$\ddot{y} = -\Omega(y)\,\dot{x} \tag{2.10.7}$$

Differentiating (2.10.6) with respect to t and using (2.10.7) we obtain the third order equation

$$\dddot{x} = -\Omega^2(y)\,\dot{X} - \dot{y}^2\,\Omega' \tag{2.10.8}$$

where $\Omega' = \partial\Omega/\partial y$. Similarly equation (2.10.7) becomes

$$\dddot{y} = -\Omega^2(y)\,\dot{y} - \dot{x}\,\dot{y}\,\Omega' \tag{2.10.9}$$

These two mutually connected non-linear differential equations are solved by the Taylor expansion of (2.10.4) and using the relations (2.3.6), (2.3.7), together with (2.10.5); these operations yield the equations

$$\dddot{x} + \Omega_0^2\,\dot{x} = -\frac{1}{2}\,u_\perp^2\,\Omega_0'\,(1 + 3\cos 2\theta), \tag{2.10.10}$$

$$\dddot{y} + \Omega_0^2\,\dot{y} = \frac{3}{2}\,u_\perp^2\,\Omega_0'\sin 2\theta \tag{2.10.11}$$

where $\theta = \Omega_0\,t - \alpha,\ u_1 = u_\perp\cos\alpha,\ u_2 = u_\perp\sin\alpha$ (2.10.12)

The non-oscillatory part of (2.10.10) is

$$\dot{x} = -\frac{u_\perp^2\,\Omega_0'}{2\Omega_0^2} \tag{2.10.13}$$

Hence a strong magnetic field in the z direction, varying parallel to the y direction, gives rise to a drift in the x direction. Vectorially the formula for this drift, called the transverse gradient drift, is

$$\mathbf{v}_G = \frac{c}{eH_0^3}\,W_\perp\,(\mathbf{H}\times\nabla)\,H \tag{2.10.14}$$

where the kinetic energy $W_\perp$ perpendicular to the magnetic field is

$$W_\perp = \frac{1}{2}\,mu_\perp^2 \tag{2.10.15}$$

(b) Curvature effects

In this case, equation (2.2.5) gives

$$\ddot{x} = -\Omega\dot{y} - \Omega_y\,\dot{z}, \tag{2.10.16}$$

$$\ddot{y} = -\Omega\,\dot{x}, \tag{2.10.17}$$
$$\ddot{z} = \Omega_y\,\dot{x} \tag{2.10.18}$$

where $\Omega_y = eH_y(z)/mc$. Neglecting squares of small quantities, equations (2.10.16) and (2.10.18) give

$$\dddot{x} + \Omega^2\,\dot{x} = -\Omega_y{}'\,v_{\|}^2, \tag{2.10.19}$$
$$\dddot{y} + \Omega^2\,\dot{y} = \Omega\,\Omega_y\,v_{\|},\ \dddot{z} = -\dot{\Omega}_y\,\dot{x}\,v_{\|} \tag{2.10.20}$$

where $\Omega_y{}' = d\Omega_y(z)/dz$. Equation (2.10.19) determines the curvature drift

$$v_R = -\frac{v_{\|}^2\,\Omega_y{}'}{\Omega^2} \tag{2.10.21}$$

In vector form it reads

$$v_R = \frac{2cW_{\|}}{eH^4}\,[\mathbf{H}\times(\mathbf{H}\,.\,\nabla)\,\mathbf{H}] \tag{2.10.22}$$

Using the differential geometry result (2.11.17), it becomes

$$v_R = \frac{2cW_{\|}}{eH^2\,R}\,(\mathbf{H}\times\hat{\mathbf{X}}) \tag{2.10.23}$$

The similar drift in the y direction, $v_{\|}\,H_y(z)/H$, is small compared to the drift $v_{\|}$ in the z direction. Locally this drift is in the plane of the curved line of force, and keeps the guiding centre motion parallel to **H** in this plane. Since R is the principal radius of curvature of the curved line of the magnetic field through the guiding centre, its motion results in the centrifugal force $mv_{\|}^2/R$. The two drifts v_G and v_R and their currents are further discussed in section 2.13.

2.11 AN ALTERNATIVE TREATMENT OF THIS MOTION

The equation of particle motion is

$$\dot{v}(P) = (v(P)\times\boldsymbol{\Omega}(P)) \tag{2.11.1}$$

Expanding $\boldsymbol{\Omega}(P)$ like the **H** of (2.9.5), and retaining only terms up to the first order of derivatives, we get

$$\boldsymbol{\Omega}(P) = \boldsymbol{\Omega}(c) - (\boldsymbol{\rho}\,.\,\nabla)\,\boldsymbol{\Omega} \tag{2.11.2}$$

where
$$\boldsymbol{\rho} = \frac{(\mathbf{v}\times\boldsymbol{\Omega})}{\Omega^2} \tag{2.11.3}$$

If the unit vectors along OX_1, OX_2 and OX_3 (section 2.9) are denoted by $\hat{\mathbf{X}}_1$, $\hat{\mathbf{X}}_2$, $\hat{\mathbf{X}}_3$, respectively, we can write equation (2.11.1) as

$$\hat{\mathbf{X}}_i\,v_i + \hat{\mathbf{X}}_i\,v_i = (\mathbf{v}\times\boldsymbol{\Omega}) - (\mathbf{v}\times(\boldsymbol{\rho}\,.\,\nabla)\,\boldsymbol{\Omega}) \tag{2.11.4}$$

where repeated index means summation over that index. Since the coordinate axes OX_1 and OX_2 rotate about OX_3, the time derivatives of $\hat{\mathbf{X}}_1$ and $\hat{\mathbf{X}}_2$ are non-zero. Now, using (2.11.3) in (2.11.4), we obtain

$$\hat{\mathbf{X}}_i\, \dot{v}_i + \dot{\hat{\mathbf{X}}}_i\, v_i = (\mathbf{v} \times \mathbf{\Omega}) - \frac{1}{\Omega^2} (\mathbf{v} \times \{((\mathbf{v} \times \mathbf{\Omega}) \,.\, \nabla\} \,\mathbf{\Omega}) \tag{2.11.6}$$

Since, at the guiding centre C,

$$\mathbf{\Omega} \equiv (0, 0, \Omega), \dot{\hat{\mathbf{X}}}_i = v_i \frac{\partial \hat{\mathbf{X}}_i}{\partial x_j} \tag{2.11.7}$$

equation (2.11.5) reduces to

$$\hat{\mathbf{X}}_i\, \dot{v}_i + v_i\, v_j \frac{\partial \hat{\mathbf{X}}_i}{\partial x_j} = \Omega\, (\hat{\mathbf{X}}_1\, v_2 - \hat{\mathbf{X}}_2\, v_1) - \frac{1}{\Omega} \hat{\mathbf{X}}_1\, v_2^2 \frac{\partial \Omega}{\partial x_1}$$

$$- \frac{v_1^2}{\Omega} \hat{\mathbf{X}}_2 \frac{\partial \Omega}{\partial x_2} + \frac{1}{\Omega} v_1\, v_2\, \hat{\mathbf{X}}_2 \frac{\partial \Omega}{\partial x_1} + \frac{v_1\, v_2}{\Omega} \hat{\mathbf{X}}_1 \frac{\partial \Omega}{\partial x_2} \tag{2.11.7}$$

where the terms having $\partial \hat{\mathbf{X}}_3 / \partial x_1$ or $\partial \hat{\mathbf{X}}_3 / \partial x_2$ are neglected in the right side. To obtain the solution of (2.11.7) as the sum of a first order term and terms of higher order, depending on the spatial derivatives of the components of the static magnetic field, we write

$$\dot{v}_i = f_i + \delta f_i \tag{2.11.8}$$

Then, separating the first and higher order terms, we get

$$f_1 = \Omega\, v_2,\; f_2 = -\, \Omega\, v_1, \tag{2.11.9}$$

$$\hat{\mathbf{X}}_i\, \delta f_i = -\, v_i\, v_j \frac{\partial \hat{\mathbf{X}}_i}{\partial x_j} - (\hat{\mathbf{X}}_1\, v_2^2 - \hat{\mathbf{X}}_2\, v_1\, v_2) \frac{1}{\Omega} \frac{\partial \Omega}{\partial x_1}$$

$$+ (\hat{\mathbf{X}}_1\, v_1\, v_2 - \hat{\mathbf{X}}_2\, v_1^2) \frac{1}{\Omega} \frac{\partial \Omega}{\partial x_2} \tag{2.11.10}$$

Denoting by $\langle x \rangle$ the average of x over the Larmor period $2\pi/\Omega$, we obtain

$$\langle v_i v_j \rangle = \frac{1}{2} u_\perp^2 \;\text{ for } i = j = (1,\, 2)$$

$$= u_\parallel^2 \quad \text{for } i = j = 3,$$

$$= 0 \quad \text{for } i \neq j \tag{2.11.11}$$

Also, if we write $\langle \delta f_i \rangle = F_i / m$, then the **F** is the time averaged force generated by the weakly inhomogeneous magnetic field on the particle. Averaging all the terms of (2.11.10) over the time period $2\pi/\Omega$, we get

$$\frac{1}{m} \mathbf{F} = -\frac{1}{2} u_\perp^2 \left(\frac{\partial \hat{\mathbf{X}}_1}{\partial x_1} + \frac{\partial \hat{\mathbf{X}}_2}{\partial x_2} \right) - u_{11}^2 \frac{\partial \hat{\mathbf{X}}_3}{\partial x_3} - \frac{1}{2} \frac{{u_\perp}^2}{H} \left(\hat{\mathbf{X}}_1 \frac{\partial H}{\partial x_1} + \hat{\mathbf{X}}_2 \frac{\partial H}{\partial x_2} \right) \tag{2.11.12}$$

To find $(\partial \hat{\mathbf{X}}_1 / \partial x_1) + (\partial \hat{\mathbf{X}}_2 / \partial x_2)$ as a linear combination of $\hat{\mathbf{X}}_1$, $\hat{\mathbf{X}}_2$, $\hat{\mathbf{X}}_3$, we write

$$\text{div}\,\mathbf{H} = 0 = \text{div}\,(\hat{\mathbf{X}}_3 H) = H\,\text{div}\,\hat{\mathbf{X}}_3 + (\hat{\mathbf{X}}_3 \,.\, \nabla)\,H$$

Hence

$$\text{div}\,\hat{\mathbf{X}}_3 = -\frac{1}{H}(\hat{\mathbf{X}}_3 \,.\, \nabla)\,H = \left(\hat{\mathbf{X}}_1 \cdot \frac{\partial \hat{\mathbf{X}}_3}{\partial x_1}\right) + \left(\hat{\mathbf{X}}_2 \cdot \frac{\partial \hat{\mathbf{X}}_3}{\partial x_2}\right) + \left(\hat{\mathbf{X}}_3 \cdot \frac{\partial \mathbf{X}_3}{\partial x_3}\right) \tag{2.11.13}$$

Writing $\hat{\mathbf{X}}_3 = (\hat{\mathbf{X}}_1 \times \hat{\mathbf{X}}_2)$, in the first and second terms, and in the third term using the fact that

$$\left(\hat{\mathbf{X}}_3 \cdot \frac{\partial \hat{\mathbf{X}}_3}{\partial x_3}\right) = \frac{\partial}{\partial x_3}\left(\frac{1}{2}\hat{\mathbf{X}}_3^2\right) = 0$$

and since $\hat{\mathbf{X}}_3^2 = 1$, we get

$$-\frac{1}{H}\frac{\partial H}{\partial x_3} = \left(\hat{\mathbf{X}}_1 \cdot \frac{\partial}{\partial x_1}(\hat{\mathbf{X}}_1 \times \hat{X}_2)\right) + \left(\hat{\mathbf{X}}_2 \cdot \frac{\partial}{\partial x_2}(\hat{\mathbf{X}}_1 \times \hat{\mathbf{X}}_2)\right)$$

$$= \left(\hat{\mathbf{X}}_1 \,.\left(\frac{\partial \hat{\mathbf{X}}_1}{\partial x_1} \times \hat{\mathbf{X}}_2\right)\right) + \left(\hat{\mathbf{X}}_1 \,.\left(\hat{\mathbf{X}}_1 \times \frac{\partial \hat{\mathbf{X}}_2}{\partial x_1}\right)\right)$$

$$+\left(\hat{\mathbf{X}}_2 \,.\left(\frac{\partial \hat{\mathbf{X}}_1}{\partial x_2} \times \hat{\mathbf{X}}_2\right)\right) + \left(\hat{\mathbf{X}}_3 \,.\left(\hat{\mathbf{X}}_1 \times \frac{\partial \hat{\mathbf{X}}_2}{\partial x_2}\right)\right) \tag{2.11.14}$$

Interchanging the elements of the scalar triple product of all terms of the right hand side with the help of the vector formula (A3.1), we get

$$\frac{1}{H}\frac{\partial H}{\partial x_3} = \left(\hat{\mathbf{X}}_3 \cdot \frac{\partial \hat{\mathbf{X}}_1}{\partial x_1}\right) + \left(\hat{\mathbf{X}}_3 \cdot \frac{\partial \hat{\mathbf{X}}_2}{\partial x_2}\right) \tag{2.11.15}$$

We can therefore write

$$\frac{\partial \hat{\mathbf{X}}_1}{\partial x_1} + \frac{\partial \hat{\mathbf{X}}_2}{\partial x_2} = \frac{1}{H}\frac{\partial H}{\partial x_3}\hat{\mathbf{X}}_3 \tag{2.11.16}$$

From the formula (A6.6) of differential geometry, we have

$$\frac{\partial \hat{\mathbf{X}}_3}{\partial x_3} = \frac{1}{R}\hat{\mathbf{X}}_1 = \frac{1}{H}\left\{(\mathbf{H}\,.\,\nabla)\left(\frac{\mathbf{H}}{|H|}\right)\right\} = \frac{1}{H^2}(\mathbf{H}\,.\,\nabla)\,\mathbf{H} \tag{2.11.17}$$

where R is the principal radius of curvature of the magnetic lines of force. Then relation (2.11.12) can be written as

$$\mathbf{F} = \sum_i (\hat{\mathbf{X}}_i \,.\, F_i) = -\frac{W_\perp}{H}\frac{\partial H}{\partial x_3}\hat{\mathbf{X}}_3 - \frac{2W_\parallel}{R}\hat{\mathbf{X}}_1 - \frac{W_\perp}{H}\nabla_\perp H \tag{2.11.18}$$

Hence resolving the force $\mathbf{F}$ into its perpendicular and parallel components with respect to the magnetic field, we get

$$\mathbf{F}_\perp = \hat{\mathbf{X}}_1 F_1 + \hat{\mathbf{X}}_2 F_2 = -\frac{2W_\parallel}{R}\hat{\mathbf{X}}_1 - \frac{W_\perp}{H}\nabla_\perp H, \; F_\parallel = -\frac{W_\perp}{H}\frac{\partial H}{\partial x_3} \quad (2.11.19)$$

Also, writing $\mathbf{F}_\perp = \mathbf{F}_G + \mathbf{F}_R$ where $\mathbf{F}_G$ is the force generated by the transverse gradient and $\mathbf{F}_R$ by the curvature, we see that

$$\mathbf{F}_G = -\frac{W_\perp}{H}\nabla_\perp H, \; \mathbf{F}_R = -\frac{2W_\parallel}{R}\hat{\mathbf{X}}_1 \quad (2.11.20)$$

$\mathbf{F}_R$ is a centrifugal force, because it is directed away from the centre of curvature of the magnetic line of force through the gyrating centre. This curvature obtains from motion generated by $\mathbf{F}_R$.

2.12 ADIABATIC INVARIANCE

The time averaged force $F_\parallel$, being static over large times, gives rise to the static velocity $v_\parallel$ which is determined from the momentum transfer relations

$$F_\parallel = mv_\parallel \frac{\partial v_\parallel}{\partial x_3} = \frac{\partial W_\parallel}{\partial x_3} = -\frac{W_\perp}{H}\frac{\partial H}{\partial x_3} \quad (2.12.1)$$

Equations (2.3.2) and (2.3.3) show that W, the total kinetic energy of the particle, is constant; hence using (2.5.8) we write

$$\frac{\partial W_\parallel}{\partial x_3} = -\frac{\partial W_\perp}{\partial x_3} = -\frac{W_\perp}{H}\frac{\partial H}{\partial x_3} \quad (2.12.2)$$

Integration with respect to x_3 shows that

$$\frac{W_\perp}{H} = \mu \text{ (constant)} \quad (2.12.3)$$

$$F_\parallel = -\frac{\partial}{\partial x_3}(\mu H) \quad (2.12.4)$$

The quantity μH is, therefore, the static potential of this force. It acts antiparallel to the direction of increasing magnetic field, and so produces a diamagnetic effect. If the particle is moving in the direction of H increasing, motion is retarded by its longitudinal gradient, resulting in the decrease of the pitch of the spiralling screw. This μ is called an adiabatic invariant if $|\delta\mu/\mu| \ll |\delta H/H|$ in distances of the order of the gyro radius ρ.

The adiabatic approximation is important in plasma physics, controlled thermonuclear fusion programmes and in the study of particle motion in space including the Van Allan radiation belt. In astrophysical conditions particles describe many free gyrations without collision in large distances. Adiabatic invariance theory therefore applies to astrophysical problems because the change $\delta\mu$ of μ over large distances (or over large times) goes to zero faster than any positive power of $\delta H/H$, over the Larmor radius ρ (or the Larmor gyration period). The adiabatic theory is an asymptotic theory and not a convergent theory.

(a) Diamagnetism in plasmas

Since $\mu = \pi \rho^2 j$, μ is also the magnetic moment of the gyrating particles where ρ is the radius of gyration, $j = e/CT$, $T = 2\pi/\Omega$. The vector μ is directed along the normal to the plane of the area of the circle described by the particle. The force $F_{\|}$ being negative, the magnetic moment vector is defined as

$$\mu = -\left(\frac{W_{\perp}}{H}\right)\left(\frac{\mathbf{H}}{H}\right) \tag{2.12.5}$$

Hence, each gyrating particle represents a microscopic current, such that the generated magnetic field is opposite to the applied field. Plasma particles therefore tend to reduce the magnetic field, and become diamagnetic. Hence, if a plasma is not in thermodynamic equilibrium in the presence of an applied magnetic field, it behaves like a diamagnetic material, and charges partially shield its interior from the field. For confinement by a magnetic field, instead of material walls the plasma pressure is macroscopically balanced by current at the boundary, because then there are no guiding centre currents at the walls to compensate for the magnetic moment produced by the particles gyrating in the interior. The single particle diamagnetic moment is added up for the plasma which then becomes diamagnetic (section 4.5 (*c*)).

(b) Magnetic mirror effect

Equation (2.12.3) shows that

$$\pi\rho^2 H = \text{constant} \tag{2.12.6}$$

So, as H increases, ρ decreases and gyrating particles proceeding towards the region of stronger fields are radially compressed. Assuming that θ is the angle between the instantaneous direction of motion and the direction of the magnetic field, we have

$$u_{\perp} = u \sin\theta, \;\; W_{\perp} = W \sin^2\theta, \tag{2.12.7}$$

and since the total kinetic energy W is constant, equation (2.12.3) gives

$$\frac{\sin^2\theta}{H} = \text{constant} \tag{2.12.8}$$

If $\theta = \theta_0$ where $H = H_0$, we obtain

$$\frac{H}{H_0} = \frac{\sin^2\theta}{\sin^2\theta_0} \tag{2.12.9}$$

When $\theta = \pi/2$ the particle is moving at right angles to the direction of the magnetic field **H**. Subsequent motion is antiparallel to **H** because this force acts in that direction. Hence the region of reflection is near $\theta = \pi/2$. If the magnetic field there is H_c, then the condition for reflection is

$$H_m > H_c \text{ or } H_m > H_0 \operatorname{cosec}^2 \theta_0$$

or
$$\sin \theta_0 > \left(\frac{H_0}{H_m}\right)^{1/2} \tag{2.12.10}$$

where H_m is the maximum field intensity. Let now

$$\sin \theta_m = \left(\frac{H_c}{H_m}\right)^{1/2}; \tag{2.12.11}$$

then all particles which lie within the loss cone of semivertical angle θ_m at the point where $H = H_0$ cannot be trapped by the magnetic field having a longitudinal gradient.

A dimensionless reflection coefficient r_0 is defined as the number of particles reflected by the magnetic mirror field divided by the total number of particles moving towards the mirror. So

$$r_0 = \frac{\int_{\theta_m}^{\pi/2} \cos\theta \, d\Omega_s}{\int_{0}^{\pi/2} \cos\theta \, d\Omega_s} \tag{2.12.12}$$

where the quantity $\cos\theta \, d\Omega_s$ is proportional to the number of particles which arrive at the mirror per second in the solid angle element $d\Omega_s$. Since $d\Omega_s = 2\pi \sin\theta \, d\theta$, we find that

$$r_0 = 1 - \frac{H_0}{H_m} \tag{2.12.13}$$

So the fraction escaping the magnetic trap is H_0/H_m. This reflection from the region of a strong magnetic field is called the magnetic mirror effect, or the magnetic bottling of charges. This property is used to trap a volume of plasma between two regions of strong magnetic field.

2.13 CURVATURE AND GRADIENT DRIFTS

The expression for the drift v_F, due to a force $\mathbf{F}$, is given by (2.7.2). So, corresponding to the forces $\mathbf{F}_G$ and $\mathbf{F}_R$ of (2.11.20), the drifts v_G and v_R can be defined as

$$v_G = \frac{cW_\perp}{eH_3} (\mathbf{H} \times \nabla_\perp H), \quad v_R = \frac{2cW_\parallel}{eH^2 H} (\mathbf{H} \times \hat{\mathbf{X}}_1) \tag{2.13.1}$$

With the help of the formula (2.11.17) we can write

$$v_R = \frac{2cW_{\|}}{eH^4} (\mathbf{H} \times (\mathbf{H} \,.\, \nabla) \,\mathbf{H}) \tag{2.13.2}$$

The drifts v_G and v_R depend on the sign of the charge e. So, for electrons and ions, these drifts are in opposite directions. The transverse gradient drift currents $\mathbf{j}_G$ and the curvature current $\mathbf{j}_R$ are given by

$$\mathbf{j}_G = \frac{c\overline{W}_{\perp}}{H^3} (\mathbf{H} \times \nabla_{\perp} H), \; \mathbf{j}_R = \frac{2c\overline{W}_{\|}}{H^2 R^2} (\mathbf{H} \times \mathbf{R}) \tag{2.13.3}$$

$\overline{W}_{\perp}$ and $\overline{W}_{\|}$ are the average densities of kinetic energy of the charges perpendicular and parallel to the direction of the magnetic field. In the static case, neglecting the longitudinal gradient of the magnetic field and volume current of charges, we have

$$\frac{\partial H_3}{\partial x_3} = 0, \;\text{curl}\, \mathbf{H} = 0 \tag{2.13.4}$$

Then, using (2.9.3), as in Jackson (1975, p. 587), we obtain

$$\nabla_{\perp} H = \hat{\mathbf{X}}_1 \frac{\partial H_3}{\partial x_1} + \hat{\mathbf{X}}_2 \frac{\partial H_3}{\partial x_2} = \hat{\mathbf{X}}_1 \frac{\partial H_1}{\partial x_3} + \hat{\mathbf{X}}_2 \frac{\partial H_2}{\partial x_3}$$

$$= \frac{\partial \mathbf{H}}{\partial x_3} = \frac{1}{H} (\mathbf{H} \,.\, \nabla) \,\mathbf{H} = \frac{H}{R} \hat{\mathbf{X}}_1, \tag{2.13.5}$$

$$v_G + v_R = \frac{c}{eRH} (W_{\perp} + 2W_{\|}) \,\hat{\mathbf{X}}_2, \tag{2.13.6}$$

$$\mathbf{F}_G + \mathbf{F}_R = -\frac{1}{R^2} (W_{\perp} + 2W_{\|}) \,\mathbf{R}, \tag{2.13.7}$$

$$| v_G + v_R | = | v_{\perp} | = \frac{1}{2\Omega R} (u_{\perp}^2 + 2u_{\|}^2) \tag{2.13.8}$$

where $|v_{\perp}|$ is the net drift perpendicular to the direction of the slowly varying magnetic field. Equation (2.13.6) shows that the drift is along the binormal to the lines of **H**. If the displacement due to this drift is δx_2, in the Larmor period $2\pi/\Omega$, then

$$\delta x_2 = | v_{\perp} | \frac{2\pi}{\Omega} = \frac{2\pi}{mR\Omega^2} (W_{\perp} + 2W_{\|}) \tag{2.13.9}$$

In the ionosphere, taking $H = 0.5$ gauss and $T = 10^3$ K, $W_{\perp} + 2W_{\|} = KT$, we find that $\Omega = 2.7 \times 10^7$ rad/sec, $\delta x_2 = 14.4/R$ cm per Larmor period. If $R = 1$ meter, and $H = 10^3$ gauss, then particles at the temperature 10^{40} K will experience the drift velocity $v_{\perp} = 1.8 \times 10^3$ cm/sec.

The radius R of the lines of geomagnetic field is of the order of or greater than the radius of earth. So δx_2 is very small. Comparing v_g of (2.7.3), with $v_G + v_R$, we find that these drifts counteract each other if **g** and **R** are in the same direction.

Exercise 2.13.1. Find the magnetic lines of force when $H \equiv H(\alpha z, \beta z, 1 + \alpha x + \beta y)$ where α and β are constants.

Answer. This **H** is both curl free and divergence free, but it has transverse gradient and curvature. If, therefore, **H** is constant along the family of surface $Q(x, y, z) = \text{constant}$, we can write $\mathbf{H} = \text{grad}\,\varphi$; then $\nabla^2 \varphi = 0$ because curl $\mathbf{H} = 0$ and

$$\varphi(x, y, z) = \int d\varphi = \int \left(\frac{\partial \varphi}{\partial x} dx + \frac{\partial \varphi}{\partial y} dy + \frac{\partial \varphi}{\partial z} d_z \right)$$

$$= H_0 z (1 + 2\alpha x + 2\beta y)$$

If $d\mathbf{r}$ ($\equiv dx, dy, dz$) is a length element along the magnetic lines normal to the surface $\varphi(x, y, z) = \text{constant}$, then the components $\partial\varphi/\partial x$, $\partial\varphi/\partial y$, $\partial\varphi/\partial z$ are the direction numbers of the normal to the surface $\varphi(x, y, z) = \text{constant}$, so

$$\frac{dx}{(\partial\varphi/\partial x)} = \frac{dy}{(\partial\varphi/\partial y)} = \frac{dz}{(\partial\varphi/\partial z)}$$

Hence the differential equations for the family of these lines are

$$\frac{dx}{\alpha z} = \frac{dy}{\beta z} = \frac{dz}{1 + \alpha x + \beta y}$$

The solutions of these equations are

$$\beta x - \alpha y + c_1 = 0, \quad \frac{2x}{\alpha} + x^2 + y^2 - z^2 = c_2$$

For different values of C_1 and C_2 these surfaces intersect to produce curves which are the magnetic lines of force.

2.14 MAGNETIC FIELD SLOWLY VARYING IN TIME

When magnetic field varies in time, a spatially varying electric field $\mathbf{E}_i$ is induced, such that

$$\text{curl}\,\mathbf{E}_i = -\frac{1}{c}\frac{\partial \mathbf{H}}{\partial t} \tag{2.14.1}$$

Rate of change of this field across an area Σ

$$= \frac{\partial}{\partial t} \iint_\Sigma (\mathbf{H} \,.\, d\boldsymbol{\Sigma}) = -c \iint_\Sigma (\text{curl}\,\mathbf{E}_i \,.\, d\boldsymbol{\Sigma}) = -c \oint_\Gamma (\mathbf{E}_i \,.\, d\mathbf{l}) \tag{2.14.2}$$

where $d\mathbf{l}$ is a line element in the curve Γ on Σ. If A is the small area covered by one gyration, then

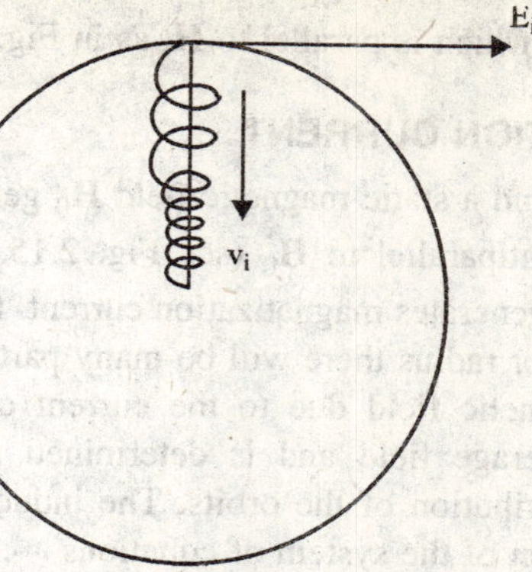

Fig. 2.14.1 Radial drift towards the axis when the axial magnetic field increases with time.

$$2\pi\,\rho E_i = -\frac{1}{c}\pi\rho^2\,\dot{H} = -\frac{1}{c}\,\mathbf{A}\,\dot{H} \tag{2.14.3}$$

where ρ is the radius of gyration. The electric field, therefore, decreases as H increases with time, and the particle loses energy in one gyration. If $\delta W_\perp$ is the gain in energy, it should be negative, and

$$\delta W_\perp = -\,2\pi\,\rho e\,E_i = \frac{e}{c}\,\dot{H}\,\pi\rho^2 \tag{2.14.4}$$

The rate of increase of this energy in one complete gyration being $\delta W_\perp/T$ where T is the Larmor period $2\pi/\Omega$, we obtain

$$\frac{\delta W_\perp}{T} = \frac{dW_\perp}{dt} = \frac{W_\perp}{H}\,\dot{H} \tag{2.14.5}$$

Integration gives

$$\frac{W_\perp}{H} = \mu = \text{constant} \tag{2.14.6}$$

Thus the magnetic moment is an adiabatic invariant for fields slowly varying in time, like fields weakly varying in space. The inducec field $\mathbf{E}_i$ is responsible for the drift given by

$$v_i = \frac{c}{H^2}\,(\mathbf{E}_i \times \mathbf{H}) \tag{2.14.7}$$

In the cylindrical coordinate system r, θ, z the z axis is taken along the direction of $\mathbf{H}$ and $\mathbf{E}$ is perpendicular to the plane $\theta = \text{constant}$. So

$$v_i = -\frac{1}{2H}\,\hat{\mathbf{X}}_r\,\rho\,\dot{H} \tag{2.14.8}$$

where $\hat{\mathbf{X}}_r$ is the unit vector in the direction of r increasing, and $\mathbf{v}_i$ is directed towards the axis OZ, which is parallel to $\mathbf{H}$, as in Fig. 2.14.1.

2.15 MAGNETIZATION CURRENT

Particles gyrating round a static magnetic field $\mathbf{H}_0$ generate their own magnetic field. This field is antiparallel to $\mathbf{H}_0$ (see Fig. 2.15.1) and the corresponding magnetic moment $\boldsymbol{\mu}$ generates magnetization current. In a volume of dimensions of the order of Larmor radius there will be many particles with random phases. The fluctuating magnetic field due to the current of these particles is small compared to the average field and is determined from the average current produced by the distribution of the orbits. The induced field $\mathbf{H}$ is in principle obtained from solution of the system of equations

$$\dot{\mathbf{v}} = (\mathbf{v} \times \boldsymbol{\Omega}) + \frac{e}{m}\mathbf{E} + \frac{e}{mc}(\mathbf{v} \times \mathbf{H}) \tag{2.15.1}$$

$$\text{curl } \mathbf{H} = \frac{1}{c}\dot{\mathbf{E}} + \frac{4\pi}{c}\mathbf{j}, \quad \text{curl } \mathbf{E} = -\frac{1}{c}\dot{\mathbf{H}} \tag{2.15.2}$$

The closed lines of induced $\mathbf{H}$ are shown in Fig. 2.15.1. An element of area Σ, bounded by a closed curve C_1 (Fig. 2.15.2) is crossed by the orbits of the

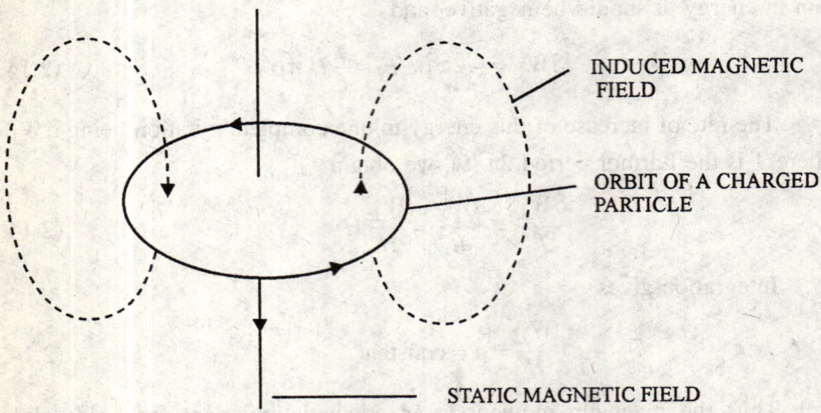

Fig. 2.15.1 Gyration of a particle and the induced magnetic field.

gyrating charges. There are orbits such as b, which penetrate twice, and so contribute nothing to the net current. Orbits like those of a and c penetrate Σ only once, and, therefore, contribute something to the current.

Let $d\mathbf{l}$ be the element of length along the curve enclosing the surface Σ. If there are N current loops per unit volume, each carrying a current I around an area $\mathbf{A}$ (vector), then the number of current loops penetrated by $d\mathbf{l}$ is $N(\mathbf{A} \cdot d\mathbf{l})$ and the current I_m, penetrating Σ, is given by

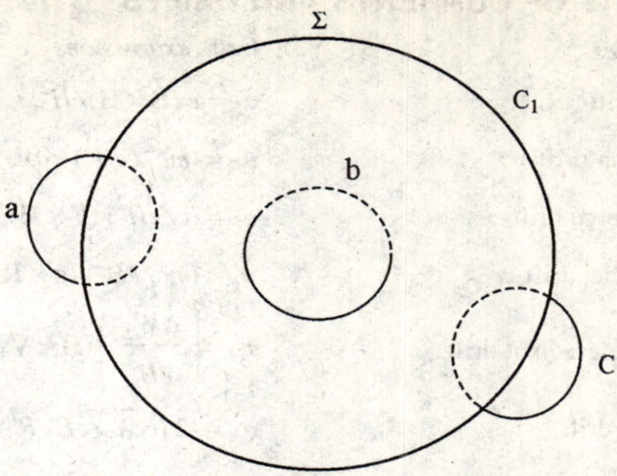

Fig. 2.15.2 Particle orbits contributing to the magnetization current.

$$\frac{I_m}{c} = \oint NI\,(\mathbf{A}\,.\,d\mathbf{l}) = \oint_{\Gamma} (\mathbf{M}\,.\,d\mathbf{l}) = \iint_{\Sigma} (\text{curl}\,\mathbf{M}\,.\,d\boldsymbol{\Sigma}) \tag{2.15.3}$$

where $$\mathbf{M} = NI\mathbf{A} = N\boldsymbol{\mu} = -(W_{\perp} N\,\mathbf{H}_0/H_0^2) \tag{2.15.4}$$

and (2.12.5) has been used; hence **M** is the magnetic moment per unit volume or simply the magnetization. Also if $\mathbf{j}_m$ is the average current of this magnetization flowing through the elementary area $d\Sigma$, then

$$I_m = \iint_{\Sigma} (\mathbf{j}_m\,.\,d\boldsymbol{\Sigma}) = c \iint_{\Sigma} (\text{curl}\,\mathbf{M}\,.\,d\boldsymbol{\Sigma})$$

Since this relation is valid for any element of area, we can write

$$\mathbf{j}_m = c\,\text{curl}\,\mathbf{M} \tag{2.15.5}$$

for the finite magnetization current $\mathbf{j}_m$ even when the guiding centres of the particles are at rest, because $\mathbf{j}_m$ acts on the particles in the eyes of the observer moving with that centre. The total current **j** is the sum of the external currents $\mathbf{j}_e$ and $\mathbf{j}_m$. The Maxwell equation containing these currents and the displacement current is

$$\text{curl}\,\mathbf{H} = 4\pi\,\text{curl}\,\mathbf{M} + \frac{4\pi}{c}\mathbf{j}_e + \frac{1}{c}\frac{\partial \mathbf{E}}{\partial t} \tag{2.15.6}$$

This relation can be written as

$$\text{curl}\,\mathbf{B}' = \frac{4\pi}{c}\mathbf{j}_e + \frac{1}{c}\frac{\partial \mathbf{E}}{\partial t} + 4\pi\nabla \times \mathbf{M} \tag{2.15.7}$$

where $$\mathbf{H} = \mathbf{B} - 4\pi\mathbf{M} \tag{2.15.8}$$

The charge density ρ_m, associated with this current, is zero because

$$\text{div}\;\mathbf{j}_m = 0 \tag{2.15.9}$$

2.16 SUMMARY OF CURRENTS AND DRIFTS

	Drift names	*Drift expressions*
1.	Electric drift	$v_E = c\,(\mathbf{E} \times \mathbf{H})/H^2$,
2.	Polarization drift	$v_P = (mc^2/eH^2)\ (\partial \mathbf{E}/\partial t)$,
3.	External force drift	$v_F = (c/eH^2)\ (\mathbf{F} \times \mathbf{H})$,
4.	Gravitational drift	$\boldsymbol{v}_g = (mc/eH^2)\ (\mathbf{g} \times \mathbf{H})$,
5.	Transverse gradient drift	$\boldsymbol{v}_G = \left(\dfrac{c\,\overline{W}_\perp}{eH^3}\right)(\mathbf{H} \times \nabla_\perp H)$
6.	Curvature drift	$\boldsymbol{v}_R = (2\,(c\,\overline{W}_\parallel/eH^2 R^2)\ (\mathbf{H} \times \mathbf{R})$

Of these v_E is identical for positve and negative charges; so $\boldsymbol{v}_E$ produces no current. The other drifts and some other factors produce the following current densities:

1.	Magnetization current	$\mathbf{j}_m = c \text{ curl } \mathbf{M},\ \mathbf{M} = -(N\,\overline{W}_\perp/H^2)\,\mathbf{H}$
2.	External force current	$\mathbf{j}_F = (e/H^2)\ ((\sum_i N_i\,\mathbf{F}_i) \times \mathbf{H})$
3.	Gravitational drift current	$\mathbf{j}_g = (\rho c/H^2)\ (\mathbf{g} \times \mathbf{H})$
4.	Polarization current	$\mathbf{j}_P = (\rho c^2/H^2)\ (\partial \mathbf{E}/\partial t)$
5.	Transverse gradient drift current	$\mathbf{j}_G = (c\,\overline{W}_\perp/H^2)\ (\mathbf{H} \times \nabla_\perp H)$
6.	Curvature current	$\mathbf{j}_R = (2c\,\overline{W}_\parallel/R^2 H^2)\ (\mathbf{H} \times \mathbf{R})$

where ρ is the sum, per unit volume, of masses of all species of charges, $\overline{W}_\perp$ and $\overline{W}_\parallel$ are similar sums of perpendicular and parallel parts, respectively, of their kinetic energy.

2.17 NET DRIFT AND FORCE

When the characteristic length L and time T of variation of **E** and **H** are much larger than the gyration radius ρ, and Larmor period $2\pi/\Omega$, respectively, the motion of the guiding centre C consists of a drift $\boldsymbol{v}_\perp$, perpendicular to **H**, given by

$$\boldsymbol{v}_\perp = \frac{c}{H^2}(\mathbf{E} \times \mathbf{H}) + \frac{c\overline{W}_\perp}{4eH^2}(\mathbf{H} \times \nabla)\,H^2 + \frac{2c\overline{W}_\parallel}{eH^2}(\mathbf{H} \times (\mathbf{H} \,.\, \nabla)\,\mathbf{H});$$

Then motion parallel to **H** is determined from the equation

$$m\frac{dv_\parallel}{dt} = \frac{1}{H}(\mathbf{H} \,.\, \nabla)\,\overline{W}_\parallel = -\frac{\overline{W}_\perp}{H^2}(\mathbf{H} \,.\, \nabla)\,H$$

If $\mathbf{j} = \mathbf{j}_m + \mathbf{j}_R + \mathbf{j}_G$, then we obtain

$$\mathbf{F} = \frac{1}{c}(\mathbf{j} \times \mathbf{H}) = \left(\mathbf{H} \times \text{curl} \left(\frac{\overline{W}_\perp \mathbf{H}}{H^2} \right) \right) + \left(\left\{ \frac{\overline{W}_\perp}{H^2} (\mathbf{H} \times \nabla) H + \frac{2\overline{W}_\parallel}{H^4} (\mathbf{H} \times (\mathbf{H} \,.\, \nabla) H \right\} \times \mathbf{H} \right)$$

$$= -\frac{\overline{W}_\perp}{H^2} (\text{curl } \mathbf{H} \times \mathbf{H} + \mathbf{H} \times \left\{ \left(\text{grad } \frac{\overline{W}_\perp}{H^2} \right) \times \mathbf{H} \right\} + \left[\left\{ \frac{\overline{W}_\perp}{H^2} (\mathbf{H} \times \nabla) H + \frac{2\overline{W}_\parallel}{H^4} (\mathbf{H} \times (\mathbf{H} \,.\, \nabla) \mathbf{H}) \right\} \times \mathbf{H} \right]$$

$$= \frac{\overline{W}_\perp}{H^2} (\mathbf{H} \times \text{curl } \mathbf{H}) + H^2 \text{ grad} \left(\frac{\overline{W}_\perp}{H^2} \right) - \mathbf{H} \left(\mathbf{H} \,.\, \text{grad } \frac{\overline{W}_\perp}{H^2} \right) + \left[\left\{ \left(\frac{\overline{W}_\perp}{H^2} (\mathbf{H} \times \nabla) H \right) + \left(\frac{2\overline{W}_\parallel}{H^4} (\mathbf{H} \times (\mathbf{H} \,.\, \nabla) \mathbf{H}) \right) \right\} \times \mathbf{H} \right]$$

For curl free magnetic fields the force

$$\mathbf{F} = H^2 \text{ grad} \left(\frac{\overline{W}_\perp}{H^2} \right) - \mathbf{H} (\mathbf{H} \,.\, \nabla) \left(\frac{\overline{W}_\perp}{H^2} \right) + (\overline{W}_\perp + 2\overline{W}_\parallel) \frac{\mathbf{R}}{R^2}$$

$$= \text{grad } \overline{W}_\perp - \left(\frac{2\overline{W}_\perp}{H^2} \right) \text{grad } H - \frac{\mathbf{H}}{\mathbf{H}^2} (\mathbf{H} \,.\, \text{grad } \overline{W}_\perp) + \frac{2\overline{W}_\perp}{H^2} \mathbf{H} (\mathbf{H} \,.\, \text{grad } H) + (\overline{W}_\perp - 2\overline{W}_\parallel) \frac{\mathbf{R}}{R^2}$$

where $\mathbf{R}$ is the vectorial radius of curvature of the magnetic line of force. Since $\nabla_\perp = \nabla - (\mathbf{H}/H^2)\ (\mathbf{H} \,.\, \nabla)$, the sum of the first and third terms on the right hand side reduces the expression for $\mathbf{F}$ to

$$\mathbf{F} = \nabla_\perp \overline{W}_\perp - \frac{2\overline{W}_\perp}{H} \text{grad } H + \frac{2\,\overline{W}_\perp}{H} \frac{\partial H}{\partial x_3} \hat{\mathbf{X}}_3 + \frac{1}{R^2} (\overline{W}_\perp + 2\overline{W}_\parallel) \mathbf{R}$$

$$= \nabla_\perp \overline{W}_\perp - \left(\frac{2\overline{W}_\perp}{H} \right) \nabla_\perp H + \frac{1}{R^2} (\overline{W}_\perp + 2\overline{W}_\parallel) \mathbf{R}$$

Again, since $\mathbf{R}/R^2 = (\nabla_\perp H)/H$ (by equation (2.13.5)), we find that

$$\mathbf{F} = \nabla_\perp \overline{W}_I + \frac{1}{R^2} (2\overline{W}_\parallel - \overline{W}_\perp) \mathbf{R}$$

But writing $\overline{W}_\perp = p_\perp$, $2\overline{W}_\parallel = p_\parallel$ where $p_\perp$ and $p_\parallel$ are the components of the anisotropic pressure tensor perpendicular and parallel to the direction of the static magnetic field, we obtain

$$\mathbf{F} = \nabla_\perp p_\perp + \frac{1}{R^2}(p_\parallel - p_\perp)\,\mathbf{R}$$

2.18 ORBIT IN PRESENCE OF OSCILLATING ELECTRIC FIELD AND CONSTANT MAGNETIC FIELD

The frequency ω of the oscillating electric field is not, in general, identical to the Larmor frequency $\Omega\ (= eH/mc)$ of gyration, due to a constant magnetic field, of a particle of charge e and mass m. So, particle motion is under the influence of two oscillations with different frequencies and forces in different directions. The effect of these two oscillations in a simple case, would be considered in this section.

Let the z-axis be parallel to the direction of the static magnetic field **H**, and let the components of the electric field (elliptically polarized perpendicular to **H**) along the three coordinate directions be

$$E_1 \cos(\omega t + \alpha),\ E_2 \sin(\omega t + \alpha),\ E_\parallel \cos(\omega t + \beta)$$

where E_1, E_2, $E_\parallel$ are real constants. The linearized equations of motion of the particle are

$$\ddot{x} = \frac{e}{m} E_1 \cos(\omega t + \alpha) + \Omega\,\dot{y}, \tag{2.18.1}$$

$$\ddot{y} = \frac{e}{m} E_2 \sin(\omega t + \alpha) - \Omega\,\dot{x} \tag{2.18.2}$$

$$\ddot{z} = \frac{e}{m} E_\parallel \cos(\omega t + \beta) \tag{2.18.3}$$

Initial position and velocity coordinates are (x_0, y_0, z_0) and (u_1, u_2, u_3) respectively. The last equation on integration gives

$$z - z_0 - u_\parallel t = -\frac{eE_\parallel}{m\omega^2}\{\cos(\omega t + \beta) - \cos\beta\} - \frac{et}{m\omega} E_\parallel \sin\beta \tag{2.18.4}$$

Hence, with time grows secularly the displacement parallel to the magnetic field, but not the kinetic energy. Equations (2.18.1) and (2.18.2) can be written as

$$(D^2 + \Omega^2)\,x' = \frac{e}{m\omega}(\omega E_1 - \Omega E_2)\cos\theta, \tag{2.18.5}$$

$$(D^2 + \Omega^2)\,y' = \frac{e}{m\omega}(\omega E_2 - \Omega E_1)\sin\theta, \tag{2.18.6}$$

where
$$\theta = \omega t + \alpha,\ D = d/dt, \tag{2.18.7}$$

$$x' = x - x_0 - \frac{u_2}{\Omega} - \frac{eE_2}{m\omega\Omega}\cos\alpha,$$

$$y' = y - y_0 + \frac{u_1}{\Omega} - \frac{eE_1}{m\omega\Omega}\sin\alpha \tag{2.18.8}$$

When $\Omega^2 \neq \omega^2$, the complete solution is a linear combination of sines and cosines of Ωt and ωt. Using the initial conditions to determine the constants of integration, the solution can be written as

$$x' = \frac{e(\omega E_1 - \Omega E_2)}{m\omega(\Omega^2 - \omega^2)}\cos\theta + A_1\cos\Omega t + B_1\sin\Omega t, \tag{2.18.9}$$

$$y' = \frac{e(\omega E_2 - \Omega E_1)}{m\omega(\Omega^2 - \omega^2)}\sin\theta + A_2\cos\Omega t + B_2\sin\Omega t \tag{2.18.10}$$

$$A_1 = -\frac{u_2}{\Omega} - \frac{e(\Omega E_1 - \omega E_2)\cos\alpha}{m\Omega(\Omega^2 - \omega^2)},\ B_1 = \frac{u_1}{\Omega} + \frac{e(\omega E_1 - \Omega E_2)\sin\alpha}{m\Omega(\Omega^2 - \omega^2)}$$

$$A_2 = \frac{u_1}{\Omega} - \frac{e(\Omega E_2 - \omega E_1)\sin\alpha}{m\Omega(\Omega^2 - \omega^2)},\ B_2 = \frac{u_2}{\Omega} - \frac{e(\omega E_2 - \Omega E_1)\cos\alpha}{m\Omega(\Omega^2 - \omega^2)} \tag{2.18.11}$$

It is evident that

$$A_1 = -B_2,\ A_2 = B_1 \tag{2.18.12}$$

When $|\Omega| << |\omega|$ we neglect $\cos\theta$ and $\sin\theta$ in (2.18.9) and (2.18.10) because their amplitudes are much less than those of $\cos\Omega t$ and $\sin\Omega t$. Then, the orbit is a circle of radius equal to $\sqrt{A_1^2 + B_1^2}$. When $|\Omega| > |\omega|$, for similar reasons, we neglect $\cos\Omega t$ and $\sin\Omega t$ and obtain

$$x' = \frac{e}{m\omega\Omega^2}(\omega E_1 - \Omega E_2)\cos\theta,\ y' = -\frac{e}{m\omega\Omega^2}(\Omega E_1 - \omega E_2)\sin\theta \tag{2.18.13}$$

So the orbit is the ellipse

$$\frac{x'^2}{(\omega E_1 - \Omega E_2)^2} + \frac{y'^2}{(\Omega E_1 - \omega E_2)^2} = \frac{e^2}{m^2\omega^2\Omega^4} \tag{2.18.14}$$

If the wave is plane polarized we put $E_2 = 0$, and find that the principal semi axes are of lengths $eE_1/m\Omega^2$ and $eE_1/m\omega\Omega$. Since $|\Omega| >> |\omega|$, the major axis (the y-axis) is in the direction perpendicular to the plane containing **E** and **H**. The lengths of the principal axes do not depend on the initial value of the velocity; they only depend upon the strength of electric and magnetic fields and the wave frequency.

When $\omega^2 = \Omega^2$, energy of electric field is resonantly absorbed by the particles and the medium gets heated. Then, for $E_1^2 \neq E_2^2$, the solution is secular, and can be written as

$$\frac{(x' - A_1 \cos \Omega t - B_1 \sin \Omega t)\, m}{e\,(E_1 - E_2)} = \frac{t}{2\,\Omega} \sin (\Omega t + \alpha), \tag{2.18.15}$$

$$\frac{(y' - A_2 \cos \Omega t - B_2 \sin \Omega t)\, m}{e\,(E_2 - E_1)} = -\frac{t}{2\Omega} \cos (\Omega t + \alpha) \tag{2.18.16}$$

Hence the parametric solutions are

$$x' = A_1 \cos \Omega t + B_1 \sin \Omega t + \frac{et}{2m\,\Omega} (E_1 - E_2) \sin (\Omega t + \alpha), \tag{2.18.17}$$

$$y' = A_2 \cos \Omega t - B_2 \sin \Omega t + \frac{et}{2m\,\Omega} (E_1 - E_2) \cos (\Omega t + \alpha) \tag{2.18.18}$$

where A_1, B_1, A_2, B_2 are given by

$$\left.\begin{aligned} A_1 &= -\frac{u_2}{\Omega} - \frac{eE_2}{m\Omega^2} \cos \alpha, \quad B_1 = \frac{u_1}{\Omega} - \frac{e}{2m\,\Omega^2} (E_1 - E_2) \sin \alpha, \\ A_2 &= \frac{u_1}{\Omega} + \frac{eE_1}{m\Omega^2} \sin \alpha, \quad B_2 = \frac{u_2}{\Omega} - \frac{e}{2m\,\Omega^2} (E_1 - E_2) \cos \alpha \end{aligned}\right\} \tag{2.18.19}$$

Equations (2.18.17) and (2.18.18) are combined to give

$$(x' - A_1 \cos \Omega t - B_1 \sin \Omega t)^2 + (y' - A_2 \cos \Omega t - B_2 \sin \Omega t)^2$$
$$= \frac{e^2 t^2}{4m^2\,\Omega^2} (E_1 - E_2)^2 \tag{2.18.20}$$

This is a circle of radius increasing with time directly and the centre oscillating at the gyration frequency Ω. The kinetic energy $m\,(\dot{x}^2 + \dot{y}^2)/2$ grows indefinitely with time, because it has a term proportional to t^2. A resonant electric field is therefore a possible means of heating a volume of plasma. In presence of the damping force, proportional to the velocity **v**, the secular growth at the resonance $\omega = \Omega$ is completely prevented. In place of (2.18.1) and (2.18.2) equation of motion is then

$$\dot{\mathbf{v}} = \frac{e}{m} \mathbf{E} + (\mathbf{v} \times \mathbf{\Omega}) - \nu \mathbf{v} \tag{2.18.21}$$

where ν is the collision frequency. Then (2.18.5) and (2.18.6) are replaced by

$$\{(D + \nu)^2 + \Omega^2\}\, v_x = \frac{e}{m} \{(D + \nu)\, E_x + \Omega\, E_y\}, \tag{2.18.22}$$

$$\{(D + \nu)^2 + \Omega^2\}\, v_y = \frac{e}{m} \{(D + \nu)\, E_y - \Omega\, E_x\} \tag{2.18.23}$$

and the field will continuously increase to an extreme value at $\omega\ (= \Omega)$. The solution of (2.18.22) and (2.18.23) for $\omega = \Omega$ is left as an exercise to the reader.

When ions, like electrons, are also mobile, equation (2.18.21) has to be supplemented by the equation of motion of ions, and in place of ν of (2.18.21) there will be two collision frequencies; one is ν_e in (2.18.21) for electrons and the other is ν_i in the equation of motion of ions. In that case two cyclotron frequencies, namely $\Omega_e\ (= -\,|eH/mc|)$ and $\Omega_i\ (= (|eH/Mc|))$ are obtained where m = electron mass and M = ion mass. There will be enhanced absorption at the two resonant frequencies Ω_e and Ω_i. Since $m << M$, $|\Omega_i| << |\Omega_e|$, and the energy distribution will be two-humped.

Exercise 2.18.1. Find the kinetic energy of an electron at $\omega = |\Omega_e|$ and of an ion at $\omega = |\Omega_i|$ and compare, in both the cases of collision free plasma and collisionally damped plasma.

2.19 LARMOR PRECESSION

Electrons orbiting round their nuclei, due to nucleas binding forces, experience a precessional rotation when moreover a homogeneous magnetic field is present. This is the Larmor precession frequency of electrons. When the central binding force is $\psi\,(r)\,\mathbf{r}/r$, the equation of motion of the charge q is

$$m\ddot{\mathbf{r}} = -\,\psi\,(\mathbf{r})\,\frac{\mathbf{r}}{r} + \frac{q}{c}\,(\dot{\mathbf{r}} \times \mathbf{H}) \tag{2.19.1}$$

where $\mathbf{r}$ is the electron position vector from the nucleus. For the force of gravitation $\psi\,(r) = MmG/r^2$, where G is the gravitational constant, and M is the ionic mass. And, for an elastic binding force, proportional to the distance $\mathbf{r}$ from the ion,

$$\psi\,(r)\,\frac{\mathbf{r}}{r} = K\mathbf{r}$$

where k is an elastic constant; the equation of motion is

$$m\ddot{\mathbf{r}} + k\mathbf{r} = \frac{q}{c}\,(\dot{\mathbf{r}} \times \mathbf{H}) \tag{2.19.1 (a)}$$

We here consider the Larmor precession effect in this simple case only. The gravitational force case is briefly treated in Exercise 2.19.3.

Before introduction of the field $\mathbf{H}$, the elastic force balances the centrifugal force of rotation. Electrons of the outermost orbits of heavy atoms are lossely bound (classically) in this manner to their respective nucleus, which being very heavy, is somewhat immobile.

If ω is the angular velocity of oscillation (harmonic) proportional to exp $(i\,\omega\,t)$, then

$$(-\,m\omega^2 + k)\,\mathbf{r} = \frac{q}{c}\,(\mathbf{v} \times \mathbf{H}) \tag{2.19.2}$$

The force of the magnetic field $q\,(\mathbf{v}\times\mathbf{H})/c$ acts perpendicular to both $\mathbf{v}$ and $\mathbf{H}$. Since $\omega = \omega_0$ introduction of the applied field, the force balance condition is

$$(-m\omega_0^2 + K) = 0$$

So the frequency of the natural orbit is

$$\omega_0 = (k/m)^{1/2} \tag{2.19.3}$$

Taking the z-axis parallel to the direction of $\mathbf{H}$, equation (2.19.2) gives

$$(-\omega^2 + \omega_0^2)\,x = \Omega\,\dot{y},\ \ (-\omega^2 + \omega_0^2)\,y = -\,\Omega\,\dot{x},\ \Omega = \frac{eH}{mc}, \tag{2.19.4}$$

$$\omega_0^2 - \omega^2 = \pm\,\omega\Omega, \tag{2.19.5}$$

for the harmonic solution exp $(i\,\omega\,t)$ for both x and y. Solving for ω we get

$$\omega = \frac{\Omega}{2} + \left\{\left(\frac{\Omega}{2}\right)^2 + \omega_0^2\right\}^{1/2} \tag{2.19.6}$$

Other solutions of (2.19.5) also exist, which are rejected because the conditions that $\omega = \Omega$ when $\omega_0 = 0$, and $\omega = \omega_0$ when $\Omega = 0$ must be satisfied.

For small magnetic fields equation (2.19.6) reduces to

$$\omega = \frac{\Omega}{2} + \omega_0 \tag{2.19.7}$$

Hence, the motion of electrons around a central nucleus, to the first order in the magnetic field $\mathbf{H}$, is the superposition of motion at the frequency ω_0 and a precessional rotation at the frequency $eH/2\,mc$, called the Larmor precession frequency. On the other hand, the quantity eH/mc (sometimes also referred to as the Larmor gyration frequency, but not as the Larmor precession frequency) is the frequency of gyration for a free electron. For a system of charges the effect of an applied magnetic field is a common precession only when their e/m has the same value under the influence of a single central force. So, the Larmor precession frequency of a system of N bound electrons is equivalent to the current

$$I = -\frac{Ne}{2\pi c}\,\frac{eH}{2mc} \tag{2.19.8}$$

The magnetic moment μ of a current loop $(= (\pi\rho^2\,I/c)$ where $\pi\rho^2$ is the loop area) for N electrons per unit volume is given by

$$\mu = -\frac{Ne^2\,H}{4\,mc^2}\,\langle\rho^2\rangle \tag{2.19.9}$$

where $\langle\rho^2\rangle$ is the mean square of the perpendicular distance of the electrons from the field axis through the nucleus. If N_1 is the number of atoms per unit volume, then the magnetic susceptibility χ is given by

$$\chi = \frac{N_1 \mu}{H} = -\frac{N N_1 e^2}{4mc^2} \langle \rho^2 \rangle \tag{2.19.10}$$

The diamagnetic field of (2.19.9), due to the Larmor precessional rotation, is the de Haas–Van Alphen effect of solid state physics.

(a) Classical notions of bound electrons

The concept of electrons, classically bound to a nucleus of positive charge, explains the Larmor precession and some other physical effects. When electrons move about a positive charge, in presence of an ambient magnetic field **H**, to the first order in H, this motion is the same as in absence of H, except for superposition of a precessional rotation at the rate of Larmor angular frequency $qH/2\,mc$, which is half of the gyration frequency of free electrons Ω_e (= qH/mc. The existence of the magnetic moment and other results from such binding forces cannot be guessed from models having only free electrons.

Orbits of electrons in atoms, distorted non-linearly by a strong monochromatic radiation, are studied as its higher harmonics, with the help of this bound electron model. Electrons are bound to positively charged nuclei, in macroscopically neutral plasmas, by a simple harmonic force, proportional to its wave induced displacement. The impact of strong radiation at a frequency which is higher than the plasma frequency, on the atom, is an electron displacement for which the quantized field concepts may be replaced by such classical models. This simple harmonic force follows from the centrifugal force of the electrons in their orbital planes. In crystals this is the one-dimensional simple harmonic force about the lattice points. So, only the transverse waves are affected thus in plasmas, whereas longitudinal waves, propagating along the directions of the unidirectional lattice vibrations, are thus affected in crystals.

Exercise 2.19.1. Show that the electric field associated with a slowly changing magnetic field generates the velocity of Larmor precession.

Answer. The electric field, associated with a magnetic field which is slowly changing from zero to the value H in time t, establishes a new equilibrium state. To show this effect we consider a charge moving in the xy plane of an inertial coordinate system, with respect to the centre of force about the origin. Since x and y are small, we can expand the electric field components E_x and E_y about their values $(E_x)_0$ and $(E_y)_0$ at the origin. Then

$$E_x = (E_x)_0 + x\left(\frac{\partial E_x}{\partial x}\right)_0 + y\left(\frac{\partial E_x}{\partial y}\right)_0 + \ldots, \tag{1}$$

$$E_y = (E_y)_0 + x\left(\frac{\partial E_y}{\partial x}\right)_0 + y\left(\frac{\partial E_y}{\partial y}\right)_0 + \ldots$$

Moreover, the momentum transfer condition gives

$$\ddot{x} = \frac{e}{m} E_x, \ \ddot{y} = \frac{e}{m} E_y \tag{2}$$

And, if Ω is the angular velocity about the z-axis,

$$\frac{d\Omega}{dt} = \frac{d}{dt}\left(\frac{x\dot{y} - \dot{x}y}{r^2}\right) = \frac{x\ddot{y} - \ddot{x}\,y}{r^2} \tag{3}$$

The Maxwell equation (1.3.7) gives

$$-\frac{1}{c}\frac{\partial H_z}{\partial t} = \frac{\partial E_x}{\partial y} - \frac{\partial E_y}{\partial x} \tag{4}$$

Since $x = r\cos\theta$, $Y = r\sin\theta$ (r is constant up to the order of approximation considered here) we obtain $d\theta/dt = \Omega$, and

$$\frac{d\Omega}{dt} = -\frac{e}{mr^2}\left[(yE_x - xE_y)_0 + \left(y^2\frac{\partial E_x}{\partial y} - x^2\frac{\partial E_y}{\partial x}\right)_0 + xy\left(\frac{\partial E_x}{\partial x} - \frac{\partial E_y}{\partial y}\right)_0 + \ldots\right] \tag{5}$$

Taking mean values (so utilizing the fact that $\langle x^2\rangle = \langle y^2\rangle = r^2/2$, and $\langle x, y, xy\rangle = 0$), with the help of (4) we find that

$$\left\langle \frac{d\Omega}{dt}\right\rangle = \frac{e}{2mc}\frac{\partial H_z}{\partial t}, \tag{6}$$

where $\langle x\rangle$ is the mean value of x. Hence the electric field of a slowly changing magnetic field generates the velocity of Larmor precession and the motion in the rotating reference system is the same as the original motion in the rest system before its application.

Exericse 2.19.2. Show that for rotation about the direction of a constant **H**, in presence of a conservative electric field, at the rate of the Larmor precession frequency, the results for non-rotating motion in the presence of **H** are recovered.

Answer. Consider the Lagrangian L of a system of N particles,

$$L = \Sigma\left[\frac{1}{2}mv^2 - q\left\{\varphi - \frac{1}{c}(\mathbf{v}\,.\,\mathbf{A})\right\}\right] \tag{1}$$

For a constant magnetic field **H**, since

$$\mathbf{H} = \frac{1}{2}\nabla \times (\mathbf{H} \times \mathbf{r}), \tag{2}$$

the vector potential is

$$\mathbf{A} = \frac{1}{2}(\mathbf{H} \times \mathbf{r}) \tag{3}$$

Hence, for a system of charges in equilibrium under a central force (say, for classical electrons bound to their respective atomic nuclei),

$$L = \Sigma \left[\frac{1}{2} mv^2 - q \{\varphi - (\mathbf{r} \,.\, (\mathbf{v} \times \mathbf{H}))\} \right] \tag{4}$$

Putting

$$\mathbf{H} = -\frac{2mc}{q} \boldsymbol{\Omega}, \tag{5}$$

in (4) we find that

$$L = \Sigma \left[\frac{1}{2} m (\mathbf{v} - (\boldsymbol{\Omega} \times \mathbf{r}))^2 - \frac{1}{2} m (\boldsymbol{\Omega} \times \mathbf{r})^2 - q\,\varphi \right] \tag{6}$$

When $\boldsymbol{\Omega}$ and $\mathbf{r}$ are so small that $|\mathbf{v}| >> |\boldsymbol{\Omega} \times \mathbf{r}|$, we neglect the second of the three terms, within the summation, and obtain

$$L = \Sigma \left[\frac{1}{2} (\mathbf{v} - (\boldsymbol{\Omega} \times \mathbf{r}))^2 - q\,\varphi \right] \tag{7}$$

Hence if $\mathbf{v}'$ is the velocity of the particle, as seen in the laboratory system, in which only the electrostatic potential force $q\,\nabla\,\varphi$ acts on the particle, evidently

$$\mathbf{v}' = \mathbf{v} - (\boldsymbol{\Omega} \times \mathbf{r}), \tag{8}$$

$$L = \Sigma \left[\frac{1}{2} mv'^2 - q\,\varphi \right] \tag{9}$$

$\mathbf{v}$ is the particle velocity in the system of coordinates in which the particle is uniformly rotating about an axis at the rate $\boldsymbol{\Omega}$. In the laboratory inertial frame K^0, equation of motion is

$$m\,\ddot{\mathbf{r}}_0 = q\,\mathbf{E} + \frac{q}{c} (\mathbf{v} \times \mathbf{H}) \tag{10}$$

where $\mathbf{r}_0$ is the radius vector from the centre of force to the electron. Let now $\mathbf{r}, \dot{\mathbf{r}}, \ddot{\mathbf{r}}$ be the values of $\mathbf{r}_0, \dot{\mathbf{r}}_0, \ddot{\mathbf{r}}_0$ in a system K which is rotating about the centre of force with uniform angular velocity $\boldsymbol{\Omega}$. Then

$$\dot{\mathbf{r}}_0 = \dot{\mathbf{r}} + (\boldsymbol{\Omega} \times \mathbf{r}) = \left(\frac{\partial}{\partial t} + \boldsymbol{\Omega} \times \right) \mathbf{r}$$

$$\ddot{\mathbf{r}}_0 = \ddot{\mathbf{r}} + 2\,(\boldsymbol{\Omega} \times \dot{\mathbf{r}}) + (\boldsymbol{\Omega} \times (\boldsymbol{\Omega} \times \mathbf{r})) \tag{11}$$

Neglecting squares of $\boldsymbol{\Omega}$ and $\mathbf{H}$, the equation of motion becomes

$$\ddot{\mathbf{r}} + 2 \left(\boldsymbol{\Omega} + \frac{q\mathbf{H}}{2mc} \right) \times \dot{\mathbf{r}} = \frac{q}{m} \mathbf{E}, \tag{12}$$

because for central forces **E** transforms in the same way as $\mathbf{r}_0$. So the equation of motion in the rotating system is the same as in the original system in absence of **H** if Ω is given by (5).

Exercise 2.19.3. Find the Larmor precession effect when the binding force is the gravitational field of the ion.

Answer. In this case the equation of motion is

$$m\ddot{\mathbf{r}} + \frac{MmG}{r^3}\mathbf{r} = \frac{q}{c}(\dot{\mathbf{r}} \times \mathbf{H})$$

After application of the field **H**, it can be written as

$$m\ddot{\mathbf{r}} = -\frac{MmG}{(r_0 + r)^3}\mathbf{r} + \frac{q}{c}(\dot{\mathbf{r}} \times \mathbf{H})$$

$$\approx -\frac{MmG}{r_0^3}\mathbf{r} + \frac{q}{c}(\dot{\mathbf{r}} \times \mathbf{H}) = -\omega_0^2\,\mathbf{r} + \frac{q}{c}(\dot{\mathbf{r}} \times \mathbf{H})$$

because the field induced displacement **r** is small in magnitude compared to $\mathbf{r}_0$, where $r = r_0$ when $H = 0$, and $\omega_0^2 = MmG/r_0^3$.

2.20 DYNAMICS OF RELATIVISTIC ELECTRONS

The relativistic equation of motion of an electron of charge $-e$ and rest mass m in a given electromagnetic field is

$$\frac{d\mathbf{p}}{dt} = -e\mathbf{E} - \frac{e}{c}(\mathbf{v} \times \mathbf{H}) \tag{2.20.1}$$

with $\mathbf{p} = m\mathbf{v}\Big/\left(1 - \dfrac{v^2}{c^2}\right)^{1/2}$. The Hamiltonian $\mathcal{H}$ of equation (1.18.15) then assumes the form

$$\mathcal{H} = \sqrt{m^2c^4 + (\mathbf{p}c + e\mathbf{A})^2} \tag{2.20.2}$$

Here φ and **A** are the scalar and vector potentials of the field, **v** is the particle velocity and **p** is the relativistic momentum; so **E** and **H** are given by (1.9.1) and (1.9.3). Then if L is the Lagrangian of (1.18.7).

$$L = -mc^2\sqrt{1 - \frac{v^2}{c^2}} - \frac{e}{c}(\mathbf{v}\,.\,\mathbf{A}) + e\,\varphi \tag{2.20.3}$$

The energy equation, obtained from (2.20.1), is

$$\frac{d}{dt}\left(\frac{mc^2}{\sqrt{1 - v^2/c^2}}\right) = \left(\mathbf{v}\cdot\frac{d\mathbf{p}}{dt}\right) = -e\,(\mathbf{E}\,.\,\mathbf{v}) \tag{2.20.4}$$

If **E** is purely due to the potential φ, then **E** = grad φ; so

$$\frac{mc^2}{\sqrt{1-v^2/c^2}} = e\,\varphi = \text{constant} \tag{2.20.5}$$

When $\varphi = 0$ **v** is constant, $\dot{v}$ is directed along the principal normal to the trajectory. So it produces the centrifugal force v^2/ρ where ρ is the principal radius of curvature. Then a time varying magnetic field cannot exist without an associated electric field and $\dot{\mathbf{v}}$ must be supported by a magnetostatic force. If **i** and **j** are unit vectors along the tangent and principal normal, respectively, to the trajectory, we have,

$$\mathbf{p} = \mathbf{i}p,\quad \frac{d\mathbf{p}}{dt} = 0 + p\frac{d\mathbf{i}}{dt} = p\,\mathbf{j}\frac{d\psi}{dt} = \frac{p\,v}{\rho}\mathbf{j}$$

where ψ is the angle made by the geometrical tangent to the trajectory at the point *P* at time *t* with a fixed line and $\psi + d\psi$ is the same at the time $t + dt$. So the equation of motion becomes

$$\frac{pv}{\rho} = \frac{|e|}{c}\,vH\sin\theta, \tag{2.20.6}$$

giving
$$\rho = \frac{v}{\Omega_0 \sin\theta\sqrt{1-v^2/c^2}} \tag{2.20.7}$$

where $\Omega_0 = |e|\,H/mc$ and θ is the angle between **v** and **H**. Hence, in a constant magnetic field **H** the motion consists of an arbitrary constant velocity parallel and a gyration perpendicular to **H** at the Larmor frequency $\Omega_0\sqrt{1-v^2/c^2}$, which is inversely proportional to the relativistic energy $mc^2/\sqrt{1-v^2/c^2}$. It is evident that the gyration radius increases by the factor $(1 - v^2/c^2)^{-1/2}$. Now, since

$$\frac{d\mathbf{p}}{dt} = m\mathbf{a},\quad \mathbf{a} = -\frac{e}{m}\left\{\mathbf{E} + \frac{1}{c}(\mathbf{v}\times\mathbf{H})\right\}, \tag{2.20.8}$$

$$\mathbf{u} = \frac{\mathbf{v}}{\sqrt{1-v^2/c^2}} \tag{2.20.9}$$

we find that **u** is a dimensionless velocity and

$$\mathbf{a} = -\frac{e}{m}\left\{\mathbf{E} + \frac{(\mathbf{u}\times\mathbf{H})}{\sqrt{1+u^2}}\right\}, \tag{2.20.10}$$

$$\sum_i \left(\frac{\partial a_i}{\partial u_i}\right) = \mathrm{div}_{\mathbf{u}}\,\mathbf{a} = -\frac{2e}{m}((\mathbf{u}\times\mathbf{H})\,.\,\mathbf{u})\cdot\frac{1}{(1+u^2)^{3/2}} = 0 \tag{2.20.11}$$

In presence of a strong, constant magnetic field **H**, neglecting the induced electric field, the relativistic electron motion is given by

$$\frac{d}{dt}\left\{\frac{m\mathbf{v}}{\sqrt{1-v^2/c^2}}\right\} = -\frac{e}{c}(\mathbf{v}\times\mathbf{H}) \tag{2.20.12}$$

Then putting $\varphi = 0$ in (2.20.5), we find that

$$\frac{m}{\sqrt{1-v^2/c^2}} = \text{constant} \tag{2.20.13}$$

Hence $m\gamma$ can be taken outside the differentiation where

$$\gamma = (1 - v^2/c^2)^{-1/2}$$

Then the equation of motion is

$$\frac{d\mathbf{v}}{dt} = -(\mathbf{v}\times\boldsymbol{\Omega}),\ \boldsymbol{\Omega} = \frac{e\mathbf{H}}{m\gamma c} \tag{2.20.14}$$

And the relations (2.12.5) and (2.13.1), are replaced by

$$\boldsymbol{\mu} = -\frac{\gamma W_\perp \mathbf{H}}{H^2},\quad v_G = \frac{c\gamma W_\perp}{eH^3}(\mathbf{H}\times\boldsymbol{\nabla}_\perp H),$$

$$v_R = \frac{2c\gamma W_\parallel}{eH^4}[\mathbf{H}\times(\mathbf{H}\,.\,\nabla)\,\mathbf{H}] \tag{2.20.15}$$

where $W_\perp = mu_\perp^2/2$, $W_\parallel = mu_\parallel^2/2$. These are the relativistically correct expressions for the magnetic moment and drifts in the presence of inhomogeneous magnetic fields. v_G and v_R are faster by the factor γ (> 1). The parallel force $F_\parallel$ of (2.12.4) is now given by

$$F_\parallel = m\gamma\frac{dv_\parallel}{dt} = -\gamma\frac{W_\perp}{B}\frac{\partial H}{\partial x_3} \tag{2.20.16}$$

The increase by the factor γ is due to the fact that the larger gyration radius subjects the particle to a greater convergence of the field lines.

Equation (2.20.1) for one particle motion should be supplemented by the Maxwell equations containing the average current $-eN\mathbf{v}$ per unit volume and the charge density $e\,(N_0 - N)$ where N and N_0 are the number densities of electrons and ions respectively (ions static). As in section 2.2, the influence of the induced fields is excluded in some cases.

Chapter 3

BASIC PLASMA PROPERTIES

3.1 PLASMA—THE FOURTH STATE OF MATTER

The word 'plasma' is of Greek origin. It means formation or configuration. The physical plasma is ionized gas, free from wall or electrode effects. It has the cohesiveness of a mould or jelly or the blood plasma of living organisms, which are solids. It, more over, has the density of gases and mobility of liquids. So, the physical plasma has features of all the three states of matter. For this reason, the plasma state is regarded as the fourth state of matter.

Coupling between classical and quantum phenomena is weak in plasmas, because the characteristic lengths or times of quantum mechanical behaviour, such as the de Broglie wavelength or the time to emit a photon, are usually short compared to the characteristic lengths and times of development of significant plasma properties. So, for study of physics of plasmas, classical electrodynamics and classical mechanics of particles, or fluids are used. Moreover, basic knowledge of kinetic theory, statistical mechanics and thermodynamics is also necessary.

Electrons being of very small mass, their motion controls the plasma behaviour in fast processes. A Lorentz plasma is a fully ionized plasma, in which electron-electron collisions are negligible, and ionic thermal motion is also negligible. In an electron plasma only electrons are mobile, ions provide a neutralizing background of positive charges.

A gas becomes electrically conducting when it contains free charged particles. Plasmas have high electrical conductivity, and mostly gaseous mechnical properties. The conductivity depends on the concentration of electrons and ions, and on their mobility. Dense, electrically conducting, fluids behave like a set of closedly packed dipoles. The plasma dielectric properties are determined by free charges, and not by dipoles. A metal, or a semiconductor can have plasma properties, although its mechanical properties of compressibility and rheology resemble those of solids.

The physical plasma of nature exists everywhere in the universe, except in the outer layer of earth of depth few kilometers containing the seven seas. The vacuum discharges, arcs and flames are examples of laboratory produced plasmas.

Even for weak oscillating electric fields, depending on the parameters, plasmas have several characteristic frequencies of collective oscillations: namely, the electron plasma frequency (also called the Langmuir frequency) ω_p; the electron gyration frequency Ω_e and the ion gyration frequency Ω_i in magnetized plasmas. The important characteristic plasma lengths are the Debye shielding radius λ_D (section 3.10), the electron gyration radius ρ_e (section 2.3), and the ion gyration radius ρ_i.

Delta functions of Dirac appear in the expressions for the discrete charges and their currents. These functions can, however, be avoided, because, for collisional processes quite slow compared with the collective ones, the individuality of the particles is of little importance, and the delta functions are replaced by smooth and continuous functions of space and time. The discrete distributions then reduce to a continuum of fluid mixtures or a distribution in multidimensional phase space. In this smearing out process the plasma frequency, Debye length, gyro frequencies and gyro radii remain fixed and the collision mean free path becomes infinite. Starting from the Liouville equation existence of several continuous distributions in phase spaces of large number of independent variables, have been proposed for study of plasma behaviour. The B B G K Y heirarchy of equations (section 8.9 (*b*)) is one such well-known formalism. Generally, an equilibrium situation is assumed where the plasma has known predetermined average properties such as charge density, temperature, magnetic field, ionization percentage, etc., and the manner in which small disturbances from this equilibrium state behave as functions of time and space coordinates is studied.

The plasma at a high pressure, surrounding a probe, can be divided into layers like those surrounding a test charge. Immediatley near the probé exists the space charge sheath, which is marked by strong deviation from quasi-neutrality of charges. Particles fall freely in this layer if the Debye shielding distance λ_D (section 3.10) is much less than the mean free path l; some particles are lost by orbital motion if probe radius is less than l but $\lambda_D \geq l$; the situation becomes complicated if $\lambda_D \simeq l$. The transition regions, beyond the space charge region, are marked by field dependent mobilities and the effect of the probe field on the plasma. After this is the diffusion region in which mobilities and diffusion coefficients are constant and the effect of the probe on the plasma is small compared to that of the density gradient. Lastly there is the undisturbed region of neutral plasma.

Plasma fields are either the applied macroscopic fields or the microscopic self fields, which are due to the individuality of the particles. If gravitational force is neglected then neutral particles are force-free in their motion between collisions with other particles. So, the forces acting on charged and neutral particles are of different origin. The Lorentz force, due to an applied electromagnetic field, is more important than the gravitational force. Forces are also either macroscopic (slowly varying), or of a short-range (fluctuating) collisional type. The macroscopic force affects all the particles in a given fluid element similarly.

The collisional force has a non-random component that tends to drive particles down the pressure gradient, and a random component that scatters the particles. In a neutral gas, the pressure gradient force and scattering forces are transmitted impulsively by hard collisions. The Coulomb collision force is a continuously acting grazing collision; so, the individual particles experience a continuous pressure gradient force, and are gradually scattered through a series of small angles.

(a) Phase transition and other transitions

The amount of heat energy introduced into a system for phase transition at a fixed temperature and pressure is the latent heat. This transition occurs from a solid to a liquid or from a liquid to a gas vapour. By further heating the thermal energy of some particles exceeds the energy binding the molecules, thus dissociating molecular gas into atomic gas. If the thermal energy is strong enough for overcoming the atomic binding energy, then some electrons become free and ionization starts; ultimately, all the atoms get stripped of their electrons until the gas becomes a mixture of freely moving electrons and nuclei if the heat applied is sufficient for that. That is the state of a fully ionized plasma. This transition from gas to plasma is gradual, and is not a phase transition in the thermodynamic sense.

(b) Time varying fields

Energy of plasma is dissipated by electrostatic Coulomb interactions of the charges and also by their interaction with electromagnetic field. The static and low frequency parts of wave fields give rise to magnetostatic interactions between the particles, and the high frequency components, being transparent to the plasma, constitute the radiation part. The motions due to electrostatic and magnetostatic interactions are identified with long range interactions; and the Debye shielding phenomenon (section 3.10) provides a natural cut-off distance for the short range force in a collision.

(c) Particles and fluids

In presence of a strong magnetic field the particle gyration period is much smaller than the mean free time t between successive collisions. Then the almost circular orbit of a particle, about a 'guiding centre' C, has a relatively steady motion through a plasma. Guiding centre theory relates the motion of C to the forces acting on the particles. But, gradients in the magnetic induction $\mathbf{B}$ cause the guiding centres of a given species to drift across the lines of force. Moreover, such gradients do not appear as forces in the fluid equation of motion. So, a conflict can arise between guiding centre theory and fluid dynamics.

(d) Collision and ionization

Ionization occurs when $(mv^2/2) > W$, where W is the ionization energy. Thus, a neutral particle absorbs so much energy that one or more electrons are lost, making it positively charged (or an ion). All collisions do not result in ionization or excitation. The energy exchange depends on certain probabilities that are very small positive fractions. Potential energy exchanged in collisions between two excited atoms, cause ionization of one of them. Collisional ionization is possible simply by exchange of kinetic energy (collisions of first order), and when a particle delivers part of its potential energy for this cause (second order collisions). A heavy particle can ionize a particle in the ground state, if its kinetic energy before impact is at least twice that of the ionization or excitation energy of the ground state particle.

(e) Temperature and Kinetic Energy

Temperatures of excitation and ionization and temperature distribution in space and in time are not necessarily the same. This temperature may be different for the different species present in non-equilibrium plasmas, or, even the concept of temperature may be meaningless.

Plasmas, encountered in nature and laboratory, are warm or cold, or hot, rarefied or dense, classical or quantum. These provide a means of transferring energy to molecules through ion collision, light absorption, or electron impact. Their internal energy consists of the kinetic energies of electrons and ions, and their Coulomb interaction energy. The ratio of average ion energy and ionization energy is much greater than unity for hot plasmas, of the order of unity for temperate plasmas, and much less than unity for cold plasmas. Cold plasmas are low electron-energy plasmas. Hot plasmas, have temperatures of many million degrees centigrade.

Kinetic theory establishes relationships between temperature (kinetic), and averaged kinetic energy. However, plasma particles also possess vibrational energy, rotational energy and the potential energy of Coulomb force and are acted on by other electromagnetic interactions. Also, all particles do not possess the same velocity. Both neutral particles and charged particles, in the absence of applied driving forces have considerable random motion, so these are generally supposed to follow the Maxwellian velocity distribution.

Electrons being smaller in size and mass than the ions and neutral particles, have much greater thermal and average velocities, and therefore, much higher temperatures than the other components of a plasma. Compared to the neutral particles, ions have higher velocities and temperatures. So the different species of particels in a plasma possess different temperatures.

3.2 SPACE FIELD OF CHARGE

The motion of a test particle is determined by its interaction with other particles in space and by processes like convection and radiation. The infinite space around a test charge is divided into the following four regions: (1) the region dominated by quantum mechanical effects, nearest to the charge; (2) the Debye shielding field region (section 3.10); (3) the fluid region and (4) the far field space. In region (1) non-elastic close range collisions with other particles, which are of quantum mechanical nature cause a net deviation through a large angle. The region (2) is a sheath of radius, few Debye lengths; which shields the test particle from other parts of the plasma. In this region the classical theory is sufficient. The field at the test particle fluctuates due to the discrete nature of the source within both regions (1) and (2). The short range interactions (chapter 6) there give rise to rapidly varying fields which are determined by the methods of study of collision cross-section. Region (3) is specified by the field of continuous media, due to prevalence of large number of charges of both signs, and their collective influence. Only macroscopic variations affect the external field region (4) which is farthest from the test charge.

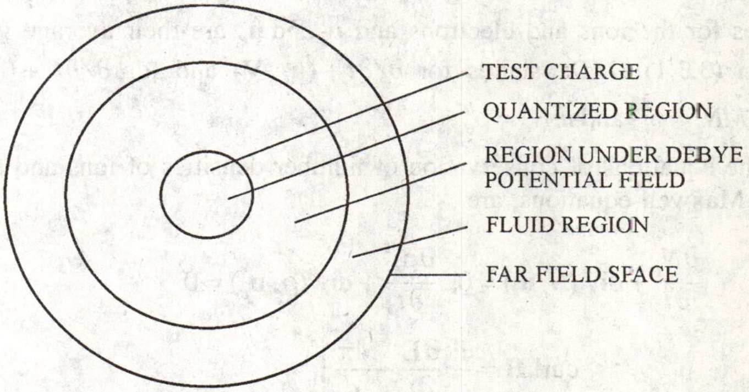

Fig. 3.2.1 Division of the space surrounding a charge into regions, according to their field features.

For electrons of gaseous plasma the de Broglie wavelength λ is about $10^{-7}\,\xi^{-1/2}$ cm, where ξ is in electron volt, and the mean interparticle distance $N^{-1/3}$ is about 10^{-5} cm at $N = 10^{15}/\text{cm}^3$. A comparison between these two lengths indicates that classical description of gaseous plasma is adequate in the Debye sphere and beyond. But, classical description is not sufficient, for study of all plasma properties.

3.3 MIXTURE OF FLUIDS OF POSITIVE AND NEGATIVE CHARGES

For simplicity we assume that a plasma is a mixture of several species of fluids, both charged and uncharged. Each such species is assumed to move independent of the other species. So, separate hydrodynamical equations of motion and continuity condition are used. Interaction among these fluid components occurs through electromagnetic effects of long range and through collisions among them. Moreover, the stresses and strains of fields of electrodynamics its pressure tensor. Also the field induced velocities of the constituents are small compared to the random velocities of thermal origin of the particles of a species.

The momentum transfer equations for a mixture of compressible fluids of positive and negative charges are

$$\frac{D}{Dt}(\rho_i^M \mathbf{u}_i) = -\nabla p_i + \mathbf{F}_i + N_i e \mathbf{E} + \frac{1}{c} N_i e (\mathbf{u}_i \times \mathbf{H}), \tag{3.3.1}$$

$$\frac{D}{Dt}(\rho_e^m \mathbf{u}_e) = -\nabla p_e + \mathbf{F}_e - N_e l \mathbf{E} - \frac{1}{c} N_e e (\mathbf{u}_e \times \mathbf{H}) \tag{3.3.2}$$

where the substantial differential operator D/Dt is the total time derivative operator, e, N_i and ρ_i^M are the charge, number density and mass density of ions respectively, $-e$, N_e, ρ_e^m are the corresponding quantities for electrons, $\mathbf{F}_i$ and $\mathbf{F}_e$ contain forces of collision, gravitation, viscosity, etc., p_i and p_e are partial pressures for the ions and electrons and $\mathbf{u}_i$ and $\mathbf{u}_e$ are their average velocities. Thus, in (3.3.1) D/Dt stands for $\partial/\partial t + (\mathbf{u}_i \,.\, \nabla)$ and for $\partial/\partial t + (u_e \,.\, \nabla)$ in (3.3.2), $MN_i = \rho_i^M$ and $mN_e = \rho_e^m$.

The equations of conservation of number densities of ions and electrons, and the Maxwell equations, are

$$\frac{\partial N_i}{\partial t} + \text{div}\,(N_i \mathbf{u}_i) = 0, \quad \frac{\partial \rho_e^m}{\partial t} + \text{div}\,(\rho_e \mathbf{u}_e) = 0 \tag{3.3.3}$$

$$\text{curl}\,\mathbf{H} = \frac{1}{c}\frac{\partial \mathbf{E}}{\partial t} + \frac{4\pi}{c}\mathbf{j}, \tag{3.3.4}$$

$$\text{curl}\,\mathbf{E} = -\frac{1}{c}\frac{\partial \mathbf{H}}{\partial t}, \tag{3.3.5}$$

$$\text{div}\,\mathbf{E} = 4\pi Q, \quad \text{div}\,\mathbf{H} = 0 \tag{3.3.6}$$

where $\mathbf{j}$ and Q are the macroscopic current and charge densities, so

$$\mathbf{j} = N_i e \mathbf{u}_i - N_e e \mathbf{u}_e, \; Q = N_i e - N_e e \tag{3.3.7}$$

If thermal change is of adiabatic nature, then the relations between the changes of pressure and density for the two components of the plasma are

$$\nabla p_i = c_i^2 \nabla \rho_i, \; \nabla p_e = c_e^2 \nabla \rho_e \tag{3.3.8}$$

where c_i and c_e are the sound velocities (section 3.4 (*a*)). These equations satisfy the conditons of a force-free neutral equilibrium state, which, macroscopically, is without any current and charge. To satisfy the charge conservation condition, N_i and N_e must be constants and equal; so

$$N_i = N_e = N \text{ (say)}, \; \mathbf{u}_i = 0, \; \mathbf{u}_e = 0, \; \nabla p_i = 0, \; \nabla p_e = 0$$

If there is no other force, then $\mathbf{F}_i$ and $\mathbf{F}_e$ vanish, and equations (3.3.1) and (3.3.2) give $0 = N_e \mathbf{E}$, $0 = -N_i \mathbf{E}$, then no static electric field can exist, and the non-trivial equations left are

$$\nabla \times \mathbf{H}_0 = 0,\ \nabla . \mathbf{H}_0 = 0, \frac{\partial \mathbf{H}_0}{\partial t} = 0 \tag{3.3.9}$$

The general solution can be written as $\mathbf{H}_0 = \nabla \psi$ where ψ satisfies the Laplace equation $\nabla^2 \psi = 0$. Thus, the equilibrium plasma can sustain a static magnetic field, the potential of which satisfies the Laplace equation.

The field variables in the disturbed state are written as

$$N_i \equiv N_0 + N_i,\ N_e \equiv N_0 + N_e,\ \mathbf{E} \equiv 0 + \mathbf{E},\ \mathbf{H} \equiv \mathbf{H}_0 + \mathbf{H},$$

$$\mathbf{u}_i \equiv 0 + \mathbf{u}_i,\ \mathbf{u}_e \equiv 0 + \mathbf{u}_e,\ \rho_i^M \equiv MN_0 + MN_i = \rho_i^0 + \rho_i^M, \tag{3.3.10}$$

$$\rho_e^m = mN_0 + mN_e = \rho_e^0 + \rho_e^m$$

where the increments due to the perturbation are functions of space and time. The linearized field equations are

$$\rho_i^0 \frac{\partial \mathbf{u}_i}{\partial t} = -c_i^2 \nabla \rho_i^M + N_0 e\mathbf{E} + \frac{1}{c} N_0 e (\mathbf{u}_i \times \mathbf{H}_0), \tag{3.3.11}$$

$$\rho_e^0 \frac{\partial \mathbf{u}_e}{\partial t} = -c_e^2 \nabla \rho_e^m - N_0 e\mathbf{E} - \frac{1}{c} N_0 e (\mathbf{u}_e \times \mathbf{H}_0), \tag{3.3.12}$$

$$\frac{\partial N_i}{\partial t} + N_0 (\nabla . \mathbf{u}_i) = 0, \frac{\partial N_e}{\partial t} + N_0 (\nabla . \mathbf{u}_e) = 0, \tag{3.3.13}$$

$$\nabla \times \mathbf{H} = \frac{1}{c} \frac{\partial \mathbf{E}}{\partial t} + \frac{4\pi}{c} N_0 e (\mathbf{u}_i - \mathbf{u}_e), \tag{3.3.14}$$

$$\nabla \times \mathbf{H} = -\frac{1}{c} \frac{\partial \mathbf{H}}{\partial t}, \tag{3.3.15}$$

$$\nabla . \mathbf{E} = 4\pi e (N_i - N_e),\ \nabla . \mathbf{H} = 0 \tag{3.3.16}$$

In the simple case of electron plasma, ions providing the neutralising static background, the linearized equations (3.3.11) – (3.3.16) simplify to

$$N_0 \frac{\partial \mathbf{u}}{\partial t} = -c_e^2 \nabla N - \frac{N_0 e}{M} \mathbf{E} - N_0 (\mathbf{u} \times \mathbf{\Omega}), \tag{3.3.17}$$

$$\frac{\partial N}{\partial t} + N_0 (\nabla . \mathbf{u}) = 0, \tag{3.3.18}$$

$$\nabla \times \mathbf{E} = -\frac{1}{c} \frac{\partial \mathbf{H}}{\partial t}, \tag{3.3.19}$$

$$\nabla \times \mathbf{H} = \frac{1}{c} \frac{\partial \mathbf{E}}{\partial t} - \frac{4\pi e}{c} N_0 \mathbf{u}, \tag{3.3.20}$$

$$\nabla . \mathbf{E} = -4\pi\, eN,\ \nabla . \mathbf{H} = 0 \tag{3.3.21}$$

where the subscript has been removed from $\mathbf{u}_e$ and N_e, and $\mathbf{\Omega} = e\mathbf{H}_0/mc$. Here N is connected to the divergence of the electric field and velocity only, and not to the curl of any field vector. All these equations are of first order in derivatives with respect to time and space coordinates, and each of them connect at least two of the field variables **u**, N, E, H. The equations, exclusively for one of the field variables **u**, N, E, H for unmagnetized plasma ($\mathbf{H}_0 = 0$) are homogeneous Klein Gordon wave equations.

For unmagnetized plasmas, take divergence of (3.3.17), and using (3.3.18) and (3.3.21), eliminate **u** and **E**, to obtain for N the equation

$$\ddot{N} - c_e^2 \nabla^2 N + \omega_\rho^2 N = 0 \tag{3.3.22}$$

where
$$\omega_\rho^2 = 4\pi N_0 e^2/m \tag{3.3.23}$$

and the time derivative is denoted by a dot. Taking curl of (3.3.20), then using (3.3.19) to eliminate **E** and **u**, the equation for **H** is obtained :

$$\ddot{\mathbf{H}} - c^2 \nabla^2 \mathbf{H} + \omega_\rho^2 \mathbf{H} = 0 \tag{3.3.24}$$

Eliminating next **H** and **u**, by taking curl of (3.3.19) and using (3.3.20) and (3.3.22), the equation for **E**,

$$\ddot{\mathbf{E}} - c^2 \nabla^2 \mathbf{E} + \omega_\rho^2 \mathbf{E} = -(c^2 - c_e^2)\, \nabla (\nabla \cdot \mathbf{E}), \tag{3.3.25}$$

and similarly for **u** are derived.

3.4 WAVES IN UNMAGNETIZED PLASMAS

(a) Electron acoustic wave

The waves of N or ρ_e^m or the components of **u** or **E**, along the direction of propagation, are essentially longitudinal in nature. So these are acoustic waves. Being free of time varying magnetic fields, these do not radiate into matter-free space.

When kinetically derived (section 9.6), $p = NK_B T$ where K_B is the Boltzmann constant, N is the number density of electrons and T is the kinetic temperature in degrees Kelvin. So, for isothermal changes, T remains unchanged and an adiabatic law gives $\delta p = K_B T\, \delta N = (K_B T/m)\, \delta\rho = c_e^2\, \delta\rho_e^m$ where c_e is the isothermal sound speed of the compressible electron fluid. It follows from the more general adiabatic law of thermodynamics $p\,\rho^{-\gamma} =$ constant, where γ is the ratio of specific heats at constant pressure and constant volume (section 4.2 (*c*)). Then $dp = (\gamma p/\rho^m)\, d\rho = c_s^2\, d\rho^m$ with $c_s^2 = \gamma p/\rho$, c_s is the adiabatic sound speed for the electron fluid. This law, leading to an instantaneous proportional change in density, is valid for only slow changes in pressure, so that sufficient time is available for density to adjust accordingly. But, for rapid fluctuations in pressure, the energy equation (9.10.13) has to be used.

Assuming the plasma to be free from collisional damping and the wave solution proportional to exp $(i\,(\mathbf{K} \cdot \mathbf{r}) - i\,\omega\, t)$ for N, with real ω and K, equation (3.3.22) gives the dispersion relation for electron acoustic waves,

$$K^2 c_e^2 = \omega^2 - \omega_\rho^2 \tag{3.4.1}$$

in unmagnetized plasmas. ω_ρ is the characteristic frequency of electron oscillation or simply the plasma frequency or the Langmuir frequency. Since $\omega_\rho/2\pi = 8920\sqrt{N_0}$, for a typical plasma ω_p is rather large; for example, $\omega_\rho = 3.6 \times 10^9$ rad/sec when $N_0 = 4 \times 10^9/\text{cm}^3$. When $\omega^2 > \omega_\rho^2$, K is real; so the waves propagate through the plasma at frequencies greater than the plasma frequency ω_ρ. At low frequencies ($\omega^2 < \omega_\rho^2$), K becomes imaginary and the waves decay as they penetrate into the plasma. These are only surface waves; their skin depth

$$d_l = \frac{c_e}{\sqrt{\omega_\rho^2 - \omega^2}} \approx \left(\frac{mc_e^2}{4\pi N_0 e^2}\right)^{1/2} \tag{3.4.2}$$

The amplitude is reduced to about 37 per cent of its initial value in this depth.

In the cold plasma approximation the pressure gradient, like the spread of electron and ion velocity distributions about their respective mean values, can be neglected. So $\nabla p_e = 0$ and (3.3.22) reduces to

$$\ddot{N} + \omega_\rho^2 N = 0 \tag{3.4.3}$$

giving $\omega^2 = \omega_\rho^2$. Hence, all the electrons of a homogeneous cold plasma simultaneously oscillate about their mean positions at the Langmuir frequency ω_ρ. Equation (3.4.3) can be written as $(\omega^2 - \omega_\rho^2)(\mathbf{\nabla} \cdot \mathbf{E}) = 0$, showing that $(\mathbf{\nabla} \cdot \mathbf{E}) = 0$ whenever $\omega^2 \neq \omega_\rho^2$, so, space charge separation, in a cold plasma, is not possible at frequencies different from the plasma frequency ω_ρ.

In a low temperature plasma (example, $N_0 = 10^{15}/\text{cm}^3$ and $T = 10^6\ ^\circ\text{K}$), or near $T = 10^8\ ^\circ\text{K}$ and $N_0 = 10^{15}/\text{cm}^3$, the particle collision number is of the order of 10^3/sec. Then electrons move without collision, from their positions of equilibrium, under the influence of the Coulomb attractive force. Upon reaching these positions they move further due to acceleration until a displacement is reached where the kinetic energy is completely converted into potential energy. The attractive Coulomb force again draws them towards their equilibrium position, causing the electrostatic oscillation at the frequency ω_ρ. This oscillatory motion persists indefinitely at the plasma frequency ω_ρ. The convection current of the electrons is exactly cancelled by the displacement current. So, there is no magnetic field in this oscillatory motion, no radiation, and no propagation of one dimensional oscillations of electrons away from their points of origin.

(b) Transverse wave

As in optically transparent dielectrics, high frequency transverse waves propagate through plasmas. Assuming the harmonic wave solution exp $(i\,(\mathbf{K}\,.\,\mathbf{r}) - i\,\omega\,t)$, the dispersion relation, obtained from the solution of (3.3.24), is

$$K^2\,C^2 = \omega^2 - \omega_\rho^2 \tag{3.4.4}$$

So the plasma has little influence on the waves at frequencies much greater than ω_ρ. The skin depth d_t, for $\omega^2 << \omega_\rho^2$, is given by

$$dt = (mC^2/4\pi\,Ne^2)^{1/2}) >> d_l$$

where d_l is the skin depth for longitudinal waves. A plane transverse wave is plane polarized if the field polarization is confined to a fixed plane; the plane wave is elliptically polarized if the locus of the field vector is an ellipse in a time period. If the locus is a circle the wave is circularly polarized. The index of refraction of transverse waves in a plasma neglecting effects of finite temperature and applied static megnetic field, is

$$n = \frac{Kc}{\omega} = \left(1 - \frac{\omega_\rho^2}{\omega^2}\right)^{1/2} \approx 1 - \frac{\omega_\rho^2}{2\omega^2}, \tag{3.4.4 (a)}$$

It holds good for low electron densities and weak field strength. This n is always less than unity for free wave propagation through a uniform plasma slab.

The phase velocity v_ρ and group velocity v_g are given by the formula

$$v_\rho = \frac{\omega}{K} = c\left(1 - \frac{\omega_\rho^2}{\omega^2}\right)^{-1/2}, \quad v_g = \frac{\partial\omega}{\partial K} = \frac{Kc^2}{\omega} = nc, \tag{3.4.4 (b)}$$

These approximately, read

$$v_\rho = c\left(1 + \frac{\omega_\rho^2}{2\omega^2}\right), \quad v_g = c\left(1 - \frac{\omega_\rho^2}{2\omega^2}\right); \tag{3.4.4 (c)}$$

so that $v_\rho > c > v_g$. Hence a sharp pulse of electromagnetic emission is dispersed such that the high frequency components reach the receiver before the low frequency components. This effect is conspicuous in whistler modes (section 11.8).

(c) Electrokinetic waves

Putting the wave solution exp $(i\,(\mathbf{K}\,.\,\mathbf{r}) - i\,\omega\,t)$ in equation (3.3.25) we obtain

$$\mathbf{K}\,(c^2 - c_e^2)\,(\mathbf{K}\,.\,\mathbf{E}) + (\omega^2 - \omega_\rho^2 - K^2\,c^2)\,\mathbf{E} = 0 \tag{3.4.5}$$

Its scalar product with $\mathbf{K}$ gives $(\mathbf{K}\,.\,\mathbf{E})\,(\omega^2 - \omega_\rho^2 - K^2\,c_e^2) = 0$. Since, in general $(\mathbf{K}\,.\,\mathbf{E}) \neq 0$, it gives the dispersion relation (3.4.1) for longitudinal waves.

The vector product of (3.4.5) with **K** gives $(\mathbf{K} \times \mathbf{E})(\omega^2 - \omega_\rho^2 - K^2 c^2) = 0$. So, the waves of $(\mathbf{K} \times \mathbf{E})$ are transverse; their dispersion is given by (3.4.4). Since **E** (or **u**) is a mixture of both longitudinal and transverse components, the equation for it, equation (3.3.25), contains both the equations for transverse and longitudinal waves. The divergence-free self-magnetic field, given by (3.3.24), is a purely transverse wave. The mixed waves are called electrokinetic waves.

Exericse 3.4.1. Assuming that both electrons and ions are mobile, in unmagnetized plasma, show that the dispersion relation for transverse waves is

$$K^2 c^2 = \omega^2 - \omega_{\rho e}^2 - \omega_{\rho i}^2$$

and for longitudinal waves is

$$(K^2 c_e^2 - \omega^2 + \omega_{\rho e}^2)/\omega_{\rho e}^2 = \omega_{\rho i}^2/(K^2 c_i^2 - \omega^2 + \omega_{\rho i}^2)$$

where

$$\omega_{pe}^2 = 4\pi N_0 e^2/m_i,\ \omega_{\rho i}^2 = 4\pi N_0 e^2/M;\ \nabla p_e = c_e^2 \nabla \rho_e,\ \nabla p_i = C_i^2 \nabla \rho_i$$

3.5 DISTINGUISHING FEATURES OF WAVES IN PLASMAS

If θ is the angle between the direction of oscillation of a vector field and its direction of propagation, then for $\theta = 0$ the dispersive wave is longitudinal or acoustic; and for $\theta = \pi/2$ it is transverse. When θ is neither zero nor $\pi/2$, the waves are called electrokinetic waves; these are a mixed system of longitudinal and transverse waves. If **A** is a field vector, then (*i*) for pure longitudinal waves $\theta = 0$, $(\mathbf{K.A}) \neq 0$, $(\mathbf{K} \times \mathbf{A}) = 0$; (*ii*) for pure transverse waves $\theta = \pi/2$, $(\nabla.\mathbf{A}) = 0$, $(\nabla \times \mathbf{A}) \neq 0$; $(\mathbf{K} \cdot \mathbf{A} = 0$, $(\mathbf{K} \times A) \neq 0$ (*iii*) for coupled waves (electrokinetic waves) $(\nabla \cdot \mathbf{A}) \neq 0$, $(\nabla \times \mathbf{A}) \neq 0$, $(\mathbf{K} \cdot \mathbf{A}) \neq 0$, $(\mathbf{K} \times \mathbf{A}) \neq 0$. For conditions (*i*) we can write $\mathbf{A} = \text{grad}\, \varphi$ where φ is a scalar field variable. The plasma equations show that $(\nabla \times \mathbf{E}) \neq 0$ and $(\nabla \cdot \mathbf{E}) \neq 0$, so the E-wave has both the longitudinal and transverse components (section 11.2). We can write $\mathbf{A} = \mathbf{A}_l + \mathbf{A}_t$ where $\mathbf{A}_l$ is the longitudinal part and $\mathbf{A}_t$ is the transverse part. So $(\nabla \cdot \mathbf{A}_l) \neq 0$, $(\nabla \times \mathbf{A}_l) = 0$, $\mathbf{A}_l = \nabla \varphi$, $(\nabla \cdot \mathbf{A}_t) = 0$, $(\nabla \times \mathbf{A}_t) \neq 0$.

In dispersive media, for a mixed field (**E** or **u**), when the two types of waves are not coupled, the dispersion relation is a product of two factors equated to zero, one of which equated to zero gives the dispersion for longitudinal waves and the other for transverse waves. For coupled waves this separation is not possible. Longitudinal waves depend on sound velocities and are excited at frequencies lower than those for transverse waves. Unless these are converted by some agencies (called the coupling factors) into transverse waves (which can propagate through vacuum) the longitudinal disturbances growing in distant regions cannot be detected by electronic laboratory devices.

(a) Temporal and spatial evolution of dispersive waves

A monochromatic wave is dispersive if it propagates at a definite phase velocity and group velocity. In spatial evolution problems the wave evolves in space at a given instant of time and so K is solved as a function of ω. In temporal evolution problems the wave evolves in time at a given point of space; so ω is solved as a function of K. Since the dispersion relations are generally not symmetric with respect to ω and K, these two categories of evolution are qualitatively different. The non-symmetric dispersion relations (3.4.1) and (3.4.4) are hyperbolic in ω and K. Equation (3.4.1), in the temporal evolution problem, gives

$$v_g = c_e \sqrt{1 - \omega_\rho^2/\omega^2}, \quad v_\rho = (\sqrt{K^2 c_e^2 + \omega_\rho^2})/K,$$

and in the spatial problem $v_g = Kc_e^2/\sqrt{K^2 c_e^2 + \omega_\rho^2}$, $v_\rho = (\sqrt{K^2 c_e^2 + \omega_\rho^2})/K$. In both cases, evidently $v_\rho > c_e$ and $v_g < c_e$.

(b) Relations between the two types of propagating fields

We first consider the case of unmagnetized plasma. So, putting $\Omega = 0$ in (3.3.17), we obtain

$$N_0 \frac{\partial \mathbf{u}}{\partial t} = - c_e^2 \nabla N - \frac{N_0 e}{m} \mathbf{E} \tag{3.5.1}$$

Writing $\mathbf{u} = \mathbf{u}_l + \mathbf{u}_t$, $\mathbf{E} = \mathbf{E}_l + \mathbf{E}_t$, and separating the longitudinal and transverse parts of (3.5.1) and (3.3.18 – 21), we obtain

$$N_0 \frac{\partial \mathbf{u}_l}{\partial t} = - c_e^2 \nabla N - \frac{N_0 e}{m} \mathbf{E}_l, \; \frac{\partial \mathbf{E}_l}{\partial t} = 4\pi N_0 e \mathbf{u}_l, \; (\nabla \cdot \mathbf{E}_l) = - 4\pi Ne,$$

$$\frac{\partial N}{\partial t} + N_0 (\nabla \cdot \mathbf{u}_l) = 0, \; (\nabla \times \mathbf{E}_l) = 0$$

for the longitudinal wave, and

$$N_0 \frac{\partial \mathbf{u}_t}{\partial t} = - \frac{N_0 e}{m} \mathbf{E}_t, \; \nabla \times \mathbf{E}_t = - \frac{1}{c} \frac{\partial \mathbf{H}}{\partial t}, \; (\nabla \cdot \mathbf{H}) = 0,$$

$$(\nabla \times \mathbf{H}) = \frac{1}{c} \frac{\partial E_t}{\partial t} - \frac{4\pi N_0 e}{c} \mathbf{u}_t, \; (\nabla \cdot \mathbf{E}_t) = 0$$

for the transverse wave. Since the fields of one subset do not appear in the equations for the other, one subset of equations can be solved ignoring the other, when the two types of waves are uncoupled, or independent.

If $\mathbf{j}_t$ is the electron current density induced by $\mathbf{E}_t$, then $\mathbf{j}_i = - N_0 e \mathbf{u}_t = - (\mathbf{E}_t N_0 ie^2/m\omega)$, so $\nabla \times \mathbf{H} = (i\omega/c)(1 - \omega_\rho^2/\omega^2) \mathbf{E}_t$. When this equation is compared with $\nabla \times \mathbf{H} = (\partial \mathbf{D}/c\,\partial t)$, for dielectrics ($\mathbf{D} = \varepsilon \mathbf{E}_t$) we find that

$$\varepsilon = 1 - \omega_\rho^2/\omega^2$$

Hence $\varepsilon < 1$ for $\omega^2 > \omega_\rho^2$, though for dielectrics $\varepsilon > 1$. Also this ε is frequency dependent though, for ordinary weak fields, in dielectrics, it is not so. Thus, the characteristic features of transverse waves in plasmas are somewhat different from those in dielectrics.

When $N \neq 0$, the equation of motion in a magnetized plasma is

$$\frac{\partial \mathbf{u}}{\partial t} = -\frac{c_e^2}{N_0} \nabla N - \frac{e}{m} \mathbf{E} - (\mathbf{u} \times \Omega)$$

The term $(\mathbf{u} \times \Omega)$ relates the component of $\partial \mathbf{u}/\partial t$ parallel to the wave vector **K** with the component of **u** perpendicular to it. Also, for the same reason, the components of $\partial \mathbf{u}/\partial t$ perpendicular to **K** are related to $\mathbf{u}_l$. So, these are electrokinetic waves.

(c) Plasmons and Phonons

Plasmons are quantas of oscillations of electron plasmas. Phonons are quantas of energy of ion acousting waves in plasmas.

3.6 LONGITUDINAL WAVES WHEN BOTH ELECTRONS AND IONS ARE MOBILE

When both electrons and ions are considered mobile and compressible, the linearized field equations are

$$\frac{\partial \mathbf{u}_e}{\partial t} = -\frac{c_e^2}{N_0} \nabla N_e - \frac{e}{m} \mathbf{E}, \frac{\partial \mathbf{u}_i}{\partial t} = -\frac{c_i^2}{N_0} \nabla N_i + \frac{e}{M} \mathbf{E}, \tag{3.6.1}$$

$$\frac{\partial N_e}{\partial t} + N_0 (\nabla \cdot \mathbf{u}_e) = 0, \ \frac{\partial N_i}{\partial t} + N_0 (\nabla \cdot \mathbf{u}_i) = 0, \ (\nabla \cdot \mathbf{E}) = 4\pi e (N_i - N_e) \tag{3.6.2}$$

Here $\mathbf{u}_e$ and $\mathbf{u}_i$ are the wave field induced perturbation velocities of electrons and ions respectively, N_e and N_i are the perturbations in their number densities, the equilibrium values of those densities are equal, and equal to N_0; c_e and c_i are the sound speeds in the fluids of electrons and ions respectively. So

$$\nabla p_e = mc_e^2 \nabla N_e, \nabla p_i = Mc_i^2 \nabla N_i \tag{3.6.3}$$

and **E** is the electric field. Taking divergence of the two equations of (3.6.1) and using (3.6.2) we get

$$\ddot{N}_e - C_e^2 \nabla^2 N_e - \omega_{\rho e}^2 (N_i - N_e) = 0, \ \ddot{N}_i - C_i^2 \nabla^2 N_i + \omega_{\rho i}^2 (N_i - N_e) = 0 \tag{3.6.4}$$

Assuming the space-time dependence $\exp(\bar{i}(\mathbf{K} \cdot \mathbf{r}) - i\omega t)$, for both N_e and N_i, two simultaneous equations are obtained; these are

$$(K^2 c_e^2 - \omega^2 + \omega_{\rho e}^2) N_e - \omega_{\rho e}^2 N_i = 0, \ (K^2 c_i^2 - \omega_{\rho i}^2) N_i - \omega\rho i^2 N_e = 0 \tag{3.6.5}$$

Eliminating N_i/N_e from those relations we get the dispersion relation

$$\frac{(K^2 c_e^2 - \omega^2 + \omega_{\rho e}^2)}{\omega_{\rho e}^2} = \frac{\omega_{\rho i}^2}{(K^2 c_i^2 - \omega^2 + \omega_{\rho i}^2)} \tag{3.6.6}$$

This is a quadratic equation in both K^2 and ω^2. From kinetic theory (section 9.7) we observe that $c_e^2 = K_B T_e/m$ and $c_i^2 = K_B T_i/M$ where K_B is the Boltzmann constant, and T_e and T_i are the kinetic temperatures of electrons and ions respectively. Then, since $c_e^2 \omega_{\rho i}^2 = (T_e/T_i) c_i^2 \omega_{\rho e}^2$, the dispersion relation (3.6.6) becomes

$$K^4 c_e^2 c_i^2 - K^2 \left[(c_e^2 + c_i^2) \omega^2 - c_e^2 \omega_{\rho i}^2 \left(1 + \frac{T_i}{T_e} \right) \right] + \omega^4 - \omega^2 (\omega_{\rho e}^2 + \omega_{\rho i}^2) = 0 \tag{3.6.7}$$

Since $c_i^2 << c_e^2$ and $\omega_{\rho i}^2 << \omega_{\rho e}^2$, this relation simplifies to

$$K^4 c_e^2 c_i^2 - K^2 c_e^2 \left\{ \omega^2 - \omega_{\rho i}^2 \left(1 + \frac{T_i}{T_e} \right) \right\} + \omega^4 - \omega^2 \omega_{\rho e}^2 = 0 \tag{3.6.7 (a)}$$

(a) High frequency longitudinal waves

When $\omega^2 > \omega_{\rho i}^2 (1 + T_i/T_e)$, we neglect the second term in the coefficient of $K^2 c^2$ and get

$$K^4 c_e^2 c_i^2 - K^2 c_e^2 \omega^2 - \omega^2 \omega_{\rho e}^2 + \omega^4 = 0 \tag{3.6.7 (b)}$$

Since $c_e^2 \omega^2 >> c_i^2 (\omega^2 - \omega_{\rho e}^2)$ when $\omega^2 > \omega_{\rho e}^2$, this equation reduces to the two dispersion relations

$$K^2 c_e^2 = \omega^2 - \omega_{\rho e}^2, \; K^2 c_i^2 = \omega^2 \tag{3.6.7 (c)}$$

The first one is the dispersion relation for the electron acoustic wave of section 3.4a; the second dispersion relation is that for a longitudinal ion wave which travels at the ion sound speed without any cut-off minimum frequency.

(b) Low frequency, long wave iength waves

Then $\omega^2 < \omega_{\rho i}^2$ and $K^2 c_i^2 < \omega_{Pi}^2$, so equation (3.6.7 (*b*)) reduces to

$$K^2 c_e^2 \omega_{Pi}^2 = \omega^2 \omega_{Pe}^2 \tag{3.6.7 (d)}$$

because $\omega_{\rho i}^2 c_e^2 = \omega_{\rho e}^2 c_S^2$ where $c_S^2 = K_B T_e/M$ (because both sides equal the quantity $(4\pi N_0 e^2) K_B T_e/Mm$), and $T_i < T_e$).

It gives the dispersion relation for ion acoustic waves,

$$K^2 c_S^2 = \omega^2, \tag{3.6.8}$$

c_S is the ion acoustic sound speed.

Unlike electron acoustic waves, the ion acoustic waves have no low frequency cut-off; and these can propagate at any low frequency. Since $c_S^2 < c_e^2$, for the same wave number, K, the frequency ω of (3.6.8) is much less than the ω of (3.4.1). Ions being slow, are accompanied by their shielding clouds. This is not possible for electrons, which are relatively much hotter. Collisions generate, in a high density neutral gas, ion acoustic waves. Their restoring force is electrical, arising from a very slight electron-ion charge separation, which is due to the existence of the long range Coulomb force. Such low frequency oscillations are highly damped in the plasma; and electrons do not behave collectively, they remain at temperatures higher than for ions; so they play the role of a neutralising moving stream of negative charges. Also, the approximation $\omega^2 < \omega_{\rho e}^2$ implies that electron inertial term is negligible for low frequencies.

Exercise 3.6.1. Find the dispersion relation (3.6.7) by starting with the quasistatic solution of the Boltzmann equation (chapter 9) for electrons, and using Gauss's theorem for the net charge and for ions the equations of momentum transfer and continuity.

Answer. The equation (9.14.3) for electrons, is

$$N_e = N_0 \exp\left(-\frac{e\varphi}{K_B T_e}\right) - N_0 \tag{3.6.9}$$

where $\mathbf{E} = \mathbf{\nabla}\,\varphi$, T_e is the kinetic temperature of electrons, and K_B is the Boltzmann constant, in the linear approximation $|\,e\varphi/K_B T_e\,| << 1$, the relation

$$N_e = -(N_0\, e/K_B\, T_e)\,\varphi \tag{3.6.10}$$

The other relevant equations, in the linear approximation are

$$\dot{N}_i + N_0\,(\mathbf{\nabla}\,.\,\mathbf{u}) = 0,\ \ \rho_0\,\dot{\mathbf{u}} = -\mathbf{\nabla}\,p_i + N_0\,e\mathbf{E},\ \mathbf{\nabla}\,.\,\mathbf{E} = 4\pi e\,(N_i - N_e) \tag{3.6.11}$$

Writing $p_i = MC_i^2\,\mathbf{\nabla} N_i$ and eliminating $\mathbf{u}$ between these relations, we get

$$M\ddot{N}_i - MC_i^2\,\nabla^2 N_i + N_0\, e\,\nabla^2\varphi = 0 \tag{3.6.12}$$

From (3.6.10) and the last relation of (3.6.11) we obtain

$$N_i = -\left(\frac{N_0\, e\varphi}{K_B\, T_e}\right) + \frac{\nabla^2\varphi}{4\pi e} \tag{3.6.13}$$

Eliminating now N_i between (3.6.12) and (3.6.13) we have a fourth order equation in φ :

$$c_i^2\,\nabla^2\,\nabla^2\,\varphi - \omega_{\rho i}^2\,\nabla^2\,\varphi - \left(\frac{c_i^2\,\omega_{Pi}^2}{c_s^2}\right)\nabla^2\,\varphi + \left(\frac{\omega_{Pi}^2}{c_s^2}\right)\ddot{\varphi} - \nabla^2\,\ddot{\varphi} = 0 \tag{3.6.14}$$

The plane wave solution $\exp\,(i\,(\mathbf{K}\,.\,\mathbf{r}) - i\,\omega t)$ which gives the dispersion relation (3.6.7), except for the term ω^4 which is negligible for low frequency waves. In terms of the Debye length $\lambda_D = (K_B\, T_e/4\pi\, N_0\, e^2)^{1/2}$ (section 3.10), in (3.6.7), we obtain the dispersion relation

$$\omega^2 = K^2 c_i^2 + \frac{\omega_{\rho i}^2}{1 + (1/\lambda_D^2 K^2)} \tag{3.6.15}$$

For long wavelength waves $\lambda_D K << 1$ and (3.16.15) reduces to

$$\omega^2 = K^2 c_i^2 + \omega_{\rho i}^2 \lambda_D^2 K^2 = K^2 c_i^2 + K^2 c_s^2 \tag{3.6.16}$$

Since $c_s^2 > c_i^2$, it approximately gives (3.6.8).

Exercise 3.6.2. Show that for the two component thermally drifting plasma the dispersion relation for a longitudinal wave can be written as

$$1 + \frac{\omega_{Pe}^2}{K^2 c_e^2 - \omega_e^2} + \frac{\omega_{Pi}^2}{K^2 c_i^2 - \omega_i^2} = 0$$

where $\omega_e = \omega - KV_\rho$, $\omega_i = \omega - KV_i$, $c_e^2 = dp_e/dP_e$, $c_i^2 = dp_i/dP_i$.

Answer. The field equations of the linearized approximation are

$$mN_0 (\dot{\mathbf{u}}_e + V_e \mathbf{u}_e') = - N_0 e\mathbf{E} - mc_e N_e'$$

$$MN_0 (\dot{\mathbf{u}}_i + V_i \mathbf{u}_i') = N_0 e\mathbf{E} - Mc_i^2 N_i$$

$$\dot{N}_e + V_e N_e' + N_0 u_e' = 0, \ \dot{N}_i + V_i N_i' + N_0 u_i' = 0$$

$$E' = - 4\pi e (N_e - N_i), \ \dot{E} = - 4\pi j = - 4\pi e N_0 (u_i - u_e) - 4\pi e (N_i V_i - N_e V_e)$$

where dot denotes the time derivative and prime denotes the derivative with respect to Z. We assume that

$$E = a \cos\theta, \ \theta = KZ - wt, \ u_e = \alpha_e \sin\theta, \ u_i = \alpha_i \sin\theta$$

$$N_e = \lambda_e \sin\theta, \ N_i = \lambda_i \sin\theta$$

Then

$$- mN_0 \omega_e \alpha_e + mc_e^2 K\lambda_e = - N_0 ea, \ - MN_0 \omega_i \alpha_i + Mc_i^2 K\lambda_i = N_0 ea,$$

$$- \lambda_e \omega_e = - N_0 K\alpha_e, \ - \lambda_i \omega_i = - N_0 K\alpha_i, \ Ka = 4\pi e (\lambda_e - \lambda_i),$$

$$\omega a = 4\pi e N_0 (\alpha_e - \alpha_i) + 4\pi e (N_e V_e - N_i V_i).$$

Therefore, $MN_0 \omega_e \alpha_e + mc_e^2 K (N_0 K/\omega_e) \alpha_e = - N_0 ea$

or $$\alpha_e \left(\frac{K^2 c_e^2}{\omega_e} - \omega_e \right) m = - ea \text{ or } \alpha_e = \frac{ea}{m} \frac{\omega_e}{K^2 c_e^2 - \omega_e^2}$$

Similarly we find that these relations give

$$\alpha_i = \frac{ea}{M} \frac{\omega_i}{K^2 c_i^2 - \omega_i^2}, \ \lambda_e = \frac{N_0 K}{\omega_e} \alpha_e = - \frac{N_0 K}{\omega_e} \frac{ea}{m} \frac{\omega_e}{K^2 c_e^2 - \omega_e^2}$$

$$= - \frac{N_0 e}{m} \frac{Ka}{K^2 c_e^2 - \omega_e^2}$$

$$\lambda_i = \frac{N_0 e}{M} \frac{Ka}{K^2 c_i^2 - \omega_i^2}$$

Hence the relation $-Ka = -4\pi e(\lambda_e - \lambda_i)$, gives

$$-1 = \frac{\omega_{Pe}^2}{(K^2 c_e^2 - \omega_e^2)} + \frac{\omega_{Pi}^2}{(K^2 c_i^2 - \omega_i^2)}$$

3.7 ENERGY TRANSPORT

In section 1.7 the energy transport problem of electrodynamics were considered with the help of the macroscopic phenomenological field equations of Maxwell and the State equations. Moreover, the relations between the energies in electric and magnetic forms, and the flux of these energies and the rate of work done by the applied field on the medium were determined. But the interaction of the Lorentz force with materials also generates kinetic energy, some forms of potential energy and the flux of mechanical energy through the boundaries. To find these, the equations of momentum transfer and continuity of mass should be considered together with the Maxwell equations. These equations, for small amplitude fields, are

$$\frac{\partial \mathbf{u}_e}{\partial t} = -\frac{1}{\rho_e^0}\nabla p_e - \frac{e}{m}\mathbf{E} - \frac{e}{mc}(\mathbf{u}_e \times \mathbf{H}_0) - \nu_e(\mathbf{u}_e - \mathbf{u}_i) - \nu_-(\mathbf{u}_e - \mathbf{u}_n), \quad (3.7.1)$$

$$\frac{\partial \mathbf{u}_i}{\partial t} = -\frac{1}{\rho_i^0}\nabla p_i + \frac{e}{M}\mathbf{E} + \frac{e}{MC}(\mathbf{u}_i \times \mathbf{H}_0) - \nu_i(\mathbf{u}_i - \mathbf{u}_e) - \nu_+(\mathbf{u}_i - \mathbf{u}_n) \quad (3.7.2)$$

$$\frac{\partial \mathbf{u}_n}{\partial t} = -\frac{1}{\rho_n^0}\nabla p_n - \nu_n^i(\mathbf{u}_n - \mathbf{u}_i) - \nu_n^e(\mathbf{u}_n - \mathbf{u}_e), \quad (3.7.3)$$

$$\frac{\partial N_e}{\partial t} + N_0(\nabla \cdot \mathbf{u}_e) = 0, \quad \frac{\partial N_i}{\partial t} + N_0(\nabla \cdot \mathbf{u}_i) = 0, \quad \frac{\partial N_0}{\partial t} + N_n^0(\nabla \cdot \mathbf{u}_n) = 0, \quad (3.7.4)$$

$$\nabla \times \mathbf{H} = \frac{1}{c}\frac{\partial \mathbf{E}}{\partial t} + \frac{4\pi e}{e}N_0(\mathbf{u}_i - \mathbf{u}_e), \quad (\nabla \times \mathbf{E}) = -\frac{1}{c}\frac{\partial \mathbf{H}}{\partial t},$$

$$(\nabla \cdot \mathbf{E}) = 4\pi e(N_i - N_e), \quad (3.7.5)$$

$$p_e = mc_e^2 N_e, \; p_i = Mc_i^2 N_i, \; p_n = M_n c_n^2 N_n, \quad (3.7.6)$$

$$\rho_e^0 = mN_0, \; \rho_i^0 = MN_0, \; \rho_n^0 = M_n N_n^0 \quad (3.7.7)$$

E is the electric field, **H** is the magnetic field, **j** is the current per unit volume, $\mathbf{u}_e$ is the average velocity of electrons, $\mathbf{u}_i$ is the same of ions, $\mathbf{u}_n$ is the same of the neutral particles, p_e, p_i, p_n are the partial pressures of the three fluids, N_0 is the number density before perturbation of positive or negative charges; N_n^0 is the same of the neutral particles, N_e, N_i, N_n are their perturbations, c_e, c_i, c_n are the speeds of sound in the three fluids and $\nu_e, \nu_i, \nu_-, \nu_+, \nu_n^i, \nu_n^e$ are the collision frequencies of momentum loss.

The Maxwell equations of (3.7.6) give the Poynting theorem (section 1.7) in the form

$$(\mathbf{E}\,.\,\mathbf{j}) = -\frac{\partial}{\partial t}(W_E + W_H) - (\nabla\,.\,\mathbf{S}^t),\ W_E = \frac{E^2}{8\pi},\ W_H = \frac{H^2}{8\pi},\ \mathbf{S}^t = \frac{c}{4\pi}(\mathbf{E}\times\mathbf{H}) \tag{3.7.8}$$

We now write $\mathbf{j} = N_0\, e\,(\mathbf{u}_i - \mathbf{u}_e)$ and obtain

$$(\mathbf{E}\,.\,\mathbf{j}) = N_0\, e\,(\mathbf{E}\,.\,\mathbf{u}_i) - N_0\, e\,(\mathbf{E}\,.\,\mathbf{u}_e) \tag{3.7.9}$$

In the first term of the right hand side of (3.7.9) the value of $\mathbf{E}$ is taken from (3.7.2), and in the second term from (3.7.1). The force terms $eN_0\,(\mathbf{u}_i\times\mathbf{H})/C$ and $eN_0\,(\mathbf{u}_e\times\mathbf{H})/c$ do not contribute to the power transfer process, because $(\mathbf{u}\,.\,(\mathbf{u}\times\mathbf{H})) = 0$, so

$$(\mathbf{E}\,.\,\mathbf{j}) = \rho_i^0\left(\frac{\partial \mathbf{u}_i}{\partial t}\,.\,\mathbf{u}_i\right) + (\mathbf{u}_i\,.\,\nabla p_i) + \rho_e^0\left(\frac{\partial \mathbf{u}_e}{\partial t}\,\mathbf{u}_e\right) + (\mathbf{u}_e\,.\,\nabla p_e) + R \tag{3.7.10}$$

where
$$R = \nu_i\,\rho_i^0\,\{u_i^2 - (\mathbf{u}_e\,.\,\mathbf{u}_i)\} + \nu_e\,\rho_e^0\,\{u_e^2 - (\mathbf{u}_e\,.\,\mathbf{u}_i)\} + + \nu_+\,\rho_i^0\,\{u_i^2 - (\mathbf{u}_i\,.\,\mathbf{u}_n)\} + \nu_-\,\rho_e^0\,\{u_e^2 - (\mathbf{u}_e\,.\,\mathbf{u}_n)\} \tag{3.7.11}$$

Defining now the kinetic energy density,

$$W_k = \frac{1}{2}(\rho_i^0\,u_i^2 + \rho_e^0\,u_e^2), \tag{3.7.12}$$

and using the vector relation

$$\{\nabla\,.\,(\mathbf{u}p)\} = p\,(\nabla\,.\,\mathbf{u}) + (\mathbf{u}\,.\,\nabla)\,p \tag{3.7.13}$$

in (3.7.10), we obtain

$$(\mathbf{E}\,.\,\mathbf{j}) = \frac{\partial W_R}{\partial t} + \{\nabla\,.\,(\mathbf{u}_e\,p_e + \mathbf{u}_i\,p_i)\} - p_i\,(\nabla\,.\,\mathbf{u}_i) - p_e\,(\nabla\,.\,\mathbf{u}_e) + R \tag{3.7.14}$$

Simplifying the third and fourth terms in the right hand side of this equation, with the help of the continuity conditions of (3.7.4), we obtain

$$(\mathbf{E}\,.\,\mathbf{j}) = \frac{\partial}{\partial t}(W_k + W_p) + (\nabla\,.\,\mathbf{S}^l) + R \tag{3.7.15}$$

with $$\frac{\partial W_p}{\partial t} = \frac{p_e}{N_0}\frac{\partial N_e}{\partial t} + \frac{p_i}{N_0}\frac{\partial N_i}{\partial t}$$

$$= \frac{Mc_i^2}{2N_0}\frac{\partial}{\partial t}(N_i^2) + \frac{mc_e^2}{2N_0}\frac{\partial}{\partial t}(N_e^2) = \frac{\partial}{\partial t}\left\{\frac{1}{2N_0}(p_i\,N_i + p_e\,N_e)\right\}, \tag{3.7.16}$$

$$\mathbf{S}^l = p_i\,\mathbf{u}_i + p_e\,\mathbf{u}_e \tag{3.7.17}$$

Here $\mathbf{S}^l$ is the energy flux per unit volume carried away by the particles and W_p is energy of compressibility per unit volume in the plasma. So, W_p is the stored potential energy density. Equating now the expressions in (3.7.8) and (3.7.15) for $(\mathbf{E}\,.\,\mathbf{j})$, we find that

$$(\mathbf{E}\,.\,\mathbf{j}) = -\frac{\partial}{\partial t}(W_E + W_H) - (\nabla\,.\,\mathbf{S}^t) = \frac{\partial}{\partial t}(W_k + W_p) + (\nabla\,.\,\mathbf{S}^l) + R \tag{3.7.18}$$

If W be the total energy and $\mathbf{S}$ be its flux, then these relations give the energy conservation condition

$$\frac{\partial W}{\partial t} + \nabla\,.\,\mathbf{S} + R = 0 \tag{3.7.19}$$

with
$$W = W_E + W_H + W_k + W_p \tag{3.7.20}$$

and
$$\mathbf{S} = \mathbf{S}^l + \mathbf{S}^t \tag{3.7.21}$$

The differential equation (3.7.19) or the corresponding integral relation obtained by integrating all the terms in a volume V is the generalized energy flux theorem. It states that the time rate of decrease of the energy $\iiint W\,dV$ equals the flux $(\mathbf{S}\,.\,\mathbf{n})$ per unit area through the boundary Σ of V (where $\mathbf{n}$ is the unit vector along the outward normal at a point of Σ) plus the amount dissipated through collisions. Averaging all the terms of (3.7.19) over the time period $2\pi/\Omega$ of a wave field, we find that it reduces to

$$(\nabla\,.\,\langle\mathbf{S}\rangle) = -\langle R\rangle \tag{3.7.22}$$

3.8 TRANSPORT THROUGH HARMONIC WAVES

For study of energy transport through harmonic waves, we consider a field in a narrow range about the mean frequency ω. Taking the wave solution proportional to $\exp\,(i\,(\mathbf{K}\,.\,\mathbf{r}) - i\,\omega\,t)$, the Maxwell equations of (3.7.5) become

$$(\mathbf{K}\times\mathbf{E}) = \frac{\omega}{c}\,\mathbf{H},\ (\mathbf{K}\times\mathbf{H}) = -\frac{\omega}{c}\,\mathbf{E} - \frac{4\pi\,i}{c}\,\mathbf{j},\ (\mathbf{K}\,.\,\mathbf{E}) = -4\pi\,ie\,(N_i - N_e) \tag{3.8.1}$$

Consider the plasma to be non-dissipative; so regard $\mathbf{K}$ and ω to be real. Then, if $\overline{\mathbf{E}}$ and $\overline{\mathbf{H}}$ are the complex conjugates $(c\,c)$ of $\mathbf{E}$ and $\mathbf{H}$ respectively, taking the scalar products of the first and second equations of (3.8.1), with $\overline{\mathbf{H}}$ and $\overline{\mathbf{E}}$ respectively, we find that

$$(\mathbf{K}\,.\,(\mathbf{E}\times\overline{\mathbf{H}})) = \frac{\omega}{c}\,(\mathbf{H}\,.\,\overline{\mathbf{H}}),\ (\mathbf{K}\,.\,(\overline{\mathbf{E}}\times\mathbf{H})) = \frac{\omega}{c}\,(\mathbf{E}\,.\,\overline{\mathbf{E}}) + \frac{4\pi i}{c}\,(\overline{\mathbf{E}}\,.\,\mathbf{j}) \tag{3.8.2}$$

Adding these two relations gives,

$$(\mathbf{K}\,.\,(\mathbf{E}\times\overline{\mathbf{H}})) + (\mathbf{K}\,.\,(\overline{\mathbf{E}}\times\mathbf{H})) = \frac{\omega}{c}\,(\mathbf{E}\,.\,\overline{\mathbf{E}}) + \frac{\omega}{c}\,(\mathbf{H}\,.\,\overline{\mathbf{H}}) + \frac{4\pi i}{c}\,(\overline{\mathbf{E}}\,.\,\mathbf{j}), \tag{3.8.3}$$

and the average of the Poynting vector $\mathbf{S}$, over the time period $2\pi/\omega$, yields

$$\langle \mathbf{S}^t \rangle = \frac{c}{8\pi} ((\mathbf{E} \times \overline{\mathbf{H}}) + (\overline{\mathbf{E}} \times \mathbf{H})) \tag{3.8.4}$$

Also, we have

$$\langle W_E \rangle = \frac{(\mathbf{E} \,.\, \overline{\mathbf{E}})}{8\pi}, \langle W_H \rangle = \frac{(\mathbf{H} \,.\, \overline{\mathbf{H}})}{8\pi}, \tag{3.8.5}$$

$$\begin{aligned}(\mathbf{E} \,.\, \overline{\mathbf{j}}) &= \rho_e^0 (\dot{\mathbf{u}}_e \,.\, \overline{\mathbf{u}}_e) + \rho_i^0 (\dot{\mathbf{u}}_i \,.\, \overline{\mathbf{u}}_i) + (\overline{\mathbf{u}}_e \,.\, \nabla p_e) + (\overline{\mathbf{u}}_i \,.\, \nabla p_i) \\ &= -i\omega \left[\rho_e^0 (\mathbf{u}_e \,.\, \overline{\mathbf{u}}_e) + \rho_i^0 (\dot{\mathbf{u}}_i \,.\, \overline{\mathbf{u}}_i) + \frac{1}{\rho_e^0 c_e^2} p_e \overline{p}_e + \frac{1}{\rho_i^0 c_i^2} p_i \overline{p}_i \right] \\ &\quad + i (\mathbf{K} \,.\, \mathbf{u}_e) \overline{p}_e + i (\mathbf{K} \,.\, \mathbf{u}_i) \overline{p}_i \end{aligned} \tag{3.8.6}$$

If two waves have frequencies ω and $\omega + \delta\,\omega$, and wave vectors $\mathbf{K}$ and $\mathbf{K} + \delta\,\mathbf{K}$, then inserting (3.8.3) in (3.8.6) to eliminate $(\mathbf{E} \,.\, \mathbf{J})$, the increments $\delta\,\omega$ and $\delta\,K$ are found to satisfy the relation

$$(\delta\,\mathbf{K} \,.\, (\langle \mathbf{S}^t \rangle + \langle \mathbf{S}^l \rangle) = \delta\omega\, [\langle W_E + W_H + W_p + W_k \rangle] \tag{3.8.7}$$

$$\left.\begin{aligned} \langle \mathbf{S}^l \rangle &= \frac{1}{2} (\mathbf{u}_e \overline{p}_e + \mathbf{u}_i \overline{p}_i), \\ \langle W_K \rangle &= \frac{1}{2} \left\{ \rho_e^0 (\mathbf{u}_e \,.\, \overline{\mathbf{u}}_e) + \rho_i^0 (\mathbf{u}_i \,.\, \overline{\mathbf{u}}_i) \right\}, \\ \langle W_p \rangle &= \frac{1}{2\rho_e^0 c_e^2} p_e \overline{p}_e + \frac{1}{2\rho_i^0 c_i^2} p_i \overline{p}_i \end{aligned}\right\} \tag{3.8.8}$$

The group velocity* $\mathbf{v}_g$ is therefore given by

$$\mathbf{v}_g = \frac{\langle \mathbf{S}^t + \mathbf{S}^l \rangle}{\langle W_E + W_H + W_K + W_p \rangle} \tag{3.8.9}$$

Any non-steady process in plasmas is, to some extent, thermodynamically irreversible. The electric and magnetic losses, however small in a variable field, always occur. For free transmission $(\omega > \omega_\rho)$ the loss is small, absorption is

---

* The group velocity of propagating wave trains is the velocity of propagation of a modulation envelope, or of a wave packet, or of wave energy. These concepts are not generally identical for anisotropic dispersive media; but, in simple cases of weak fields through linear dielectrics, the differences in these definitions do vanish. The amplitude curve of the wave train moves with the group velocity and the wave forerunner actually races ahead of a signal travelling even through dispersive media with the vacuum velocity C of light. The received signal, even from very small amplitude waves, remains small until the main wave packet appears travelling at the group velocity. The modulation frequency $\delta\,\omega$ impressed on the carrier wave at frequency ω results in a wave train moving with the group velocity $\mathbf{v}_g = \lim_{\delta K \to 0} (\delta\omega/\delta\mathbf{K})$. In that sense the velocity $\mathbf{v}_g$ of (3.8.9) is the velocity at which energy propagates. These two velocities coincide only in the linear approximation. A strong monochromatic wave becomes quasimonochromatic because of intensity induced effects; then the group velocity $\partial\omega/\partial\mathbf{K}$ differs from the energy carrier velocity $\langle \mathbf{S} \rangle / \langle W \rangle$.

negligible and the concept of internal energy of the plasma in the electromagnetic field can be introduced with the help of the definition of dielectric and conductivity tensors. Writing $\mathbf{j}_\alpha = \sigma_{\alpha\beta} E_\beta$ (where, by repeated index summation over that index is denoted) and introducing it in the right hand side of (3.8.3), we obtain

$$(\delta \mathbf{K} \,.\, \langle \mathbf{S}^t \rangle) = \delta\omega \left[\langle W_E + W_H \rangle + \frac{i}{2} \langle \overline{E}_i \,.\, E_j \rangle \frac{\partial}{\partial \omega} \sigma_{ij} \right] \tag{3.8.10}$$

because $\langle (\overline{\mathbf{E}}\,(\omega + \delta\omega) \,.\, \mathbf{j}\,(\omega + \delta\omega) \rangle - \langle (\overline{\mathbf{E}}\,(\omega) \,.\, \mathbf{j}\,(\omega)) \rangle$

$$= \langle (\overline{E}_\alpha E_\beta) \rangle \{ \sigma_{\alpha\beta}(\omega + \delta\omega) - \sigma_{\alpha\beta}(\omega) \} = \langle \overline{E}_\alpha E_\beta \rangle \left(\frac{\partial}{\partial \omega} \sigma_{\alpha\beta} \right) \delta\omega \tag{3.8.11}$$

because $\langle \overline{E}_\alpha E_\beta \rangle$ is independent of ω. Hence, for electromagnetic fields, we can write

$$\mathbf{v}_g = \frac{\langle \mathbf{S}^t \rangle}{\langle W_E + W_H + W_1 \rangle} \tag{3.8.12}$$

with
$$\langle W_1 \rangle = \frac{i}{2} \langle \overline{E}_\alpha E_\beta \rangle \frac{\partial}{\partial \omega} \sigma_{\alpha\beta} \tag{3.8.13}$$

It is now evident that $\overline{E}_\alpha E_\beta \, \partial \sigma_{\alpha\beta} / \partial \omega$ has the dimension of energy density and $\overline{E}_\alpha E_\beta \, \partial \sigma_{\alpha\beta} / \partial \mathbf{K}$ is that of its flux.

(a) The case of waves in unmagnetized plasmas

To find the dependence of $\sigma_{\alpha\beta}$ on ω we first consider the simple case of electron motion in the presence of electromagnetic field in cold and unmagnetized plasma. In this case, the momentum change equation is $i\omega\mathbf{u} = e\mathbf{E}/m$ and, for propagation parallel to the z-axis, we have

$$\sigma_{xx} = \sigma_{yy} = \frac{i\omega_\rho^2}{4\pi}, \quad \sigma_{zz} = 0, \quad \sigma_{ij} = 0 \text{ if } i \neq j, \tag{3.8.14}$$

$$\begin{aligned} \langle W_1 \rangle &= \frac{i}{2} \left[\langle \overline{E}_x E_x \rangle \frac{\partial}{\partial \omega} \sigma_{xx} + \langle \overline{E}_y E_y \rangle \frac{\partial}{\partial \omega} \sigma_{yy} \right] \\ &= [\langle \overline{E}_x E_x \rangle + \langle \overline{E}_y E_y \rangle] \; (\omega_\rho^2 / 8\pi\, \omega^2), \end{aligned} \tag{3.8.15}$$

$$\begin{aligned} \langle W_E + W_H + W_1 \rangle &= \frac{1}{8\pi} \left(\langle \overline{E}_x E_x \rangle + E_y \overline{E}_y \rangle \left(1 + \frac{K^2 c^2 + \omega_\rho^2}{\omega^2} \right) \right) \\ &= \frac{1}{4\pi} (\langle \overline{E}_x E_x \rangle + \langle E_y \overline{E}_y \rangle), \end{aligned} \tag{3.8.16}$$

$$\langle S_z^t \rangle = \frac{Kc^2}{4\pi\omega} (\langle E_x \overline{E}_x \rangle + \langle E_y \overline{E}_y \rangle) \tag{3.8.17}$$

The dispersion relation $K^2 c^2 = \omega^2 - \omega_\rho^2$ gives $v_g = Kc^2/\omega = \partial\omega/\partial K$, so the group velocity $\partial\omega/\partial K$ equals the energy carrier velocity $\langle \mathbf{S} \rangle / \langle W \rangle$.

(b) The case of waves in magnetized plasmas

For plasma in the bed of a constant magnetic field $\mathbf{H}_0$, parallel to the direction of propagation of waves, the relations of (3.13.4) give

$$\sigma_{xx} = \sigma_{yy} = \frac{i\omega\,\omega_\rho^2}{4\pi(\omega^2 - \Omega^2)}, \quad \sigma_{xy} = -\sigma_{yx} = \frac{\omega_\rho^2\,\Omega}{4\pi(\omega^2 - \Omega^2)},$$

$$\sigma_{zz} = 0 = \sigma_{xz} = \sigma_{zx} = \sigma_{yz} = \sigma_{zy}, K^2 c^2/\omega^2 = 1 - \omega_\rho^2/\omega(\omega \mp \Omega), \tag{3.8.18}$$

$$\langle W_1 \rangle = \frac{i}{2}\left[\langle E_x \overline{E}_x + E_y \overline{E}_y \rangle \frac{\partial}{\partial\omega}\sigma_{xx} + \langle \overline{E}_x E_x - E_y \overline{E}_y \rangle \frac{\partial}{\partial\omega}\sigma_{xy} \right]$$

$$= (\langle E_x \overline{E}_x + E_y \overline{E}_y \rangle)\, \omega_\rho^2/8\pi(\omega \mp \Omega)^2, \tag{3.8.19}$$

$$\langle \Omega_E + W_H + W_1 \rangle = \frac{1}{8\pi}(\langle E_x \overline{E}_x + E_y \overline{E}_y \rangle)\left(1 + \frac{K^2 c^2}{\omega^2} + \frac{\omega_\rho^2}{(\omega \mp \Omega)^2}\right) = \frac{\partial\omega}{\partial K} \tag{3.8.20}$$

$$v_g = \frac{\langle E_x \overline{E}_x + E_y \overline{E}_y \rangle}{4\pi\langle W_E + W_H + W_1 \rangle}\frac{Kc^2}{\omega} = \left(\frac{2Kc^2}{\omega}\right) \Big/ \left[1 + \frac{K^2 c^2}{\omega^2} + \frac{\omega_\rho^2}{(\omega \mp \Omega)^2}\right] = \frac{\partial\omega}{\partial K} \tag{3.8.21}$$

The group velocity $\partial\omega/\partial K$, evaluated with the help of the dispersion relation (3.8.18), equals the energy carrier velocity.

(c) Energy absorption in unmagnetized dissipative plasmas

The loss of energy of the wave field to the plasma is evaluated by including the retarding force $-N_0\,\nu\mathbf{u}$ per unit volume in the momentum balance conditon. For electron motion, static ions, transverse waves and no static magnetic field, the guiding equations

$$\left.\begin{aligned} m\frac{\partial \mathbf{u}}{\partial t} &= -e\mathbf{E} - m\nu\,\mathbf{u}, \ \text{curl}\,\mathbf{E} = -\frac{1}{c}\frac{\partial \mathbf{H}}{\partial t}, \\ \text{curl}\,\mathbf{H} &= \frac{1}{c}\frac{\partial \mathbf{E}}{\partial t} - \frac{4\pi N_0 e}{c}\mathbf{u}, \end{aligned}\right\} \tag{3.8.22}$$

give

$$\left.\begin{aligned} \mathbf{j} &= -N_0 e \mathbf{u} = \frac{N_0 e^2 (\nu + i\omega)}{m(\omega^2 + \nu^2)} \mathbf{E}, \ \varepsilon = 1 - \frac{\omega_p^2 (\omega - i\nu)}{\omega(\omega^2 + \nu^2)}, \\ \sigma &= \frac{N_0 e^2 (\nu + i\omega)}{m(\omega^2 + \nu^2)} \end{aligned}\right\} \tag{3.8.23}$$

$$K^2 c^2 = \omega^2 - \omega_p^2 \left(1 - \frac{i\nu}{\omega}\right) \frac{1}{(1 + \nu^2/\omega^2)} \tag{3.8.24}$$

For small ν, this relation becomes

$$\omega^2 \approx K^2 c^2 + \omega_p^2 \left(1 - \frac{i\nu}{\omega}\right) \tag{3.8.25}$$

Since K is complex, for real values of ω, the wave is damped inside the plasma and it will be reflected when it reaches the region where $\omega = \omega_p$. Putting, $K = K_1 + iK_2$ into this relation (with K_1, K_2, ω real), for evaluating the rate of absorption of electromagnetic power by a plasma, we write.

$$\mathbf{E} = \mathbf{E}_0 \exp(iKx - i\omega t) = \mathbf{E}_1 \exp(iK_1 x - i\omega t) \tag{3.8.26}$$

with $\mathbf{E}_1 = \mathbf{E}_0 \exp(-K_2 x)$. Real values of (3.8.23) for $\theta = K_1 x - \omega t$ give

$$\mathbf{j} = N_0 e^2 \mathbf{E}_1 (\nu \cos\theta + \omega \sin\theta)/m(\omega^2 + \nu^2) \tag{3.8.27}$$

$$\text{Then } P = \langle(\mathbf{E}\,.\,\mathbf{j})\rangle = \frac{E_1^2 N_0 e^2 \nu}{2m(\omega^2 + \nu^2)} \tag{3.8.28}$$

where P is the power averaged over one time period. There will be no absorption when $\nu = 0$.

In high frequency discharges energy is transferred from the electric field by electrons generated by ionization of neutral particles. These electrons gain energy from the field through collisions with other particles. The collisions convert the oscillatory motion of the electrons into a random motion. To restore the ordered oscillatory motion, the field induces a network on the electrons and thus enhances their energy. When the collision frequency ν lies between 10^9 and 10^{11} collisions per second, for waves with frequencies of the order of 10^7 oscillations/sec (radio frequency waves), since $\nu >> \omega$, the effect of the driving frequency disappears. The electrons, generated by ionization of gas disappear from the discharge zone by diffusion to the walls of the container. A steady state is reached when the rates of production and loss are balanced.

Free electrons, oscillating in an alternating electric field, cannot derive power from the field, since their motion is $\pi/2$ out of phase with the field. This may be seen by putting $\nu = 0$ in (3.8.27), then $\mathbf{j} = N_0 e^2 \omega \mathbf{E}_0 \sin\theta$ and the real value of **E** is $\mathbf{E}_0 \cos\theta$. They acquire energy only because collisions with neutral particles alter their phase relationship with the field while, at the same time, lose a small fraction of the energy thus gained through the collision process. When $\omega >> \nu$, the maximum displacement x of an electron, obtained by integrating the first equation of (3.8.22), is given by

$$x = \frac{2eE}{m\omega^2} \tag{3.8.29}$$

Exercise 3.8.1. For a transverse wave of small amplitude in a two-component, collisionally damped, unmagnetized plasma, find expressions for the dielectric constant ε, electrical conductivity σ and the static part of the rate of work done by the wave on the plasma.

Answer. It is easily seen that

$$\varepsilon = 1 - \frac{\omega_{\rho e}^2 + \omega_{\rho i}^2}{\omega^2 + (\nu_e + \nu_i)^2} + \frac{i\,(\omega_{\rho e}^2 + \omega_{\rho i}^2)\;(\nu_i + \nu_e)}{\omega\,\{\omega^2 + (\nu_i + \nu_e)^2\}},$$

$$\sigma = \frac{(\omega_{\rho e}^2 + \omega_{\rho i}^2)\;(\nu_e + \nu_i + i\,\omega)}{4\pi\,\{\omega^2 + (\nu_i + \nu_e)^2\}}$$

The static part of the rate of work done is the time averaged part of $\langle(\mathbf{E}\,.\,\mathbf{j})\rangle$. Evidently

$$\langle(\mathbf{E}\,.\,\mathbf{j})\rangle = \frac{E_1^2}{8\pi}\,\frac{(\omega_{\rho e}^2 + \omega_{\rho i}^2)\,(\nu_i + \nu_e)}{\{\omega^2 + (\nu_i + \nu_e)^2\}}$$

where ν_i and ν_e are the collision frequencies of ions and electrons respectively.

Exercise 3.8.2. Find $\langle \mathbf{E}\,.\,\mathbf{j}\rangle$ for the bound electrons of the Larmor precession effect (section 2.19).

3.9 SPACE CHARGE SEPARATION EFFECT

When the volume dimensions are much greater than the Debye length (section 3.10), charges of both sign intermix intimately because of strong attraction; so, there is no drastic deviation from statistical equilibrium in a plasma; and the difference in potential energy of particles at various points in the space-charge field can, at most, be of the order of the mean random energy. The long range Coulomb interactions among charges, generate electrostatic force inside the plasma, for restoring quasi-neutrality.

For longitudinal oscillations in cold plasmas (section 3.4), we have

$$\ddot{N}_e + \omega_\rho^2 N_e = 0 \tag{3.9.1}$$

In a typical plasma $N_0 = 10^{14}/\text{cm}^3$, and so $\omega_\rho = 6 \times 10^{11}$ rad/sec. Hence ω_ρ is usually large. For estimation of electrostatic field due to a departure from charge neutrality in a large volume, let ions of uniform number density N_i reside at the centre of a sphere of radius r cm. Electrons of uniform number density N_e are supposed to be distributed symmetrically around the centre. Then, if ψ is the potential on a unit negative charge and f is the acceleration of an electron, we find that

$$\psi = \frac{e}{r}\left(\frac{4\pi r^3}{3}\right)(N_e - N_i)$$

$$f = \frac{e}{r^2}\left(-\frac{e}{m}\right)\left(\frac{4\pi r^3}{3}\right)(N_i - N_e) \text{ cm/sec}^2$$

If $N_e = 10^{11}/\text{cm}^3$, then for 1 per cent deviation from charge neutrality, near a sphere of radius 1 cm, $\psi = 2$ esu of potential,

$$f = (4.8 \times 10^{-10})^2 \times 10^9 \times 25/(9.1 \times 10^{-28} \times 6) = 1.2 \times 10^{18} \text{ cm/sec}^2.$$

This f being enormous, accelerations due to 1 per cent or more deviation from charge neutrality give rise to a very rapidly growing instability; so, in a stable plasma the quasi-neutrality condition $|N_e - N_i| << N_e$ holds good. The electrons and ions then are bound to each other collectively by the forces of space charge separation; these, therefore, are constrained to move without separating appreciably in macroscopic motion of plasma.

Exercise 3.9.1. If $N_e = 10^{13}/\text{cm}^3$, and only 1 per cent separation of charges has occurred, then show that the potential $\psi = 600$ volts/cm and find the acceleration f near a sphere of radius 1 cm.

Answer. $f = 1.16 \times 10^{20}$ cm/sec^2.

Exercise 3.9.2. Show that for $N_e = 10^{15}/\text{cm}^3$ and for a departure of only one part in a million from charge neutrality, the field potential is 600 volts/cm near a sphere of radius 1 cm.

3.10 DEBYE SHIELDING POTENTIAL

Since plasma electrons have considerable random thermal motion, charge separation, even if to a very small extent, occurs, and the difference between electron and ion charge densities decreases rapidly with increasing distance from an ion charge. For an ion the random thermal motion of the neighbouring charges makes the surrounding region sometimes excess in ions and sometimes excess in electrons. But, in an average time, the electron number will slightly exceed the ion number because of the Coulomb attraction and repulsion. So, every charge is surrounded by an atmosphere of opposite charges. This is the Debye screening (or shielding) of a charge by opposite charges and the thickness of this screening atmosphere is the Debye shielding distance or the Debye-Huckel radius. The charge neutrality is maintained over distances large compared to this screening distance. If the random thermal motion is so weak that electrostatic attraction between opposite charges cannot be overcome within the Debye radius, as in dielectrics, only dipole polarization occurs. However, at high temperatures (> 10^6 °K) the Debye shielding distance is modified by quantum effects.

(a) A simple derivation of the Debye screening radius

For a rough estimation of this distance, equate the potential energy gained by a charge in moving through this distance with the kinetic energy per particle in one direction. Two parallel conducting planes are separated by a distance $2x_0$, and are held at the same potential φ. The electric field **E** is everywhere parallel to the x-axis. If N is the uniform number density of electrons, each of charge $-e$, and if ions, each of charge e, are removed from the region between the planes, then

$$\mathbf{E} = \nabla\varphi = (\partial\varphi/\partial x, 0, 0),$$

and by Gauss's law

$$\frac{d^2\varphi}{dx^2} = -4\pi Ne \tag{3.10.1}$$

Integrating this relation twice and assuming that $\varphi = \varphi_0$ and $d\varphi/dx = 0$ when $x = 0$, we get

$$\varphi - \varphi_0 = -2\pi Ne\, x^2 \tag{3.10.2}$$

If W is the potential energy of an electron, then the energy required to transfer a particle from the surface to the mid plane is its change δW across the thickness x_0 of the slab.

$$\delta W = -e\,(\varphi - \varphi_0) = 2\pi Ne^2\, x_0^2$$

If the particles have thermal motion, characterized by a kinetic temperature T, then the kinetic energy per particle in unidirectional motion is $K_B T/2$ where K_B is the Boltzman constant (section 9.5). Then, equating δW with $K_B T/2$, we get

$$x_0 = \lambda_D = \left(\frac{K_B T}{4\pi Ne^2}\right)^{1/2} = 6.9\left(\frac{T}{N}\right)^{1/2} \text{cm} \tag{3.10.3}$$

where T is in absolute degree Kelvin, and λ_D is a measure of the distance in which the collective space charge effect is of the order of the random behaviour due to thermal motion. For $N = 10^{15}/\text{cm}^3$ and $T = 10$ KeV, we get $\lambda_D \approx 7.5 \times 10^{-3}$ cm. For a fusion plasma, at 10 KeV and electron density $10^{16}/\text{cm}^3$, it is found that $\lambda_D \approx 10^{-3}$ cm.

(b) Derivation using distribution function

The distribution function in six-dimensional phase space, of velocity and position coordinates (chapter 9), obtained from the steady state solution of the Boltzmann equation, is

$$f(\mathbf{r},\ \mathbf{v}) = N\left(\frac{n}{2\pi K_B T}\right)^{3/2} \exp\left[-\frac{e\varphi}{K_B T} - \frac{mv^2}{2K_B T}\right]$$

where m is the electronic mass, T is the kinetic temperature, K_B is the Boltzmann constant, φ is the potential of the Coulomb electric field and $-e$ is the electronic charge. Integrating $f(\mathbf{r}, \mathbf{v})$ in the velocity space we ge

$$N_e(\mathbf{r}) = N \exp(= e\varphi/K_B T_e).$$

Similarly, the ionic number density $N_i(\mathbf{r}) = N \exp(e\varphi/K_B T_i)$. Then, following the test charge e at the origin, the total charge density is found to be

$$e\delta(\mathbf{r}) + eN \exp(e\varphi/K_B T_i) - eN \exp(-e\varphi/K_B T_e)$$

where $\delta(\mathbf{r})$ is the delta function of Dirac (Appendix 1). From the Poisson equation we obtain

$$\nabla^2\varphi = 4\pi e\delta(\mathbf{r}) + 4\pi eN[\exp(e\varphi/K_B T_i) - \exp(-e\varphi/K_B T_e)] \quad (3.10.4)$$

When the potential energy of the electromagnetic field is much less than the kinetic energy per particle of thermal motion, we have $|e\varphi| << K_B T$; then keeping terms up to the first power of φ, for the linear approximation, we get

$$\nabla^2\varphi - \varphi/\lambda_D^2 = 4\pi e\,\delta(\mathbf{r})$$

where

$$\lambda_D^2 = \frac{K_B T_e T_i}{4\pi Ne^2 (T_i + T_e)} \quad (3.10.5)$$

For radial symmetry φ satisfies the equation

$$\frac{1}{r^2}\frac{d}{dr}\left(r^2\frac{d\varphi}{dr}\right) = \frac{4\pi Ne^2}{K_B}\left(\frac{1}{T_e} + \frac{1}{T_i}\right)\varphi$$

Imposing the boundary condition $\varphi \to 0$ as $r \to \infty$, the solution

$$\varphi(r) = -\frac{e}{r}\exp\left(-\frac{r}{\lambda_D}\right) \quad (3.10.6)$$

is obtained. Hence the potential due to the charge e is reduced by the factor $\exp(-r/\lambda_D)$ from the value it has in vacuum. If ions provide a static background of positive charges of constant density, then $N_i(\mathbf{r}) = N$ (= constant) and relation (3.10.3) shows that

$$\lambda_D^2 = \frac{K_B T}{4\pi N e^2}$$

Actually λ_D approximately estimates the distance from the charge e beyond which its electrical potential falls off rapidly, so the numerical factor may be 4 (for $T_e >> T_i$) or 8 (for $T_e = T_i$). If N_D is the number of electrons or ions in the Debye sphere of radius λ_D, then

$$N_D = \frac{4\pi\lambda_D^3}{3} N \tag{3.10.7}$$

Using the values of N and T of section 3.10a, we find that $N_D = 1.8 \times 10^9$ particles of each sort. This being a large number, within the Debye sphere, a particle interacts with large number of particles, of both sign. Therefore, the plasma region must be greater in size than λ_D; λ_D should be large compared to the average interparticle distance and small compared to the characteristic distance of macroscopic changes. In equilibrium, the point charge attracts enough charges of opposite sign (and repels enough particles with charges of the same sign), so that the potential is reduced by the factor exp $(-r/\lambda_D)$ from the value it would have in vacuum at a point.

(c) Relation between the screening distance and cut-off frequency

In a plasma the charges interact with each other due to long range forces (section 6.7), and the electric and magnetic fields induced by them add together in a coherent way, generating collective modes which do not exist when molecules interact with short range forces only. Since $\omega_\rho \lambda_D = \sqrt{K_B T/m}$ where ω_ρ is the cut-off frequency $\sqrt{4\pi N_0 e^2/m}$ (sections 3.3 and 3.4), the average time required by the particles, having thermal motion to move through the distance λ_D, is of the order of the time elapsed for one complete oscillation with frequency ω_ρ. But the time of one complete oscillation of an external field at frequency ω is about $1/\omega$. If this time is greater than $1/\omega_\rho$, the particles prevent the wave from propagating through the plasma. So a wave at $\omega < \omega_\rho$ cannot penetrate into the plasma, but if $\omega > \omega_\rho$, the coherent response of the plasma does not affect its propagation. Also, a wave propagates through a plasma if the displacement current $E\,\omega/c$ exceeds the plasma current $4\pi N_0\, eu/c$ where u $(= eE/m\omega)$ is the wave induced electron velocity, that is, if $\omega^2 > \omega_\rho^2$.

3.11 GENERALIZED OHM'S LAW

The plasma electrical conductivity is either *DC* or *AC*. For electrostatic fields the conductivity is *DC* and for waves it is of *AC*. Sometimes, a strong applied wave field gives rise to a non-linearly induced *DC* conductivity in addition to an *AC* conductivity. Analytical expressions for the components of the *AC* conductivity tensor obtain for waves in plasmas. Here, for simplicity, the plasma is regarded as a mixture of cold fluids of positive and negative charges, in the presence of a constant and uniform magnetic field, both species are mobile and suffer dissipative losses due to collisions. Ohm's law, in a stationary electrically conducting medium, is

$$\mathbf{j} = \sigma \mathbf{E} \tag{3.11.1}$$

where σ is the constant electrical conductivity of the medium. If the medium is moving with velocity $\mathbf{v}$, this law reads

$$\mathbf{j} = \sigma \left\{ \mathbf{E} + \frac{1}{c}(\mathbf{v} \times \mathbf{B}) \right\} \tag{3.11.2}$$

To find the generalised Ohm's law, consider the linearized momentum transfer equations for electron and ions:

$$\dot{\mathbf{u}}_e = -\frac{e}{m}\mathbf{E} - \frac{e}{mc}(\mathbf{u}_e \times \mathbf{H}) - \nu_1 (\mathbf{u}_e - \mathbf{u}_i), \tag{3.11.3}$$

$$\dot{\mathbf{u}}_i = \frac{e}{M}\mathbf{E} + \frac{e}{Mc}(\mathbf{u}_i \times \mathbf{H}) - \nu_2 (\mathbf{u}_i - \mathbf{u}_e) \tag{3.11.4}$$

where ν_1 and ν_2 are the collision frequencies. The wave field induced velocity $\mathbf{u}$, the current density $\mathbf{J}$ and the mass density ρ^m are

$$\mathbf{u} = \frac{(MN_i \mathbf{u}_i + mN_e \mathbf{u}_e)}{(MN_i + mN_e)} \tag{3.11.5}$$

$$\mathbf{j} = e(N_i \mathbf{u}_i - N_e \mathbf{u}_e), \tag{3.11.6}$$

$$\rho^m = MN_i + mN_e \tag{3.11.7}$$

For simplicity we put $N_e = N_i = N$, and since $M >> m$, find that

$$\mathbf{u} = \mathbf{u}_i + \frac{m}{M}\mathbf{u}_e, \; \rho = MN, \tag{3.11.8}$$

$$\mathbf{j} = Ne(\mathbf{u}_i - \mathbf{u}_e) = Ne\left\{ \mathbf{u} - \left(\frac{m}{M} + 1 \right) \mathbf{u}_e \right\} \tag{3.11.9}$$

Hence

$$\mathbf{u}_e = \mathbf{u} - \frac{\mathbf{j}}{Ne}, \tag{3.11.10}$$

$$\mathbf{u}_i = \mathbf{u} + \frac{m}{\rho e}\mathbf{j}, \tag{3.11.11}$$

$$\mathbf{u}_e - \mathbf{u}_i = -\frac{\mathbf{j}}{Ne} \tag{3.11.12}$$

From equations (3.11.3), (3.11.10) and (3.11.12) we obtain

$$\dot{\mathbf{u}} - \frac{1}{Ne}\frac{\partial \mathbf{j}}{\partial t} = -\frac{e}{m}\mathbf{E} - \frac{e}{mc}(\mathbf{u} \times \mathbf{H}) + \frac{1}{mNC}(\mathbf{j} \times \mathbf{H}) \times \frac{\nu_1}{Ne}\mathbf{j} \tag{3.11.13}$$

Equations (3.11.4), (3.11.11) and (3.11.12), therefore, give

$$\dot{\mathbf{u}} + \frac{m}{\rho e}\frac{\partial \mathbf{j}}{\partial t} = \frac{e}{M}\mathbf{E} + \frac{e}{Mc}(\mathbf{u}\times\mathbf{H}) + \frac{m}{\rho Mc}(\mathbf{j}\times\mathbf{H}) - \frac{\nu_2}{Ne}\mathbf{j} \tag{3.11.14}$$

Subtraction of the terms of (3.11.14) from (3.11.13) approximately gives

$$\frac{1}{Ne}\frac{\partial \mathbf{j}}{\partial t} = \frac{e}{m}\mathbf{E} + \frac{e}{mc}(\mathbf{u}\times\mathbf{H}) - \frac{1}{mNe}(\mathbf{j}\times\mathbf{H}) - \frac{\nu}{Ne}\mathbf{j} \tag{3.11.15}$$

where $\nu = \nu_1 + \nu_2$. This equation, when written in the form

$$\mathbf{j} = \frac{Ne^2}{m\nu}\left\{\mathbf{E} + \frac{1}{C}(\mathbf{u}\times\mathbf{H})\right\} - \frac{e}{m\nu c}(\mathbf{j}\times\mathbf{H}) - \frac{1}{\nu}\frac{\partial \mathbf{j}}{\partial t}, \tag{3.11.16}$$

is the generalized Ohm's law.

(a) The Hall effect

For a stationary plasma $u = 0$ and

$$\mathbf{j} = \frac{Ne^2}{m\nu}\mathbf{E} - \frac{e}{m\nu}(\mathbf{j}\times\mathbf{H}) - \frac{1}{\nu}\frac{\partial \mathbf{j}}{\partial t} \tag{3.11.17}$$

In the steady state since $\partial\mathbf{j}/\partial t = 0$, this equation becomes

$$\mathbf{j} = \frac{Ne^2}{m\nu}\mathbf{E} - \frac{e}{m\nu c}(\mathbf{j}\times\mathbf{H}) \tag{3.11.18}$$

It can be written as

$$\mathbf{j} = \sigma\mathbf{E} + \mu(\mathbf{j}\times\mathbf{H}) \tag{3.11.19}$$

where σ is the electrical conductivity in the presence of a constant electric field and zero magnetic field. So

$$\sigma = \frac{Ne^2}{m\nu}, \quad \mu = -\frac{e}{m\nu c} \tag{3.11.20}$$

Resolving (3.11.19) into the components we get

$$j_X = \frac{\sigma(E_X + \mu H E_Y)}{1+\mu^2 H^2}, \quad j_Y = \frac{\sigma(E_Y - \mu H E_X)}{1+\mu^2 H^2} \tag{3.11.21}$$

$$j_z = \sigma E_z \tag{3.11.22}$$

where **H** is parallel to the z-axis. The short circuit condition is $E_Y = 0$ so we obtain

$$j_X = \frac{\sigma E_x}{1+\mu^2 H^2}, \quad j_y = -\frac{\sigma\mu H E_x}{1+\mu^2 H^2} \tag{3.11.23}$$

Hence if $j_x = \sigma_{\parallel} E_x$, $j_y = \sigma_{\perp} E_x$, we find that

$$\sigma_{\parallel} = \frac{\sigma}{1+\mu^2 H^2}, \quad \sigma_{\perp} = -\frac{\sigma\mu H}{1+\mu^2 H^2} \tag{3.11.24}$$

The more familiar forms for $\sigma_{\|}$ and $\sigma_{\perp}$ are

$$\sigma_{\|} = \frac{\sigma\nu^2}{\nu^2 + \Omega^2} = \frac{\sigma}{1 + \Omega^2 T^2}, \quad \sigma_{\perp} = \frac{\Omega\nu\sigma}{\nu^2 + \Omega^2} = \frac{\sigma T \Omega}{1 + \Omega^2 T^2}. \tag{3.11.25}$$

where $T = 1/\nu$ and $\Omega = eH/mc$. So T, a characteristic time, is of the order of the average time between successive collisions.

The magnetized plasma therefore has one value of electrical conductivity parallel to the electric field and another value perpendicular to the plane of the crossed fields. In the absence of boundaries, the transverse electric current $\mathbf{j}_{\perp}$ gives rise to an electric field of charge separation that acts on this current. A potential difference in the Y direction, thus obtained, is the Hall effect.

For high magnetic fields, in the range $| \ll \Omega T \ll M/m$ (where M is the ion mass and m is the electron mass), we find that

$$\sigma_{\|} = \frac{\sigma}{\Omega^2 T^2}, \quad \sigma_{\perp} = \frac{\sigma}{\Omega T} = \frac{N_0 e^2}{m\Omega},$$

Since $\sigma_{\perp} >> \sigma_{\|}$, the current is perpendicular to the magnetic field and the electrical conductivity does not depend on the collision frequency. The Hall effect is important in the study of electrical conductivity of materials in presence of an applied DC magnetic field. Some types of MHD generators (sections 5.6 and 5.7) where magnetic fields of the order of 4 tesla or more are applied, exhibit this effect.

3.12 WAVES IN MAGNETIZED PLASMA

If a constant magnetic field $\mathbf{H}_0$ is superposed on the plasma, electromagnetic waves are coupled to the longitudinal waves and the wave process is complicated. To study the propagation of small amplitude waves in cold magnetized plasma, we consider the following linearized equations of a two-fluid plasma:

$$\frac{\partial \mathbf{u}_e}{\partial t} = -\frac{c_e^2}{N_0} \nabla N_0 - \frac{e}{m} \mathbf{E} - (\mathbf{u}_e \times \mathbf{\Omega}_e), \tag{3.12.1}$$

$$\frac{\partial \mathbf{u}_i}{\partial t} = -\frac{c_i^2}{N_0} \nabla N_i + \frac{e}{M} \mathbf{E} + (\mathbf{u}_i \times \mathbf{\Omega}_i), \tag{3.12.2}$$

$$N_0 (\mathbf{\nabla} \cdot \mathbf{u}_e) + \frac{\partial N_e}{\partial t} = 0, \quad N_0 (\mathbf{\nabla} \cdot \mathbf{u}_i) + \frac{\partial N_i}{\partial t} = 0, \tag{3.12.3}$$

$$\text{curl } \mathbf{E} = -\frac{1}{c}\frac{\partial \mathbf{H}}{\partial t}, \quad \text{curl } \mathbf{H} = \frac{1}{c}\frac{\partial \mathbf{E}}{\partial t} + \frac{4\pi N_0 e}{c} (\mathbf{u}_i - \mathbf{u}_e), \tag{3.12.4}$$

$$\text{div } \mathbf{E} = 4\pi e (N_i - N_e), \text{ div } \mathbf{H} = 0 \tag{3.12.5}$$

where $\Omega_e = e\mathbf{H}_0/mc$, $\Omega_i = e\mathbf{H}_0/Mc$ and other symbols are obvious. From these the following set of nine second order partial differential equations are obtained in terms of the mixed vectors $\mathbf{E}$, $\mathbf{u}_e$, $\mathbf{u}_i$:

$$\ddot{\mathbf{u}}_e = c_e^2 \nabla (\nabla . \mathbf{u}_e) - \frac{e}{m} \dot{\mathbf{E}} - (\dot{\mathbf{u}}_e \times \Omega_e), \tag{3.12.6}$$

$$\ddot{\mathbf{u}}_i = c_i^2 \nabla (\nabla . \mathbf{u}_i) + \frac{e}{M} \dot{\mathbf{E}} + (\dot{\mathbf{u}}_i \times \Omega_i) \tag{3.12.7}$$

$$c^2 \text{ curl curl } \mathbf{E} + \ddot{\mathbf{E}} + 4\pi N_0 e (\dot{\mathbf{u}}_i - \dot{\mathbf{u}}_e) = 0 \tag{3.12.8}$$

Equation (3.12.6) is obtained by differentiating (3.12.1) with respect to time and eliminating N_e with the help of (3.12.3). Equation (3.12.7) similarly follows from (3.12.2) and (3.12.3); (3.12.8) is deduced by taking curl of the first equation of (3.12.4) and using that to eliminate $\mathbf{H}$ in the second equation.

Since particle velocities perpendicular and parallel to $\mathbf{H}$ are different, the plasma becomes anisotropic in presence of static magnetic field. As a result the conductivity tensor σ_{ij} and the dielectric tensor ε_{ij} have some non-diagonal elements. Since

$$\nabla \times \mathbf{H} = -\frac{i\omega}{c} \hat{\mathbf{X}}_i \varepsilon_{ij} E_j = -\frac{i\omega}{c} \hat{X}_i E_i + \frac{4\pi}{c} \hat{X}_i \sigma_{ij} E_j = \frac{i\omega}{c} \hat{\mathbf{X}}_i \left(\delta_{ij} + \frac{4\pi i}{\omega} \sigma_{ij} \right) E_j$$

we find that

$$\varepsilon_{ij} = \delta_{ij} + \frac{4\pi i}{\omega} \sigma_{ij} \tag{3.12.9}$$

Here $\hat{\mathbf{X}}_i$ is the unit vector in the ith direction, use of repeated index means summation over that index, δ_{ij} is the kronecker delta (see equation (1.8.7 (*a*)) and the time dependence $\exp(-i\omega t)$ is assumed. Moreover, if the space dependence is $\exp(ikz)$, the connections between the longitudinal components u_{ez}, u_{iz}, E_z and the transverse components u_{ex}, u_{ey}, u_{ix}, u_{iy}, E_x, E_y are obtained from equations (3.12.3) and (3.12.7). Wave propagation parallel to the direction of the magnetic field $\mathbf{H}_0$, is further discussed in sections 3.13, 3.14, and 11.2-11.9.

3.13 ELECTROMAGNETIC WAVES, PROPAGATING PARALLEL TO A CONSTANT MAGNETIC FIELD

(a) Basic equations and dispersion relations

When a constant magnetic field acts parallel to the direction of propagation (z-axis) of a transverse wave in cold non-dissipative plasma, using the plane wave solution $\exp(ikz - i\omega t)$, equations (3.12.1)–(3.12.5) reduce to

$$\omega u_{ex} + i\,\Omega_e u_{ey} = -\frac{ie}{m} E_x, \quad -i\,\Omega_e u_{ex} + \omega u_{ey} = -\frac{ie}{m} E_y, \tag{3.13.1}$$

$$\omega\, u_{ix} - i\Omega_i\, u_{iy} = \frac{ie}{M} E_x,\quad i\Omega_i\, u_{ix} + \omega\, u_{iy} = \frac{ie}{M} E_y, \tag{3.13.2}$$

$$\frac{\omega_{\rho e}^2 (u_{ex} - u_{ix})}{K^2 c^2 - \omega^2} = \frac{ie}{m} E_x,\quad \frac{\omega_{\rho e}^2\, \omega\, (u_{ey} - u_{iy})}{K^2 c^2 - \omega^2} = \frac{ie}{m} E_y, \tag{3.13.3}$$

where
$$\omega_{\rho e}^2 = \frac{4\pi N_0 e^2}{m},\quad \omega_{\rho i}^2 = \frac{4\pi N_0 e^2}{M},\quad \boldsymbol{\Omega}_e = \frac{e\mathbf{H}_0}{mc},\quad \Omega_i = \frac{\epsilon \mathbf{H}_0}{Mc} \tag{3.13.4}$$

When $\omega^2 \neq \Omega_e^2$ and $\omega^2 \neq \Omega_i^2$, equations (3.13.1) and (3.13.2) become

$$u_{ex} = -\frac{ie}{m}\frac{(\omega E_x - i\Omega_e E_y)}{\omega^2 - \Omega_e^2},\quad u_{ey} = -\frac{ie}{m}\frac{(\omega E_y + i\Omega_e E_x)}{\omega^2 - \Omega_e^2}$$

$$u_{ix} = \frac{ie}{M}\frac{(\omega E_x + i\Omega_i E_y)}{\omega^2 - \Omega_i^2},\quad u_{iy} = \frac{ie}{M}\frac{(\omega E_y - i\Omega_i E_x)}{\omega^2 - \Omega_i^2} \tag{3.13.5}$$

The resonant cases, for which $\omega = |\Omega_e|, |\Omega_i|$, cannot be determined from (3.13.5). In section 11.9 these relations have been discussed for collisionally damped plasma, due to appearance in denominator of zero, which is avoided by inclusion of collisional loss in force law. To find u_{ex} and u_{ix} equations (3.13.1) and (3.13.2) are used to eliminate u_{ey} and u_{iy}. Then with the help of (3.13.3), we get

$$i\left\{\omega^2 - K^2 c^2 - \frac{\omega^2 \omega_{\rho e}^2}{\omega^2 - \Omega_e^2} - \frac{\omega^2 \omega_{\rho i}^2}{\omega^2 - \Omega_i^2}\right\} E_x = \left\{\frac{\omega\, \Omega_e\, \omega_{\rho e}^2}{\omega^2 - \Omega_e^2} - \frac{\omega\, \Omega_i\, \omega_{\rho i}^2}{\omega^2 - \Omega_i^2}\right\} E_y \tag{3.13.6}$$

Similarly, eliminating from (3.13.1) and (3.13.2) the quantities u_{ex} and u_{ix}, then finding u_{ey} and u_{iy} and finally putting these values in the second equation of (3.13.3), we get

$$i\left\{\omega^2 - K^2 c^2 - \frac{\omega^2 \omega_{\rho e}^2}{\omega^2 - \Omega_e^2} - \frac{\omega^2 \omega_{\rho i}^2}{\omega^2 - \Omega_i^2}\right\} E_y = -\left\{\frac{\omega\, \Omega_e\, \omega_{\rho e}^2}{\omega^2 - \Omega_e^2} - \frac{\omega\, \Omega_i\, \omega_{\rho i}^2}{\omega^2 - \Omega_i^2}\right\} E_x \tag{3.13.7}$$

These two relations can also be written as

$$iAE_X - BE_Y = 0,\quad BE_X + iAE_Y = 0 \tag{3.13.8}$$

with

$$A = \omega^2 - K^2 c^2 - \frac{\omega^2 \omega_{\rho e}^2}{\omega^2 - \Omega_e^2} - \frac{\omega^2 \omega_{\rho i}^2}{\omega^2 - \Omega_i^2},\quad B = \frac{\omega\, \Omega_e\, \omega_{\rho e}^2}{\omega^2 - \Omega_e^2} - \frac{\omega\, \Omega_i\, \omega_{\rho i}^2}{\omega^2 - \Omega_i^2} \tag{3.13.9}$$

Elimination of E_x/E_y gives the dispersion relation

$$A^2 = B^2; \tag{3.13.10}$$

Explicitly it is

$$\left\{\omega^2 - K^2 c^2 - \frac{\omega^2 \omega_{\rho e}^2}{\omega^2 - \Omega_e^2} - \frac{\omega^2 \omega_{\rho i}^2}{\omega^2 - \Omega_i^2}\right\}^2 = \left\{\frac{\omega \Omega_e \omega_{\rho e}^2}{\omega^2 - \Omega_e^2} - \frac{\omega \Omega_i \omega_{\rho i}^2}{\omega^2 - \Omega_i^2}\right\}^2 \tag{3.13.11}$$

The two values of K^2 are given by

$$\frac{C^2 (K_R^2, K_L^2)}{\omega^2} = 1 - \frac{\omega_{\rho e}^2}{\omega (\omega \mp \Omega_e)} - \frac{\omega_{\rho i}^2}{\omega (\omega \pm \Omega_i)} \tag{3.13.12}$$

Since $\omega_{\rho e}^2 \Omega_i = 4\pi N_0 e^3 H_0/Mmc = \omega_{\rho i}^2 \Omega_e$ we obtain

$$\frac{K_R^2 c^2}{\omega^2} = 1 - \frac{\omega_{\rho e}^2 + \omega_{\rho i}^2}{(\omega - \Omega_e)(\omega + \Omega_i)}, \quad \frac{K_L^2 c^2}{\omega^2} = 1 - \frac{\omega_{\rho e}^2 + \omega_{\rho i}^2}{(\omega + \Omega_e)(\omega - \Omega_i)}, \tag{3.13.13}$$

$K_R^2 = K_L^2$ when $\mathbf{H}_0 = 0$ and $K_R^2 \neq K_L^2$ when $\mathbf{H}_0 \neq 0$. Hence, the wave splits into two waves of different wave numbers in a magnetized plasma. This is bi-refringence in a magnetized plasma. If $\mathbf{E}_R$ and $\mathbf{E}_L$ are electric fields of the two waves,

$$E_{RY} = iE_{RX}, \; E_{LY} = -\, iE_{LX} \tag{3.13.14}$$

These are two dispersion relations for circularly polarized waves, one right and the other left. The real values of the field components are

$$\left.\begin{aligned} E_{RX} &= |E_R| \cos \varphi_R, \; E_{RY} = |E_R| \cos\left(\varphi_R + \frac{\pi}{2}\right), \; \varphi_R = (K_R Z - \omega t + \alpha_R), \\ E_{LX} &= |E_L| \cos \varphi_L, \; E_{LY} = |E_L| \cos\left(\varphi_L - \frac{\pi}{2}\right), \; \varphi_L = (K_L Z - \omega t + \alpha_L) \end{aligned}\right\} \tag{3.13.15}$$

where α_R and α_L are the constant wave phases on $Z = 0$ at time $t = 0$. Let $\theta = \tan^{-1} (E_y/E_x)$. These relations show that φ_R increases with time and φ_L decreases at a given space point, if the wave is propagating anti-parallel to $\mathbf{H}_0$. Hence $\mathbf{E}_R$ rotates anticlockwise, and $\mathbf{E}_L$ clockwise. The $\mathbf{E}_R$ wave is right circularly polarized (or the 'extraordinary'), and $\mathbf{E}_L$ is the left circularly polarized (or the 'ordinary' wave). Electrons appear to gyrate anticlockwise, (that is, in the sense of the extraordinary wave); and ions gyrate clockwise, (in the sense of the ordinary wave (section 2.3)). When $\omega = |\Omega_e|$ the $\mathbf{E}_R$ field resonates with electron gyration, and when $\omega = |\Omega_i|$, the $\mathbf{E}_L$ wave resonates with ion gyration. At these resonances the associated charges gyrate in phase with the electric field and gain energy indefinitely in the absence of collisional dissipation. Thus, an electric field, at the ion cyclotron frequency, can be applied to heat the ions in a plasma preferentially.

When $\omega^2 \neq (\Omega_e^2, \Omega_i^2)$, equations (3.13.13) can be written as

$$K^2 c^2 = \omega^2 - \frac{(\omega_{\rho e}^2 + \omega_{\rho i}^2)}{(i \mp \Omega_e/\omega)\,(1 \pm \Omega_i/\omega)} \tag{3.13.16}$$

Hence $K^2 > 0$ if $\omega^2 > (\omega_{\rho e}^2 + \omega_{\rho i}^2)\,\omega^2/(\omega \mp \Omega_e)\,(\omega \pm \Omega_i)$. So, the cut-off frequency ω_c is given by $(\omega_c \mp \Omega_e)\,(\omega_e \pm \Omega_i) = \omega_{\rho\,e}^2 + \omega_{\rho\,i}^2$; when $\mathbf{H}_0 = 0$ we find that $\omega_c = \sqrt{\omega_{\rho e}^2 + \omega_{\rho i}^2}$. When ion motion is completely ignored, the dispersion relation is

$$K^2 c^2 = \omega^2 - \frac{\omega\,\omega_\rho^2}{\omega \mp \Omega} \tag{3.13.17}$$

where $\Omega = eH_0/mc$ and ω_p stands for $\omega_{\rho e}$. For large magnetic field, at low frequencies, we have $\omega << \Omega$. Then $K_R^2 c^2 = \omega^2 + \omega_\rho^2\,\omega/\Omega$ and $K_L^2 c^2 = \omega^2 - \omega_\rho^2\,\omega/\Omega$. In this case K_R, the wave number of the extraordinary wave has no cut-off though the wave number K_L of the ordinary wave has the cut-off value ω_ρ^2/Ω. These facts justify the names ordinary wave and extraordinary wave to the two resolved waves.

Exercise 3.13.1. Deduce the dispersion relation (3.13.11) by using the complex fields $E_\pm$ given by $E_\pm = (E_x \pm iE_y)$ in place of E_x and E_y in (3.13.1)–(3.13.3).

(b) Dispersion relation for Alfven waves

At the low frequency limit, $\omega^2 < \Omega^2$, the dispersion relation for Alfven waves (section 1.8) can be obtained from (3.13.13).

$$\frac{K^2 c^2}{\omega^2} = 1 + \frac{\omega_{\rho e}^2 + \omega_{\rho i}^2}{\Omega_e\,\Omega_i} = 1 + \frac{4\pi\,\rho c^2}{H_0^2} \tag{3.13.17 (a)}$$

where $\rho = N_0\,(M + m)$. Hence $K^2 c^2/\omega^2 = 1 + c^2/c_A^2$ because $c_A = (H_0^2/4\pi\rho)^{1/2}$. Again, since $c^2 >> c_A^2$, the Alfven wave condition $K^2\,c_A^2 = \omega^2$ for the dispersion of low frequency electromagnetic waves is obtained. It is due to the bending of the magnetic field lines very slightly by motion (or, due to interaction between the electromagnetic field and a moving conducting medium) and does not depend on the cyclotron rotation of the plasma particles.

(c) Wave characteristics depending on the frequency range

Actually the dispersion relations of (3.13.15) represent propagation characteristic of three categories of waves, depending on the range of wave frequencies ω relative to the characteristic frequencies Ω_e and Ω_i. For $|\omega| > \Omega_e$, the two circular polarizations in opposite senses have slightly different dispersion rates, giving rise to a Faraday rotation (section 3.14). For the intermediate range, $\Omega_i < |\omega| < \Omega_e$, the dispersion relations approximately are

$$\frac{K_R^2 c^2}{\omega^2} = 1 - \frac{(\omega_{\rho e}^2 + \omega_{\rho i}^2)}{\omega(\omega - \Omega_e)}, \quad \frac{K_L^2 c^2}{\omega^2} = 1 - \frac{(\omega_{\rho e}^2 + \omega_{\rho i}^2)}{\omega(\omega + \Omega_e)} \tag{3.13.18}$$

These also produce Faraday rotation and reduce to (3.13.17) because $M >> m$. In the low frequency range of $\omega^2 < \Omega_i^2$, as in equation (3.13.17 (*a*)), the distinction between the two circular polarizations vanishes.

(e) The temporal evolution problem

Spatial evolution of waves in magnetized plasma shows the bifurcation in the wave length for a given period of time. Temporal evolution indicates the multiplicity of time periods of different modes having the same fixed wave length. The two problems are somewhat different and the temporal problem is more complicated, because (3.13.11) is of eight degree in ω and of fourth degree in K.

For mobile electrons and immobile ions, the dispersion relations (3.13.17) for right and left circular polarizations are quadratic in K and cubic in ω. So, for given K, we write the dispersion relation as

$$K^2 c^2 = \omega_R^2 - \frac{\omega_R\, \omega_\rho^2}{\omega_R - \Omega}, \quad K^2 c^2 = \omega_L^2 - \frac{\omega_L\, \omega_\rho^2}{\omega_L + \Omega} \tag{3.13.19}$$

where ω_R and ω_L are respectively the wave frequencies. Since no dissipation mechanism is included in the process, the real roots only are accepted. In the resonant case, ($\omega_R = \Omega$), for collisionally damped plasma, that value of K is useful for which the real part of one of the values of ω_R equals Ω.

3.14 FARADAY ROTATION AND OTHER EFFECTS OF WAVES IN MAGNETIZED PLASMAS

(a) Faraday rotation

If the resultant electric field $\mathbf{E}_R + \mathbf{E}_L$ (section 3.13) is received by a linearly polarized antenna, the plane of polarization is found to rotate through the Faraday rotation angle θ. It was originally observed in crystals as a magneto-optical effect; the rotation is doubled on passing the same optically active crystal backward and forward.

Consider a plasma slab between the planes $Z = 0$ and $Z = l$, embedded in a magnetic field $\mathbf{H}_0$ parallel to the z-axis. A plane linearly polarized electromagnetic wave, with electric field components ($a \cos \varphi$, $b \cos \varphi$, 0), is incident normally to the plane $Z = 0$. In $Z < 0$ the electric field is written in the complex form

$$E_x + iE_y = r \cos \varphi, \quad \varphi = KZ - \omega t \tag{3.14.1}$$

where $r = a + ib$. So, the angle θ, which the plane of polarization makes with the ZOX plane, is given by

$$\theta = \tan^{-1}\left(\frac{E_y}{E_x}\right) = \tan^{-1}\left(\frac{b}{a}\right) \tag{3.14.2}$$

If, upon advancement of the wave front through one centimeter into the slab, the plane of polarization is rotated through the angle χ then

$$\frac{E_y}{E_x} = \tan(\theta + \chi z) \tag{3.14.3}$$

and the electric field can be written as

$$E_x + iE_y = r \exp(i \chi Z) \cos \varphi \tag{3.14.4}$$

If χ is a function of Z, then χZ should be replaced by the integral $\int_0^z \chi(Z)\, dZ$. Actually, in the region $0 < Z < l$, the right and left circular polarization components split due to unequal wave numbers K_R and K_L. So, as suggested in (3.13.15), the expression for electric field should be

$$E_x + iE_y = \frac{r}{2}\{\exp(-iK_R z + i\omega t) + \exp(iK_L z - i\omega t)\} \tag{3.14.5}$$

For $\mathbf{H}_0 = 0$, this relation yields (3.14.1); and, by putting $K_R = K + \chi$, $K_L = K - \chi$ we obtain (3.14.4). Now separating the $\mathbf{E}_R$ and $\mathbf{E}_L$ fields, and equating the real and imaginary parts from both sides, we get

$$E_{RX} = \frac{1}{2}(a \cos \varphi_R + b \sin \varphi_R), \quad E_{RY} = \frac{1}{2}(-a \sin \varphi_R + b \cos \varphi_R),$$

$$E_{LX} = \frac{1}{2}(a \cos \varphi_L - b \sin \varphi_L), \quad E_{LY} = \frac{1}{2}(a \sin \varphi_L + b \cos \varphi_L) \tag{3.14.6}$$

where $$\varphi_R = K_R z - \omega t, \quad \varphi_L = K_L z - \omega t \tag{3.14.7}$$

If $K_L = K_R = K$ say, then the resultant of E_+ and E_- is always E_x, as in Fig. 3.14.1. But, if $K_L \neq K_R$, as in equation (3.13.12), the resultant is no longer E_x, as may be seen from Fig. 3.14.2, owing to Faraday rotation. A simple geometrical analysis of these figures shows that

$$\psi = \varphi_L - \frac{1}{2}(\varphi_L + \varphi_R) = \frac{1}{2}(K_L - K_R) z$$

where φ_L and φ_R are given by (3.14.7). Then, using the two values of K from (3.13.17), we find that

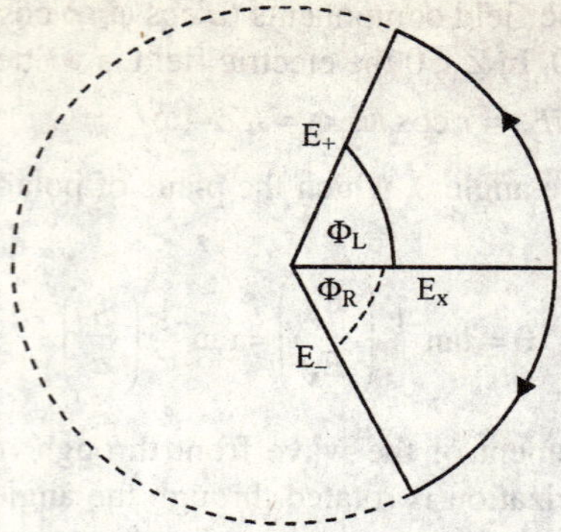

Fig. 3.14.1 The plane polarized wave E_x as the resultant of the LCP and RCP waves.

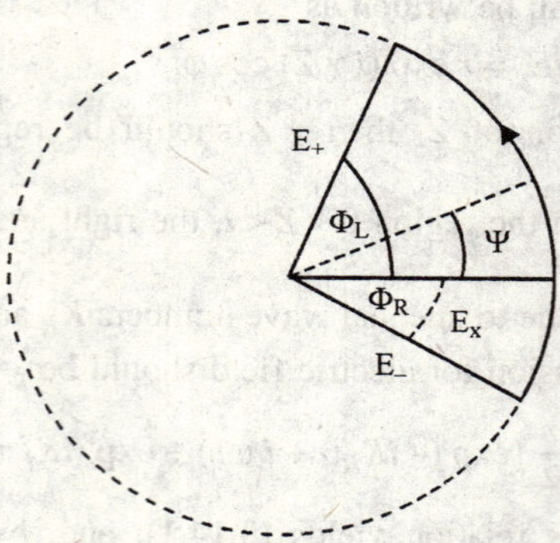

Fig. 3.14.2 Faraday rotation through an angle ψ.

$$\frac{d\psi}{dz} = \frac{1}{2}(K_L - K_R) = \frac{1}{2c}\left\{\left(\omega^2 - \frac{\omega\,\omega_\rho^2}{\omega+\Omega}\right)^{1/2} - \left(\omega^2 - \frac{\omega\,\omega_\rho^2}{\omega-\Omega}\right)^{1/2}\right\} \quad (3.14.8)$$

Hence ψ is proportional to the length traversed by the wave. For weak magnetic fields, $\omega >> \Omega$, and (3.13.17) yields

$$\frac{K_R\,c}{\omega} \approx \left(1 - \frac{\omega_\rho^2}{\omega^2} - \frac{\omega_\rho^2\,\Omega}{\omega^3}\right)^{1/2} \cong 1 - \frac{\omega_\rho^2}{2\,\omega^2} - \frac{\omega_\rho^2\,\Omega}{2\,\omega^3},$$

$$\frac{K_L\,c}{\omega} \approx 1 - \frac{\omega_\rho^2}{2\,\omega^2} + \frac{\omega_\rho^2\,\Omega}{2\omega^3},$$

yielding

$$\psi \approx \frac{\omega_\rho^2\,\Omega\,z}{2c\,\omega^2} \quad (3.14.9)$$

Hence this rotation is anticlockwise, and is directly proportional to the magnetic field $\mathbf{H}_0$. For strong magnetic fields at low frequencies, $(\omega << \Omega)$, in equation (3.14.8) we replace $1/(\omega \pm \Omega)$ by $1/(\pm \Omega)$ and obtain

$$\frac{d\psi}{dz} = \frac{\omega}{2c}\left[\left(1 - \frac{\omega_\rho^2}{\Omega\,\omega}\right)^{1/2} - \left(1 + \frac{\omega_\rho^2}{\Omega\,\omega}\right)^{1/2}\right]$$

If, moreover, $\omega_\rho^2 << \Omega\,\omega$, (which is valid in rarefied plasmas), we find that

$$\psi \approx -\frac{z\,\omega_\rho^2}{2\Omega\,c} \tag{3.14.10}$$

In this case, Faraday rotation is inversely proportional to the static magnetic field $\mathbf{H}_0$, and since the signs of ψ in (3.14.9) and (3.14.10) are opposite, the rotations in the two cases are in opposite senses. If $d\,\psi$ is the change of ψ over the spectral width $d\,\omega$ of the signal, then

$$\frac{d\psi}{d\omega} = \frac{1}{2}\left(\frac{dK_L}{d\omega} - \frac{d\,K_R}{d\,\omega}\right) \tag{3.14.11}$$

When the two values of K of (3.13.16) or (3.13.17) are complex, we write them as $K_R + i\chi_R$ and $K_L + i\chi_L$, because, the dissipation in the medium differs for the two polarized modes, and since $|E_R| \neq |E_L|$, the resultant is elliptically polarized, because one mode is more damped than the other; then the calculations become complicated.

In laboratory, this rotation is measured by rotating the receiving horn and also by using polarization diplexer to compare the relative magnitude of the x-component and y-component of the electric field. When the phase shift and Faraday rotation are combined, the amplitude fluctuations are superimposed on the phase fringes. Each half-rotation gives a phase reversal of 180°. Faraday rotation ceases when one of the circular polarizations goes to the cut-off, then both the x and y components behave similarly. The left and right circular polarizations can be studied independently with the help of circularly polarized antennas.

Exercise 3.14.1. Faraday rotation of a beam of wavelength 8×10^{-3} cm in a uniform plasma slab of thickness 100 cm in 1000 gauss magnetic field is 90°. Find the density of the plasma.

Answer. This λ corresponds to the frequency

$$\omega = c/\lambda = (3/8) \times 10^{13} \text{ rad/sec.}$$

Since $H = 1000$ gauss, $\Omega = (5/27) \times 10^{11}$ rad/sec. So $\Omega < \omega$ and the appropriate formula for use is (3.14.9). Since $\psi = \pi/2$, the electron density N_0 is given by

$$N_0 = (\pi/2) \times (m/4\pi e^2) \times (mc/eHz) \times 2c\omega^2 = 2 \times 10^{16} \text{ cm}^{-3}.$$

Exercise 3.14.2. If $\omega = (3/8) \times 10^{11}$ rad/sec, $H_0 = 10000$ gauss and other parameters of exercise 3.14.1 remain unchanged, find the plasma density.

Answer. In this case formula (3.14.10) gives

$$N = 5 \times 10^{10} \text{ cm}^{-3}.$$

Exercise 3.14.3. Find ψ and $d\psi$ of (3.14.11) in the ionsphere taking

$$\omega = 10^8 \text{ rad/sec}, \; z = 3 \times 10^7 \text{ cm}, \; H_z = 0.3 \text{ gauss},$$

$$N_0 = 10^6 \text{ cm}^{-3} \text{ and } d\,\omega/\omega = 1\%$$

Answer. $\omega_\rho = 5.5 \times 10^7$ rad/sec, $\Omega = 5 \times 10^6$ rad/sec,

$$K_R - K_L = 5 \times 10^{-5}/\text{cm}, \; \psi = -7.5 \times 10^2 \text{ rad and } d\psi = 5 \times 10^{-7} \text{ rad.}$$

(b) Energy transport

The electric energy density components of the RCP and LCP waves are

$$W_{E_R} = \frac{1}{8\pi}(E_{RX}^2 + E_{RY}^2) = \frac{1}{32\pi}(a^2 + b^2) = W_{E_L} \tag{3.14.12}$$

For the total field $\mathbf{E}_l + \mathbf{E}_R$, the electrical energy density is

$$W_E = \frac{1}{16\pi}(a^2 + b^2)\;\{1 + \cos(\varphi_L + \varphi_R)\} + \frac{1}{16\pi}ab\sin(\varphi_L + \varphi_R)$$

Hence $W_{E_l} = W_{E_R} = \frac{1}{2}\langle W_E\rangle$ where $\langle W_E\rangle$ is the average value of W_E over all $\varphi_L + \varphi_R$. The components of $\mathbf{H}$, obtained from the first equation of (3.12.4), are

$$H_x = -\frac{c}{\omega}(K_R E_{Ry} + K_L E_{Ly}), \; H_y = \frac{c}{\omega}(K_R E_{Rx} + K_L E_{Lx})$$

Hence the magnetic energy density

$$\langle W_H\rangle = \frac{c^2}{2\omega^2}(K_R^2 + K_L^2)\;\langle W_E\rangle,$$

because $$W_{H_R} = \left(\frac{K_R^2 c^2}{\omega^2}\right) W_{E_R}, \; W_{H_L} = \left(\frac{K_L^2 c^2}{\omega^2}\right) W_{E_l}$$

If W_e and W_i be the kinetic energy densities of electrons and ions respectively, then

$$W_e = \frac{1}{2} m N_0 (u_{ex}^2 + u_{ey}^2), \; W_i = \frac{1}{2} M N_0 (u_{ix}^2 + u_{iy}^2)$$

Using (3.14.6) in the equation of motion $\dot{\mathbf{u}}_e + \mathbf{u}_e \times \mathbf{\Omega}_e = -e\mathbf{E}/m$, we find that

$$u_{ex} = -\frac{e}{2m}\left[\frac{(-a \sin \varphi_R + b \cos \varphi_R)}{\omega - \Omega_e} - \frac{(a \sin \varphi_L + b \cos \varphi_L)}{\omega + \Omega_e}\right],$$

$$u_{ey} = -\frac{e}{2m}\left[-\frac{(a \cos \varphi_R + b \sin \varphi_R)}{\omega - \Omega_e} - \frac{(a \cos \varphi_L - b \sin \varphi_L)}{\omega + \Omega_e}\right]$$

Hence we have

$$\langle W_e \rangle = \frac{1}{2}\,\omega_{\rho e}^2 \langle W_E \rangle \left[\frac{1}{(\omega - \Omega_e)^2} + \frac{1}{(\omega + \Omega_e)^2}\right],$$

$$\langle W_i \rangle = \frac{1}{2}\,\omega_{\rho i}^2 \langle W_E \rangle \left[\frac{1}{(\omega + \Omega_i)^2} + \frac{1}{(\omega - \Omega_i)^2}\right],$$

$$\langle W_k \rangle = \langle W_e + W_i \rangle = \frac{1}{2} \langle W_E \rangle \left[\frac{\omega_{\rho e}^2}{(\omega + \Omega_e)^2} + \frac{\omega_{\rho e}^2}{(\omega - \Omega_e)^2} + \frac{\omega_{\rho i}^2}{(\omega + \Omega_i)^2} + \frac{\omega_{\rho i}^2}{(\omega - \Omega_i)^2}\right]$$

Since the Poynting flux

$$s_z^t = \frac{c}{4\pi}\{(E_{Rx} + E_{Lx})(H_{Ry} + H_{Ly}) - (E_{Ry} + E_{Ly})(H_{Rx} + H_{Lx})\}$$

$$= \frac{c^2}{16\pi\,\omega}(K_R + K_L)(a^2 + b^2)\,\{1 + \cos(\varphi_R + \varphi_L)\}$$

we find that $\langle S_z^t \rangle = \langle W_E \rangle (K_R + K_L)\, c^2/\omega$. If W is the total energy density, then $W = W_E + W_H + W_k$, and

$$\langle W \rangle = \frac{1}{2} \langle W_E \rangle \left[1 + \frac{c^2}{2\omega^2}(K_R^2 + K_L^2) + \frac{\omega_{\rho e}^2}{2(\omega + \Omega_e)^2} + \frac{\omega_{\rho e}^2}{2(\omega - \Omega_e)^2} + \frac{\omega_{\rho i}^2}{2(\omega + \Omega_i)^2} + \frac{\omega_{\rho i}^2}{2(\omega - \Omega_i)^2}\right] = \langle W_E \rangle \left[1 - \frac{\omega_{\rho e}^2\,\Omega_e^2}{(\omega^2 - \Omega_e^2)^2} - \frac{\omega_{\rho i}^2\,\Omega_i^2}{(\omega^2 - \Omega_i^2)^2}\right]$$

(3.14.13)

The static part of the energies in different forms, namely the quantities $\langle W_E \rangle$, $\langle W_H \rangle$, $\langle W_i \rangle$, $\langle W_e \rangle$ can now be compared for information on the distribution of energies in different forms in magnetized plasmas.

(c) Wave induced magnetic moment (Inverse Faraday Effect)

In the DC field of circularly polarized waves in plasmas electrons gyrate in orbits of small radii which depend on radiation intensity, wave frequency and electron density. This gyration of electrons induces a magnetic moment field which, for left circular polarization, is parallel to the direction of wave propagation and for right circular polarization it is antiparallel. When dispersion rates of the two components are equal, the two magnetizations cancel each other; but there is a residual induced magnetization, including a zero harmonic (DC) field in a homogeneous plasma when these rates differ. This induced magnetic field was originally called the inverse Faraday effect (IFE in short). The direct Faraday rotation effect is the rotation of the plane of polarization of a plane polarized wave about the direction of the applied constant magnetic field, and is a consequence of the acquisition of angular momentum by the radiation field in interaction with a gyrotropic medium.

Consider a particle of charge e and rest mass m in the field of circularly polarized waves. The motion is such that the particle velocity is perpendicular to the electric field and parallel to the magnetic field of these waves. So $(\mathbf{K} \cdot \mathbf{v}) = 0$ where $\mathbf{K}$ is the wave vector of the wave and $\mathbf{v}$ is the particle velocity. This condition ensures that no work is done in maintaining the rotatory motion. Therefore, due to interaction with the radiation, the charged particle acquires the angular momentum

$$\mathbf{L} = (\mathbf{r} \times \mathbf{p}) \tag{3.14.14}$$

where $\mathbf{r}$ is the position vector of the particle relative to a point in the axis of its rotation and $\mathbf{p}$ is its momentum. Each orbiting electron contributes a magnetic dipole of moment

$$\boldsymbol{\mu} = \frac{1}{2c}(\mathbf{r} \times \mathbf{j}) \tag{3.14.15}$$

where $\mathbf{j}$ is the current of its charge. So for non-relativistic electron motion. $\mathbf{j} = -e\,\mathbf{v}$ and $\boldsymbol{\mu} = -e\mathbf{L}/2me$ and the magnetic induction $\mathbf{B}_i$ specified by the electron number density N per cm^3, is given by

$$\mathbf{B}_i = 4\pi N \boldsymbol{\mu} = 4\pi\,\mathbf{M} \tag{3.14.16}$$

where $\mathbf{M}$ is the magnetization per unit volume. The total electric field $\mathbf{E}$ is the sum of the fields of right and left circular polarizations of a wave propagating parallel to the direction of a constant magnetic field. So, we can write

$$\mathbf{E} = \mathbf{E}_R + \mathbf{E}_L \tag{3.14.17}$$

If the direction of wave propagation is parallel to OZ, from (3.14.6) we have

$$E_{Rx} = \frac{1}{2}(a\cos\varphi_R + b\sin\varphi_R),\ E_{Ry} = \frac{1}{2}(-a\sin\varphi_R + b\cos\varphi_R),$$

$$E_{Lx} = \frac{1}{2}(a\cos\varphi_L - b\sin\varphi_L),\ E_{Ly} = \frac{1}{2}(a\sin\varphi_L + b\cos\varphi_L) \tag{3.14.18}$$

But, in place of (3.14.7) for φ_R and φ_L, we now write

$$\varphi_R = Kz - \omega_R t,\ \varphi_L = Kz - \omega_L t$$

where we use (3.13.19) for ω_R and ω_L. This is valid for the temporal evolution problems for waves of the same given wavelength; equation (3.14.19) is then solved as a function of time. Here a and b are the constant amplitudes of the fields. When $\mathbf{H}_0 = 0$, we obtain $\varphi_R = \varphi_L = \varphi$ and $E_x = a \cos \varphi$, $E_y = b \cos \varphi$; the field is polarized in the plane which makes the angle $\tan^{-1}(b/a)$ with the ZOX plane.

Equation of motion for an electron, in the linearized approximation, is

$$m\ddot{\mathbf{r}} = -e\,(\mathbf{E}_R + \mathbf{E}_L) - \frac{e}{c}(\mathbf{v} \times \mathbf{H}_0) \tag{3.14.19}$$

This equation gives

$$u_x = -\frac{e}{2m}\left[\frac{(-a \sin \varphi_R + b \cos \varphi_R)}{\omega_R - \Omega} - \frac{(a \sin \varphi_L + b \cos \varphi_L)}{\omega_L + \Omega}\right],$$

$$u_y = -\frac{e}{2m}\left[\frac{(-a \cos \varphi_R - b \sin \varphi_R)}{\omega_R - \Omega} + \frac{(a \cos \varphi_L - b \sin \varphi_L)}{\omega_L + \Omega}\right],$$

$$x = -\frac{e}{2m}\left[-\frac{(a \cos \varphi_R + b \sin \varphi_R)}{\omega_R\,(\omega_R - \Omega)} - \frac{(a \sin \varphi_L + b \cos \varphi_L)}{\omega_L\,(\omega_L + \Omega)}\right], \text{ etc.}$$

Hence

$$\langle x u_y - y u_x \rangle = -\frac{e^2\,(a^2 + b^2)}{4m^2}\left[\frac{1}{\omega_R\,(\omega_R - \Omega)} - \frac{1}{\omega_L\,(\omega_L + \Omega)}\right] \tag{3.14.20}$$

Since $\langle E_x^2 + E_y^2 \rangle = (a^2 + b^2)/2$, we find that

$$B_i = \frac{\pi N e^3}{m^2\,c}\langle E^2 \rangle \left[\frac{1}{\omega_R\,(\omega_R - \Omega)^2} - \frac{1}{\omega_L\,(\omega_L + \Omega)^2}\right]$$

If we write $B_i = \alpha\,(E_R^2 - E_L^2)$, then

$$\alpha E_R^2 = \frac{\pi N e^3 \langle E^2 \rangle}{m^2\, c\, \omega_R\,(\omega_R - \Omega)^2},\quad \alpha\, E_L^2 = \frac{\pi N e^3 \langle E^2 \rangle}{m^2\, c\, \omega_L\,(\omega_L + \Omega)},$$

and so $B_i = 0$ when $H_0 = 0$. Instead of solving the difficult temporal evolution problem for ω_R and ω_L, we can estimate B_i by solving (3.13.17) for the simpler spatial problem for K_R and K_L, and use the dielectric conditions $K_R\, C = \omega_R$ and $K_L\, C = \omega_L$ to find approximately ω_R and ω_L for the induced magnetization.

In the non-linearly correct equation of motion

$$\ddot{\mathbf{r}} = -\frac{e}{m}\mathbf{E} - (\mathbf{v} \times \mathbf{\Omega}) - \frac{e}{mc}(\dot{\mathbf{r}} \times \mathbf{H}) \tag{3.14.21}$$

the third term in the right hand side gives an induced longitudinal acceleration and a Doppler shift of the wave frequency, in times so small that the longitudinal electron velocity can be neglected; and the orbit is still approximately in a plane and circular. The total energy of the particle in circular motion (kinetic + potential) ε (= ωL_z where L_z is the angular momentum of (3.14.14)), is also the energy of a circularly polarized electromagnetic wave. Thus, when energy is transferred from field to particle, angular momentum is also transferred.

When the radiation field is switched on, the work done on the induced currents can be interpreted in terms of the energy stored or propagating in the medium. The total electric and magnetic fields are **E** and **B** and the sum of the energy density terms, which appear in the Poynting theorem, is

$$\frac{1}{8\pi}(E^2 + B^2) + \int (\mathbf{E} \,.\, (\mathbf{j} + \langle \mathbf{j} \rangle))\, dt$$

where **j** is the true external current and $\langle \mathbf{j} \rangle$ is the induced current, some time written as $(\partial \mathbf{P}/\partial t) + (\nabla \times \mathbf{M})$. In the medium this term is the stored energy density; it is usually included in the quantities **D** and **H** and in the total electromagnetic energy density.

Exercise 3.14.4. Find the induced magnetic moments when, in place of (3.13.18), the field components are given by $E_{Rx} = \frac{1}{2}(a+b)\cos\theta_R$,

$$E_{Lx} = \frac{1}{2}(a-b)\cos\theta_L,\ E_{Ry} = \frac{1}{2}(a+b)\sin\theta_R,\ E_{Ly} = \frac{1}{2}(a-b)\sin\theta_L.$$

Note that both $\mathbf{E}_R$ and $\mathbf{E}_L$ are circularly polarized waves, but the sum $\mathbf{E}_R + \mathbf{E}_L$ is not so. In fact, if $\theta_R = \theta_L$, the sum $\mathbf{E}_R + \mathbf{E}_L$ reduces to an elliptically polarized wave in unmagnetized plasmas.

3.15 RADIATION FROM PLASMA

Radiation from a plasma depends on its thermal condition and the degree of ionization. It is the only source of information on the processes that take place in interstellar and solar plasma.

Radiation from electrons due to their acceleration through the collision process is called bremsstrahung. The radiation when electrons are subjected to acceleration due to their cylotron motion (that is, gyration in the presence of static magnetic field) the resulting radiation, is called the cyclotron or synchrotron radiation. Radiation also occurs through the Cerenkov process, in which the particle velocity is faster than the velocity of light in a medium.

Radio sources emit a continuous spectrum of waves. The conditions in them cannot be determined uniquely by studying their radation. Synchrotron radiation is believed to be the most important process; bremsstrahlung and inverse Compton effect are less important processes of radiation from astrophysical bodies. Bremsstrahlung is the main source of radiation from a hot and fully ionized plasma at very high frequencies and free from any strong applied static magnetic field.

Since electrons repel each other, their accelerations are equal and opposite in magnitude. So, to a first approximation, the radiations of two electrons closely situated and both accelerated, cancel each other.

Changes in vibration, rotation, and electronic states, of di- or polyatomic gaseous matter, on heating, cause the emission of band spectra. Further heating dissociates the molecules into atoms. At MHD power generation temperatures (see sections (5.6) and (5.7)), the radiation is mainly from changes in vibration and from rotational motion of the molecules, and is of quantum mechanical origin.

We can write

$$\sum_i (e_i \dot{\mathbf{v}}_i) = \frac{d^2}{dt^2} \sum_i \left\{ \left(\frac{e_i}{m_i} \right) m_i \mathbf{r}_i (t) \right\} \tag{3.15.1}$$

$$= \left(\frac{e}{m} \right) \left(\frac{d^2}{dt^2} \right) \sum_i \{ m_i \mathbf{r}_i (t) \} = \frac{e}{m} \frac{d^2}{dt^2} (\mathbf{R} M) \tag{3.15.1 (a)}$$

if the ratio e / m of charge and mass of the system has the same value, where $\sum_i m_i \mathbf{r}_i = \mathbf{R} (\sum_i m_i) = \mathbf{R}M$; so the total mass is M, and $\mathbf{R}$ is the position vector of the centre of gravity ($c\,g$) of the system. If the cg is at rest or in uniform motion, we obtain

$$\sum_i (e_i \dot{\mathbf{v}}_i) = 0 \tag{3.15.1 (b)}$$

and, for radiation power, equation (1.14.9) gives $P = 0$. Then, moreover, the dipole radiation power also vanishes, which can be shown by Fourier analysis and Parseval's theorem for the Fourier components of a function.

Including the radiation damping term $(2e^2/3c^3)\, \ddot{\mathbf{v}}$ of equation (1.15.4), in a Lorentz force field, the equation for electron motion is

$$m \dot{\mathbf{v}} = - e\mathbf{E} - \frac{e}{c} (\mathbf{v} \times \mathbf{H}) + \frac{2e^2}{3c^3} \ddot{\mathbf{v}} \tag{3.15.1 (c)}$$

In the classical model of a bound electron, executing a simple harmonic oscillation of angular frequency ω_0, and under the influence of an applied electric field $\mathbf{E}$, the equation of motion

$$m \ddot{\mathbf{r}} = - m\, \omega_0^2\, \mathbf{r} + \left(\frac{2e^2}{3c^3} \right) \dddot{\mathbf{r}} - e\mathbf{E} \tag{3.15.2}$$

has been studied in section 6.11 to find the Rayleigh scattering cross-section.

(a) The bremsstrahlung

Bremsstrahlung is the radiation emitted by decelerated charged particles. It includes radiation when velocity of particles changes due to Coulomb interaction with particles of the neighbourhood. The motion is slowed down by the emission of a gamma photon. When electrons are distributed uniformly, and isotropically, they approach the heavy ions from all directions with all possible speeds. This radiation is isotropic, and the polarization is random. Bremsstrahlung in plasma is the radiation from electrons under the influence of electric field of the nearby positive charges, and due to encounters between electrons and atoms. The former is dominant in highly ionized plasmas, and the latter in weakly ionized plasmas. We only consider classically the former type of radiation here, for non-relativistic energies in low density plasmas in the absence of strong applied static magnetic fields. The minimum distance of impact is the de Broglie wavelength of the particles.

The acceleration f of an electron by an ion of equal charge at a distance R is given by

$$\dot{v} = f = \frac{e^2}{mR^2} \tag{3.15.3}$$

Putting this value of $\dot{v}$ in the expression for the radiated power P for non-relativistic energies, namely equation (1.14.9), we obtain

$$P = \frac{2e^6}{3m^2 c^3 R^4} \tag{3.15.4}$$

Let N_e and N_i be the number densities of electrons and ions respectively. Multiplying (3.15.4) by $4\pi R^2 \, dR \, N_e$ we obtain the radiation from electron plasma. Integrating that quantity with respect to R, the total radiation P_1 from the electrons is obtained.

$$P_1 = \int_{R_{\min}}^{\infty} P \, . \, 4\pi R^2 N_e \, dR = \frac{2\pi}{3} \frac{N_e e^6}{c^3 m^2 R_{\min}} \tag{3.15.5}$$

Here $R_{\min}$ is the distance up to which the classical approximation holds good. We can take

$$R_{\min} = \frac{\hbar}{p} = \lambda_1 \tag{3.15.6}$$

where λ_1 is the de Broglie wavelength, $\hbar$ is the Planck constant and p is the particle momentum; we write $p = \sqrt{m\kappa T}$ (κ is the Boltzmann constant (Appendix 7) and T is the kinetic temperature of electrons). For considerable electron energies λ_1 is larger than the collision parameter p_0 of minimum impact (section 6.2).

The principal source of radiation from a hot and fully ionized plasma, free from any strong static applied magnetic field, is the bremsstrahlung from acceleration of free electrons by ions. It is N_i times P_1 of (3.15.5), and is the total radiation per unit volume of plasma, P_T. So,

$$P_T = \frac{8\pi}{3} \frac{e^6 N_e N_i}{m^2 c^3 R_{min}} \tag{3.15.7}$$

In thermonuclear reactors, the energy loss by bremsstrahlung is considerable, compared to the rate of energy production by nuclear fields; this loss imposes certain restrictions to their operation.

(b) Synchrotron radiation

Charged particles have spiral-like ordered velocity around static magnetic fields, which is an accelerated motion. The associated radiation is known as synchrotron radiation (or the cyclotron or betatron radiation, or magnetic bremsstrahlung). In the presence of an applied constant magnetic field **H**, the relativistic equation of motion (section 2.20) is

$$\frac{d}{dt}(\gamma \mathbf{v}) = -(\mathbf{v} \times \mathbf{\Omega}) \tag{3.15.8}$$

where $\mathbf{\Omega}_0 = e\mathbf{H}/mc$ and $\gamma^2 = 1/(1 - v^2/c^2)$. When the motion is in the plane perpendicular to the direction of the magnetic field **H**, $(\dot{\boldsymbol{\beta}} . \boldsymbol{\beta}) = 0$ and this equation reduces to

$$\dot{\boldsymbol{\beta}} = -\frac{1}{\gamma}(\boldsymbol{\beta} \times \mathbf{\Omega}_0), \quad \left(\boldsymbol{\beta} = \frac{\mathbf{v}}{\mathbf{c}}\right) \tag{3.15.9}$$

Putting this value of $\dot{\boldsymbol{\beta}}$ in (1.14.17) we obtain

$$P = \frac{2e^2}{3c} \gamma^4 \frac{(\boldsymbol{\beta} \times \mathbf{\Omega}_0)^2}{\gamma^2} = \frac{2e^2 \Omega_0^2 \beta^2}{3c(1 - \beta^2)} \tag{3.15.10}$$

for the power radiated.

3.16 STREAM INSTABILITIES IN COLD PLASMAS

Consider a beam of electrons penetrating a plasma, when their velocity is considerably higher than the thermal velocity of the plasma electrons, and the beam electron plasma frequency is much smaller than the plasma electron frequency. The beam is decelarated due to the scattering of its electrons by the electrons and ions of the plasma. Also the beam can be modulated; in that case an amplification at certain frequencies can occur. Deceleration also develops when the temperature of the beam electrons (ejected from cathode surfaces, for instance) equals the temperature of the electrons of the gas discharge into which the beam penetrates.

When the mean velocity of the electrons differs from the mean velocity of ions, we assume that all the electrons travel with the drift velocity **V** with respect to the ions. Also we assume that the plasma is quasi neutral and cold (both temperature of electrons and ions is zero); so their pressures are constant. For both transverse and longitudinal waves of small amplitudes propagating along arbitrary directions with respect to the direction of flow, instabilities (called the Buneman instability) arise even in the absence of a steady magnetic field. Then the dispersion relation for propagation of a plane polarized electromagnetic wave is of sixth degree in the wave frequency.

The wave vector **K** of the wave is assumed to be along the *X*-axis of the rectangular cartesion coordinate system OXYZ; the other coordinate axes are chosen such that the stream velocity **V** is in the *XY* plane, making angle θ with the direction of propagation (the *X*-axis). So the components of **V** are $(V_x, V_y, 0)$ or simply $V(\cos\theta, \sin\theta, 0)$. Since the stream velocity **V** is constant in magnitude and direction, the linearized equations of motion for electrons of the beam are

$$\frac{\partial \mathbf{u}_e}{\partial t} + (\mathbf{V}\,.\,\nabla)\,\mathbf{u}_e = -\frac{e}{m}\left(\mathbf{E} + \frac{1}{c}\,\mathbf{V}\times\mathbf{B}\right) \tag{3.16.1}$$

In the linearized approximation the continuity condition

$$\frac{\partial N}{\partial t} + \nabla\,.\,(N_e\,\mathbf{V} + N_0\,\mathbf{u}_e) = 0 \tag{3.16.2}$$

reduces to

$$\frac{\partial N_e}{\partial t} + N_0\ \text{div}\,\mathbf{u}_e + (\mathbf{V}\,.\,\nabla)\,N_e = 0 \tag{3.16.3}$$

where N_0 is the equilibrium value of the number density of electrons and N_e is its fluctuation. The harmonic space-time dependence of the fluctuating field quantities is $\exp(iKx - i\omega t)$. Using this dependence in (3.16.1) and (3.16.3), we get

$$i\omega'\,\mathbf{u}_e = \left(\frac{e}{m}\right)\left(\mathbf{E} + \frac{1}{c}(\mathbf{V}\times\mathbf{B})\right), \tag{3.16.4}$$

$$N_e\,\omega' = N_0\,K u_{ex} \tag{3.16.5}$$

where ω′ is the Doppler shifted frequency which differs from ω due to the existence of a component of the stream velocity along the direction of propagation. So

$$\omega' = \omega - (\mathbf{K}\,.\,\mathbf{V}) = \omega - KV_x \tag{3.16.6}$$

Since the stream is a beam of electrons only, for the ion motion we can put $\mathbf{V} = 0$ and get

$$i\,\omega\,\mathbf{u}_i = -\frac{e}{M}\,\mathbf{E}, \tag{3.16.7}$$

$$N_i\,\omega = N_0\,K u_{ix} \tag{3.16.8}$$

The Maxwell equation curl $\mathbf{E} = -\,\partial \mathbf{H}/c\,\partial t$ gives

$$B_x = 0,\; B_y = -\frac{Kc}{\omega}E_x,\; B_z = \frac{Kc}{\omega}E_y \tag{3.16.9}$$

because div $\mathbf{B} = 0$. Solving the other Maxwell equation, curl $\mathbf{H} = (\partial \mathbf{E}/c\,\partial t) + (4\pi\,\mathbf{j}/e)$, where

$$\mathbf{j} = N_0\,e\,\mathbf{u}_i - e\,(N_e\,\mathbf{V} + N_0\,\mathbf{u}_e), \tag{3.16.10}$$

with the help of (3.16.5), (3.16.7) and (3.16.8) we obtain

$$u_{ex} = \frac{(\omega^2 - \omega_{\rho i}^2)\,(\omega - KV_x)}{4\pi\,i\,N_0\,e\omega^2}\,E_x, \tag{3.16.11 (a)}$$

$$u_{ey} = \frac{i}{4\pi\,N_0\,e}\left[\left(\frac{K^2\,c^2}{\omega} - \omega + \frac{\omega_{\rho i}^2}{\omega}\right)E_y + \left(\frac{KV_y}{\omega}\right)(\omega^2 - \omega_{\rho i}^2)\,E_x\right], \tag{3.16.11 (b)}$$

$$u_{ez} = \frac{i}{4\pi\,N_0\,e}\left[\left(\frac{K^2\,c^2}{\omega}\right)E_x - \left(\omega + \frac{\omega_{\rho i}^2}{\omega}\right)E_z\right] \tag{3.16.11 (c)}$$

where
$$\omega_{\rho i}^2 = \frac{4\pi\,N_0\,e^2}{M},\; \omega_{\rho e}^2 = \frac{4\pi\,N_0\,e^2}{m} \tag{3.16.12}$$

Eliminating now **B** from (3.16.4) with the help of (3.16.9) we get

$$i\omega'\,mu_{ex} = e\left\{E_x + \left(\frac{KV_y}{\omega}\right)E_y\right\},\; i\omega'\,mu_{ey} = eE_y\left(1 - \frac{KV_x}{\omega}\right),$$

$$i\omega'\,m\,u_{ez} = e\left(E_z - \frac{KV_x}{\omega}E_x\right) \tag{3.16.13}$$

These equations also relate the components of **E** with those of $\mathbf{u}_e$. Eliminating the components of **E** and $\mathbf{u}_e$ between the six linear simultaneous equations of (3.16.11) and (3.16.13), we obtain the determinantal relation.

$$\begin{vmatrix} A_{11} & A_{12} & 0 \\ A_{21} & A_{22} & 0 \\ 0 & 0 & A_{33} \end{vmatrix} = 0 \tag{3.16.14}$$

This equation gives the following two dispersion relations

$$A_{11}\,A_{22} - A_{12}\,A_{21} = 0,\; A_{33} = 0 \tag{3.16.15}$$

where $A_{11} = \dfrac{(\omega - KV_x)}{\omega_{\rho e}^2 \, \omega} - 1, \; A_{12} = \dfrac{KV_y}{\omega},$

$$A_{21} = -\frac{KV_y\,(\omega - KV_x)}{\omega^2\,\omega_{\rho e}^2}, \; A_{22} = -\frac{(\omega - KV_x)\;(K^2 c^2 + \omega_{\rho e}^2 + \omega_{\rho i}^2 - \omega^2)}{\omega\,\omega_{\rho e}^2},$$

$$A_{33} = 0 = -\frac{(\omega - KV_x)}{\omega_{\rho e}^2}\left(\frac{K^2 c^2}{\omega} - \omega + \frac{\omega_{\rho i}^2 + \omega_{\rho e}^2}{\omega}\right) \tag{3.16.16}$$

The equation for A_{33} reduces to the dispersion relation for the transverse E_z wave (relative to the propagation direction):

$$\omega^2 = K^2 c^2 + \omega_{\rho e}^2 + \omega_{\rho i}^2 \tag{3.16.17}$$

The propagation features of this transverse wave are unaffected by the electron beam. The other dispersion relation is of sixth degree in ω. It is the first relation of (3.16.15) and can be written with the help of the expressions for $A_{11}, A_{12}, A_{21}, A_{22}$ in (3.16.16):

$$\begin{aligned}\Omega^6 - 2p\,\Omega^5 \cos\theta - (q^2 + 2 + 2r - p^2\cos^2\theta)\,\Omega^4 + 2p\,(1 + q^2 + 2r)\,\Omega^3\cos\theta - \\ - [(1 + q^2 + r)\,(p^2\cos^2\theta - 1 - r) + p^2\,(r^2\cos^2\theta + \sin^2\theta)]\,\Omega^2 - 2pr \\ (1 + q^2 + r)\,\Omega\cos\theta + rp^2\,[(1 + q^2 + r)\cos^2\theta + \sin^2\theta] = 0\end{aligned} \tag{3.16.18}$$

where $\Omega = \omega/\omega_{\rho e},\; p = KV/\omega_{\rho e},\; q = Kc/\omega_{\rho e},\; r = \omega_{\rho i}^2/\omega_{\rho e}^2$ (3.16.19)

When **V** is parallel to **K** we put $\theta = 0$ in (3.16.18) and get

$$\Omega^2 - (1 + q^2 + r) = 0, \tag{3.16.20}$$

$$\Omega^4 - 2p\,\Omega^3 + (p^2 - r - 1)\,\Omega^2 + 2pr\,\Omega - rp^2 = 0 \tag{3.16.21}$$

Equation (3.16.20) is identical to (3.16.17); and (3.16.21) has the familiar form,

$$\frac{\omega_{\rho e}^2}{(\omega - KV)^2} + \frac{\omega_{\rho i}^2}{\omega^2} = 1 \tag{3.16.22}$$

For $p = 0$ and $r = 0$ equation (3.16.18) reduces to

$$\Omega^4 - (q^2 + 2)\,\Omega^2 + (1 + q^2) = 0 \tag{3.16.23}$$

It gives the two dispersion relations

$$\omega^2 = K^2 c^2 + \omega_{\rho e}^2,\; \omega^2 = \omega_{\rho e}^2 \tag{3.16.24}$$

Since equation (3.16.22) gives complex roots of ω, streams make a plasma unstable.

When the stream is perpendicular to the direction of wave propagation, we put $\theta = \pi/2$ in (3.16.18) and obtain

$$\Omega^4 - (q^2 + 2)\,\Omega^2\,(1 + q^2 - p^2) = 0 \tag{3.16.25}$$

The roots are given by $2\Omega^2 = q^2 + 2 \pm (q^4 + 4p^2)^{1/2}$ (3.16.26)

For instability $(q^2 + 2)^2 < q^4 + 4p^2$ or $q^2 + 1 < p^2$ or $c \leq KV/\omega_{pe}$. This is not relativistically possible, so the plasma cannot be made unstable thus.

Chapter 4

MAGNETOHYDRODYNAMICS

4.1 INTRODUCTION

Magnetohydrodynamics (or MHD) is concerned with the interaction of electrically conducting fluids in motion and fields. (Often, a dense volume of electrically conducting fluid is referred to as a plasma.) Mechanical force producing motion, and external agents inducing magnetic fields, couple the electrical properties with those of fluid motion. So, in MHD theory, the difficulties of fluid mechanics are multiplied by the difficulties of classical electrodynamics. The MHD materials have large densitites. So the characteristic lengths, such as the mean free path (Appendix 12b) and the Debye length (section 3.10) are of the order of molecular dimensions. The positive and negative charges remain so close to each other that space charge separation, due to internal thermal motion, becomes negligible, and their individual motion, depending on their mass and charge, is not important in the MHD model for a plasma. Moreover, wavelengths of applied fields and volume lengths of the materials are larger than these characteristic lengths. Also, the velocities are small compared with the speed of light; the electromagnetic changes are instantaneous as compared to the time scale of MHD effects; low frequency phenomena including ohm's law are important and the electromagnetic radiation field is filtered out, because the displacement current in the Maxwell equations is then negligible.

The pre-Maxwell equations, based on the laws of Coulomb, Ampere and Faraday, and the charge conservation condition, are sufficient for study of MHD (sections 4.2–4.4). So this subject could have been developed before the discovery of the displacement current by Maxwell.

In the adiabatic process (1) The regions of compression and rarefaction of the gas do not have time to exchange energy during propagation of the acoustic wave; (2) Compressions are so slow that the reactions are reversible; (3) As wavelength of a wave is increased, the adiabaticity condition is better satisfied; (4) The particles must hit the boundary walls several times; (5) But the compression also must not be so slow that the particles find time to scatter in different directions. When the collision frequency is very large, the plasma remains isotropic, and pressure remains scalar. In the collisionless limit, ion gyration radius and Debye shielding length are smaller than the characteristic lengths specifying the problem, and the magnetic moment is an adiabatic invariant of motion of each particle.

In the intermediate double-adiabatic theory (or the CGL theory, section 4.12) there will be two different equations of state for the transverse and longitudinal pressure tensor components, with respect to the field direction, because collisions are not strong enough to keep the pressure isotropic, but these are sufficiently strong to make heat flow negligible. So the term adiabaticity has different implications in the adiabatic and double-adiabatic theories.

4.2 MHD EQUATIONS

Equations for conservation of mass and momentum, the phenomenological Maxwell equations of electrodynamics and the thermal and caloric equations of state are the starting equations. Both the flow variables and the electromagnetic field variables are defined as functions of position coordinates $\mathbf{r}$ (x, y, z) and time t, as in the Eulerian kind of specifications. So the substantial time derivative operator is

$$\frac{d}{dt} = \frac{\partial}{\partial t} + (\mathbf{v} \cdot \nabla)$$

where $\mathbf{v}$ is the velocity of the fluid element. Actually the derivative $\partial\varphi/\partial t$ of a field variable $\varphi\,(\mathbf{r}, t)$ is the instantaneous rate of change of φ at the space time point $\mathbf{r}, t$. The correct expression for the time rate of change of φ may be determined by noting that an element at the position $\mathbf{r}$ at time t is at position $\mathbf{r} + \mathbf{v}\,\delta t$ at time $t + \delta t$, so

$$\varphi\,(\mathbf{r} + \mathbf{v}\,\delta t,\ c + \delta t) - \varphi\,(\mathbf{r},\ t)$$

$$= \delta t \left\{ \frac{\partial \varphi}{\partial t} + (\mathbf{v} \cdot \nabla)\,\varphi \right\} + \delta\,(\delta\,t^2)$$

Hence $$\frac{d\varphi}{dt} = \frac{\partial \varphi}{\partial t} + (\mathbf{v} \cdot \nabla)\,\varphi$$

(a) Mass conservation condition

The mass conservation condition is

$$\frac{d\rho}{dt} + \rho \operatorname{div} \mathbf{v} = 0 \text{ or } \frac{\partial \rho}{\partial t} + \operatorname{div}(\rho\,\mathbf{v}) = 0 \tag{4.2.1}$$

where $\rho\,(\mathbf{r}, t)$ is the mass per unit volume of the fluid element.

(b) Momentum transfer condition

If $q\,(\mathbf{r}, t)$ is the net free charge density, then the momentum transfer condition is

$$\frac{d\mathbf{v}}{dt} = -\frac{1}{\rho} \nabla p + \frac{q}{\rho} \mathbf{E} + \frac{1}{\rho\,c} (\mathbf{j} \times \mathbf{B}) + \frac{\nu}{\rho} \nabla^2 \mathbf{v}$$

$$+ \frac{1}{\rho} (\delta + \frac{1}{3} \nu)\, \nabla\,(\nabla \cdot \mathbf{v}) + \frac{1}{\rho} \mathbf{F}_0 \tag{4.2.2}$$

where $\mathbf{E}$ is the electric field, $\mathbf{B}$ is the magnetic induction, $\mathbf{j}$ is the current density, p is the pressure of the fluid, $\mathbf{F}_0$ represents the other forces, ν is the coefficient of viscosity and ζ is the coefficient of bulk viscosity; generally $\zeta \ll \nu$. When both $\nu = 0$ and $\rho = 0$, (4.2.2) is called the Euler equation; when $\nu \neq 0$ and $\rho \neq 0$, (4.2.2) is called the Navier. Stokes equation (4.2.2), with the help of (4.2.1), can be written as

$$\frac{d}{dt}(\rho\, v_i) = -\frac{\partial}{\partial x_j} P_{ij} + F_i', \tag{4.2.3}$$

or as

$$\frac{\partial}{\partial t}(\rho\, v_i) = -\frac{\partial}{\partial x_j} \pi_{ij} + F_i' \tag{4.2.4}$$

where

$$\pi_{ij} = \rho\, v_i\, v_j + P_{ij}, \tag{4.2.5}$$

$$P_{ij} = p\, \delta_{ij} - \sigma_{ij}, \tag{4.2.6}$$

$$\sigma_{ij} = \nu \left(\frac{\partial v_i}{\partial x_j} + \frac{\partial v_j}{\partial x_i} - \frac{2}{3} \delta_{ij} \frac{\partial v_k}{\partial x_k} \right) + \zeta\, \delta_{ij} \frac{\partial v_k}{\partial x_k}, \tag{4.2.7}$$

$$\mathbf{F}' = \mathbf{F}_0 + q\,\mathbf{E} + \frac{1}{c}(\mathbf{j} \times \mathbf{B}) \tag{4.2.8}$$

Here δ_{ij} is the kronecker delta ($\delta_{ij} = 1$ if $i = j$, otherwise $\delta_{ij} = 0$) and repeated index means summation over that index. The Lorentz force $\mathbf{F}$ $(= q\mathbf{E} + (\mathbf{j} \times \mathbf{B})/c)$ has been reduced in section 1.8 to the form

$$\mathbf{F} = \hat{\mathbf{X}}_i \frac{\partial T_{ij}}{\partial x_j} - \frac{\partial}{\partial t}\left(\frac{\varepsilon\mu}{c^2} \mathbf{S}^t \right) \tag{4.2.9}$$

where $\hat{\mathbf{X}}_i$ is the unit vector in the ith direction. With the help of these transformations, equation (4.2.4) can be written as

$$\frac{\partial}{\partial t}\left(\rho v_i + \frac{\varepsilon\mu}{c^2} S_i^t \right) = -\frac{\partial}{\partial x_j}(\pi_{ij} - T_{ij}) + F_{oi} \tag{4.2.10}$$

(c) Phenomenological Maxwell equations

These are

$$\left.\begin{aligned} &\text{curl}\,\mathbf{H} = \frac{1}{c}\frac{\partial \mathbf{D}}{\partial t} + \frac{4\pi}{c}\mathbf{j},\ \text{curl}\,\mathbf{E} = -\frac{1}{c}\frac{\partial \mathbf{B}}{\partial t}, \\ &\text{div}\,\mathbf{D} = 4\pi\, q,\ \text{div}\,\mathbf{B} = 0 \end{aligned}\right\} \tag{4.2.11}$$

(d) Electrodynamic state relations

These are Ohm's law,

$$\mathbf{j} = \sigma \left\{ \mathbf{E} + \frac{1}{c}(\mathbf{v} \times \mathbf{B}) \right\} + q\mathbf{v}, \tag{4.2.12}$$

and the constitutive relations

$$\mathbf{B} = \mu\mathbf{H},\ \mathbf{D} = \varepsilon\mathbf{E} \tag{4.2.13}$$

Here σ is the electrical conductivity, μ is the magnetic permeability, ε is the dielectric constant, and $q\,\mathbf{v}$ is the advective current*, which depends on the motion of the net charge q. It is different from convection current $\mathbf{j}$, which is transport of a conducting fluid due to heat.

(e) Thermodynamic state relations

The state of a compressible fluid is defined macroscopically by the pressure P, the entropy S, the internal energy ξ, the absolute temperature T (in degrees Kelvin), the mass density ρ and the specific volume λ of the fluid (the volume per unit mass $= 1/\rho$). The medium is two-parametric if all the thermodynamic variables are functions of two of these quantities (say functions of ρ (or λ) and S). For the two-parameter media the thermodynamic law is

$$\rho\,(d\xi + p d\lambda) = \rho T ds, \tag{4.2.14}$$

then

$$T ds = \frac{\partial \xi}{\partial s} ds + \left(\frac{\partial \xi}{\partial \lambda} + p\right) d\lambda \tag{4.2.15}$$

Equating the coefficients of $d\lambda$ and ds from both sides we have

$$T = \frac{\partial \xi}{\partial s},\ p = -\frac{\partial \xi}{\partial \lambda} \tag{4.2.16}$$

In an ideal gas (temperatures not too high, and the range of stream conditions not unduly severe, as in most real gases), with negligible viscosity and thermal conductivity,

$$p = R\rho T \tag{4.2.17}$$

where R is the universal gas constant. Assuming the local processes to be adiabatic and isentropic (this excludes the shock regions), for polytropic gases we have

$$\xi = c_v T,\ c_p = c_v + p\frac{\partial \lambda}{\partial T} \tag{4.2.18}$$

where c_v is the specific heat at constant volume $(= T(\partial s/\partial T) = \partial \xi/\partial T)$ and c_p is the same at constant pressure. For $P =$ constant, relation (4.2.17) yields

$$p\frac{d\lambda}{dT} = R,\ R = c_p - c_v \tag{4.2.19}$$

Equations (4.2.14) and (4.2.18) give

$$T ds = c_v\, dT + p d\lambda \tag{4.2.20}$$

Hence evidently

$$ds = c_v\left(\frac{dT}{T}\right) + \left(\frac{p}{T}\right) d\lambda = c_v\left(\frac{dp}{p} + \frac{d\lambda}{\lambda}\right) + \left(\frac{R}{\lambda}\right) d\lambda$$

* Advection is transport of momentum and energy of a fluid due to motion.

$$= c_v \frac{dp}{p} + c_p \frac{d\lambda}{\lambda} \tag{4.2.21}$$

where we have used in the first term of the second step the logarithmic differentiation of (4.2.17), in the second term equation (4.2.17) itself and in the final step the second relation of (4.2.19). Solution of (4.2.21) can be written as

$$p = \left(\frac{\rho^{\gamma} p_0}{\rho_0^{\gamma}} \right) \exp\left(\frac{s - s_0}{c_v} \right), \ \gamma = \frac{c_p}{c_v} \tag{4.2.22}$$

where the initial values of p, ρ and S are denoted by p_0, ρ_0 and s_0 respectively. For isentropic gases equation (4.2.22) gives the range of possibilities of entropy variation without the help of the field equations for molecular transport processes. When a fluid element changes its state isentropically, for negligible dissipation the adiabatic state equation

$$\frac{d}{dt}(p\rho^{-\gamma}) = 0 \tag{4.2.23}$$

yields the solution $(p/\rho_0) = (\rho/\rho_0)^{\gamma} = (T/T_0)^{\gamma/(\gamma - 1)}$. (4.2.24)

The constant $\gamma = c_p/c_v$; c_p and c_v are the specific heats at constant pressure and volume respectively. If d is the number of degrees of freedom in the system, then $\gamma = (2 + d)/d$. Monatomic gases have only translational motions in all directions; so for these $d = 3$ and $\gamma = 5/3$. In a two-dimensional compression (example : collisional plasmas, in presence of a static magnetic field) there is no energy transfer to the third spartial degree of freedom and $\gamma = 2$.

(f) Energy transport

The law of conservation of energy is equivalent to the first law of thermodynamics, which states that the specific change in the internal energy ξ of the particles per unit mass is due to compression by the surrounding medium, and the energy generated by the external sources. Mathematically, this law can be expressed as

$$d\xi = T\,ds - p\,d\lambda \tag{4.2.25}$$

and

$$dQ = Tds + dp/\rho \text{ or } \nabla p = \rho \nabla Q - \rho\lambda \nabla s \tag{4.2.26}$$

where the heat function per unit mass

$$Q = \xi + \frac{1}{\rho} p \tag{4.2.27}$$

and $\lambda = 1/\rho$. Since the entropy per unit mass, s, for adiabatic motion, is conserved,

$$\frac{ds}{dt} = 0 \text{ or } \frac{\partial s}{\partial t} + (\mathbf{v} \cdot \nabla)\, s = 0 \tag{4.2.28}$$

The energy per unit volume of the medium is the sum of the kinetic energy density of random thermal motion, $\rho v^2/2$, and the internal energy density $\rho\xi$. Differentiating this sum partially with respect to time we get

$$\frac{\partial}{\partial t}(\rho v^2 + \rho\xi) = -\frac{1}{2}v^2 \operatorname{div}(\rho \mathbf{v}) + \rho\left(\mathbf{v}\cdot\frac{\partial \mathbf{v}}{\partial t}\right) + \xi\left(\frac{\partial \rho}{\partial t}\right) + \rho\left(\frac{\partial \xi}{\partial t}\right) \tag{4.2.29}$$

where, in the first term of the right hand side, we have used (4.2.1). The right hand side of this equation is

$$= -\frac{1}{2}v^2\,\nabla\,.\,(\rho\mathbf{v}) - \rho\,(\mathbf{v}\,.\,(\mathbf{v}\,.\,\nabla)\,\mathbf{v}) - (\mathbf{v}\,.\,\nabla)\,p + \nu\,(\mathbf{v}\,.\,\nabla^2\,\mathbf{v})$$

$$+\left(\zeta + \frac{\nu}{3}\right)(\mathbf{v}\,.\,\nabla)(\nabla\,.\,\mathbf{v}) + (\mathbf{F}_0\,.\,\mathbf{v}) + \xi\frac{\partial\rho}{\partial t} + \rho\left(T\frac{\partial s}{\partial t} - p\frac{\partial\lambda}{\partial t}\right) \tag{4.2.30}$$

where we have put the value of $\partial\mathbf{v}/\partial t$ in the second term from (4.2.2), used (4.2.25) and put $\mathbf{F} = q\,\mathbf{E} + (\mathbf{j}\times\mathbf{B})/c$. The second term of this result $= -\rho\,(\mathbf{v}\,.\,\nabla)\,(v^2/2)$. On using equation (4.2.26), the third term

$$= -\rho\,(\mathbf{v}\,.\,\nabla)\,Q + \rho\,\lambda\,(\mathbf{v}\,.\,\nabla)\,s$$

The sum of these two terms $= -\rho\,(\mathbf{v}\,.\,\nabla)\left(\frac{v^2}{2} + Q\right) + \rho\lambda\,(\mathbf{v}\,.\,\nabla)\,s$ and the sum of the fourth and fifth terms $= \nabla\,.\,\mathbf{A}$ where

$$\mathbf{A} = \nu\sum_{(i,\,k)\,=\,1}^{3}\left(\frac{\partial v_i}{\partial x_k} + \frac{\partial v_k}{\partial x_i} - \frac{2}{3}\,\delta_{ik}\operatorname{div}\mathbf{v}\right)v_i\,\hat{\mathbf{X}}_k$$

If χ is the thermal conductivity and T is the temperature, then the sixth term in the right hand side of (4.2.30) $= (\mathbf{F}_0\,.\,\mathbf{v}) = \operatorname{div}(\chi\,\operatorname{grad}\,T)$. The seventh term, $(\mathbf{F}\,.\,\mathbf{v})$, has been studied in section 1.7. If (4.2.27) is used, then the sum of the eighth, ninth and tenth terms

$$= \xi\left(\frac{\partial\rho}{\partial t}\right) + \rho T + \frac{\partial s}{\partial t} + \frac{p}{\rho}\,\frac{\partial\rho}{\partial t} = Q\,\frac{\partial\rho}{\partial t} + \rho T\frac{\partial s}{\partial t};$$

Moreover, if equations (4.2.1) and (4.2.28) are employed, then it reduces to $-\,Q\nabla\,.\,(\rho\mathbf{v}) - \rho\lambda\,(\mathbf{v}\,.\,\nabla)\,s$. Hence we find that

$$\frac{\partial}{\partial t}\left(\frac{1}{2}\rho\,v^2 + \rho\,\xi\right) = -\frac{v^2}{2}\,\nabla\,.\,(\rho\,\mathbf{v}) - \rho\,(\mathbf{v}\,.\,\nabla)\left(\frac{v^2}{2} + Q\right) + \rho\,\lambda\,(\mathbf{v}\,.\,\nabla)\,s +$$

$$+\,\nabla\,.\,(\mathbf{A} + \chi\,\nabla T) - Q\,\nabla\,(\rho\mathbf{v}) - \rho\lambda\,(\mathbf{v}\,.\,\nabla)\,s + (\mathbf{F}\,.\,\mathbf{v})$$

$$= \nabla\cdot\left\{-\,\rho\mathbf{v}\left(\frac{v^2}{2} + Q\right) + \mathbf{A} + \chi\,\nabla T\right\} + (\mathbf{F}\,.\,\mathbf{v})$$

If, with the help of (4.2.27), φ is replaced by $\xi + (p/\rho)$ then this equation reduces to

$$\frac{\partial}{\partial t}\left(\rho\xi + \frac{1}{2}\rho v^2\right) + \text{div}\left\{\rho\mathbf{v}\left(\xi + \frac{v^2}{2}\right) + p\,\mathbf{v}\right\}$$

$$= \text{div}\,(\mathbf{A} + \chi\,\text{grad}\,T) + (\mathbf{F}\,.\,\mathbf{v}) \qquad (4.2.30\,(a))$$

Hence the energy per unit volume at a point is due to (1) the energy flux to or from the volume during fluid motion, (2) the work of pressure forces and (3) the energy applied from external sources. The heat increase per unit volume equals the sum of (1) the heat flux div **A** due to viscous dissipation, (2) the heat flux div (χ grad T) due to thermal conduction and (3) the rate of work done, (**F** . **v**), by the Lorentz force. More explicit form of equation (4.2.30) is

$$\frac{d}{dt}\left(\rho\xi + \frac{1}{2}\rho v^2\right) = \text{div}\,(\chi\,\text{grad}\,T) + \frac{\partial}{\partial x_i}(\sigma_{ij}\,v_j) - \frac{\partial}{\partial x_i}(pv_i) \quad (4.2.31)$$

where σ_{ij} is given by (4.2.7).

The expression for the viscous force obtained from the fourth term in the right hand side of (4.2.2), is $\nu\nabla\times(\nabla\times\mathbf{v})$. Then the rate of work done against this forces is $\nu\,(v\,.\,\nabla\times(\nabla\times\mathbf{v}))$. If the fluid is in a perfectly rough container of volume V and surface Σ, then using the boundary condition $\mathbf{v}=0$ on Σ after integration over V, gives the total rate of dissipation of energy

$$\nu\iiint_V (\mathbf{v}\,.\,\nabla\times(\nabla\times\mathbf{v}))\,dV.$$

Since, div ($\mathbf{v}\times$ curl $\mathbf{v}$) = (curl **v** . curl **v**) – (**v** . curl curl **v**), formula 7 of Appendix 3, for an incompressible fluid, this integral is replaced by

$$\nu\iiint_V (\text{curl}\,\mathbf{v})^2\,dV - \nu\iint_\Sigma (\mathbf{v}\times\text{curl}\,\mathbf{v})\,.\,d\boldsymbol{\Sigma};$$

and, using the boundary condition, by $\nu\iiint_V (\text{curl}\,\mathbf{v})^2\,dV$.

4.3 APPROXIMATIONS OF MHD FROM DIMENSIONAL CONSIDERATIONS

In electromagnetics, the vacuum speed of light, c ($\equiv 3\times10^{10}$ cm/sec), appears in the field equations as a universal constant, so, in plasmas and MHD processes, comparing the typical speeds (material speeds and speed-like material parameters) with the speed of light, one can know which terms to keep and which to ignore. But, in mechanics of uncharged particles so universal consant exists, in the guiding equations; so, comparison conclusion is not possible. In dense media the mean free path (Appendix 12b) and the Debye shielding distance (section 3.10) are of the order of molecular distances, and the macroscopic motion does not differ effectively from thermal motion. Then, high frequency electromagnetic waves (a consequence of the existence of the displacement current in the Maxwell equations) are filtered out because these waves propagate at infinite phase velocity and interaction of fields with fluid motion is mainly through the conduction current.

The speeds of interest in MHD processes are v, C_A and C_S and v is a typical speed of material flow, C_A is the Alfven speed (equation 1.8.12) and C_S is the sound speed in the medium. The ratios v^2/c^2, C_A^2/c^2, C_s^2/c^2 are the measures of macroscopic kinetic energy, magnetic energy and thermal energy, in terms of the rest energy of the fluid. In gas dynamics the flow is supersonic if $M > 1$, subsonic if $M < 1$ and transonic if $M = 1$, where the Mach number $M = v/c_s$. Similarly, in MHD flows, the Alfven number A $(= v/C_A)$ determines the nature of steady flow of electrically conducting fluids in channels in the presence of axial magnetic field. An MHD flow is super-Alfvenic if $A > 1$, sub-Alfvenic if $A < 1$ and trans-Alfvenic if $A \approx 1$.

The effectivity of a physical quantity depends on its order of magnitude relative to other quantities of importance in it. The statement that a quantity P is of the order of the magnitude of Q, or symbolically $P \approx Q$, means that P and Q have the same dimension and that $P = \lambda Q$ where λ is of the order of unity, say, lying between 1 and 1/10. For the evaluation of the order of magnitude of a quantity, with respect to the time scale and length scale of a phenomenon, its time derivative and space derivative (along the z direction, say) can be expressed as

$$\frac{\partial \psi}{\partial t} = \frac{\psi}{T}, \quad \frac{\partial \psi}{\partial z} = \frac{\psi}{L_z},$$

where L_z and T are the characteristic length (along the z direction) and time respectively, of its variation in that effect. This means that ψ varies appreciably in times T and distances L_z and not much less or much more than these magnitudes. Then we write $L_z/T = v$; v stands for a characteristic speed of the process. The time scale of variation of an electromagnetic wave field is l/c where c is the vacuum speed of light. For bulk motion the time L/v is much greater than the time L/c, because the maximum v is the sound speed C_s, which is of the order of 330 cm/sec. So, if $Q \approx (v^2/c^2)\, P$, then

$$P + Q \approx P \tag{4.3.1}$$

The second equation of (4.2.11) can be written as

$$\frac{E}{L} = \frac{B}{CT} \tag{4.3.2}$$

The displacement current $\partial \mathbf{D}/\partial t$ is negligible when compared with $(\nabla \times \mathbf{H})$,

$$\frac{1}{c}\left|\frac{\partial D}{\partial t}\right| = \frac{E}{CT} = \frac{Bl}{C^2 T^2} = |\nabla \times \mathbf{H}| \frac{v^2}{c^2} \tag{4.3.3}$$

Consequently,

$$\nabla \times \mathbf{H} = \frac{4\pi}{c}\mathbf{j},\ \nabla \cdot \mathbf{j} = 0,\ j = CH/L \tag{4.3.4}$$

The third equation of (4.2.11) gives $q \approx E/l$. Next, to compare the two terms of the Lorentz force, we write

$$|q\mathbf{E}| \approx \frac{E^2}{l} \approx \left(\frac{B^2}{l}\right)\left(\frac{v^2}{c^2}\right) = \left|\frac{1}{c}(\mathbf{j} \times \mathbf{B})\right|\left(\frac{v^2}{c^2}\right) \tag{4.3.5}$$

Hence the body force $q\,\mathbf{E}$ is negligible when compared with $(\mathbf{j} \times \mathbf{B})/c$. The momentum transfer equation (4.2.2) therefore simplifies to

$$\rho \frac{d\mathbf{v}}{dt} = -\nabla p + \frac{1}{c}(\mathbf{j} \times \mathbf{B}) + \mathbf{F}_0 + \nu \nabla^2 \mathbf{v} + \left(\zeta + \frac{1}{3}\nu\right)\nabla(\nabla \cdot \mathbf{v}) \tag{4.3.6}$$

The viscous force $\nu \nabla^2 \mathbf{v}$, on comparison with the force $(\mathbf{j} \times \mathbf{B})/C$, gives

$$|\nu \nabla^2 \mathbf{v}| \approx \left|\frac{1}{c}(\mathbf{j} \times \mathbf{B})\right| (\nu v/B^2 l) \tag{4.3.7}$$

Putting $H = 100$ gauss, $l = 100$ cm, $v = 1$ cm/sec and $\nu = 10^{-2}$ gm/cm sec (mercury), we get

$$\frac{\nu v}{B^2 L} = 10^{-7} \tag{4.3.8}$$

In astrophysical problems l is much greater, so this coefficient is still smaller. Only in some cases of high speed laboratory experiments, the viscous force is not neglected. When the pressure gradient is also large, compared with the viscous term, we have $|\nabla p| > |\nu \nabla^2 \mathbf{v}|$, so $|\nu\, v/l| < p$. Since

$$|q\mathbf{v}| \approx \left(\frac{Ev}{l}\right) \approx \left(\frac{Bv}{cT}\right) \approx \left(\frac{v^2}{c^2}\right) j, \tag{4.3.9}$$

the state relation (4.2.12) simplifies to Ohm's law

$$\frac{\mathbf{j}}{\sigma} = \mathbf{E} + \frac{1}{c}(\mathbf{v} \times \mathbf{B}) \tag{4.3.10}$$

and excludes the possibility of plasma oscillations. Relation (4.3.2) gives

$$\frac{E^2}{8\pi} \approx \left(\frac{v^2}{c^2}\right)\left(\frac{B^2}{8\pi}\right) \text{ or } W_E \approx \left(\frac{v^2}{c^2}\right) W_B \tag{4.3.11}$$

Hence the electrical energy density is negligible in comparison with the magnetic energy density. But the relativistic nature of the dispersive magnetic field (section 1.16) is such that non-relativistically the Faraday law of induction is $\nabla \times \mathbf{E} = 0$; and the electric and magnetic fields are given by the two relations of (1.16.8). From these we find that $W_H = (v^2/c^2)\, W_E$, which is different from (4.3.11). Hence, the MHD process is basically different from the high frequency electromagnetic processes involving the displacement current.

4.4 REDUCED MHD EQUATIONS

With the help of (4.3.4), the Lorentz force term of (4.3.6) reduces to

$$\frac{(\mathbf{j} \times \mathbf{B})}{c} = \left(\frac{\mu}{4\pi}\right) ((\nabla \times \mathbf{H}) \times \mathbf{H}) = \nabla \left(-\frac{\mu H^2}{8\pi}\right) + \left(\frac{\mu}{4\pi}\right) (\mathbf{H} \,.\, \nabla)\, \mathbf{H} \quad (4.4.1)$$

Hence if p^* is the total pressure in the fluid, we find

$$p^* = p + \frac{\mu H^2}{8\pi} \quad (4.4.2)$$

So a static magnetic field, acting like a second fluid, with energy $\mu H^2/8\pi$ per unit volume, modifies the compressibility of the fluid.

Since $\nabla \,.\, \mathbf{B} = 0$, Ohm's law, equation (4.3.10), gives

$$\frac{4\pi}{c} \nabla \times \mathbf{j} = -\nabla^2 \mathbf{H} = \frac{4\pi\sigma}{c} \left\{ (\nabla \times \mathbf{E}) + \frac{1}{c} \nabla \times (\mathbf{v} \times \mathbf{B}) \right\}$$

$$= -\frac{4\pi\sigma}{c} \left(\frac{\partial \mathbf{B}}{\partial t}\right) + \frac{4\pi\sigma}{c^2} \{(\mathbf{B} \,.\, \nabla)\, \mathbf{v} - (\mathbf{v} \,.\, \nabla)\, \mathbf{B} - \mathbf{B}\, (\nabla \,.\, \mathbf{v})\} \quad (4.4.3)$$

$$\text{Hence } \frac{\partial \mathbf{B}}{\partial t} + (\mathbf{v} \,.\, \nabla)\, \mathbf{B} - (\mathbf{B} \,.\, \nabla)\, \mathbf{v} + \mathbf{B}\, (\nabla \,.\, \mathbf{v}) = \eta \nabla^2 \mathbf{B} \quad (4.4.4)$$

where η is the electrical resistivity of the medium. Eliminating $(\nabla \,.\, \mathbf{v})$ from this equation, with the help of (4.4.1), and replacing $(\partial/\partial t) + (\mathbf{v} \,.\, \nabla)$ by d/dt we get

$$\frac{d\mathbf{H}}{dt} - (\mathbf{H} \,.\, \nabla)\, \mathbf{v} - \left(\frac{\mathbf{H}}{\rho}\right) \frac{d\rho}{dt} = \eta \nabla^2 \mathbf{H}$$

Dividing both sides of this equation by ρ and further simplifying we get

$$\frac{(\eta\, \nabla^2 \mathbf{B})}{\rho} = \frac{d}{dt}\left(\frac{\mathbf{B}}{\rho}\right) - \frac{1}{\rho} (\mathbf{B} \,.\, \nabla)\, \mathbf{v}$$

Hence the reduced MHD field equations are

$$\frac{\partial}{\partial t} (\rho \mathbf{v}) - \frac{\partial}{\partial x_j} (\rho v_j\, \mathbf{v}) = -\nabla \left(p + \frac{\mu H^2}{8\pi}\right) + \frac{\mu}{4\pi} (\mathbf{H} \,.\, \nabla)\, \mathbf{H}$$

$$+ \mathbf{F} + \nu\, \nabla^2 \mathbf{v} + \left(\zeta + \frac{\nu}{3}\right) \nabla\, (\nabla \,.\, \mathbf{v}), \quad (4.4.5)$$

$$\rho\,(\nabla\,.\,\mathbf{v})+\frac{d\rho}{dt}=0,\ \frac{\eta\nabla^2 B}{\rho}=\frac{d}{dt}\left(\frac{\mathbf{B}}{\rho}\right)-\frac{1}{\rho}(\mathbf{B}\,.\,\nabla)\,\mathbf{v},\ (\nabla\,.\,\mathbf{B})=0 \quad (4.4.6)$$

When the flow of an infinitely conducting fluid is in the XY plane and $\mathbf{H}$ is parallel to OZ, the second equation of (4.4.6) gives $d\,(\mathbf{H}/\rho)/dt=0$, so $H=\alpha\rho$ where α is a constant. Then the relation (4.4.2) gives

$$p^*=p+\frac{\mu\alpha^2\rho^2}{8\pi}\ \text{and}\ \frac{dp^*}{d\rho}=C_S^{*2}=C_S^2+C_A^2 \quad (4.4.7)$$

where C_S is the sound speed in the absence of the magnetic field, C_A is the Alfven speed and C_S^* is the sound speed in such a magnetized medium.

(a) Energy conservation

The rate of work done per unit volume, by the electrical forces on the medium, is

$$(\mathbf{F}\,.\,\mathbf{v})\equiv(\mathbf{E}\,.\,\mathbf{j})=-\frac{\partial W}{\partial t}H-(\nabla\,.\,\mathbf{S}^t)=\frac{1}{\sigma}j^2-\frac{1}{c}(\mathbf{v}\,.\,(\mathbf{j}\times\mathbf{B})) \quad (4.4.8)$$

where W_H $(=\mu\,H^2/8\pi)$ is the magnetic energy density and

$$\mathbf{S}^t=\frac{c}{4\pi}(\mathbf{E}\times\mathbf{H})=\frac{\mu}{4\pi}(\mathbf{H}\times(\mathbf{v}\times\mathbf{H}))-\frac{c^2}{16\pi^2\sigma}(\mathbf{H}\times(\nabla\times\mathbf{H})) \quad (4.4.9)$$

$\mathbf{S}^t$ is obviously the Poynting vector. The identity

$$\frac{d}{dt}\left(\rho\xi+\frac{1}{2}\rho v^2\right)\equiv\frac{\partial}{\partial t}\left(\rho\xi+\frac{1}{2}\rho v^2\right)+\nabla\,.\left(\rho\xi\mathbf{v}+\frac{1}{2}\rho v^2\,\mathbf{v}\right) \quad (4.4.10)$$

reduces the energy transfer condition (4.3.31) to

$$\frac{\partial W}{\partial t}+(\nabla\,.\,\mathbf{S})=0 \quad (4.4.11)$$

where
$$W=\rho\xi+(\rho v^2/2)+(\mu H^2/8\pi), \quad (4.4.12)$$

$$\mathbf{S}=\mathbf{v}\left(\rho\xi+\frac{1}{2}\rho v^2+p\right)+\mathbf{S}^t-\chi\,(\nabla T)-\boldsymbol{\sigma}\,.\,\mathbf{v} \quad (4.4.13)$$

The elements of the tensor $\boldsymbol{\sigma}$ are to be obtained from the right hand side of (4.2.7). The total energy density is W, and its flux is $\mathbf{S}$. Equations (4.2.1), (4.4.5), (4.4.6) and (4.4.8) form a complete set, for study of MHD processes.

(b) Some case studies

(*i*) If p_i and $\mathbf{H}_i$ are respectively the pressure and magnetic field inside a sun spot region, and p_e, $\mathbf{H}_e$ are those outside, for pressure balance at the interface we have

$$p_i + \frac{\mu H_i^2}{8\pi} = p_e + \frac{\mu H_e^2}{8\pi}$$

Moreover, if $H_i = 2000$ to 3000 gauss and $H_e = 0$, then

$$p_e - p_i \approx \frac{\mu H_i^2}{8\pi} = 0.16 \times 10^6 \text{ to } 0.36 \times 10^6 \text{ dynes/cm}^2.$$

(*ii*) In the formula $p_e - p_i = N\kappa\,(T_e - T_i)$ where κ (the Boltzmann constant of kinetic theory (chapter 9)) $= 1.4 \times 10^{-16}$ erg/°K, we put $N = 10^{17}$ particle/cm^3, and find that

$$T_e - T_i \approx \frac{0.16 \times 10^6}{10^{17} \times 1.4 \times 10^{-16}}\,^{\circ}\text{K to } \frac{0.36 \times 10^6}{10^{17} \times 1.14 \times 10^{-16}}\,^{\circ}\text{K}$$

Putting $T_e \approx 5700$°K, $T_i \approx 4600$°K, we find that

$$T_e - T_i = 1100\,^{\circ}\text{K} = \{(p_e - p_i)/N\kappa\}\,^{\circ}\text{K}.$$

Hence $p_e - p_i \approx 1.5 \times 10^4$ dynes/cm^2. Equating this quantity with the magnetic pressure $\mu\, H_i^2/8\pi$, the sun spot magnetic field is estimated.

4.5 SOME PROPERTIES

(a) Generalization of Bernoulli's theorem

Ignoring the bulk viscosity coefficient ζ in (4.3.6) we get

$$\rho \frac{\partial \mathbf{v}}{\partial t} - \rho\,(\mathbf{v} \times \Omega) = -\nabla \left(p + \frac{\mu H^2}{8\pi} + \frac{1}{2}\rho v^2 + \psi \right) + \nu \nabla^2\, \mathbf{v} + \frac{\mu}{4\pi}\,(\mathbf{H}\,.\,\nabla)\,\mathbf{H} \tag{4.5.1}$$

where $\Omega = \nabla \times \mathbf{v}$, the vector formula (A.3.10) has been used for $(\mathbf{v}\,.\,\nabla)\,\mathbf{v}$ and ψ is the potential of the force $\mathbf{F}_0$, assumed conservative here.

In steady flow $\partial/\partial t = 0$ and the scalar product of (4.5.1) with $\mathbf{v}$ gives

$$(\mathbf{v}\,.\,\nabla) \left(p + \frac{1}{2}\rho v^2 + \frac{\mu H^2}{8\pi} + \psi \right) = \frac{\nu}{\rho}\,(\mathbf{v}\,.\,\nabla^2\,\mathbf{v}) = \frac{\mu}{4\pi}\,(\mathbf{v}\,.\,(\mathbf{H}\,.\,\nabla)\,\mathbf{H}) \tag{4.5.2}$$

When $\nu = 0$ and $\mathbf{H} = 0$ this relation yields the classical Bernoulli theorem, namely, $p + (\rho v^2/2) + \psi$ is constant along a stream line. If $(\mathbf{H}\,.\,\nabla)\,\mathbf{H} = 0$, but $\mathbf{H} \neq 0$ along a stream line, then for $\nu = 0$,

$$p + \frac{\rho v^2}{2} + \psi + \frac{\mu H^2}{8\pi} = \text{constant} \tag{4.5.2 (a)}$$

So any change in the fluid pressure is balanced by an opposite change in magnetic pressure; and, if fluid pressure in a region decreases, the magnetic pressure must increase to keep the total pressure same.

For hydrostatic equilibrium, in the absnece of externally applied force fields, this relation gives

$$p + \frac{\mu H^2}{8\pi} = \text{constant}; \qquad (4.5.2\ (b))$$

So the sum of plasma pressure and magnetic pressure remains constant. And, if p increases then H decreases, and the plasma behaves as a diamagnetic material (section 2.12 (*a*)). Diamagnetism is negligible in rarefied plasmas, which satisfy everywhere, the condition

$$p << \frac{\mu H^2}{8\pi} \qquad (4.5.2\ (c))$$

(b) Generalization of Kelvin's theorem

To study evolution of vorticity we take curl of equation (4.5.1) and obtain

$$\frac{\partial \mathbf{\Omega}}{\partial t} = \nabla \times (\mathbf{v} \times \mathbf{\Omega}) + \frac{\mu}{4\pi\rho} \nabla \times (\mathbf{H} \,.\, \nabla)\,\mathbf{H} + \frac{\nu}{\rho} \nabla^2 \mathbf{\Omega} \qquad (4.5.3)$$

Now, using the vector identity (A.3.9) to simplify the first term on the right hand side and the equation of continuity in (4.4.6) we get

$$\frac{d\mathbf{\Omega}}{dt} = (\mathbf{\Omega} \,.\, \nabla)\,\mathbf{v} - \frac{\mathbf{\Omega}}{\rho}\frac{d\rho}{dt} = \frac{\mu}{4\pi\,\rho} \nabla \times (\mathbf{H} \,.\, \nabla)\,\mathbf{H} + \frac{1}{\rho} \nu \nabla^2 \mathbf{\Omega}$$

This equation can also be written as

$$\frac{d}{dt}\left(\frac{\mathbf{\Omega}}{\rho}\right) - \frac{1}{\rho}(\mathbf{\Omega} \,.\, \nabla)\,\mathbf{v} = \frac{\mu}{4\pi\rho^2} \nabla \times (\mathbf{H} \,.\, \nabla)\,\mathbf{H} + \frac{\nu}{\rho^2} \nabla^2 \mathbf{\Omega} \qquad (4.5.4)$$

In incompressible, non-viscous, but electrically conducting, fluids, therefore, we have

$$\frac{d\mathbf{\Omega}}{dt} - (\mathbf{\Omega} \,.\, \nabla)\,\mathbf{v} = \frac{\mu}{4\pi\rho} \operatorname{curl} (\mathbf{H} \,.\, \nabla)\,\mathbf{H} \qquad (4.5.5)$$

Hence, vorticity is created and not conserved by magnetic field. The circulation theorem of Kelvin (the flux of vorticity is carried away by the stream lines) follows from (4.5.5), when the magnetic field is zero. Hydromagnetic analogue of Kelvin's theorem in integral form, when the magnetic lines follow the stream lines, has been considered.

Exercise 4.5.1. For hydromagnetics, prove analogue of Kelvin's theorem, J. De. The analogue of Kelvins theorem in Hydromagnetics. Naturwissen Schaften **44**, 256-257 (1957).

Answer. (Proof) Kelvin's theorem in hydrodynamics is: The circulation of a fluid element, $\Gamma_0 \left(= \oint_{\Gamma} (\mathbf{v} \,.\, d\mathbf{l})\right)$, along a closed stream line (which equals the flux of vorticity through the surfaces Σ enclosed by the curve Γ) is a constant of motion for conservative forces. The analogue for hydromagnetics is: The circulation Γ_0 along a closed magnetic line of force is a constant of motion when the field lines are frozen with the electrically conducting fluid.

Consider the MHD equations for infinitely electrically conducting, incompressible and invised media:

$$\frac{\partial H}{\partial t} = \nabla \times (\mathbf{v} \times \mathbf{H}) \tag{1}$$

$$\nabla . H = 0 \tag{2}$$

$$\frac{\partial \mathbf{v}}{\partial t} - \mathbf{v} \times (\nabla \times \mathbf{v}) = \frac{1}{4\pi\rho} \; ((\nabla \times \mathbf{H}) \times \mathbf{H}) \tag{3}$$

$$(\nabla . \mathbf{v}) = 0 \tag{4}$$

$$\text{Then} \oint_{\Gamma} ((\dot{\mathbf{v}} - \mathbf{v} \times (\nabla \times \mathbf{v})) . d\mathbf{l}) = \frac{1}{4\pi\rho} \oint_{\Gamma} (((\nabla \times \mathbf{H}) \times \mathbf{H}) . d\mathbf{l}) \tag{5}$$

$$\text{RHS} = \frac{1}{4\pi\rho} \oint_{\Gamma} (\nabla \times H) . (\mathbf{H} \times d\mathbf{l}) = 0$$

because **H** is frozen with the streamlines in the fluid, and so H is parallel to the line element $d\mathbf{l}$ of a streamline. Hence, in this case, circulation of electrically conducting fluid along stream lines is a constant of motion.

(c) Diamagnetic drifts and currents

For magnetohydrostatic equilibrium, ignoring motion and the electric field, or using the reference frame moving with the v_E drift (section 2.6), in which $\mathbf{E} = 0$, we find that, if the magnetic lines are straight and parallel, the momentum transfer condition reduces to

$$P_{\perp} + \frac{B^2}{8\pi} = \text{constant} = \frac{B_0^2}{8\pi} \tag{4.5.6}$$

where $p_{\perp}$ denotes the pressure created in planes perpendicular to the field lines, and $\mathbf{B}_0$ is the magnetic field outside the plasma. Since $p_{\perp} > 0$, we find that $B^2 < B_0^2$ and the field inside the plasma is weaker than outside. This means that the medium becomes diamagnetic (section 2.12 (*a*)); so it experiences a net force in the direction of decreasing magnetic field. In the interior of a plasma the microscopic current due to gyration of charges overlap, however, and cancel each other. But, in regions where the magnetic field is inhomogeneous, the pressure increment $\delta B^2/8\pi$ is governed by (4.5.6). There is a net circulation of the magnetization current which is from the curl of the magnetic moment per unit volume.

For small drifts in the gyration period $2\pi/\Omega$, we neglect the acceleration $mN\,\partial\mathbf{v}/\partial t$ and the non-linear term $(\mathbf{v} . \nabla)\,\mathbf{v}$; then assuming that **E** and **B** are uniform, and that both the number density N and pressure P have gradients, we get

$$0 = EN \left[(\mathbf{E} \times \mathbf{B}) + \frac{1}{c} (\mathbf{v} \times \mathbf{B}) \times \mathbf{B} \right] - (\nabla p) \times \mathbf{B},$$

or
$$0 = EN \left[(\mathbf{E} \times \mathbf{B}) + \frac{1}{c} \mathbf{B} \, (\mathbf{v}_\perp \, . \, \mathbf{B}) - \frac{1}{c} B^2 \mathbf{v}_\perp \right] - (\nabla \, p) \times \mathbf{B} \tag{4.5.7}$$

where we have taken the cross product of the momentum transfer relation with **B**. Now, since $(\mathbf{v}_\perp \, . \, \mathbf{B}) = 0$, we obtain

$$\mathbf{v}_\perp = \frac{c}{B^2} (\mathbf{E} \times \mathbf{B}) - \frac{c \, (\nabla p \times \mathbf{B})}{eNB^2} = \mathbf{v}_E + \mathbf{v}_D \tag{4.5.8}$$

where
$$\boldsymbol{v}_E = \frac{c}{B^2} (\mathbf{E} \times \mathbf{B}), \;\; \boldsymbol{v}_D = -\frac{c}{eNB^2} (\nabla p \times \mathbf{B}), \tag{4.5.9}$$

and $\boldsymbol{v}_D$, called the diamagnetic drift, is independent of mass but depends on the sign of the charge. Since $\boldsymbol{v}_D$ is perpendicular to the direction of the gradient, the neglect of $(\mathbf{v} \, . \, \nabla) \, \mathbf{v}$ is justified. The thermodynamic equation of state relating the pressure p with the particle number N is $(\nabla p)/p = \gamma \, (\nabla N)/N$ where γ is the ratio of the specific heats. For isothermal compression

$$\nabla p = C_s^2 \, \nabla \rho \tag{4.5.10}$$

where C_s is the sound speed in the medium. Using this relation in (4.5.9) we obtain

$$\boldsymbol{v}_D = -\frac{CC_s^2 \, (\nabla \rho \times \mathbf{B})}{eNB} = \frac{mcC_s^2}{eBN} (\hat{\mathbf{z}} \times \nabla N) \tag{4.5.11}$$

The electron drift and the diamagnetic current are

$$\boldsymbol{v}_{DE} = -\frac{cc_e^2}{eBN} m \, (\hat{\mathbf{z}} \times \nabla N) \tag{4.5.12}$$

$$\mathbf{j}_D = Ne \, (\boldsymbol{v}_{Di} - \boldsymbol{v}_{De}) = \frac{c}{B} (\kappa T_e + \kappa T_i) \;\; (\hat{\mathbf{z}} \times \nabla n) \tag{4.5.13}$$

where κ is the Boltzmann constant, T_e and T_i are the kinetic temperatures, and $\boldsymbol{v}_{Di}$ is the drift of the ions.

From an equilibirum plasma configuration, energy flows out due to drift from the radial decrease of the magnetic field. This is a consequence of the diamagnetic repulsion of the plasma from the region of strong magnetic field and cooling by expansion. This transfer is important in the ejection of plasma streams from the sun.

4.6 ANALOGY WITH HYDRODYNAMICS. MAGNETIC REYNOLD'S NUMBER

Equation (4.4.4) can be written as

$$\frac{\partial \mathbf{B}}{\partial t} - \nabla \times (\mathbf{v} \times \mathbf{B}) = \eta \nabla^2 \mathbf{B}, \ \eta = \frac{c^2}{4\pi\mu\sigma}; \tag{4.6.1}$$

for $\mathbf{H} = 0$ equation (4.5.3) becomes

$$\frac{\partial \Omega}{\partial t} - \nabla \times (\mathbf{v} \times \Omega) = \frac{\nu}{\rho} \nabla^2 \Omega \tag{4.6.2}$$

Comparison of these two equations shows that the magnetic lines of force in a conducting fluid resemble the vortex lines of force in an incompressible fluid and that the kinetic viscosity ν is analogous to the symbol η which inversely depends on magnetic permeability μ and electrical conductivity σ. From this similarity η can be called the magnetic diffusivity of the medium, because the magnetic **B** lines slip through the material like the vortex lines in incompressible fluids. Therefore, the finite electrical conductivity of the medium introduces a dissipative loss which is similar to dissipation in non-conducting fluids.

Following the definition of the ordinary Reynolds number $R\,(= vl/\nu_0$ where $\nu_0 = \nu/\rho$, **v** is the flow speed and l is a characteristic length of variation of the medium) the magnetic Reynolds number R_m is defined as

$$R_m = \frac{vl}{\eta} = \frac{4\pi\mu}{c^2} \sigma vl \tag{4.6.3}$$

R_m is actually the ratio of the order of magnitude of the terms curl $(\mathbf{v} \times \mathbf{B})$ and $\eta \nabla^2 \mathbf{B}$ of (4.6.1); this magnitude decides whether magnetic lines move with the fluid or diffuse through it. If R_m is large, η is small and so (4.6.1) approximately becomes

$$\frac{\partial \mathbf{B}}{\partial t} = \nabla \times (\mathbf{v} \times \mathbf{B}), \ R_m >> 1; \tag{4.6.4}$$

When R_m is small we obtain the magnetic diffusion equation in a stationary conductor in the form

$$\frac{\partial \mathbf{B}}{\partial t} = \eta \nabla^2 \mathbf{B}, \ R_m << 1 \tag{4.6.5}$$

It shows that the **B** lines decay in the diffusion time l^2/η if η is real.

That the steady flow of electrically conducting fluids is similar to the steady vortex motion in incompressible non-viscous fluids can be seen from the following scheme :

Equations for the kinetics of vortex motion	Magnetohydrostatic equations
$\nabla . \mathbf{v} = 0,$	$\nabla . \mathbf{H} = 0,$
$\nabla \times \mathbf{v} = \Omega$	$\nabla \times \mathbf{H} = \frac{4\pi}{c} \mathbf{j},$
$(\Omega \times \mathbf{v}) = \nabla \left(\frac{p}{\rho} + \frac{v^2}{2} \right)$	$\frac{1}{c} (\mathbf{j} \times \mathbf{B}) = \nabla p$

Though a vortex sheet should be analogous to a current sheet and a current ring in a perfect conductor to a vortex ring, and though, in steady flow of a fluid its vorticity corresponds to the magnetic field, conclusions on the stability of one system cannot, however, be transferred to their analogue in the other system, because, for stability study, one must consider unsteady flows in which time derivatives of field variables cannot be ignored. This correspondence, therefore, cannot be extended to unsteady flows. For example, the pinch effect (section 5.4), which is a plasma string with a current, is unstable though a single vortex filament in a fluid is a stable system.

(a) Decay time estimation

Since l is the characteristic length of variation, equation (4.6.5) can be written as $\partial \mathbf{B}/\partial t = -\eta \mathbf{B}/l^2$. The solution being $\mathbf{B} = \mathbf{B}_0 \exp(-\eta t/l^2)$, the decay time T is equal to l^2/η $(= 4\pi\mu\sigma l^2/c^2)$.

For mercury since $\sigma = 9.4 \times 10^{15}$ sec, $\rho = 13.5$ gm/cm^3, $\nu_0 = \nu_1/\rho = 1.2 \times 10^{-3}$ cm^2/sec, $\mu = 1$, we find that

$$T = 1.3 \times 10^{-4} \times l^2 \text{ sec},\ \nu_0/\eta = \nu/\eta\rho = 1.5 \times 10^{-7},\ R_m = 10^{-4} \times l \times v.$$

For copper $\sigma = 4.5 \times 10^{17}$/sec, so $1/\eta = 0.36$ sec cm^2. Taking, for the core of earth, $\sigma = 10^{15}$/sec, and radius = 5000 km, we find that $T = 2.5 \times 10^{12}$ sec $\approx 10^5$ years. This decay time is too short to conclude that the earth had initially, during its formation, a much stronger magnetic field which decayed with age. In laboratory plasmas R_m is small because l is of the order of few centimeters and v is also of the order of few cm/sec. So, there will be considerable diffusion and insignificant transport of magnetic field. But, in astrophysical conditions, and within the core of the earth, l and v being large, the situation will be just the reverse.

(b) Alfven Theorem for the Frozen-in Effect

Since the magnetic induction field satisfies (4.6.11), and since

$$4\pi\mu \nabla \times \mathbf{j} = (\nabla \times (\nabla \times \mathbf{B}) = -c\nabla^2 \mathbf{B},$$

we write (A 8.5) as

$$\frac{dF}{dt} = \eta \iint_\Sigma (\nabla^2 \mathbf{B} \,.\, d\Sigma) = -\frac{4\pi\mu\eta}{c} \iint_\Sigma (\nabla \times \mathbf{j}) \,.\, d\Sigma,\ \mathbf{F} = \iint_\Sigma (\mathbf{B} \,.\, d\Sigma) \qquad (4.6.6)$$

With the help of Stoke's theorem, this equation for flux conservation becomes

$$\frac{dF}{dt} = -\frac{c}{\sigma} \oint (\mathbf{j} \,.\, d\mathbf{l}) \qquad (4.6.7)$$

It shows that the rate of change of the flux of **B** lines through a surface Σ is determined by the circulation of the conduction current **j** through a closed curve Γ on Σ and moving with the fluid. For infinite conductivity the flux of **B** lines through Σ remains unchanged as Σ moves with the streamlines. So, this flux is attached to the particles and is carried away by the streamlines. In other words, the magnetic lines are frozen with the fluid. The result (4.6.7) is called Alfven's theorem (section 1.8).

Since fluid particles move with their streams, streamlines of all particles lying on a material surface at an instant will remain on it afterwards also. Similarly, a magnetic surface is not penetrated by the magnetic lines of force, and these lines are embedded in such surfaces. if the magnetic diffusivity is negligible, then flux of **B** lines through a magnetic surface is possible; and if viscosity is negligible, there will be no flux of particles through a material surface. The magnetic field of space plasma may slow down the rotation of a star due to the interaction of its magnetic field with the surrounding space plasma.

For $\sigma = \infty$ equation (4.6.7) is equivalent to

$$\mathbf{B}\, d\Sigma = \text{constant} \tag{4.6.8}$$

where $d\Sigma$ is the cross-section of a tube of force. Also the mass element bounded by two normal cross-sections formed by fluid particles is $\rho l\, d\Sigma = \text{constant}$ where l is the length of the small cylindrical element. For incompressible fluids ρ is constant; so

$$l\, d\Sigma = \text{constant} \tag{4.6.9}$$

Equations (4.6.8) and (4.6.9) give

$$B/l = \text{constant} \tag{4.6.10}$$

So the magnetic field lines are lengthened by fluid motion, or, in other words, the magnetic field intensity increases directly with increase in length of the streamlines. Since (4.6.7) is valid for the sum $\mathbf{v}_1 + \mathbf{v}_2$, where $\mathbf{v}_1$ is any velocity satisfying the condition curl $(\mathbf{v} \times \mathbf{B}) = 0$, it is evident that more than one type of motion of particles can be attached to the same field lines.

4.7 PLASMAS OF EARTH AND OTHER SPACE BODIES

(a) Space Plasma

Transformation relations for *EM* fields,

$$\mathbf{E}' = \mathbf{E} + \frac{1}{c}(\mathbf{v} \times \mathbf{B}), \ \mathbf{B}' = \mathbf{B},$$

show that strong electric fields are produced by motion across strong magnetic fields, where (**E**, **B**) are the fields in the laboratory frame (*LF*) and (**E**′, **B**′) are the fields in the frame moving with relative uniform velocity **v**. These secondary electric fields are much weaker than the magnetic fields. The magnetic fields are independent of choice of coordinate systems in relative inertial motion. But these electric fields depend on choice of coordinate systems in which they are measured. A 10^{10} gauss cosmic magnetic field in a medium drifting at the rate 3×10^5 cm/sec, generates 10^5 esu of electric field.

All celestial bodies, on account of their rotation, are electrically polarized when they are observed from a system at rest, though these are electrically neutral or nearly so on astronomical scales. On the atomic scale the Coulomb attraction and repulsion far exceed gravitational attraction. So, it is the other way round, and the *EM* theory and that of gravitation can be studied separately. On large spatial scales the universe is homogeneous, and isotropic. Cosmic (or global) magnetic field, violates the isotropy, but not the homogeneity. In astrophysics non-sustained discharges carry large currents because external ionizers are of high temperature. These ionize matter rather completely due to the enormous currents generated. In presence of such high currents the plasma is pinched by *EM* forces. The pinch constriction generates heat, and *EM* radiation which gives information on plasmas of chromospheric flares from giant explosions on the sun's surface affecting processes in space close to earth, in super novae and in quasars havings masses 10^8 times the solar mass. The laboratory analogue of cosmic space is not vacuum in a tank, but a highly ionized and dense gas and large magnetic fields.

(b) Solar Plasma

The sun is a sphere of radius 7×10^{10} cm, mass 10^{33} gms, mass density 142 gm/cm^3 and surface gravitational acceleration 2.73×10^4 cm/sec^2. The period of its rotation is 2.2×10^6/sec. Gravitational pressure at the centre is greater than 10^{11} atmosphere and temperature is 1 to 2 KeV. Probably hydrogen exists there at more than one thousand times its liquid density. Diameter of the solar core is about one-third of solar diameter. Temperature is 5500°C on the solar surface and 1.4×10^7 °C in the solar core.

From the innermost core to its surface the Sun is in a plasma state. At its surface the magnetic field is 1 to 2 gauss and the sound speed is 10^3 cm/sec. The photosphere is a layer of thickness 1000 km on the radius 7×10^5 km, at the temperature 6000°K and mass density 10^{-7} gm/cm^3. The Alfven wave phase velocity there is about 10^3 cm/sec. In the deep interior thermo nuclear fusion reaction is possible because the temperature is about 16 million degrees °K. The solar energy is mainly radiated from the photosphere. The solar wind is the flux of ions at a steady rate from the solar corona which is a low density high temperature (= 10^6 °K) region in solar surface. It radiates *UV* waves.

Thermo nuclear fusion reaction involving hydrogen ions is the source of stellar radiation. But this reaction is less efficient than the reaction with $D\_$ and

T_- ions. Since the reaction volume in stars is very large, the temperature is of the order often million degrees. This temperature is considerably lower than that of laboratory thermonuclear plasma for a more favourable reaction.

Solar magnetic field is about 25 gauss from dipole moment of strength 4.2×10^{33} gauss cm^2. Solar magnetic field decreases rapidly towards the solar corona from the photophere. Sunspots are MHD phenomena due to magnetic field of 4000 gauss. The greater part of the solar energy exists in the range of radiation wave length from 200 to 1200 nanometers (= 200×10 A to 1200×10 A). The spectral density 12 W/m^{-2}/nm, and the total energy flux is about 1.2×10^3 W/m^2. The pressure of radiation for Sun is 4.5×10^{-5} dynes/cm^2. For the whole earth the force of this pressure is 5.7×10^7 kgm wt and the gravitational attraction is 3.6×10^{21} kgm wt per unit volume. Sun loses its gravitational mass at the rate of 2.5×10^9 kgm per minute or 1.21×10^{15} kgm per year. This number, though very large, numerically, is small compared with the total mass of the Sun. If the sun radiates the same constant energy, it will take 11 million years for its annihilation. The solar energy is produced in the core of sun. Solar energy reaching the entire surface of the earth is about one by a billion of the energy radiated by Sun.

Solar core does not collapse inspite of the huge pressure exerted by a billion tons of over lying matter on every square meter, because the core temperature is very high compared to the surface temperature. Hence particles in the core collide at great speed, knocking out orbital electrons and creating favourable conditions for fusion, overcoming the force of mutual electrostatic repulsion.

The solar plasma consists mainly of free electrons and ions of hydrogen nuclei. Usually four protons fuse together in such fusion reactions creating the α particle which is a helium nucleus. Though probability of protons fusing together in Sun is very small, the number of fusion reactions per second is sufficient to keep the sun burning.

The two processes causing fusion reaction are (1) Proton-Proton cycle; (2) Carbon cycle. In (1) two protons fuse together, and the intermediate products are deuterium nuclei (deuterons) and helium 3 nuclei. In (2) the carbon acts as a catalyst. In both (1) and (2) a considerable amount of energy is liberated, because the mass of an alpha particle is about 99.3 percent of the sum total of the masses of four protons, the missing mass is converted into energy according to the Einstein formula $\xi = mc^2$. In sun thus 657 million tons of hydrogen are converted to 652.5 million tons of helium every second. The missing 4.5 million tonnes of mass are transformed into energy which is an enormous amount of energy.

(c) The earth and its Plasma

The earth is approximately a sphere of radius 6380 km, mass 6×10^{24} kgm;

mean mass density is 5.5 gm/cm^3, mean surface gravitational attraction is 981 cm/sec^2, mean distance from Sun is 150×10^6 km, period of rotation about Sun is one year, escape velocity is 11.2 km/sec, second of rotation about the north-south axis is 23.93 hours and orbital eccentricity is 0.017. It has solid crust of thickness 35 km, a mantle of thickness 2900 km and a core of radius 3500 km. The velocity of earth round the sun is 30 km/sec. The radiation pressure of sun light is 0.45×10^{-4} dynes/cm^2. The force of this pressure on the whole earth is 5.7×10^7 kg wt. The force of gravitational attraction of Sun on earth is 3.6×10^{21} kg wt/cm^3. The force of the radiation pressure on earth is therefore small compared to the force of gravitation[1].

The inner core, upto radius 1200 km, consists of a solid alloy of iron and nickel at 2 to 3 million atmosphere, and temperature 5000°k. The electrical conductivity of the core and the mantle are 3×10^{15}/sec and 3×10^{11}/sec. An intermediate liquid core, between 1200 and 3500 km, is mainly of molten iron, and some amount of sulpher and silicon. The other solid mantle from the radial distance 3500 km upto the outer surface consists of solid ferromagnesium silicate. The magnetic field in the inner core is 10 gauss, at the core surface 4 gauss, on the earth surface at equator it is 0.3 gauss, at the two poles it is 0.6 gauss and at the other surface of earth it is 0.5 gauss.

The distribution of the magnetic field of earth is that of a dipole moment of strength 8.1×10^{25} gauss cm^3, situated at the centre of earth when measured at distance greater than 10^5 km. The axis of this magnetic moment intersects at the earth's surface in antipodel points situated at the latitude 78.5°5, longitude 111°E and at the latitude 78.5°N, longitude 69°W.

The plasma ejected from the solar surface is captured by the magnetic field to generate some radiation belts around the earth. The farthest such belt is situated between 20,000 km to 30,000 km, and the nearest one is between 70 km and few thousand km. In the inner belt, the radiation intensity is about 10^4 times greater than due to cosmic radiation; and the ion energies are about 0.1 MeV to a few MeV, and the electron energies are about 40 KeV.

(d) Earth's rotation and precession

The acceleration by Sun's gravitational field of the orbit of the earth is about 0.5 cm/sec^2 and equals the centrifugal acceleration due to earth's orbital motion. The effect of moon on the acceleration of earth is about 4×10^{-3} cm/sec^3. The angular velocity of a point in the earth about an axis pointed towards the polaris is $(2\pi/24 \times 60/64)$ rad/sec. An object at rest on the surface of earth is acted

* The radiation pressure is a surface force and gravity is a volume force. The ratio of the two forces becomes larger when the volume density is smaller. The deflection of a comet, when it comes near the sun, can be explained thus.

upon by centrifugal force from the diurnal rotation of the earth about its axis, the annual rotation of earth about Sun and the monthly rotation of the centre of earth about the centre of gravity of the earth-moon system. In the liquid metallic core of the rotating earth the torque of coriolis force can balance the torque of the Lorentz force, thereby generating slowly oscillating hydromagnetic waves.

The angular velocity Ω of the solid mantle of earth undergoes slow changes, both in direction and magnitude. The axis of rotation has a slow precession Q_p, about the normal to the plane of rotation about the sun, $\Omega_p \approx 7.7 \times 10^{-12}$ rad/sec; and the time period is 2.6×10^4 years. The corresponding cone has the semi vertical angle 23.5°. This precession is due to the net torque exerted by Sun and moon on the equitorial protrusion of earth, and provides a non-rigid rotation of the intermediate liquid core. Since the ellipticities of the two regions differ, the mean density of the core is much greater than that of the mantle. So, the core and the mantle can have very different directions of precessional rotation after several thousand years, due to the uncoupling of their motion, and large relative velocities between the core and the mantle are possible. But the core and the mantle are strongly coupled together by viscous and electromagnetic means, keeping the precessional velocities of the two regions same, both in direction and magnitude. This precessional torque generates core motion which generates the magnetic field of earth. Other sources of magnetization are the convection due to buoyancy forces of thermal or sedimentary origin and turbulent flow in the core of earth due to instability of the boundary layer. This angular momentum is transferred from mantle to core generating secondary flows.

4.8 FERRARO'S LAW OF ISOROTATION

Stellar bodies rotate, have symmetry about the axis of rotation, and are magnetized. The **B** lines are frozen-in with the streamlines. For magnetized rotating bodies the surfaces of constant magnetic field value are also surfaces of constant density and vice-versa. In other words, material layers and matter all along a field line also rotate with the same angular velocity, at every point of a magnetic line of force. This is Ferraro's law of isorotation.

For rotation about the z-axis, all the layers at different distances (axial) will have different angular velocity ω, so this is non-rigid rotation (also called differential rotation or non-uniform rotation (section 13.14)), because ω, then, is a function of r and Z. Velocity components $(0, r\omega\,(r, Z), 0)$ at the point (r, θ, Z) (in cylindrical polar coordinates) are independent of the polar angle θ in the steady state because both the motion and the **B** lines are symmetrical about the z-axis. Then, in infinitely conducting materials, the MHD equation (4.6.1) gives

$$\nabla \times (\mathbf{v} \times \mathbf{B}) = 0, \tag{4.8.1}$$

Since $\quad (\mathbf{v} \times \mathbf{B}) = (r\omega\, B_z,\ 0,\ -r\,\omega\, B_r),$

$$\text{curl}\,(\mathbf{v} \times \mathbf{B}) = \hat{\mathbf{X}}_\theta \left[\frac{\partial}{\partial z}(r\omega\, B_z) + \frac{\partial}{\partial r}(r\omega\, B_r) \right] \tag{4.8.2}$$

where $\hat{\mathbf{X}}_\theta$ is the unit vector parallel to the direction of θ increasing at (r, θ, z), Using the vector formula (A2.1), we find that

$$\nabla \times (\mathbf{v} \times \mathbf{B}) = \hat{\mathbf{X}}_\theta \left[r\,\omega \left\{ \frac{1}{r}(r\,B_r) + \frac{\partial}{\partial z} B_z \right\} + r\left(B_r \frac{\partial \omega}{\partial r} + B_z \frac{\partial \omega}{\partial z} \right) \right]$$

$$= \hat{\mathbf{X}}_\theta \left[r\,\omega\,(\nabla \,.\, \mathbf{B}) + r\,(\mathbf{B} \,.\, \nabla)\,\omega \right] = 0 \tag{4.8.3}$$

Hence, $(\mathbf{B} \,.\, \nabla)\,\omega = 0,$ (4.8.4)

ω is constant along a line of force of **B** and so on a surface of constant **B**. The period of rotation at the equator of a magnetized revolving star is different from that of other latitudes. This theorem is applicable only for electrically highly conducting materials, like dense cosmic bodies in which the magnetic lines of force are frozen properly.

All points of a surface traced out by a family of magnetic lines of force having the same value of B must rotate with the same average angular velocity. Otherwise, strong electric currents of destabilization would be induced. Moreover, if the angular velocity of a certain part of a magnetic line of force differs from that of other parts, the line becomes twisted. A twisted line of force tries to straighten itself by sending out an Alfven wave. Although, during a short time, different parts of a line of force may move with different velocities, for propagation of the Alfven waves, and to eliminate strong induced electric currents of destabilization, the average over a sufficiently long time must have the same angular velocity for all parts of it. The time necessary for establishing the same average angular velocity of all parts of a line of force is given by the time of travel of the Alfven waves.

4.9 LINEARIZATION OF MHD EQUATIONS

In the uniform unperturbed state the fluid is assumed at rest at pressure P and mass density ρ; and embedded in the homogeneous magnetic field $\mathbf{B}_0$. A small amplitude perturbation introduces oscillatory motion into the medium; the field variables then can be written as $\mathbf{v}$, $\mathbf{E}$, $\mathbf{B}_0 + \mathbf{B}$, $\mathbf{j}$, $\rho_0 + \rho$, $p_0 + p$. Up to the first order of perturbed quantities, the field equations are

$$\rho_0 \frac{\partial \mathbf{v}}{\partial t} = -\nabla p + \frac{1}{c}(\mathbf{j} \times \mathbf{B}_0) + \nu \nabla^2 \mathbf{v}, \quad \frac{\partial \rho}{\partial t} + \rho_0 \,\mathrm{div}\, \mathbf{v} = 0,$$

$$\nabla \times \mathbf{H} = \frac{4\pi}{c}\mathbf{j}, \quad \nabla \times \mathbf{E} = -\frac{1}{c}\frac{\partial \mathbf{B}}{\partial t}, \quad (\nabla \,.\, \mathbf{B}) = 0 \tag{4.9.1}$$

$$\nabla p = C_S^2 \nabla \rho, \quad \mathbf{j} = \sigma \left\{ \mathbf{E} + \frac{1}{c}(\mathbf{v} \times \mathbf{B}) \right\}$$

where C_S is the speed of propagation of sound in the medium. In an infinitely conducting medium, Ohm's law reduces to

$$\mathbf{E} + \frac{1}{c}(\mathbf{v} \times \mathbf{B}) = 0 \tag{4.9.2}$$

If the perturbations in pressure and density over their equilibrium values p_0 and ρ_0 are denoted by δp and $\delta \rho$, the adiabatic law, the sixth equation of (4.9.1), is

$$\delta p = C_S^2 \, \delta \rho; \tag{4.9.3}$$

but if these two quantities are denoted by p and ρ, then obviously this law reads

$$p = C_S^2 \, \rho \tag{4.9.4}$$

4.10 MHD WAVES IN INCOMPRESSIBLE MEDIA

These are some low frequency modes, longitudinal or transverse or both, mutually coupled, of the MHD approximation. Effects of polarization of fields of these are not important. In some cases of small amplitude wave motion in infinitely conducting and non-viscous media, the effects of compressibility can be ignored. Slightly more general are the waves in viscous and finitely conducting media.

(a) Waves in infinitely conducting and non-viscous media

In this case the first equation of (4.9.1) becomes

$$\rho_0 \frac{\partial^2 \mathbf{v}}{\partial t^2} = -\nabla \left(\frac{\partial p}{\partial t} \right) + \frac{1}{4\pi} \left(\nabla \times \frac{\partial \mathbf{H}}{\partial t} \right) \times \mathbf{B}_0 \tag{4.10.1}$$

Using (4.6.1) and putting $\sigma = \infty$, the second term in the right hand side becomes $(1/4\pi\mu)\,[\nabla \times \{\nabla \times (\mathbf{v} \times \mathbf{B}_0)\} \times \mathbf{B}_0]$; with the help of the vector identity (A3.9) it can be written as

$$\frac{1}{4\pi\mu} [\{\nabla \times (\mathbf{B}_0 \,.\, \nabla)\, \mathbf{v}\} \times \mathbf{B}_0] \tag{4.10.2}$$

Then, taking curl of (4.10.1), we obtain

$$\ddot{\mathbf{\Omega}} = \frac{1}{4\pi\mu\,\rho_0} [\nabla \times (\{\nabla \times (\mathbf{B}_0 \,.\, \nabla)\, \mathbf{v}\} \times \mathbf{B}_0)]$$

$$= \frac{1}{4\pi\mu\,\rho_0} (\mathbf{B}_0 \,.\, \nabla)\, \nabla \times (\mathbf{B}_0 \,.\, \nabla)\, \mathbf{v} = \frac{1}{4\pi\,\mu\,\rho_0} (\mathbf{B}_0 \,.\, \nabla)^2\, \mathbf{\Omega} \tag{4.10.3}$$

where curl $\mathbf{v} = \mathbf{\Omega}$ and the operators curl and $(\mathbf{B}_0 \,.\, \nabla)$ have been interchanged. Assuming in (4.10.3) the wave solution $\exp\,[i\,(\mathbf{K} \,.\, \mathbf{r}) - i\omega t]$ for $\mathbf{\Omega}$ we find the Alfven wave dispersion relation (section 1.8 (*b*)):

$$\omega^2 = (\mathbf{K} \,.\, \mathbf{C}_A)^2, \quad \mathbf{C}_A = \frac{\mathbf{B}}{\sqrt{4\pi\,\rho_0\,\mu}} \tag{4.10.4}$$

The phase speed $(\mathbf{K} \,.\, \mathbf{C}_A)/|K|$ depends on the angle between $\mathbf{K}$ and $\mathbf{B}_0$. Since $\nabla \,.\, \mathbf{B} = 0$, the wave is transverse. This type of waves, propagating along a magnetic field line (which acts like a stretched string, loaded by the plasma mass per unit length, attached to the field line), does not occur in hydrodynamics, because no elastic force can counteract the sliding of the neighbouring layers of non-conducting fluids with respect to each other. In MHD, this sliding stretches the magnetic lines of force; since these lines are attached to the fluid elements, the stretching is counteracted by the tension of the magnetic field $\mathbf{B}_0$.

Like optical waves at the interface of two different dielectics, Alfven waves are reflected and transmitted at the interface of two conducting fluids of different mass densities. Alfven waves transfer energy to hot plasma because these are (1) of low frequency, and (2) are strongly coupled to the plasma. These are absorbed over a wide range of frequencies, in theta-pinch plasma and in toroidal geometry of several stellarator type devices. Launching structures for heating by Alfven waves are often large since in these it is necesary to generate waves whose lengths are comparable to the dimensions of the expeirment. Alfven wave frequencies are quite high in strongly magnetized plasmas.

In mercury $\rho = 10^{-7}$ gm/cm^3; so $C_A = B_0/13$. In Sun's photosphere $\rho = 10^{-7}$ gm/cm^3 and B_0 is about 1 or 2 Gauss, so $C_A \approx 10^3$ to 2×10^3 cm/sec; this speed is much less than the sound speed there, about 10^6 cm/sec. Frequency range of Alfven waves is 100 KH_z to 10 MH_z.

(b) Waves in viscous and finitely conducting media

In viscous, finitely conducting and incompressible media the guiding equations are

$$\rho_0 \frac{\partial \mathbf{v}}{\partial t} = -\nabla p + \frac{1}{4\pi\mu}((\nabla \times \mathbf{B}) \times \mathbf{B}_0) + \nu \nabla^2 \mathbf{v}, \tag{4.10.5}$$

$$\frac{\partial \mathbf{B}}{\partial t} = \nabla \times (\mathbf{v} \times \mathbf{B}_0) + \frac{c^2}{4\pi\mu\sigma}\nabla^2 \mathbf{B} \tag{4.10.6}$$

Since $\partial/\partial t = -i\omega$ and $\nabla = i\mathbf{k}$, these equations can be written as

$$\rho_0 \frac{\partial \mathbf{v}}{\partial t} = \frac{-\nabla p + ((\nabla \times \mathbf{B}) \times \mathbf{B}_0)/4\pi\mu}{1 + iK^2 \nu/(\rho_0 \omega)}, \quad \frac{\partial \mathbf{B}}{\partial t} = \frac{\nabla \times (\mathbf{v} \times \mathbf{B}_0)}{1 + iK^2 c^2/(4\pi\mu\sigma\omega)} \tag{4.10.7}$$

When treated in a manner similar to the first case they reduce to an equation in $\mathbf{\Omega}\,(= \nabla \times \mathbf{v})$, the left side of which is identical to that of (4.10.3) and the right side is $1/[(1 + iK^2 \nu/\rho_0 \omega)\,\{1 + (ik^2 c^2/4\pi\mu\sigma\omega)\}]$ times that of (4.10.3). So the dispersion relation is

$$(\mathbf{K} \,.\, \mathbf{C}_A)^2 = \omega^2 \left(1 + \frac{iK^2 c^2}{4\pi\mu\omega\sigma}\right)\left(1 + \frac{i\nu K^2}{\rho_0 \omega}\right) \tag{4.10.8}$$

After taking the square root of this equation we expand the right hand side, then, retaining terms only up to the first power in ν and $1/\sigma$, we get

$$\pm \frac{(\mathbf{K} \cdot \mathbf{C}_A)}{\omega} = 1 + \frac{iK^2}{2\,\omega}\left(\frac{C^2}{4\pi\,\mu\,\sigma} + \frac{\nu}{\rho_0}\right) \tag{4.10.9}$$

Hence the wave is damped both by viscosity and finite electrical conductivity.

Exercise 4.10.1. If the displacement current $\partial\,\mathbf{D}/\partial t$ is also included, find the dispersion relation. In that case, if $\nu = 0$, $\sigma = \infty$ and $\mathbf{K}$ is parallel to $\mathbf{B}_0$, show that $\omega/K = CC_A/\sqrt{C^2 + C_A^2}$.

4.11 MAGNETOSONIC WAVES

In compressible fluids (or compressional Alfven waves) density varies. This variation is related to pressure variation. Propagation of disturbance through inviscid, infinitely conducting compressible plasma in presence of homogeneous magnetic, field will be studied in this section.

An energy principle exists for investigation of instabilities in plasma. It depends on a Lagrangian displacement of a fluid element to find a small deviation from the equilibrium state. The method used for wave solutions in chapter 3 can also be adopted here.

The field variables are functions of the initial location $\mathbf{r}$ of a fluid element and of time t. If $\mathbf{R}$ is its displaced position at time t, then $\mathbf{R}(t) = \mathbf{r}(t) + \boldsymbol{\xi}(\mathbf{r}, t)$ with $|\mathbf{r}| >> |\boldsymbol{\xi}|$. The displacement $\boldsymbol{\xi}$ vanishes initially. So $\boldsymbol{\xi}(\mathbf{r}, 0) = 0$. Also $\boldsymbol{\xi}(0,\ t) = 0$. The macroscopic (average) fluid velocity in terms of $\boldsymbol{\xi}$ is given by

$$\mathbf{v}(\mathbf{R},\ t) = \frac{d\,\mathbf{R}}{dt} = \mathbf{v}(\mathbf{r} + \boldsymbol{\xi},\ t)$$

$$= \mathbf{v}(\mathbf{r},\ t) + (\boldsymbol{\xi} \cdot \nabla)\,\mathbf{v}'(\mathbf{r},\ t) + \frac{1}{2}(\boldsymbol{\xi} \cdot \nabla)^2\,\mathbf{v} + \ldots$$

Since the displacement amplitude is small we neglect $(\boldsymbol{\xi} \cdot \nabla)\,\mathbf{v}$ and still higher order terms, and find that $\mathbf{v}(\mathbf{R},\ t) \equiv \mathbf{v}(\mathbf{r},\ t)$. To the lowest order $\boldsymbol{\xi}$ is regarded as given at a point, but to higher orders, it is left undefined through the relation

$$\mathbf{v}(\mathbf{r},\ t) = \frac{\partial \boldsymbol{\xi}(\mathbf{r},\ t)}{\partial t} \tag{4.11.1}$$

This Lagrangian description gives the time history of the velocity through the derivative $d\boldsymbol{\xi}/dt$ of the displacement; the other (Eulerian) description gives the time history of the velocity at a point as fluid elements pass through it. Relation (4.11.1) shows that, up to the first order of small quantities, the two descriptions are identical, and that the Eulerian velocity $\mathbf{v}(\mathbf{R},\ t)$ can be replaced by the derivative of the Lagrangian displacement $\boldsymbol{\xi}(\mathbf{r},\ t)$.

Differential operators satisfy the relations

$$\frac{\partial}{\partial \mathbf{R}} = \frac{\partial \mathbf{r}}{\partial \mathbf{R}} \cdot \frac{\partial}{\partial \mathbf{r}} = \left(\frac{\partial}{\partial \mathbf{R}}(\mathbf{R} - \boldsymbol{\xi}) \cdot \frac{\partial}{\partial \mathbf{r}}\right) = \frac{\partial}{\partial \mathbf{r}} - \frac{\partial \boldsymbol{\xi}}{\partial \mathbf{R}} \cdot \frac{\partial}{\partial \mathbf{r}},$$

and
$$\frac{d\xi}{dt}=\frac{\partial \xi}{\partial t}+\left(\frac{\partial \xi}{\partial \mathbf{r}}\cdot\frac{\partial \mathbf{r}}{\partial t}\right)$$

We now set $\nu = 0$ and $\sigma = \infty$ in the system of equations (4.9.1); then putting the value of **v** from (4.11.1) in the second equation of (4.9.1), we find that

$$\rho = -\rho_0 \operatorname{div} \xi \tag{4.11.2}$$

The magnetic induction **B** is evaluated by first eliminating **E** between $\nabla \times \mathbf{E} = -\partial \mathbf{B}/c\,\partial t$ and (4.9.2), and then, by integrating the result with respect to time,

$$\mathbf{B} = \nabla \times (\xi \times \mathbf{B}_0) = (\mathbf{B}_0 \,.\, \nabla)\,\xi - \mathbf{B}_0\,(\nabla \,.\, \xi) \tag{4.11.3}$$

Hence

$$\frac{4\pi\mu}{c}\,\mathbf{j} = \nabla \times \nabla \times (\xi \times \mathbf{B}_0) = \nabla \times (\mathbf{B}_0 \,.\, \nabla)\,\xi - \nabla\,(\nabla \,.\, (\xi \times \mathbf{B})) \tag{4.11.4}$$

$$p = -\rho_0\, C_S^2\,(\nabla \,.\, \xi) \tag{4.11.5}$$

where (4.9.4) has been used. Then the momentum transfer equation of (4.9.1) becomes

$$\ddot{\xi} = c^2\,\nabla\,(\nabla \,.\, \xi) + \frac{1}{4\pi\,\rho_0\,\mu}\,(\{\nabla \times (\mathbf{B}_0 \,.\, \nabla)\,\xi + (\mathbf{B}_0 \times \nabla\,(\nabla \,.\, \xi))\} \times \mathbf{B}_0) \tag{4.11.6}$$

Expanding now the vector triple product in the last term in the right hand side, dividing both sides by C_A^2 and putting the value of C_A from (4.10.4), we obtain

$$\frac{\ddot{\xi}}{C_A^2} - (1+\alpha)\,\nabla\,(\nabla \,.\, \xi) + (\mathbf{h} \,.\, \nabla)\,(\nabla \,.\, \xi)\,\mathbf{h} - (\{\nabla \times (\mathbf{h} \,.\, \nabla)\,\xi\} \times \mathbf{h}) \tag{4.11.7}$$

where $\mathbf{h} = \mathbf{B}_0/|\,B_0\,|$; **h** is the unit vector along $\mathbf{B}_0$ and

$$\alpha = C_S^2/C_A^2 \tag{4.11.8}$$

Assuming ξ to be proportional to $\exp\,[i\,(\mathbf{K} \,.\, \mathbf{r}) - i\omega t]$ and setting $V = \omega/K$, $\mathbf{K}_0 = \mathbf{K}/|\,K\,|$, we find the equatin for ξ.

$$\left[\frac{V^2}{C_A^2} - (\mathbf{K}_0 \,.\, \mathbf{h})^2\right]\xi + [(\mathbf{K}_0 \,.\, \mathbf{h})\ (\mathbf{h} \,.\, \xi) - (1+\alpha)\ (\mathbf{K}_0 \,.\, \xi)]\,\mathbf{K}_0 + (\mathbf{K}_0 \,.\, \mathbf{h})\ (\mathbf{K}_0 \,.\, \xi)\,\mathbf{h} = 0 \tag{4.11.9}$$

This is a linear equation in ξ of the type $a\,\xi + b\mathbf{h} + c\mathbf{K}_0 = 0$ with the component ξ_2 perpendicular to the plane of **h** and $\mathbf{K}_0$, ξ_1 is perpendicular to **h** in this plane and ξ_3 is parallel to **h**. When $\mathbf{K}_0$ is parallel to **h** we obtain

$$\left(\frac{V^2}{C_A^2}-\alpha\right)\xi_3=0,\ \left(\frac{V^2}{C_A^2}-1\right)(\xi_1,\ \xi_2)=0 \tag{4.11.10}$$

Since, in general, $\xi \neq 0$, we find that $V^2 = C_S^2$ for ξ_3 and $V^2 = C_A^2$ for ξ_1 and ξ_2. The displacement ξ_3 is not connected to ξ_1 or ξ_2, so the change in fluid pressure generating the sound speed C_S does not couple to the change in magnetic pressure which is responsible for the generation of the Alfven wave. When $\mathbf{K}_0$ is not parallel to $\mathbf{h}$, we take the scalar product of (4.11.9) with $\mathbf{K}_0$ and $\mathbf{h}$; and obtain

$$\left[\frac{V^2}{C_A^2}-(1+\alpha)\right](\mathbf{K}_0\,.\,\xi)=-(\mathbf{K}_0\,.\,\mathbf{h})\ (\mathbf{h}\,.\,\xi), \tag{4.11.11}$$

$$(V^2/C_A^2)\ (\mathbf{h}\,.\,\xi)=\alpha\,(\mathbf{K}_0\,.\,\mathbf{h})\ (\mathbf{K}_0\,.\,\xi) \tag{4.11.12}$$

Eliminating now $(\mathbf{K}_0\,.\,\xi)$ and $(\mathbf{h}\,.\,\xi)$ from these two equations we get

$$V^2-(C_A^2+C_S^2)\,V^2-C_A^2\,C_S^2\cos^2\theta=0 \tag{4.11.13}$$

where θ is the angle between $\mathbf{K}_0$ and $\mathbf{h}$. This is the dispersion relation for the waves of ξ_1 and ξ_2. Since $(\mathbf{K}\,.\,\xi)=K_1\,\xi_1+K_3\,\xi_3$ and $(\mathbf{K}\,.\,\mathbf{h})=K_3$, $(\mathbf{h}\,.\,\xi)=\xi_3$, equation (4.11.11) yields

$$\frac{K_1\,\xi_1}{K_3\,\xi_3}=-\frac{(V^2-C_S^2)}{V^2-C_A^2-C_S^2} \tag{4.11.14}$$

Hence ξ_1 and ξ_3 are connected, if $\theta \neq 0$. Equation (4.11.13) gives

$$V_\pm^2=\frac{1}{2}\ [C_A^2+C_S^2\pm\sqrt{(C_A^2+C_S^2)^2-4C_A^2\,C_S^2\cos^2\theta}\,] \tag{4.11.15}$$

The quantity under the root sign is always positive; its maximum and minimum values are $(C_A^2+C_S^2)^2$ and $(C_A^2-C_S^2)^2$. Hence V^2 lies between $C_A^2+C_S^2$ and zero. These non-decaying waves are called magnetosonic or magneto acoustic waves, because the displacement vector has both a longitudinal component ξ_3 and a transverse component ξ_1 with respect to the static magnetic field $\mathbf{H}_0$. Since $V_+^2 > V_-^2$, the wave with phase velocity V_+ is called the fast wave and the V_- wave is called the slow wave. For $\mathbf{B}_0 = 0$ these coincide and become a wave with phase velocity C_S. When $\theta = \pi/2$, we find that $V_- = \theta$, $V_+ = \sqrt{C_A^2 + C_S^2}$. The phase velocity is therefore increased by the dc magnetic field.

To find ξ_2 we take the scalar product of all the terms of (4.11.9) with the unit vector $\hat{\mathbf{X}}_2$ (perpendicular to the plane of $\mathbf{h}$ and $\mathbf{K}$) and get

$$\left\{\frac{V^2}{C_A^2}-(\mathbf{K}_0\,.\,\mathbf{h})^2\right\}\xi_2=0 \tag{4.11.16}$$

Hence ξ_2, growing with the perturbed pressure, propagates at the Alfven speed $C_A \cos\theta$.

(c) Phase relations between the variables

Using for $(\mathbf{h}\,.\,\boldsymbol{\xi})$ the value from (4.11.11) into (4.11.9), and of ρ from (4.11.2) we get

$$\boldsymbol{\xi}=\left[\left(C_A^2+C_S^2-\frac{C_A^2\,C_S^2\cos^2\theta}{V^2}\right)\mathbf{K}_0-\mathbf{h}\,C_A^2\cos\theta\right]\frac{i\rho}{K\rho_0\,(V^2-C_A^2\cos^2\theta)} \tag{4.11.17}$$

Vector product of $\boldsymbol{\xi}$ with $\mathbf{K}_0$ transforms (4.11.17) into

$$(\mathbf{K}\times\boldsymbol{\xi})=\cdot\,\frac{i\rho\,C_A^2\cos\theta\,(\mathbf{K}_0\times\mathbf{h})}{K_0\,(V^2-C_A^2\cos^2\theta)} \tag{4.11.18}$$

So, for the two values of V^2 there will be two vlaues of $\boldsymbol{\xi}$ and ρ, denoted by ξ_+, ξ_-, and ρ_+, ρ_- respectively. For these quantities

$(\mathbf{K}_0\times\boldsymbol{\xi})=i\,\rho_+\,C_A^2\cos\theta\,(\mathbf{K}_0\times\mathbf{h})/K\,\rho_0\,(V_+^2-C_A^2\cos^2\theta)$, etc. Hence

$$(\mathbf{K}_0\times(\boldsymbol{\xi}_++\boldsymbol{\xi}_-))=\frac{iC_A^2}{K\rho_0}-(\mathbf{K}_0\times\mathbf{h})\cos\theta\left[\frac{\rho_+}{V_+^2-C_A^2\cos^2\theta}+\frac{\rho_-}{V_-^2-C_A^2\cos^2\theta}\right] \tag{4.11.19}$$

To find the relation between ρ_+ and ρ_- we take the scalar product of all terms of (4.11.19) with $\boldsymbol{\xi}_++\boldsymbol{\xi}_-$ and get

$$\frac{\rho_-}{\rho_+}=\frac{V_-^2-C_A^2\cos^2\theta}{V_+^2-C_A^2\cos^2\theta} \tag{4.11.20}$$

If δp_B be the perturbation in the magnetic pressure, then

$$8\pi\mu\,\delta p_B=(\mathbf{B}_0+\mathbf{B})^2-\mathbf{B}_0^2.$$

So, up to the first power of $\mathbf{B}$,

$$\delta p_B=\frac{(\mathbf{B}_0\,.\,\mathbf{B})}{4\pi\mu} \tag{4.11.21}$$

Putting, in this result, the value of $\mathbf{B}$ from (4.11.3) we get

$$\delta p_B=\frac{1}{4\pi\,\mu}\,\{(\mathbf{B}_0\,.\,(\mathbf{B}_0\,.\,\nabla)\,\boldsymbol{\xi}-B_0^2\,(\nabla\,.\,\boldsymbol{\xi})\}$$

$$= \frac{iB_0^2}{4\pi\mu} \{(\mathbf{K}_0 \cdot \mathbf{h})(\mathbf{h} \cdot \boldsymbol{\xi}) - (\mathbf{K}_0 \cdot \boldsymbol{\xi})\} \tag{4.11.22}$$

Now, elimination of $(\mathbf{K}_0 \cdot \mathbf{h})$ $(\mathbf{h} \cdot \boldsymbol{\xi})$, with the help of (4.11.11), gives

$$\delta p_B = -\frac{i\,K\,B_0^2}{4\pi\mu}\left(\frac{V^2}{c_A^2} - \alpha\right)(\mathbf{K}_0 \cdot \boldsymbol{\xi}) = (V^2 - C_S^2)\,\rho$$

$$= \frac{\rho}{2}\,[C_A^2 - C_s^2 \pm \sqrt{(C_A^2 + C_S^2)^2 - 4C_A^2\,C_S^2 \cos^2\theta}\,] \tag{4.11.23}$$

Hence $\delta p_B/\rho$ is positive for the fast wave, and negative for the slow wave; for the fast wave δp_B and ρ are in phase, and for the slow wave they are 180° out of phase. Also, the relation (4.11.14) shows that ξ_1 and ξ_2, are in phase for the fast wave and out of phase by 180° for the slow wave.

4.12 DOUBLE-ADIABATIC THEORY

In magnetized media a pressure $\mu H^2/8\pi$ and a tension $\mu H^2/4\pi$ parallel to the field direction act at all points. So the pressure components $p_{\|}$ and $p_{\perp}$ parallel and perpendicular to $\mathbf{B}$ are different. To study the time behaviour in a local cartesian coordinate system, of this pressure anisotropy, we write

$$p_{ij} = p_{\|}\, l_i\, l_j + p_{\perp}\,(\delta_{ij} - l_i\, l_j) \tag{4.12.1}$$

$$\varepsilon = \varepsilon_{\|} + \varepsilon_{+},\ \varepsilon_{\|} = \frac{p_{\|}}{2\rho},\ \ \varepsilon_{\perp} = \frac{p_{\perp}}{\rho} \tag{4.12.2}$$

where l_i, l_j, l_k are the direction cosines of $\mathbf{B}$ with respect to these axes, ρ is the mass density, and ε is the internal energy per unit mass. When collisions are not very frequent, $\varepsilon_{\|}$ and $\varepsilon_{\perp}$ are not interdependent. By the first law of thermodynamics,

$$d\varepsilon = -p\,d\lambda \tag{4.12.3}$$

where $\lambda\ (= 1/\rho)$ is the specific volume (section 4.2 (*e*)). For longitudinal and transverse components of the pressure tensor, this law gives

$$d\,\varepsilon_{\|} = -p_{\|}\,d\lambda_{\|},\ \ d\,\varepsilon_{\perp} = -p_{\perp}\,d\,\lambda_{\perp} \tag{4.12.4}$$

Let l be a length element parallel to $\mathbf{B}$, and let an element of area perpendicular to $\mathbf{B}$ be Σ. Then the change dl due to a small displacement $\boldsymbol{\xi}$ is given by

$$d\,\mathbf{l} = (\hat{\mathbf{X}} \cdot \boldsymbol{\xi}(\mathbf{r} + \mathbf{R})) - (\hat{\mathbf{X}} \cdot \boldsymbol{\xi}(\mathbf{r})) = (\hat{\mathbf{X}} \cdot \nabla)(\boldsymbol{\xi} \cdot \mathbf{R}),$$

$$\mathbf{R} = l\,\hat{\mathbf{X}},\ \ \hat{\mathbf{X}} = \frac{\mathbf{B}}{B} \tag{4.12.5}$$

The continuity condition $(d\rho/dt) + \rho\,(\nabla \cdot \mathbf{v}) = 0$, for total change in volume, gives

$$d\lambda = \lambda (\nabla . \mathbf{v})\, dt \tag{4.12.6}$$

Taking $\boldsymbol{\xi} = \mathbf{v}\, dt$, this change due to the total change $d\lambda_{\|}$ is

$$d\lambda_{\|} = \Sigma\, dt = \Sigma\, (\hat{\mathbf{X}} . \nabla)(\mathbf{v} . \hat{\mathbf{X}})\, l\, dt = \lambda\, (\hat{\mathbf{X}} . \nabla)(\mathbf{v} . \hat{\mathbf{X}})\, dt \tag{4.12.7}$$

Hence $$d\lambda_{\perp} = d\lambda - d\lambda_{\|} = \lambda\, \{(\nabla . \mathbf{v}) - (\hat{\mathbf{X}} . \nabla)(\mathbf{v} . \hat{\mathbf{X}})\}\, dt, \tag{4.12.8}$$

and with the help of (4.12.2) and (4.12.4) we have

$$d\varepsilon_{\|} = \frac{1}{2} d\, (p_{\|} \lambda) = -\, p_{\|}\, d\lambda_{\|} \tag{4.12.9}$$

$$d\varepsilon_{\perp} = l\, (p_{\perp} \lambda) \tag{4.12.10}$$

These relations can be writen as

$$\lambda\, dp_{\|} = -\, p_{\|}\, d\lambda - 2p_{\|}\, d\lambda_{\|} \tag{4.12.11}$$

$$= -\, p_{\|}\, dt\, \lambda\, (\nabla . \mathbf{v}) - 2p_{\|}\, \lambda\, (\hat{\mathbf{X}} . \nabla)\ (\mathbf{v} . \hat{\mathbf{X}})\, dt, \tag{4.12.12}$$

where the values of $d\lambda$ and $d\lambda_{\|}$ are taken from (4.12.6) and (4.12.7) respectively. So the equation for $p_{\|}$ is

$$\frac{d\, p_{\|}}{dt} + p_{\|}\, (\nabla . \mathbf{v}) = -\, 2p_{\|}\, (\hat{\mathbf{X}} . \nabla)\ (\hat{\mathbf{X}} . \mathbf{v}) = -\, 2p_{\|}\, l_i\, l_j \frac{\partial v_i}{\partial x_j} \tag{4.12.13}$$

Similarly proceeding with $p_{\|}$,we have $-\, p_{\perp}\, d\lambda_{\perp} = \lambda\, dp_{\perp} + p_{\perp}\, d\lambda$; so

$$\lambda\, dp_{\perp} = -\, p_{\perp}\, \lambda\, (\nabla . \mathbf{v})\, dt - p_{\perp}\, \{\lambda\, (\nabla . \mathbf{v}) - \lambda\, (\hat{\mathbf{X}} . \nabla)\, (\mathbf{v}\hat{\mathbf{X}})\}\, dt \tag{4.12.14}$$

The differential equation for $p_{\perp}$, therefore, is

$$\frac{dp_{\pm}}{dt} + 2p_{\perp}\, (\nabla . \mathbf{v}) = p_{\perp}\, (\hat{\mathbf{X}} . \nabla)\ (\hat{\mathbf{X}} . \mathbf{v}) \tag{4.12.15}$$

The MHD equation $\partial \mathbf{B}/\partial t = \nabla \times (\mathbf{v} \times \mathbf{B})$ gives

$$\frac{d\mathbf{B}}{dt} + \mathbf{B}\, (\nabla . \mathbf{v}) = (\mathbf{B} . \nabla)\, \mathbf{v} \tag{4.12.16}$$

Its scalar product with $\mathbf{B}/B^2$ gives the relation

$$\frac{1}{B} \frac{dB}{dt} + (\nabla . \mathbf{v}) = l_i\, l_j \frac{\partial v_i}{\partial x_j} \tag{4.12.17}$$

for the convective rate of change of B, where the Einstein convention (summation over a repeated index) is assumed. Using the continuity condition and eliminating $l_i\, l_j\, (\partial v_i/\partial x_j)$ between (4.12.13) and (4.12.17), we get the double-adiabatic equations of state

$$\frac{d}{dt} \left(\frac{B^2}{\rho} p_{\|} \right) = 0 \tag{4.12.18}$$

$$\frac{d}{dt} \left(\frac{p_{\perp}}{\rho_B} \right) = 0 \tag{4.12.19}$$

where (4.12.15) has also been used. These are called the CGL (Chew-Goldberger-Low) approximations for the pressure tensor, section 9.10 (*c*).

4.13 MHD SHOCK WAVES

A wave front (a moving surface, on one side of which a disturbance exists, and the other side is undisturbed) is a steady transition region of small thickness and variable curvature, between the known region of the undisturbed fluid (the region 2) and that through which the disturbance has already passed (region 1). Thickness of this surface of discontinuity of some flow parameters is of the order of few mean free paths (Appendix 12 (*b*)). The solution ahead of the front is different from that behind it. These surfaces are of weak discontinuity, when the variables themselves are continuous, but their derivatives along the normal to the wave front are discontinuous.

A shock is an abrupt, non-adiabatic change of fluid variables for propagation of disturbance. The shock front is any transition layer which propagates through the medium; causing a change of state. On the average, it is stationary in the rest frame. Any change in the state of the fluid is due to the conversion of ordered flow energy into the random thermal energy. The energy flux through a shock front is zero, because the entropy must increase as the front is crossed. The local velocity of propagation being higher in the compressive part, the density gradients steepen with time and eventually form a discontinuous pressure front. In plasmas ions transport momentum and energy of electrons and excite viscous effects; electrons cause thermal conduction. So, even in the absence of any static magnetic field, the joule heat and other viscous forces dissipate at different rates.

Ordinary fluids of non-interacting particles have only one (longitudinal) degree of freedom; so these allow only one mode of propagation at the sound speed C_S. MHD materials, however, can support one longitudinal and two transverse dispersive modes. Hence, these have three degrees of freedom; the three propagation speeds are known as fast, intermediate and slow wave speeds (sections 4.10 and 4.11). Intermediate waves being purely transverse, do not steepen to form shock waves. The fast and slow modes contain both transverse and longitudinal components, and the compression waves give rise to shocks. High speed, high altitude flights of missiles and reentering satellites, under low pressure, in the atmosphere, generate shocks which may dissociate the air molecules. If flow kinetic energy ahead of a shock is converted into thermal energy behind, then controlled producton of gases of high temperature might be possible in laboratory by passing a shock wave of suitable strength through a volume of gas. In the MHD shock theory, the macroscopic equations of conservation, and the Maxwell equations are integrated across the shock front, for obtaining a set of equations which are independent of the shock structure; these relate the fluid properties on either side of the front.

For irrotational flow the equation of motion is

$$\frac{\partial \mathbf{v}}{\partial t} + \nabla\left(\frac{v^2}{2}\right) = \frac{1}{c\,\rho}(\mathbf{j} \times \mathbf{B}) - \frac{1}{\rho}\nabla p$$

When the flow speed is comparable to C_S and C_A (C_S is the sound speed and C_A is the Alfven phase speed), the non-linear term $\nabla(v^2/2)$ acts effectively like an additional gradient of pressure. At its maximum the wave front becomes steepest and flattens the wave behind. In shocks energy is transferred successively from a given initial mode to modes with shorter and shorter wavelengths. Thus, ultimately, a narrow shock zone develops which depends on the energy distribution through finite heat conduction, viscosity, and other irreversible heat dissipation mechanisms. For adiabatic changes

$$C_S^2 = \frac{\gamma}{\rho} p = \gamma \rho^{\gamma - 1} = \gamma p^{(\gamma - 1)}/\gamma$$

If, on one side of the front $\rho = \rho_1$, and on the other $\rho = \rho_2$, then

$$C_{S1}^2 > C_{S2}^2 \text{ if } \rho_1 > \rho_2 \text{ and } \gamma > 1,$$

and matter in medium 1 soon overlaps that in medium 2. The sound velocity at the peak of compression is greater than at lower densities. The peak density has the tendency to fall upon that part which is ahead of it. For rapid density changes, due to viscous forces and heat conduction, the wave front becomes steep. For waves of infinitesimal amplitudes, this type of heat dissipation is ignored.

MHD shock theory is meant for understanding (1) general properties of shock waves and, the influence of magnetic field on these; (2) the thickness of a shock wave, and (3) the stability of shocks from small perturbation.

(a) Relations across the shock front

If $\mathbf{v}$ be the flow velocity and $\mathbf{u}$ the shock front velocity, the starting shock relations across the wave front, following (A9.9), are

$$[\rho u_n - \rho\, v_n] = 0 \tag{4.13.1}$$

$$[(u_n - v_n)\,\rho \mathbf{v}] = \left[p + \frac{\mu H^2}{8\pi}\right]\mathbf{n} - \frac{\mu}{8\pi}[\mathbf{H}\, H_n], \tag{4.13.2}$$

$$[(u_n - v_n)\,\mathbf{B} + \mathbf{v}\, B_n] = 0, \tag{4.13.3}$$

$$[B_n] = 0 \tag{4.13.4}$$

$$\left[(u_n - v_n)\left(\frac{\rho v^2}{2}\right) + \rho\varepsilon + \frac{\mu H^2}{8\pi}\right] = \left[p v_n + \frac{\mu H^2}{8\pi} v_n - \frac{\mu}{4\pi}(\mathbf{v}\,.\,\mathbf{H})\, H_n\right] \tag{4.13.5}$$

Here (4.13.1) obtains from the continuity condition of (4.4.6), (4.13.2) from (4.4.5) with $\mathbf{F} = 0$, $\rho = 0$, $\nu = 0$, the magnetic diffusion equation (4.4.5) yields (4.13.3); the condition $\nabla \cdot \mathbf{B} = 0$ gives (4.13.4) and (4.13.5) obtains from the energy conservation condition (4.4.11). The fluid velocity relative to the shock front, $\bar{v}_n = v_n - u_n$; so these shock conditions reduce to

$$[\rho \bar{v}_n] = 0, \tag{4.13.6}$$

$$[\rho\, \bar{v}_n\, \mathbf{v}] + [p^* \mathbf{n}] = \frac{\mu}{4\pi} [\mathbf{H}\, H_n], \tag{4.13.7}$$

$$[\mathbf{v}\, B_n - \bar{v}_n\, \mathbf{B}] = 0, \ [B_n] = 0, \tag{4.13.8}$$

$$\left[\bar{v}_n \left(\frac{\rho v^2}{2} + \rho \varepsilon + \frac{\mu H^2}{8\pi} \right) \right] + [p^* \, v_n] = \frac{\mu}{8\pi} [(\mathbf{v} \cdot \mathbf{H})\, H_n] \tag{4.13.9}$$

in which
$$p^* = p + \frac{\mu H^2}{8\pi} \tag{4.13.10}$$

If the subscripts 1 and 2 are used to denote a variable behind and ahead of the shock front, then

$$\rho_1\, \bar{v}_{n_1} = \rho_2\, \bar{v}_{n_2} = m \tag{4.13.11}$$

where m is the rate of flow of mass across the front. Since the speed u_n is continuous at the shock surface,

$$[\bar{v}_n] = [v_n] \tag{4.13.12}$$

Equations (4.13.6) and (4.13.11) then give

$$m\, [\lambda] = [v_n] \tag{4.13.13}$$

where λ $(= 1/\rho)$ is the specific volume. Hence (4.13.9) becomes

$$\frac{m}{2} [v^2] + m\, [\varepsilon] + \left(\frac{m\mu}{8\pi} \right) [\lambda H^2] + [v_n\, p] + \frac{\mu}{8\pi} [v_n\, H^2] = \frac{\mu H_n}{4\pi} [(\mathbf{v} \cdot \mathbf{H})] \tag{4.13.14}$$

(b) MHD Rankine-Hugoniot relation

Using the thermodynamic state relations

$$R = c_p - c_v, \ \gamma = \frac{c_p}{c_v}, \ \varepsilon = c_v\, T = \frac{p\, \lambda}{R}\, c_v = \frac{\lambda}{\gamma - 1} \tag{4.13.15}$$

in (4.13.14) we get

$$\frac{m}{2} [v^2] + \frac{m}{\gamma - 1} [p\lambda] + \frac{m\, \mu}{8\pi} [\lambda H^2] + [p v_n] + \frac{\mu}{8\pi} [v_n\, H^2] = \frac{\mu H_n}{4\pi} [(\mathbf{v} \cdot \mathbf{H})] \tag{4.13.16}$$

where T is the temperature in degrees Kelvin. We now write

$$\frac{1}{2}(P_1 + P_2) = \langle P \rangle, \; P_2 - P_1 = [P], \tag{4.13.17}$$

$$\langle P \rangle [Q] + \langle Q \rangle [P] \equiv [PQ], \; \langle P^2 \rangle \langle P \rangle^2 \equiv \frac{1}{4}[P]^2 \tag{4.13.18}$$

These relations are valid for scalar products of pairs of vectors and for the ordinary products of two scalar quantities. Scalar product of (4.13.7) with $\langle \mathbf{v} \rangle$, therefore, gives

$$([\rho \, \overline{v}_n \, \mathbf{v}] \,.\, \langle \mathbf{v} \rangle) + ([p^* \, \mathbf{n}] \,.\, \langle \mathbf{v} \rangle) = \frac{\mu}{4\pi}([\mathbf{H} \, H_n] \,.\, \langle \mathbf{v} \rangle)$$

or
$$\frac{m}{2}[v^2] + [p] \langle v_n \rangle + \frac{\mu}{8\pi}[H^2] \langle v_n \rangle = \frac{\mu H_n}{4\pi}([\mathbf{H}] \,.\, \langle \mathbf{v} \rangle) \tag{4.13.19}$$

If (4.13.17) is used in this relation, then

$$\frac{m}{2}[v^2] \; [pv_n] - m \langle p \rangle [\lambda] + \frac{\mu}{8\pi}[H^2] \langle v_n \rangle = \frac{\mu H_n}{4\pi}([\mathbf{H}] \,.\, \langle \mathbf{v} \rangle) \tag{4.13.20}$$

Subtracting both sides of (4.13.20) from the corresponding sides of (4.13.16) we obtain

$$\frac{m}{\gamma - 1}[p\lambda] + m \langle p \rangle [\lambda] + \frac{m\mu}{8\pi}[\lambda H^2] + \frac{\mu}{8\pi}[v_n] \langle H^2 \rangle - \frac{\mu H_n}{4\pi}$$
$$(\langle \mathbf{H} \rangle \,.\, [\mathbf{v}]) = 0 \tag{4.13.21}$$

Scalar product of (4.13.7) with $\langle \mathbf{H} \rangle$ gives

$$B_n([\mathbf{v}] \,.\, \langle \mathbf{H} \rangle) = \mu \, [v_n] \langle H \rangle^2 + \mu \langle v_n \rangle ([\mathbf{H}] \,.\, \langle \mathbf{H} \rangle)$$

$$= \mu m \, [\lambda] \left(\langle H^2 \rangle - \frac{1}{4}[H]^2 \right) + \mu \langle v_n \rangle ([\mathbf{H}] \,.\, \langle \mathbf{H} \rangle) \tag{4.13.22}$$

Using (4.13.21) in (4.13.22) we obtain

$$m[\varepsilon] + m \langle p \rangle [\lambda] + \frac{m\mu}{8\pi}[\lambda H^2] + \frac{m\mu}{8\pi}[\lambda]$$

$$\langle H^2 \rangle - \frac{m\mu}{4\pi}[\lambda] \left(\langle H^2 \rangle - \frac{1}{4}[H]^2 \right)$$

$$= \frac{\mu}{4\pi} \langle v_n \rangle \; ([\mathbf{H}] \,.\, \langle \mathbf{H} \rangle)$$

Using now (4.13.17) and (4.13.18) in the right side of this equation for $P = Q = \mathbf{H}$ and the fact that $\langle v_n \rangle = m \langle \lambda \rangle + u_n$, we get

$$m[\varepsilon + \lambda \langle p \rangle] + \frac{m\,\mu}{16\pi}[\lambda][H]^2 + \frac{m\,\mu}{8\pi} \langle \lambda \rangle [H^2] = \frac{m\,\mu}{8\pi} \langle \lambda \rangle [H^2] + \frac{\mu u_n}{8\pi}[H^2]$$

Hence, $u_n = 0$ if the observer follows the shock; so

$$[\varepsilon + \lambda \langle p \rangle] + \frac{\mu}{16\pi} [\lambda] \; [H^2] = 0 \qquad (4.13.23)$$

This is a form of the MHD Rankine-Hugoniot shock relation; for $\mathbf{H} = 0$; when $\mathbf{H}$ is perpendicular to the wave front, $[\mathbf{H}] = \mathbf{n}\,[H_n] = 0$; then equation (4.13.23) resembles that of hydrodyna-mics. Putting $\varepsilon = p\,\lambda/(\gamma - 1)$ in (4.13.23) we get

$$\frac{1}{\gamma - 1}\left(\frac{p_1}{\rho_1} - \frac{p_2}{\rho_2}\right) + \frac{1}{2}(p_1 + p_2)\left(\frac{1}{\rho_1} - \frac{1}{\rho_2}\right) = -\frac{\mu}{16\pi}\left(\frac{1}{\rho_1} - \frac{1}{\rho_2}\right)[H]^2 \qquad (4.13.24)$$

For evaluation of the ratio p_1/p_2 in terms of ρ_2, ρ_1/ρ_2 and $[H]^2$ we write (4.13.24) as

$$\left(\frac{p_1}{p_2} - \frac{\rho_1}{\rho_2}\right)\frac{1}{(\gamma - 1)} + \left(\frac{p_1}{p_2} - 1\right)\frac{1}{2}\left(1 - \frac{\rho_1}{\rho_2}\right) = -\frac{\mu}{16\pi\, p_2}\left(1 - \frac{\rho_1}{\rho_2}\right)[H]^2 \qquad (4.13.25)$$

Putting $r = \rho_1/\rho_2 = \lambda_2/\lambda_1$ in (4.13.25) we get

$$\frac{p_1}{p_2} = \frac{r(\gamma + 1) - (\gamma - 1)}{(\gamma + 1) - r(\gamma - 1)} + \frac{\mu\,[H]^2\,(\gamma - 1)\,(r - 1)}{8\pi\, p_2\,\{(\gamma - 1) - r(\gamma - 1)\}} \qquad (4.13.26)$$

It can be further simplified to

$$\frac{p_1}{p_2} = \frac{(1 + K^2)\,\{(\gamma - 1)\,r - \gamma + 1\} - 2K^2 r}{(\gamma + 1) - r(\gamma - 1)}, \quad K^2 = \frac{\mu\,[H]^2}{8\pi\, p_2} \qquad (4.13.27)$$

Hence the strength of the shock can be obtained when all the variables in front of the shock are specified together with the magnitude of a single one behind the front. Since, in any physical problem, the movement of the shock front is known, the jumps in pressure and density and the permissible range of values of r can be determined.

4.14 COLLISIONLESS SHOCKS AND SHOCK THICKNESS

If collision times are long compared to the other times of interest, then classical binary collision effects are unimportant, and collisionless shocks in low density plasma are produced. These have thickness less than the smallest collision mean free path and hydrodynamic dissipative processes are unimportant in these. Motion of charges is then restricted more by magnetic field, than by collisions, and the mechanisms of change of the plasma state by the passage of the shock front are rather complex. Landau damping is an example of collisionless absorption of fields, generating shocks in plasmas. Collisionless shocks can heat ions rapidly to temperatures of thermonuclear reactions. The pinch effect is often used to compress an initial plasma, and then to generate a cylindrically imploding shock. Ahead of such shocks the plasma should be sufficiently homogeneous and reproducible. To avoid cylindrical effects, the shock must separate from the piston and become stationary before it gets near to the discharge axis. For large shock velocity, thickness of a collisionless shock is about ten times the ratio c/w_p where c is the vacuum speed of light and ω_p is the characteristic electron plasma frequency. For very high shock velocities, the shock front broadens to scales of the order of c/w_{pi} where $w_{pi} = (m/M)\, w_p$, M and m being the mass of an ion and electron respectively.

Earth's magnetosphere lies in a supersonic solar wind, near the earth; but it appears to be protected from this wind by a bow shock. The average speed (of the order of 10 km/sec) is much greater than the typical satellite speeds; the shock thickness is of the order of 50 to 1000 km, which is also of the order of the distance c/w_{pi}. In the solar wind, near the earth, the particle number density is about 5 cm^{-3}, the wind velocity u is about 400 km/sec, the ion temperature $T_i = 5$ ev, the electron temperature $T_e = 10T_i$, $H = 5 \times 10^{-5}$ Gauss. Alfven speed $c_A = 50$ Km/sec and $u/c_A = 8$.

Within the thickness of a discontinuity surface (of the order of a few mean free path) the dissipation rates per unit area, determined by the conservation equations, become extremely large. When the Larmor radius is small, compared to the mean free path, the dissipation process becomes complicated. The cases of plasmas of high density, low temperature, and small mean free path, compared to the Larmor radius, are simple to work out. When the collision frequency between electrons and ions is high (higher than the gyro frequency of electrons and ions), the shock thickness, depending on the density of the plasma and on the mean free path of the particle, is also of the order of the mean free path.

The shock dissipates within a distance of the order of the shock thickness δ; so, in (4.14.2), the gradient operator is replaced by $1/\delta$. Then the total rate of dissipation, within the shock front, due to the viscous force, is δ times the rate of work done, per unit volume, for dissipation (say, for joule heating). The dissipation rate is $d\,\varepsilon/dt$, within the thickness δ where ε is the density of energy lost. So

$$dt = \frac{\delta}{v_1} \tag{4.14.1}$$

where v_1 is the velocity behind the shock front. From dimensional considerations of the transfer equations, the rate of dissipation is equal to $(\Delta\,\mathbf{v}\,.\,\nu\nabla^2\,(\Delta\,\mathbf{v}))$ where $\Delta\,\mathbf{v}$ is the velocity ahead of the shock relative to that behind, and ν is the coefficient of viscosity. Hence

$$\frac{d\varepsilon}{dt} = (\Delta\,\mathbf{v}\,.\,\nu\nabla^2\,(\Delta\,\mathbf{v})) \tag{4.14.2}$$

We find that $(v_1/\delta)\,d\varepsilon \approx \nu\,((\mathbf{v}_1 - \mathbf{v}_2)/\delta)^2$, because quantitatively

$$\nabla^2\,(\Delta\mathbf{v}) \approx \frac{\Delta\mathbf{v}}{\delta^2} \tag{4.14.3}$$

Taking $\varepsilon \approx \rho_1 v_1^2/2$, the shock thickness is approximately given by

$$\delta \approx \frac{\nu (v_1 - v_2)^2}{v_1 (\rho_1 v_1^2/2)} \tag{4.14.4}$$

Relations (4.14.2) and (4.14.3) show that the rate of dissipation is proportional to the square of the velocity gradient, the volume rate varies inversely as δ^2, and the total rate of dissipation within the shock front (approximately equal to $\delta (d\varepsilon/dt)$) is inversely proportional to δ. When the loss is mainly through joule heating,

$$\frac{d\varepsilon}{dt} \approx (\mathbf{E} \cdot \mathbf{j}) \approx \frac{1}{\sigma} (\mathbf{j}_1 - \mathbf{j}_2)^2 \approx \frac{c^2}{16\pi^2 \mu^2 \sigma} [\nabla \times (\mathbf{B}_1 - \mathbf{B}_2)]^{-2}$$

$$\approx \frac{c^2}{16\pi^2 \mu^2 \sigma} \left\{ \frac{(\mathbf{B}_1 - \mathbf{B}_2)}{\delta} \right\}^2,$$

so

$$\delta \approx \frac{c^2 (\mathbf{B}_1 - \mathbf{B}_2)^2}{16\pi^2 \sigma \mu^2 \nu_1 (\rho_1 v_1^2/2)}$$

Hence, joule heating depends on the square of the currents; current being equivalent to the gradient of the magnetic field, this heating varies inversely as the square of δ. The time for a fluid particle to pass through the shock layer is $t_1 \approx \delta/c_S$ where c_S is the sound speed in the layer. For viscous forces if t_2 is the corresponding time, then $F \approx \nu \nabla^2 v = \nu v/\delta^2 \approx v/t_2$ and $t_2 \approx \delta^2/\nu$. Equating t_1 and t_2 we get $\delta = \nu/c_S$. For oscillations this length is of the order of the mean free path. When dissipation is due to Ohmic heating the corresponding time t_3 can be estimated from (4.6.5); it is $t_3 \approx \delta^2/\eta \approx \delta/c_s$ where $\eta = c^2/4\pi \mu\sigma$ because (4.6.5) gives $Bt_3 \approx \eta B/\delta^2$.

4.15 MHD STABILITY THEORY

The terms stable and unstable systems are associated with the behaviour of dynamical systems towards small perturbations about their equilibrium states. If the perturbation gives rise to forces which tend to restore the system to its equilibrium state, then the system is stable. If the displacement increases indefinitely, the system is unstable.

The self-consistent linearized field equations, in the normal mode technique, determine the frequency ω and the wave number K in terms of macroscopic parameters for small displacements from the equilibrium state. A particular mode is stable when ω (or the wave length $\lambda = 1/K$) is real, and unstable when it is imaginary.

In the energy principle technique, a variational formulation of the equations of motion is done. Any perturbation which decreases the potential energy from its equilibrium value is the stable perturbation. The system is unstable or stable according as the normal component of the gradient of $p^* \ (= p + \mu H^2/8\pi)$ acts in the direction of the normal component of displacement at the boundary, or in the opposite direction. Only those displacements are considered which obtain from the assumption that the system is in stable equilibrium, and then extraneous forces are introduced which displace the system away from equilibrium. For some types of complicated equilibrium configurations, this technique is preferred to the normal mode method.

Mathematically, the starting point is the Lagrangian displacement of a fluid element, section 4.11 (*a*). Initially one can start with the assumption that both $\boldsymbol{\xi}$ and $\mathbf{v}$ have independent values. In simple cases the following MHD equations for infinitely conducting fluids are used:

$$\frac{d\mathbf{v}}{dt} = -\frac{1}{\rho}\nabla p + \frac{1}{c\rho}(\mathbf{j}\times\mathbf{B}),\ \frac{\partial\rho}{\partial t} + \nabla\cdot(\rho\,\mathbf{v}) = 0,$$

$$\mathbf{E} + \frac{1}{c}(\mathbf{v}\times\mathbf{B}) = 0,\ \frac{d}{dt}(p\rho^{-\gamma}) = 0,\ \nabla\times\mathbf{E} = -\frac{1}{c}\frac{\partial\mathbf{B}}{\partial t},$$

$$\nabla\times\mathbf{H} = \frac{4\pi}{c}\mathbf{j},\ (\nabla\cdot\mathbf{B}) = 0 \tag{4.15.1}$$

For small departure from equilibrium, the field variables can be written as $\rho_0 + \rho,\ 0 + \mathbf{v},\ p_0 + p,\ \mathbf{j}_0 + \mathbf{j},\ \mathbf{B}_0 + \mathbf{B},\ 0 + \mathbf{E}$ where the first part of the sum for each field variable is its equilibrium state value and the second part is the increment due to perturbation. The static field equations are

$$\nabla p_0 = \frac{1}{c}(\mathbf{j}\times\mathbf{B}_0),\ p_0\,\rho_0^{-\gamma} = \text{constant}, \tag{4.15.2}$$

$$\nabla\times\mathbf{H}_0 = \frac{4\pi}{c}\mathbf{j}_0,\ (\nabla\cdot\mathbf{B}_0) = 0$$

Putting the value $\partial\boldsymbol{\xi}/\partial t$ for $\mathbf{v}$ from (4.11.1) in the second relation of (4.15.1) we get

$$\rho = -\nabla\cdot(\rho_0\,\boldsymbol{\xi}) \tag{4.15.3}$$

Similarly the third and fifth equations of (4.15.1) give

$$\mathbf{B} = \nabla\times(\boldsymbol{\xi}\times\mathbf{B}_0), \tag{4.15.4}$$

and the fourth equation on integration becomes

$$p\rho_0 = \gamma\, p_0\,\rho \ \text{ or } \ p = -c_S^2\,\nabla\cdot(p_0\,\boldsymbol{\xi}) \tag{4.15.5}$$

where $c_S\ (= \sqrt{\gamma p_0/\rho_0}\,)$ is the speed of sound in the medium; the fourth equation of (4.15.1) can be written as

$$\nabla p_0 = \frac{\gamma}{\rho_0}\,p_0\,\nabla\rho_0 \tag{4.15.6}$$

Putting this relation in (4.15.5) we get

$$p = -\gamma p_0 (\nabla . \xi) - (\xi . \nabla) p_0 \tag{4.15.7}$$

Then the first equation of (4.15.1) is transformed into

$$\rho_0 \frac{\partial^2 \xi}{\partial t^2} = \nabla \{\gamma p_0 (\nabla . \xi) + (\xi . \nabla) p_0\} + \frac{1}{4\pi\mu} [\{\nabla \times \nabla \times (\xi \times \mathbf{B}_0)\} \times \mathbf{B}_0]$$

$$+ \frac{1}{4\pi\mu} \{(\nabla \times \mathbf{B}_0) \times \nabla \times (\xi \times \mathbf{B}_0)\} \tag{4.15.8}$$

Solution of this equation to determine the displacement ξ and its growth or decay is sought in the form

$$\xi = \sum_n [\xi_n \exp (i \omega_n t)], \tag{4.15.9}$$

together with suitable boundary conditions. If all ω_n are real, ξ is bounded. For complex frequencies, ω_n has a negative imaginary part, which gives an indefinite growth of ξ if $\xi_n \neq 0$.

Chapter 5

APPLICATIONS OF MHD AND PLASMA PROCESSES

5.1 INTRODUCTION

Plasma occurs both in space and in laboratory. Applications of plasma and MHD are many and of various types. We here discuss Hartman flow, Couette flow, pinch effect, MHD generation of power, plasma of nature and laboratory, including magnetically confined fusion plasma.

Diagnostic techniques are known for probing the electrical, thermal transport properties of a plasma. Spectroscopic methods determine its chemistry and temperature. Microwave interferometry, Langmuir probe and optical interferometry assess the electrical properties. The fluid properties of continuous jets of arc are determined by probes, photographic techniques and optical instruments. Low gas densities and high level of luninosity of plasma create problems. Streak cameras or photomultiplier tubes, if placed at proper positions along the shock tube, can measure luminosity of the shock front in electromagnetic shock tubes.

(a) Plasma accelerators

A plasma accelerator, by its electric power, drives a plasma in a given direction. To show that assume that the flow velocity **V** is parallel to the *x*-axis, the applied magnetic field **B** is parallel to the *z*-axis, the applied electric field is parallel to the *y*-axis and the unit vectors parallel to the coordinate axes are $\hat{\mathbf{X}}$, $\hat{\mathbf{Y}}$, $\hat{\mathbf{Z}}$. The induced electric field $\mathbf{E}_i$ is given by $\mathbf{E}_i = (\mathbf{V} \times \mathbf{B})/c = -(VB/c)\,\hat{\mathbf{Y}}$. The induced current $\mathbf{j} = -(\sigma\, BV/c)\,\hat{\mathbf{y}}$, and the conduction current $\mathbf{j}_c = \sigma\,\mathbf{E}a$. The force parallel to the direction of the flow is

$$\mathbf{F} = \frac{1}{c}(\mathbf{j}_c + \mathbf{j}_i) \times B = \left(\frac{B}{c}\right)\left(E_a - \frac{VB}{c}\right)\hat{\mathbf{X}}$$

Hence **F** will be in the sense of **V** if $E_a > (VB/c)$ and opposite if $E_a < (VB/c)$.

A plasma in rapid rotation (Ixion) can be produced between two coaxial cylindrical electrodes (cylindrical condensors), surrounded by coils which produce an axial magnetic field, Lehnert (1971). A radial electric field is set-up by connecting the electrodes with a condensor bank and the plasma starts rotating with a velocity **v** determined from the relation $\mathbf{E} + (\mathbf{V} \times \mathbf{B})/c = 0$. Then $\mathbf{v} = c\,(\mathbf{E} \times \mathbf{B})/B^2$ if $(\mathbf{v} \,.\, \mathbf{B}) = 0$. Since the device has cylindrical symmetry **E** is in the radial direction, **B** is in the axial z direction and so

$$V_\theta = -\,cB_z\,E_r/B^2.$$

(b) MHD flow through channels

These are flows of electrically conducting, incompressible viscous fluids through insulated channels of rectangular or other types of given cross-sections in the presence of homogeneous magnetic field along a direction perpendicular to the flow. An electric field, perpendicular to both the magnetic field and the direction of flow, is also applied in some cases. The force $(\mathbf{j} \times \mathbf{B})/c$, due to the conduction current $\mathbf{j} = \sigma\,\{\mathbf{E} + (\mathbf{v} \times \mathbf{B})/c\}$, moves the fluid against a pressure gradient. In other problems no current flows and voltage of the induced field is used to regulate the flow. For these problems the magnetic Reynolds number (section 4.6) is

$$R_m = \left(\frac{4\pi\mu\,\sigma}{c}\right)(au) \tag{5.1.1}$$

where a is a typical length in the channel cross-section and u is a typical velocity associated with the flow.

Let b be the width, l the length and $2a$ be the thickness of the channel. When $b >> 2a$ and $l >> 2a$, the problems reduce to those of flows between infinite parallel plates. These, then, can be regarded as one-dimensional flows, because the flow variables no longer depend on the two coordinates corresponding to the larger dimensions b and l. When $b = 2a$ the end effects complicate the study and introduce secondary flows. But these are not considered here. Sometimes the flow between coaxial cylinders is also studied. In these flows the distance between cylindrical surfaces should be small compared to their radii and length.

In general the problem is to find the velocity, current per unit volume and the magnetic field profiles in terms of electrical conductivity, channel material and external circuits in presence of electrostatic fields.

5.2 HARTMAN FLOW

In the Hartman problem the flow is maintained by the pressure gradient, which is the pressure drop between the two ends of the channel divided by the channel length l. The flow is parallel to the x-axis and the z-axis is normal to the plates; the boundary planes are $z = \pm\, a$. The fluid motion and the longitudinal magnetic field are assumed to depend only on the z coordinate. The pressure p depends also on x, because the constant pressure gradient $P\ (= \partial p/\partial x)$ maintains the flow parallel to the x-axis.

The condition $\nabla \,.\, \mathbf{H} = 0$ gives

$$H_z = \text{constant} = H_0\ \text{(say)} \tag{5.2.1}$$

Since v_x depends only on z, the incompressibility condition $(\nabla \cdot \mathbf{v}) = 0$ is automatically satisfied. Neglecting gravity the momentum transfer condition in the steady state gives

$$0 = -\nabla p + \frac{1}{c}(\mathbf{j} \times \mathbf{B}) + \nu \nabla^2 \mathbf{v} \tag{5.2.2}$$

The conduction current $\mathbf{j} = \sigma \{\mathbf{E} + (\mathbf{v} \times \mathbf{B})/c\}$ (5.2.3)

The electric current $\mathbf{j}_y$ of an electric field $\mathbf{E}$, applied between the walls in the direction of the y-axis, is given by

$$j_y = \sigma \left(E - \frac{B_0 v_x}{c} \right) \tag{5.2.4}$$

Putting this relation in the X component of (5.2.2) we get

$$0 = -P + \frac{\sigma B_0}{c} \left(E - \frac{1}{c} B_0 v_x \right) + \nu \frac{\partial^2 v_x}{\partial z^2}, \tag{5.2.5}$$

and the second order equation

$$\frac{\partial^2 v_x}{\partial z^2} - \left(\frac{M}{a}\right)^2 v_x = P_0 - \left(\frac{M}{a}\right)^2 \left(\frac{CE}{B_0}\right) \tag{5.2.6}$$

with $$\left(\frac{M}{a}\right)^2 = \frac{\sigma B_0^2}{\nu c^2}, \; P_0 = \frac{P}{\nu} \tag{5.2.7}$$

where $2a$ is the distance between the two parallel plates. The dimensionless quantity M, depending on the ratio σ/ν, is called the Hartman number. For mercury $\sigma = 9.4 \times 10^{15}$/sec, $\nu = 1.5 \times 10^{-2}$ gm/cm sec, $\rho = 13.5$ gm/cm^3; so

$$M = (B_0 a/c)\sqrt{(\sigma/\nu)} = 2.6 \times 10^{-2} \times B_0 \times a, \; R_m \approx 10^{-4} \times a \times u$$

If $B_0 = 10^4$ gauss and $a = 1$ cm, then $M \approx 250$. Since electrical conductivity is responsible for the distortion of magnetic lines of force, viscosity dominates over the magnetic freezing if M is small and magnetic diffusion dominates over the viscous loss if M is large.

The boundary conditions for viscous flow are $v_x = 0$ at $z = a$ and $z = -a$. The tangential magnetic field should be continuous at the boundaries and $H_x = 0$ at $z = \pm a$. Otherwise, the magnetic field outside the flow should have a component H_x for the sake of continuity; but that is contrary to the given fact that $H = H_0$ in both $z > a$ and $z < -a$.

With the help of these boundary conditions the solution of (5.2.6) becomes

$$v_x = \left(\frac{CE}{B_0} - \frac{P_0 a^2}{M^2} \right) \left\{ 1 - \frac{\cosh (Mz/a)}{\cosh M} \right\} \tag{5.2.8}$$

Hence, if $M << 1$, we find that

$$v_x = \left(\frac{CE}{B_0} - \frac{P_0 a^2}{M^2} \right) \left(\frac{M^2}{2} \right) \left(1 - \frac{z^2}{a^2} \right) \tag{5.2.9}$$

The coefficients CE/B_0 and $P_0 a^2/M^2$ have obviously the dimensions of velocity. Using (5.2.4) and the Maxwell equation $\nabla \times \mathbf{H} = 4\pi \, \mathbf{j}/c$, the differential equation for H_x is obtained;

$$\frac{\partial H_x}{\partial z} = \frac{4\pi \, \sigma}{c} \left(E - \frac{1}{c} B_0 \, v_x \right) \tag{5.2.10}$$

Hence
$$H_x = \frac{4\pi \, \sigma}{c} \left\{ Ez - \frac{B_0}{c} \int v_x \, dz + A \right\} \tag{5.2.11}$$

where A, the constant of integration, is chosen such that $H_x = 0$ at $z = \pm a$. Putting the value of v_x from (5.2.8) in (5.2.11) we find that

$$H_x = \frac{4\pi\sigma}{c} \frac{B_0}{c} \left[\frac{cE}{B_0} z - \left(\frac{cE}{B_0} - \frac{P_0 a^2}{M^2} \right) \left\{ z - \frac{a}{M} \frac{\sinh (Mz/a)}{\cosh M} \right\} + \frac{CA}{B_0} \right] \tag{5.2.12}$$

If electrodes are placed on either side of the channel, an external circuit, connected across them, together with the fluid, makes a continuous circuit. The field E is then determined by the current flowing in the circuit. If E is so adjusted that the total electric current $\int_{-a}^{a} j_y \, dz$ vanishes, then

$$2aE = \frac{B_0 \, a}{cM} \left\{ 2 \left(\frac{cE}{B_0} - \frac{P_0 a^2}{M^2} \right) (M - \tanh M) \right\} \tag{5.2.13}$$

This relation simplifies to

$$\left(\frac{cE}{B_0} \frac{P_0 a^2}{M^2} \right) = - P_0 \frac{a^2}{M^2} M \coth M \tag{5.2.14}$$

To eliminate E from (5.2.14) we use (5.2.12) and get

$$H_x = \frac{4\pi\sigma}{c^2} B_0 \left[P_0 \left(\frac{a^2}{M^2} \right) \left\{ z - \frac{a \sinh (Mz/a)}{\sin M} \right\} + \frac{CA}{B_0} \right] \tag{5.2.15}$$

If $A = 0$ then the boundary condition $H_x = 0$ at $z = \pm a$ is satisfied by (5.2.15); so

$$H_x = \left(\frac{4\pi\,\sigma}{c^2}B_0\right)\left(\frac{P_0\,a^2}{M^2}\right)\left\{z - \frac{a\sinh(Mz/a)}{\sinh M}\right\} \tag{5.2.16}$$

Putting (5.2.14) in (5.2.8) we find that

$$v_x = \frac{P_0\,a^2}{M^2}\,\frac{M}{\sinh M}\left\{-\cosh M + \cosh\left(\frac{Mz}{a}\right)\right\} \tag{5.2.17}$$

The average value V of v_x is given by

$$V = \frac{1}{2a}\int_{-a}^{a} v_x\,dz = \frac{P_0\,a^2}{M^2\sinh M}(\sinh M - M\cosh M) \tag{5.2.18}$$

In case of plane Poisseuile flow $M << 1$,

$$\sinh M \approx M + M^3/6,\ \cosh M \approx 1 + M^2/2$$

etc., and therefore

$$H_x = \frac{4\pi\sigma}{c^2 M^2}B_0\,P_0\,a^2\left(1 - \frac{z^2}{a^2}\right),\quad v_x = \frac{P_0}{2}a^2\left(1 - \frac{z^2}{a^2}\right) \tag{5.2.19}$$

When $M >> 1$, on the other hand, $\cosh M = \sinh M = (\exp M)/2$, etc., and so

$$H_x = \left(\frac{4\pi\,\sigma}{c^2}B_0\,P_0\right)\left(\frac{a^2}{M^2}\right)a\left[\frac{z}{a} - \exp\left\{M\left(\frac{z}{a} - 1\right)\right\}\right]$$

Hence $H_x \to 0$ as $M \to \infty$ because $|z| \leq a$. For the u of (5.1.1) the average velocity V of (5.2.18) can be used. When R_m is small H_x is also small compared to H_0 because P is small, and there is no significant transport of the lines of force. The induced magnetic field increases with increasing Reynolds number; but, at the same time, decreases with increasing M.

If electrodes, placed on either side of the duct, are short circuited, the action of the applied magnetic field is that of a brake. When current is maximum, the external resistance is zero and the electromgnetic breaking action is maximum. The flow energy in the brake is dissipated totally within the fluid. In MHD generators (section 5.6) electric power is extracted from the fluid because of finite external resistance. When $E = B_0\,V/c$, the average current vanishes and the fluid flux may be determined by estimating E, taking for velocity the average V of (5.2.18). When the external circuit contains a power supply source, the power is used to accelerate the flow in electromagnetic pumps (accelerators). In nuclear reactors, the force $(\mathbf{j} \times \mathbf{B})/c$ drives liquid alkali metals for absorption of heat.

For mercury $\sigma = 9.4 \times 10^{15}/\text{sec}$, and $\nu = 1.5 \times 10^{-2}$ poise. If we take $p = 10^6$ dynes/cm^2 (= one atmosphere), $a = 0.1$ cm, $l = 1$ cm, then $P_0 = 10^6$. From (5.2.9) we obtain $v_x = 0.5 \times 10^4/\text{cm sec}$. Hence $M = 174$ if $B_0 = 5000$ gauss. In (5.2.9), if we take $E = P_0\, a^2\, B_0/M^2\, c$, then $v_x = 0$; this is the minimum electric field required to produce a positive flow. Thus

$$E_{\min} = 9.6 \times 10^{-6} \text{ esu} = 28.8 \times 10^{-2} \text{ V/cm}.$$

5.3 COUETTE FLOW

This is a shear flow of electrically conducting and viscous fluids, between two infinite plates moving relative to each other in their own planes. If a boundary is moving it is the open fluid surface, and a stationary boundary is any real wall. The initial fluid motion is caused by the motion of boundary wall. The boundary conditions are : $v_x = V_1$ at $z = a$ and $v_x = V_2$ at $z = -a$. Other conditions being similar to those of Hartman flow, the solution of (5.2.6) is

$$v_x = \frac{1}{2}(V_1 + V_2)\frac{\cosh(Mz/a)}{\cosh M} + \frac{1}{2}(V_1 - V_2)\frac{\sinh(Mz/a)}{\sinh M} + \left(\frac{cE}{B_0}\right)\left\{1 - \frac{\cosh(Mz/a)}{\cosh M}\right\} \quad (5.3.1)$$

Differentiating this relation with respect to Z, we get

$$\frac{\partial v_x}{\partial z} = \left(\frac{V_1 + V_2}{2} - \frac{cE}{B_0}\right)\frac{M}{a}\,\frac{\sinh(Mz/a)}{\cosh M} + \frac{M}{2a}(V_1 - V_2)\frac{\cosh(Mz/a)}{\sinh M}$$

If, relative to the boundary $z = -a$, the velocity of $z = a$ is V_0, then putting $V_z = 0$, $V_1 = V_0$ and $z = -a$ in this relation, we get

$$\frac{\partial v_x}{\partial z} \pm \left(\frac{V_0}{z} - \frac{CE}{B_0}\right)\frac{M}{a}\,\frac{\sinh(Mz/a)}{\cosh M} + \frac{V_0 M}{2a}\,\frac{\cosh(Mz/a)}{\sinh M}$$

Hence we find that

$$\left(\frac{\partial v_x}{\partial z}\right)_{z=\pm a} = \frac{V_0 M}{2a}\coth M \pm \left(\frac{V_0}{2} - \frac{cE}{B_0}\right)\frac{M}{a}\tanh M$$

Then since $(V_0/2) > (cE/B_0)$, we obtain $(\partial v_x/\partial z)_{z=a} > (\partial v_x/\partial z)_{z=-a}$. So the velocity gradient is increased at the moving wall ($z = a$) and decreased at the stationary wall ($z = -a$). In hydrodynamic flows of non-conducting fluids $M = 0$; so

$$\lim_{M \to 0}\,(M \coth M) = 1.$$

Hence $(\partial v_x/\partial z)_{z=\pm a} = V_0/2a$ and the velocity gradient remains constant. Since (section 5.2) $H_x = 0$ at $z = \pm a$, corresponding to (5.2.15) we find that

$$H_x = \frac{4\pi}{c}\sigma B_0 \left[\frac{cE}{B_0} z - \frac{1}{2}(V_1 + V_2)\left(\frac{a}{M}\right)\frac{\sinh(Mz/a)}{\cosh M} - \frac{cE}{B_0}\left\{ z - \frac{a}{m}\frac{\sinh(Mz/a)}{\cosh M}\right\} - \frac{1}{2}(V_1 - V_2)\left(\frac{a}{M}\right) \left\{\frac{\cosh(Mz/a)}{\sinh M}\right\} + \frac{c}{B_0}A \right] \quad (5.3.2)$$

The condition, that the total current $\int_{-a}^{a} j_y\, dz$ vanishes, gives

$$2aE = \frac{B_0}{c}\frac{a}{M}\left[2\left(\frac{V_1 + V_2}{2}\right)\tanh M + 2\left(\frac{cE}{B_0}\right)(M - \tanh M)\right] \quad (5.3.3)$$

Cancelling the left hand side with the third term in the right hand side we get

$$\frac{cE}{B_0} = \frac{1}{2}(V_1 + V_2) \quad (5.3.4)$$

Then (5.3.2) reduces to

$$H_x = \frac{4\pi}{c^2}\sigma B_0 \left[\frac{1}{2}(V_1 - V_2)\frac{a}{M}\frac{\cosh(Mz/a)}{\sinh M} + \frac{c}{B_0}A\right] \quad (5.3.5)$$

Putting now $H_x = 0$ at $z = \pm a$ in this equation we get

$$A = \frac{B_0}{2c}(V_1 - V_2)\frac{a}{M}\coth M, \quad (5.3.6)$$

$$H_x = \frac{4\pi}{c^2}\sigma B_0 \cdot \frac{1}{2}(V_1 - V_2)\left(\frac{a}{M}\right)\frac{\{\cosh M - \cosh(Mz/a)\}}{\sinh M} \quad (5.3.7)$$

$$v_x = \frac{1}{2}(V_1 + V_2) + \frac{1}{2}(V_1 - V_2)\frac{\sinh(Mz/a)}{\sinh M} \quad (5.3.8)$$

If V is the average value of v_x, then

$$V = \frac{1}{2a}\int_{-a}^{a} v_x\, dz = \frac{1}{2}(V_1 + V_2) \quad (5.3.9)$$

When $M << 1$, we obtain

$$H_x = -\frac{4\pi}{c^2}\sigma B_0 \cdot \frac{1}{2}(V_1 - V_2)\frac{a}{2}\left(1 - \frac{x^2}{a^2}\right), \quad (5.3.10)$$

and when $M >> 1$

$$H_x = \frac{4\pi}{c^2}\sigma B_0 \frac{1}{2}(V_1 - V_2)\frac{a}{M}\ [1 - \exp\{M((Z/a) - 1)\}] \qquad (5.3.11)$$

In 5.1.1 we now put $u = (V_1 - V_2)/2$, and find that when R_m is small H_x will be small compared to H_0, because $V_1 + V_2$ will be small. The effect of MHD forces on the boundary layers can be studied when the layer thickness is small compared to other dimensions. In a compressible fluid the interplay between viscous dissipation and heat transfer can thus be studied. Couette flow experiments are employed to measure the flow of liquid metals and of blood through capillaries.

Exercise 5.3.1. Find the general solution of (5.2.6) and show that the Hartman flow solution is recovered from that by putting the wall velocities equal to zero and the Couette flow results by neglecting the pressure gradient P_0.

Answer. The general solution is

$$v_x = \left(\frac{cE}{B_0} - \frac{P_0 a^2}{M^2}\right)\left\{1 - \frac{\cosh(Hz/a)}{\cosh M}\right\} + \frac{1}{2}(V_1 + V_2)\frac{\cosh(Mz/a)}{\cosh M}$$

$$+ \frac{1}{2}(V_1 - V_2)\frac{\sinh(Mz/a)}{\sinh M}$$

5.4 PINCH EFFECT

In the pinch effect a pinching pressure is exerted on a plasma in a long circular cylinder, by its own circular magnetic field. A current at the surface of the plasma column produces this circular magnetic field. Moreover, both the plasma and space outside are subjected to an axial dc magnetic field, and the plasma is surrounded by a conducting wall of the discharge tube of radius R_0 (ranges from 2.5 to 5 cm), with electrodes at the ends; or it is a torus formed by bending a straight tube and joining the two ends. If plasma is tenuous and hot, the mean free path is greater than or equal to the pinch radius. Then free motion of the particles, due to the Lorentz force, becomes predominant. When collision-mean-free-path is comparatively smaller, shock waves develop. The circular (or the toroidal, for an endless tube) magnetic field is produced by a current sheet at the plasma surface. An electric field, running down the tube, is produced. Currents exist only on surface of the plasma, because the rarefied gas, acting like a conductor, is assumed to be intially ionized at a temperature so high that collisions become relatively unimportant. The magnetic radial compression very quickly collapses the plasma with a bang. The resulting shock waves are reflected off the cylinder axis and move outwards, striking the interface and retarding its inward motion or even reversing it. The radial compression supports confinement of a plasma by its own magnetic field.

A moderately high axial current j_z produces the azimuthal magnetic field B_ϕ. The resulting radial pressure $j_z B_\phi / C$ pinches the plasma radially towards the axis into a filament. The azimuthal field B_ϕ forces the charged particles to describe orbits in planes through the axis. These orbits create axial current, which affects the magnetic field itself. Since the gyration radius of ions is much greater than that of electrons, the ions tend to penetrate deeper into the region of the magnetic field, and create space charge, which, in turn, produces a radial electric field E_r ; in the surface layer an azimuthal field B_ϕ, falling from its value on the vacuum side $(r > R(t))$ to zero in $r < R(t)$, arises. E_r drops to zero both in $r > R(t)$ and $r < R(t)$, but is non-zero on the surface $r = R(t)$. Here $R(t)$ is the pinch radius at time t and r is any radial distance from the pinch axis.

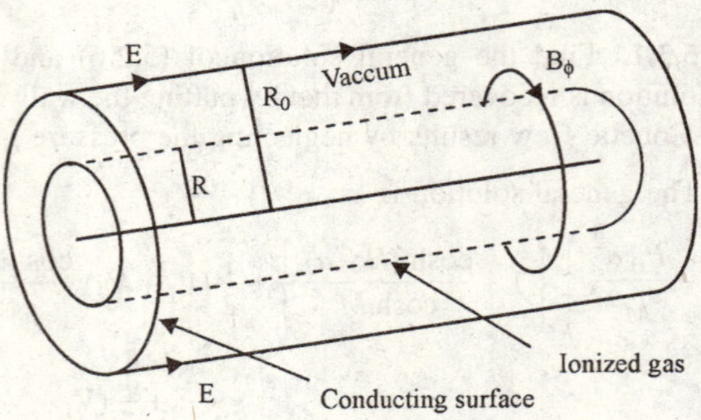

Fig. 5.4.1 The Pinch effect.

Field equations for infinitely conducting non-viscous fluids can be used. Neglecting gravitation these are

$$\frac{d\mathbf{v}}{dt} = -\frac{1}{\rho}\nabla p + \frac{1}{4\pi\mu\rho}(\nabla \times \mathbf{B} \times \mathbf{B}), \quad \frac{\partial \mathbf{B}}{\partial t} = \nabla \times (\mathbf{v} \times \mathbf{B}),$$

$$\frac{\partial \rho}{\partial t} + \nabla \cdot (\rho\, \mathbf{v}) = 0, \quad (\nabla \cdot \mathbf{B}) = 0, \quad \frac{d}{dt}(p\rho^{-\gamma}) = 0 \tag{5.4.1}$$

Here $\mathbf{v}$ is the fluid velocity, p is the scalar pressure, ρ is the mass density and γ is the ratio of the specific heats.

(a) Pinch statics : Bennet's relation

The condition of pressure balance across the sheath of surface current is

$$\frac{\mu H_\varphi^2}{8\pi} + \frac{\mu H_{ze}^2}{8\pi} = p + \frac{\mu H_{zi}^2}{8\pi} \text{ at } r = R \tag{5.4.2}$$

Here B_{zi} is the axial field inside, B_{ze} is that in the surrounding vacuum and p is the pressure at the interface with vacuum. On $r = R(t)$ we put

$$B_\varphi = B_0 \frac{R}{r}, \quad B_{ze} = \alpha\, B_0, \quad B_{zi} = \alpha_i\, B_0 \tag{5.4.3}$$

and find that (5.4.2) reduces to

$$\frac{p}{p_H} = 1 + \alpha_e^2 - \alpha_i^2 \tag{5.4.4}$$

where $p_H = \mu H_0^2/8\pi$. So p_H is the pressure due to the azimuthal magnetic field on the interface. Since p is a positive definite, α_e and α_i must satisfy the condition

$$\alpha_i^2 << 1 + \alpha_e^2 \tag{5.4.5}$$

Assuming that the variables depend on the radial distance r only, the Maxwell equation

$$\oint (\mathbf{B} \, . \, d\mathbf{l}) = \frac{4\pi\mu}{c} \iint (\mathbf{j} \, . \, d\mathbf{\Sigma})$$

reduces to

$$\int_{\varphi=0}^{2\pi} B_\varphi \, r d\varphi = \frac{4\pi\mu}{c} \int_{\varphi=0}^{2\pi} \int_0^r j_z \, r d\varphi \, dr = \frac{4\pi\mu}{c} I \tag{5.4.6}$$

where I is the volume current in the pinched plasma. If the cylinder $r = R$ is filled up with plasma, the induced magnetic field $B_\varphi (r)$ is given by

$$B_\varphi (r) = \frac{2I\mu}{cr}, \quad r > R(t) \tag{5.4.7}$$

with
$$I = \int_0^R 2\pi r j_z \, dr \tag{5.4.8}$$

I is the total current flowing in the cylinder. From the momentum transfer condition of (5.4.1), for the static pinch, we get

$$\frac{dp}{dr} = -\frac{B_\varphi}{4\pi\mu r} \frac{d}{dr}(r B_\varphi) = -\frac{1}{8\pi\mu r^2} \frac{d}{dr}(rB_\varphi)^2 \tag{5.4.9}$$

Integration of this relation with respect to r yields

$$p(r) - p_0 = -\frac{1}{8\pi\mu} \int_0^r \frac{1}{r^2} \frac{d}{dr}(rB_\varphi)^2 \, dr \tag{5.4.10}$$

where p_0 is the plasma pressure on the z-axis. Since $p = 0$ at $r = R(t)$, we find that

$$p_0 = \frac{1}{8\pi\mu} \int_0^R \frac{1}{r^2} \frac{d}{dr}(r^2 B_\varphi^2) \, dr$$

Hence
$$p = \frac{1}{8\pi\mu} \int_r^R \frac{1}{r^2} \frac{d}{dr}(r^2 B_\varphi^2) \, dr \tag{5.4.11}$$

We now define the average pressure $\langle p \rangle$ by

$$\langle p\rangle = \frac{1}{\pi R^2} \int_0^R 2\pi r p \, dr \tag{5.4.12}$$

Integrating the right hand side by parts we obtain

$$\frac{\langle p\rangle \pi R^2}{2\pi} = -\frac{1}{2} \int_0^R r^2 \frac{dp}{dr} dr$$

Using (5.4.9), the average pressure $\langle p \rangle$ is found to be related to the total current and pinch radius,

$$\langle p\rangle = \frac{1}{8\pi\mu R^2} \int_0^R \frac{d}{dr}(r^2 B_\varphi^2)\, dr = \frac{B_\varphi^2(R)}{8\pi\,\mu} = \frac{I^2 \mu}{2\pi R^2 c^2} \tag{5.4.13}$$

If electrons and ions are at the same temperature T, and the number density of electrons per unit length is N_e, then

$$N_e = \int_0^R 2\pi r N dr = \frac{2\pi}{2\kappa T} \int_0^R r p \, dr = -\frac{\pi}{2\kappa T} \int_0^R r^2 \frac{dp}{dr} dr$$

$$= \frac{\pi}{2\kappa T} \frac{1}{8\pi\mu} \int_0^R \frac{d}{dr}(r^2 B_\varphi^2)\, dr = \frac{[R^2 B_\varphi^2(R)]}{16\kappa T\,\mu} = \frac{\mu I^2}{4\kappa T c^2}$$

where (section 9.8), $p = N\kappa(T_e + T_i) = 2N\kappa T$, and N is the plasma number density, κ is the Boltzmann constant, T_e and T_i are the kinetic temperatures for electrons and ions respectively, and here $T_e = T_i = T$. Hence we obtain, the Bennet relation

$$I^2 = \frac{4\kappa T N c^2}{\mu}, \tag{5.4.14}$$

for the linear pinch. It shows that the kinetic temperature is proportional to the square of the total current I flowing in the discharge tube and inversely to the line density N of the particles. Equation (5.4.13) shows that the average pressure of matter is equal to the magnetic pressure at the surface of the cylinder. When

$$T = 10^8\,{}^\circ\text{K},\ N = 10^{15}\ \text{cm}^{-3},$$

we have $p = N\kappa T = 1{\cdot}4 \times 10^7$ dynes/cm^2 (= 14 atmosphere). Magnetic fields of the order of 1.9×10^4 gauss can produce this amount of pressure. The current I, inducing this field, is given by (5.4.7). $I = Bc\, r/2\,\mu = r \times 28.5 \times 10^{13}$ stat amperes if $B = 1.9 \times 10^4$ gauss. So a rather high current is necessary for confinement of a plasma column by its own azimuthal magnetic field. Actually the

Bennet relation gives the current discharged through the column to contain a plasma at a specific temperature and a given number of particles per unit length of the column. Since radiation losses increase with increasing current, a plasma cannot be confined by a sufficiently large current through the column at any specific temperature. So the radiation loss limits the amount of current that can be discharged. The assumption that the pinched plasma is at a specific constant temperature and stable are rather not true. For this reason the validity of this theory is doubted, though it is largely accepted.

(b) Pinch dynamics

Since the pinch is assumed infinitely conducting, the electric field in the laboratory system is given by

$$\mathbf{E}' = \frac{\mathbf{j}}{\sigma} = 0 = \mathbf{E} + \frac{1}{c}(\mathbf{v} \times \mathbf{B}) \text{ in } r < R(t) \tag{5.4.15}$$

and

$$\mathbf{E}' = \mathbf{E} \text{ in } r > R(t) \tag{5.4.16}$$

In the Maxwell equation $\oint (\mathbf{E} \,.\, d\mathbf{l}) = -\frac{1}{c}\frac{d}{dt}\iint_{\Sigma} (\mathbf{B} \,.\, d\boldsymbol{\Sigma})$ the loop of the line integral in the left hand side is taken to be a rectangle in any axial plane, whose two parallel sides are the lines of intersection of the plane with the surfaces $r = R_0$ and $r = R(t)$. If l is the axial length, then $d\boldsymbol{\Sigma} = \hat{\mathbf{X}}_{\varphi}\, l dr$ where $\hat{\mathbf{X}}_{\varphi}$ is the unit vector along the azimuthal direction. Since E is parallel to the axis,

$$EX(-l) + 0 \times l = -\frac{l}{c}\frac{d}{dt}\int_{R(t)}^{R_0} B_{\varphi}\, dr$$

because, in the left hand side the only contribution comes from that side of the loop which lies in the conducting well. Hence

$$E = \frac{1}{c}\frac{d}{dt}\int_{R}^{R_0} B_{\varphi}\, dr \tag{5.4.17}$$

In this way the applied field E is estimated in terms of the rate of change of flux of B around the plasma. Putting (5.4.7) in (5.4.17) we obtain a relation between current and pinch radius for a given electric field

$$E = -\frac{d}{dt}\left(\frac{2I\mu}{c^2}\log\frac{R_0}{R}\right) \tag{5.4.18}$$

When the plasma is pinched by the azimuthal field, the boundary moves in and sweeps off the particles with it. A surface layer is formed which is at rest in a coordinate system moving with the boundary. This boundary sheet is regarded here as a plane. The sheet velocity is generally very large compared to the thermal motion of the individual particles. These particles move towards the surface with the velocity $-\dot{R}$ and are reflected back with velocity $\dot{R}$. So the change in momentum per particle is $2M\dot{R}$ where M is the mass per particle. The number of particles striking unit area of the sheet per unit time is $N\dot{R}$. The momentum transfer per unit area per unit time, therefore, is $2M\dot{R} \,.\, N\dot{R}$. Since this transfer is due exclusively to the pressure of the pinching field, we can write

$$\frac{B_\varphi^2}{8\pi\mu} = 2MN\dot{R}^2 \tag{5.4.19}$$

Putting the value of B_φ from (5.4.7) in this relation we get $I^2 = (4\pi\rho c^2/\mu)(R\dot{R})^2$ where $\rho = MN$. We find that

$$I = -\left(\frac{4\pi}{\mu}\rho c^2\right)^{1/2} (R\dot{R}) \tag{5.4.20}$$

because, relative to the boundary, $\dot{R} < 0$. Eliminating I from this relation, with the help of (5.4.18), the equation of motion of the interface between the pinched plasma and vacuum is obtained :

$$2R\dot{R}\log\left(\frac{R_0}{R}\right) = -c\left(\frac{\mu}{4\pi\rho}\right)^{1/2} \int_0^t E\,dt \tag{5.4.21}$$

because $\dot{R} = 0$ at $t = 0$. Assuming the electric field constant in time, integrating this relation, we get

$$\log\left(\frac{R_0}{R}\right)\left(\frac{dR^2}{dt}\right) = -cEt\left(\frac{\mu}{4\pi\rho}\right)^{1/2}$$

Further integrating between the limits R_0 (upper) and R (lower), we obtain

$$1 - \frac{R^2}{R_0^2} - 2\left(\frac{R}{R_0}\right)^2 \log\left(\frac{R_0}{R}\right) = \frac{CE}{R_0^2} t^2 \left(\frac{1}{4\pi\mu\rho}\right)^{1/2} \tag{5.4.22}$$

Since both sides of this relation are dimensionless quantities, the factor $(C^2E^2/4\pi\mu\rho)^{1/2}$ in the right hand side has the dimensions of speed. When, for example, the amount $1/30$ esu of electric field is applied in a plasma, having 10^{16} particles/cm^3, this velocity is about 104×10^6 cm/sec. For $R = 10$ cm using this value of velocity for $\dot{R}$, equation (5.4.20) gives $I = 10^{15}$ stat amperes.

The parameters specifying the pinch are the line density N and discharge current I. Time of development of the pinch is of the order of the collision time, which is inversely proportional to the collision frequency, and these parameters are independent of its linear dimensions. The pinch configuration is often used to compress an initial plasma, and consequently to generate a cylindircally imploding shock propagating perpendicular to the magnetic field. The filling gas in the cylindrical discharge tube is often pre-ionized by *UV* radiation from auxilliary sources at both ends. A slowly rising magnetic field penetrates into the plasma and also compresses it. The pinch devices are used in fusion research and as sources of electromagnetic and neutron radiation.

5.5 PINCH STABILITY

In cylindrical polar coordinates the external and internal magnetic fields in the unperturbed state are

$$\mathbf{B}_{0e} \equiv (0,\ B_\varphi,\ 0),\ \mathbf{B}_{0i} \equiv (0, 0, B_z) \tag{5.5.1}$$

The axial magnetic field B_z inside the plasma is constant, the azimuthal external field B_φ is inversely proportional to the distance r from the axis and the unperturbed state field variables p_0, ρ_0, etc., are constant. Assume that the time dependence of $\boldsymbol{\xi}$ is exp $(i\omega t)$ and find that (4.15.8) reduces to

$$-\omega^2 \boldsymbol{\xi} = \frac{1}{4\pi\mu\rho_0}\ [\{\nabla \times \nabla \times (\boldsymbol{\xi} \times \mathbf{B}_{0i})\} \times \mathbf{B}_{0i}] + C_S^2 \nabla (\nabla \cdot \boldsymbol{\xi}) \tag{5.5.2}$$

Also we assume that $\boldsymbol{\xi}$ has the form

$$\boldsymbol{\xi}(\mathbf{r},\ t) = [\xi_r(r),\ \xi_\varphi(r),\ \xi_z(r)] \exp[i(kz \pm m\varphi + \omega t)] \tag{5.5.3}$$

where m is the azimuthal number and K is the wave number in the z direction. The general perturbation is the sum of perturbations at all frequencies ω and all azimuthal numbers m, and is an integral over K. Since the governing equation (4.15.8) is linear in $\boldsymbol{\xi}$, each mode can be investigated separately and independent of the other modes.

The displacement is independent of the azimuthal angular coordinate for perturbations which are symmetric about the axis; so the configurations are of necking-off type or the sausage type, which are obtained for $m = 0$. For $m = 1$ the pinch is deformed into a kink or screw of helical shape but retaining a circular cross-section. Though the series of perturbations at $m = 2, 3$, etc., are theoretically possible, the modes at $m = 0$ and $m = 1$ are mainly studied for simplicity. We here consider only the $m = 0$ mode. Equation (5.5.2), therefore, yields the following three equations

$$-\omega^2 \xi_r \exp(ikz) = C_S^2 \frac{\partial}{\partial r}(\nabla \cdot \boldsymbol{\xi}) + \frac{B_z P_\varphi}{4\pi\mu\rho}, \tag{5.5.4}$$

$$-\omega^2 \xi_\varphi \exp(ikz) = -\frac{B_z P_r}{4\pi\mu\rho}, \tag{5.5.5}$$

$$-\omega^2 \xi_z \exp(ikz) = C_S^2\, ik\, (\nabla \cdot \boldsymbol{\xi}) \tag{5.5.6}$$

Here P_r and P_φ are the radial and azimuthal components of $\mathbf{P}$ where

$$\mathbf{P} = \nabla \times \nabla \times (\boldsymbol{\xi} \times \mathbf{B}_0) = \nabla \times \{(\mathbf{B}_{0i}\, \nabla)\, \boldsymbol{\xi} - \mathbf{B}_{0i}\, (\nabla \cdot \boldsymbol{\xi})\}$$

$$= B_z \nabla \times \left(\frac{\partial \xi}{\partial t} \right) - (\nabla (\nabla . \xi)) \times \mathbf{B}_{0i} = \frac{1}{r} B_z \left\{ - \hat{\mathbf{X}}_r \, r \frac{\partial^2 \xi_\varphi}{\partial z^2} \right.$$

$$+ \hat{\mathbf{X}}_\varphi \, r \left(\frac{\partial^2 \xi_r}{\partial z^2} - \frac{\partial^2 \xi_z}{\partial r \, \partial z} \right) + \hat{\mathbf{X}}_z \frac{\partial^2}{\partial z \, \partial r} (r \, \xi_\varphi) \left. \vphantom{\frac{\partial^2}{\partial z^2}} \right\} +$$

$$+ B_z \frac{\partial}{\partial r} \left\{ \frac{1}{r} \frac{\partial}{\partial r} (r \xi_r) + \frac{\partial \xi_z}{\partial z} \right\} \quad (5.5.7)$$

where $\hat{\mathbf{X}}_r$, $\hat{\mathbf{X}}_\varphi$, $\hat{\mathbf{X}}_z$ are the unit vectors along the radial, azimuthal and axial directions, respectively. Hence, equations (5.5.4)–(5.5.6) become

$$- \omega^2 \, \xi_r = \exp(-ikz) \, C_S^2 \frac{\partial}{\partial r} (\nabla . \xi) + C_A^2 \left(\frac{\partial^2 \xi_r}{\partial z^2} + \frac{1}{r} \left\{ \frac{1}{r} \frac{\partial}{\partial r} (r \, \xi_r) \right\} \right), \quad (5.5.8)$$

$$- \omega^2 \, \xi_\varphi = 0 + C_A^2 \frac{\partial^2 \xi_\varphi}{\partial z^2}, \quad (5.5.9)$$

$$- \omega^2 \, \xi_z = \exp(-ikz) \, C_S^2 \, ik \, (\nabla . \xi) \quad (5.5.10)$$

where obviously $C_A^2 = B_z^2 / 4\pi\mu \, \rho_0$. These relations simplify to

$$(K^2 \, C_A^2 - \omega^2) \, \xi_r = C_S^2 \frac{\partial}{\partial r} \left[\frac{1}{r} \left\{ \frac{\partial}{\partial r} (r \, \xi_r) + \frac{\partial}{\partial z} (r \, \xi_z) \right\} \right]$$

$$+ C_A^2 \frac{\partial}{\partial r} \left\{ \frac{1}{r} \frac{\partial}{\partial r} (r \, \xi_r) \right\}$$

$$= (C_S^2 + C_A^2) \frac{\partial}{\partial r} \left\{ \frac{1}{r} \frac{\partial}{\partial r} (r \, \xi_r) \right\} + ikC_S^2 \frac{\partial \xi_z}{\partial r}, \quad (5.5.11)$$

$$(K^2 \, C_A^2 - \omega^2) \, \xi_\varphi = 0, \quad (5.5.12)$$

$$\frac{K^2 \, C_S^2}{iK} \left[\frac{1}{r} \left\{ \frac{\partial}{\partial r} (r \, \xi_r) \right\} + iK \, \xi_z \right] - \omega^2 \, \xi_z = 0 \quad (5.5.13)$$

Since, generally $\xi_\varphi \neq 0$, equation (5.5.12) shows that the dispersion relation for the wave of ξ_φ is

$$K^2 \, C_A^2 = \omega^2 \quad (5.5.14)$$

Equations (5.5.11) and (5.5.13) give for ξ_r and ξ_z the following second degree equations

$$\frac{\partial^2 \xi_r}{\partial r^2}+\frac{1}{r}\frac{\partial \xi_r}{\partial r}+\left\{\frac{(\omega^2-k^2C_S^2)(\omega^2-K^2C_A^2)}{\omega^2(C_A^2+C_S^2)-K^2C_A^2C_S^2}-\frac{1}{r^2}\right\}\xi_r=0, \quad (5.5.15)$$

$$\frac{\partial^2 \xi_z}{\partial z^2}+\frac{1}{r}\frac{\partial \xi_z}{\partial r}+\frac{(\omega^2-K^2C_S^2)(\omega^2-K^2C_A^2)}{\omega^2(C_S^2+C_A^2)-K^2C_A^2C_S^2}\xi_z=0 \quad (5.5.16)$$

Solution of (5.5.16) is

$$\xi_z = I_0(\chi r) \quad (5.5.17)$$

where
$$\chi^2=\frac{(\omega^2-K^2C_S^2)(\omega^2-K^2C_A^2)}{\omega^2(C_A^2=C_S^2)-K^2C_S^2C_A^2} \quad (5.5.18)$$

and I_0 is the modified Bessel function of the first kind of order zero. Putting $r\chi = x$ in (5.5.12) we obtain the Bessel equation

$$\xi_r''+\frac{1}{x}\xi_r'+\left(1-\frac{1}{x^2}\right)\xi_r=0 \quad (5.5.19)$$

where $\xi_r' = d\xi_r/dx$, etc. Putting the value of ξ_z from (5.5.17) into (5.5.13) we get

$$\xi_r=\frac{ix\{\omega^2C_S^2-C_A^2(k^2C_S^2-\omega^2)\}}{K(K^2C_A^2-\omega^2)C_S^2}I_0'(\chi r) \quad (5.5.20)$$

(a) Boundary conditions

At the plasma-vacuum interface, both before and after the perturbation, the two boundary conditions are (1) the continuity of tangential component of the electric field and (2) the continuity of pressure. If **n** is the unit vector normal to the plasma surface, then $\mathbf{n}=\hat{\mathbf{X}}_r$. The tangential component of the electric field $\mathbf{E}'$ is $(\mathbf{n}\times\mathbf{E}')$; so within the pinched plasma

$$\mathbf{E}_i'=\mathbf{E}_i+(\mathbf{v}\times\mathbf{B}_0)/c \quad (5.5.21)$$

and outside it

$$\mathbf{E}_c'=\mathbf{E}_e+(\mathbf{v}\times\mathbf{B}_{0e})/c \quad (5.5.22)$$

because $\mathbf{E}_e$ is the outside field. If **A** is the vector potential of the perturbation in the vacuum field, then

$$\mathbf{E}_e=-\frac{1}{c}\frac{\partial \mathbf{A}}{\partial t},\ \mathbf{B}_e=\nabla\times\mathbf{A},\ \mathbf{v}=\frac{\partial \boldsymbol{\xi}}{\partial t} \quad (5.5.23)$$

The plasma is assumed to be infinitely conducting. So, from (5.5.21), we obtain

$$\mathbf{E}_i'=0=\mathbf{E}_i+(\mathbf{v}\times\mathbf{B}_{0i})/c \quad (5.5.24)$$

But the continuity of tangential component of the electric field gives

$$(\mathbf{n} \times \mathbf{E}_i') = (\mathbf{n} \times \mathbf{E}_e') \tag{5.5.25}$$

and

$$(\mathbf{n} \times \mathbf{E}_e) = -[\mathbf{n} \times (\mathbf{v} \times \mathbf{B}_{0e})]/c \tag{5.5.26}$$

Putting, in this relation, the value of $\mathbf{E}_e$ from (5.5.23) we get

$$(\mathbf{n} \times \mathbf{A}) = [\mathbf{n} \times (\boldsymbol{\xi} \times \mathbf{B}_{0e})] \tag{5.5.27}$$

Since $\mathbf{B}_{0e} = \hat{\mathbf{X}}_\varphi B_\varphi$ this relation reduces to

$$(\mathbf{n} \times \mathbf{A}) = -\hat{\mathbf{X}}_\varphi \xi_r B_\varphi \tag{5.5.28}$$

The second set of boundary conditions is the continuity of pressure both in equilibrium and in the perturbed state. These conditions give

$$P_0 + \frac{B_{0i}^2}{8\pi\mu} = \frac{B_{0e}^2}{8\pi\mu}, \tag{5.5.29}$$

$$P_0 + \delta P + \frac{1}{8\pi\mu}(\mathbf{B}_{0i} + \delta \mathbf{B}_i)^2 = \frac{1}{8\pi\mu}(\mathbf{B}_{0e} + \delta \mathbf{B}_e)^2 \tag{5.5.30}$$

Here

$$\delta P = -\rho_0 C_S^2 (\nabla \cdot \boldsymbol{\xi}),\ \delta \mathbf{B}_i = \mathbf{B}_i + (\boldsymbol{\xi} \cdot \nabla)\mathbf{B}_{0i}, \tag{5.5.31}$$

$$\delta \mathbf{B}_e = \mathbf{B}_e + (\boldsymbol{\xi} \cdot \nabla)\mathbf{B}_{0e}$$

δP is the perturbation in p of the relations of (4.15.5), $\mathbf{B}_i$ and $\mathbf{B}_e$ are the perturbations in the magnetic field due to oscillatory fluctuations.

(b) Stability condition

Using (5.5.29) and (5.5.30) and keeping terms up to first power of $\boldsymbol{\xi}$ and the field increments, we get

$$-C_S^2(\nabla \cdot \boldsymbol{\xi}) + \frac{1}{4\pi\mu\rho_0}(\mathbf{B}_{0i} \cdot \delta \mathbf{B}_i) = \frac{1}{4\pi\mu\rho_0}(\mathbf{B}_{0e} \cdot \delta \mathbf{B}_e) \tag{5.5.32}$$

Since $\mathbf{B}_{0i} = \hat{\mathbf{X}}_z B_z$, $\mathbf{B}_{0e} = \hat{\mathbf{X}}_\varphi B_\varphi$ and since B_φ is a function of r only, this equation becomes

$$-C_S^2(\nabla \cdot \boldsymbol{\xi}) + \frac{B_z}{4\pi\mu\rho_0}(B_{iz} + 0) = \frac{B_\varphi}{4\pi\mu\rho_0}\left\{B_{e\varphi} + \xi_r \frac{\partial B_\varphi}{\partial r}\right\} \tag{5.5.33}$$

Also $(\nabla \cdot \boldsymbol{\xi}) = \frac{1}{r}\left\{\frac{\partial}{\partial r}(r\xi_r) + 0 + \frac{\partial}{\partial z}(r\xi_z)\right\}$

Putting the value of $\partial(r\xi_r)/r\partial r$ from (5.5.10) in this relation we obtain

$$(\nabla \cdot \boldsymbol{\xi}) = -\frac{\omega^2 \xi_z}{ikC_S^2}\exp(ikz), \tag{5.5.34}$$

and (5.5.33) reduces to

$$\left(\frac{\omega^2}{ik}\xi_z\right)\exp(ikz)+\frac{B_z B_{iz}}{4\pi\mu\rho_0}=\left(\frac{B_\varphi}{4\pi\mu\rho_0}\right)\left(B_{e\varphi}+\xi_r\frac{\partial B_\varphi}{\partial r}\right) \qquad (5.5.35)$$

Here B_{iz} and $B_{e\varphi}$ are the perturbed increments of internal and external magnetic fields. Since $\mathbf{B}_i=\nabla\times(\boldsymbol{\xi}\times\mathbf{B}_{0i})$, we find that $B_{iz}=-(B_z/r)\,\partial(r\xi_r)/\partial r$. Moreover, the induced external azimuthal magnetic field $B_{e\varphi}$ is intially zero. Equation (5.5.35), therefore, becomes

$$\frac{\omega^2\xi_z}{ik}-\frac{B_z^2}{4\pi\mu\rho_0}\frac{1}{r}\frac{\partial}{\partial r}(r\xi_r)=\frac{B_\varphi\xi_r}{4\pi\mu\rho_0}\frac{\partial B_\varphi}{\partial r}$$

Putting the value of $\partial(r\xi_r)/r\,\partial r$ from (5.5.13) in this equation we get

$$\frac{\xi_z}{K}\left\{\omega^2+\frac{B_z^2(\omega^2-k^2C_S^2)}{4\pi\mu\rho_0 C_S^2}\right\}=\frac{\xi_r B_\varphi^2}{4\pi\mu\rho_0 r}$$

because B_φ is inversely proportional to r and so $\partial B_\varphi/\partial r=-B_\varphi/r$. Since $c_A^2=B_z^2\,4\pi\mu\rho_0$, this equation becomes

$$\frac{\xi_z}{k}\left\{\omega^2+\frac{c_A^2}{C_S^2}(\omega^2-k^2c_s^2)\right\}=\frac{\xi_r B_\varphi^2}{4\pi\mu\rho_0 r}$$

We now put the value of ξ_z from (5.5.17) and of ξ_r from (5.5.20) in this equation. Then cancelling the factor $\omega^2C_S^2+C_A^2(\omega^2-k^2C_S^2)$ from both sides we get

$$\frac{I_0(\chi r)}{kC_S^2}=-\frac{B_\varphi^2\chi I_0'(\chi r)}{4\pi\mu\rho_0 rK(-K^2C_A^2+\omega^2)C_s^2}$$

$$\text{Hence}\quad \omega^2-K^2C_A^2=-\frac{B_\varphi^2\chi I_0'(\chi r)}{4\pi\mu\rho_0 rI_0(\chi r)}=-\frac{B_\varphi^2}{B_z^2}\frac{C_A^2\chi I_0'(\chi r)}{rI_0(\chi r)}$$

From solutions of the Bessel equation it can be shown that $\chi rI_0'(\chi r)/I_0(\chi r)$ is positive for real χ. So instability will occur if $\omega^2<0$, that is, if $B_z^2<(B_\varphi^2/k^2r^2)\{\chi rI_0'(\chi r)\}/I_0(\chi r)$. Hence if $B_z^2>(B_\varphi^2/k^2r^2)\{\chi rI_0'(\chi r)\}/I_0(\chi r)$, the pinch will remain stable. Thus, for stability, a finite minimum axial magnetic field is necessary. But the condition of equilibrium, (5.5.29). must be satisfied. So a maximum value of B_z also exists because $B_z^2<B_\varphi^2$.

A self-constricted cylindrical pinch is therefore unstable with respect to the bending, which distorts the straight current path into a serpentine series of kinks.

The curvature can increase to bring the discharge into contact with the chamber walls.

5.6 MHD GENERATOR

In MHD power generators electrically conducting working fluids (gas or liquid) flow in a transverse magnetic field, producing, in the direction perpendicular to both of flow and the magnetic field, an induced emf, which can be used to derive a current via external load. In conventional power generation schemes the electric current of the power is produced according to Faraday's law of electromagnetic induction; however, in MHD power generation it is Ohm's law which gives that current. Particularly, the Hall current is used in some cases. The metal conductor of conventional generators is replaced in MHD generators by a current of plasma moving in a transverse magnetic field which is transmitted through a channel of conventional steam turbine cycle producing electric power, Fig. 5.6.1. However, as in other low frequency MHD processes, radiation losses are not important. MHD power generation principles, and those of the conventional power generators, differ in some respects and also concur in some other respects.

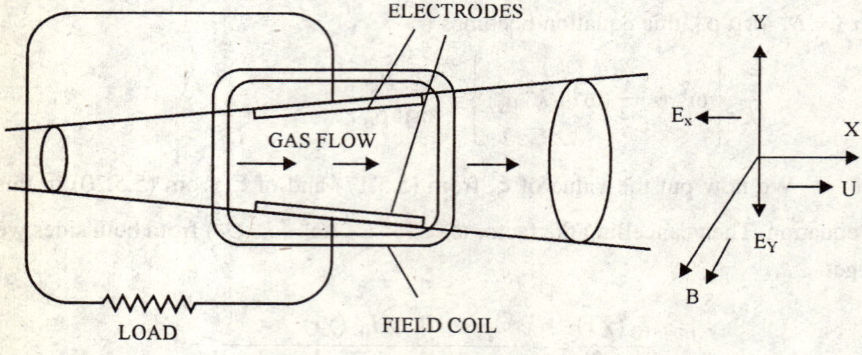

Fig. 5.6.1 MHD generator.

The electrical conductivity is enhanced through additives, called seed materials, having low ionization potentials. MHD flow can be a linear flow, vortex or radial flow. Linear channel flow geometry is widely studied and its fabrication technology is at an advanced level. Ionization and energy transfer occur mainly through interparticle collisions. Inelastic collisional losses are considerable from interaction between electrons and impurity atoms or electrons and molecules of the working fluid; so that the plasma is at equilibrium even in some regions near the electrodes. Concentration of neutral particles is 10^4 to 10^5 times that of electrons or ions, so the fluid is electrically conducting but neutral to a high degree; in other words, it is only partially ionized through radiation and seeding, and not through pure heating. Electron concentration is greater than about 10^{12} per CC. Applied dc magnetic field is about 2 tesla (20,000 gauss); so the induced magnetic field is totally negligible.

The working fluid, generally a monatomic inert gas, like argon, neon or helium or some combustion product of hydrogen—oxygen—air system like water gas, for example, is considered as a mixture of two fluids at two different temperatures, namely the electron fluid, consisting of both bound and free electrons at the electron temperature T_e, and the bulk of the working fluid consisting mainly of the heavy particles at the gas temperature T_g (≈ 2500°C to 3000°C). Since $T_e >> T_g$, the MHD working fluid is guided by the phenomena of two temperature ionization. Moreover, the thermal energy of the gas current is in the neutral gas. Flow speed is at two or more Mach numbers. Pressure at the exit is one atmosphere. Power is produced as the gas moves itself expanding through the nozzle in presence of very large dc magnetic field. Pinch effect is negligible in the flow because current is not large enough for that purpose. Gas discharge effects are also negligible because these occur only at very low pressures.

As in homopolar rotating devices, only dc power is produced. To directly convert the dc power to ac power, direct electrodeless conversion is employed with the help of sinusoidally varying magnetic field; which is extracted from the magnetic coils themselves or from subsidiary coils and which travels in the flow direction at a velocity far less than the flow velocity. Such a field, travelling down the duct, allows electrodes to be eliminated and generator power output to be induced into the magnetic circuit. So the conventional circuit switching or alternating the magnetic field are not employed for the conversion.

The plasma flow study in MHD generators is complicated by the presence of boundary layers near the walls of both insulator and electrodes. Plasma parameters in the boundary layer regions determine quantities like heat losses, friction, voltage drop near electrode, reversal of currents. The channel cross-section is divided into three regions : (1) the core region, (2) the electrode boundary layer region, namely the anode and cathode layers and (3) insulator boundary layer region, namely the upper and lower insulator layers. The plasma is in thermodynamic equilibrium there and radiation losses are negligible.

Maintenance of a steady flow of conducting gas and a steady value of the conductivity are some of the major problems responsible for the late appearance of MHD power generators even in this era. Presently it is cost, and not the efficiency, which is the prohibitive factor for extensive use of these types of power plants.

(a) Seeding effect

Most of the noble gases and combustion products, which serve as working fluids in MHD generators, have ionization potentials more than 12 ev (1 ev means 11600°K). These gases, therefore, get ionized only at temperatures far in excess of 4000°K. So, at the temperatures necessary for operation of MHD generators, thermal ionization and thermally produced electrical conductivity are completely negligible. Though the degree of ionization is small at these temperatures, the magnitude of the resulting electrical conductivity is sufficient for use in MHD power generation in commercial scale. Even 0.1 to 1 per cent of the particles of the seed material, by volume, gives sufficient energy of ionization to raise the electrical conductivity σ to values up to about 5×10^{12} sec^{-1} at 3000°K and from 5×10^{9} to 5×10^{10} sec^{-1} at 2000°K at equilibrium.

Cesium is much used as a seed material. Helium, mixed with Cesium vapour, works well as the working fluid. Maximum electrical conductivity is obtained when the Cesium concentration is about 1.4 atomic per cent; but, to keep the partial pressure of cesium in the system low, a smaller concentration is preferred, because the molecular weight of cesium is high. So, a large doze will increase the overall molecular weight and thereby will reduce the speed of sound in the working fluid. About 0.5 per cent of cesium in helium eliminates most of the difficulties without appreciably decreasing the electrical conductivity. The partial pressure of seed vapour is about 10^{-2} atmosphere in the working fluid at one atmosphere. The amount of seeding material is decided by optimization, the electrical conductivity being a very important parameter in this decision.

(b) Closed-cycle generators

In closed cycle MHD generators the seeded gas is repeatedly raised to the necessary temperature in a special heater. Even nuclear heaters or thermonuclear reactors can be employed for heating the gas to temperatures of the order of 2500° to 3500°K in a reactor with solid elements. Joule heating appreciably increases the electron temperature. Energy transfer is possible through elastic collisions between electrons and inert gas atoms. The electron concentration is determined by the ionization equilibrium at the electron temperature.

The physical principles of open-cycle generators have been developed; but there are no corresponding high temperature gas-cooled nuclear reactors which can consume the heat under high temperature conditions of these generators.

5.7 OPEN-CYCLE GENERATORS

All MHD power plants in operation are open-cycle generators. So their theory and fabrication techniques have been developed.

In open-cycle MHD generators thermal ionization is responsible for the generation of electrical conductivity of the working fluid. The hot gas possessing adequate conductivity pass through a channel of rectangular cross-section. One pair of walls facing the poles of the electromagnet forms the insulating walls, while the other pair is for the electrodes. The combustion products are directly passed through the channel only once and then are discharged. This hot discharge gas is generally cycled to increase efficiency in thermal power plants providing unlimited power for limited durations.

These generators can be used in space vehicles and in mobile ground units where high power from light apparatus is required. A heat exchanger to the energy source is required. A nuclear reactor can be used as the energy source.

Pulsed MHD generators employing the combustion products of solid fuels with alkali metal mixture have been planned.

Assuming both charge neutrality and incompressibility conditions and ignoring the effects of electric current varying in time (*i.e.*, neglecting $\partial \mathbf{j}/\partial t$) and ion collision frequency; the simplified generalized Ohm's law then is

$$\frac{1}{\sigma}\mathbf{j} = \mathbf{E} + \frac{1}{c}(\mathbf{u} \times \mathbf{B}) - \frac{M}{\rho e c}(\mathbf{j} \times \mathbf{B}) \tag{5.7.1}$$

where $\mathbf{j}$ is the electric current density, ρ is the mass density, $\mathbf{u}$ is the average velocity, M is the ionic mass, $\mathbf{E}$ is the applied electric field, e is the electronic charge per particle and σ is the electrical conductivity. Evidently

$$\sigma = \frac{Ne^2}{m\nu}, \quad \rho = MN \tag{5.7.2}$$

where ν is the collision frequency, N is the electron number density, and m is the electron mass.

The efficiency β of the MHD generators is defined by

$$\beta = \frac{\text{work output}}{\text{work input}} \tag{5.7.3}$$

For the Lorentz force,

$$\beta = \frac{(\mathbf{E} \cdot \mathbf{j})}{\frac{1}{c}((\mathbf{j} \times \mathbf{B}) \cdot \mathbf{u})} \tag{5.7.4}$$

This β is called the generator loading coefficient. Another quantity k, called the load factor, is the ratio of the load resistance to total resistance of the circuit. So, for the geometries adopted here,

$$E_y = K\frac{u}{c}B \tag{5.7.5}$$

Another efficiency parameter η is defined in the relation (5.7.39). The load factor k and the efficiency β are identical in many cases. Assume the flow to be approximately one-dimensional; then take the flow velocity $\mathbf{u}$ parallel to the X-axis and the applied magnetic field parallel to the Z direction. The induced electric field is

$$\mathbf{E}_i = \frac{1}{c}(\mathbf{u} \times \mathbf{B}) = -\frac{u}{c}B\,\hat{\mathbf{Y}} \tag{5.7.5 (a)}$$

where $\hat{\mathbf{X}}$, $\hat{\mathbf{Y}}$, $\hat{\mathbf{Z}}$ are the unit vectors parallel to the coordinate axes. The corresponding induced current is

$$\mathbf{j} = \sigma \mathbf{E}_i = -(\sigma/c)\,uB\,\hat{\mathbf{Y}} \tag{5.7.5 (b)}$$

The body force is the reduced Lorentz force

$$\mathbf{F}_i = \frac{1}{c}(\mathbf{j} \times \mathbf{B}) = -\frac{\sigma}{c^2}uB^2\,\hat{\mathbf{X}} \tag{5.7.5 (c)}$$

This force is opposite to the directions of flow of the working fluid. So the plasma is decelerated and the fluid pressure upstream of the channel is greater than that at the down stream. Hence energy is extracted from the moving fluid at the expenses of a pressure drop. The lost kinetic energy induces direct current in the external circuit. The internal dissipation ($= j^2/\sigma$, per unit volume), due to the current produced, can induce non-equilibrium ionization. The induced electric field, $-(\mathbf{U} \times \mathbf{B})/C$, can produce electron concentration much in excess of that in thermodynamic equilibrium at the gas temperature. The influence of non-equilibrium ionization on electrical conductivity and Hall effect is considerable.

Ohm's law, equation (5.7.1), gives

$$j_x = \frac{\sigma}{1+\alpha^2}\left[E_x - \alpha\left(E_y - \frac{1}{c}uB\right)\right]$$

$$= \sigma_{\|} E_x - \sigma_{\perp}\left(E_y - \frac{u}{c}B\right), \tag{5.7.6}$$

$$j_y = \frac{\sigma}{1+\alpha^2}\left[E_y - \frac{u}{c}B + \alpha E_x\right] = \sigma_{\|}\left(E_y - \frac{u}{c}B\right) + \sigma_{\perp} E_x \tag{5.7.7}$$

where

$$\alpha = \frac{\sigma M}{\rho e}B = \frac{\sigma B}{Ne} = \frac{Ne^2 B}{m\nu\, Nec} = \frac{\Omega}{\nu},\quad \Omega = \frac{eB}{mc} \tag{5.7.8}$$

$$\sigma_{\|} = \frac{\sigma}{1+\alpha^2} = \frac{Ne^2}{m}\,\frac{\nu}{\nu^2+\Omega^2}, \tag{5.7.9}$$

$$\sigma_{\perp} = \frac{\sigma\alpha}{1+\alpha^2} = \frac{Ne^2}{m}\,\frac{\omega}{\nu^2+\Omega^2} \tag{5.7.10}$$

Ω is the Larmour frequency of gyration of the electrons. Eliminating E_y between (5.7.6) and (5.7.7) we obtain

$$j_x = -\alpha j_y + \sigma E_x \tag{5.7.11}$$

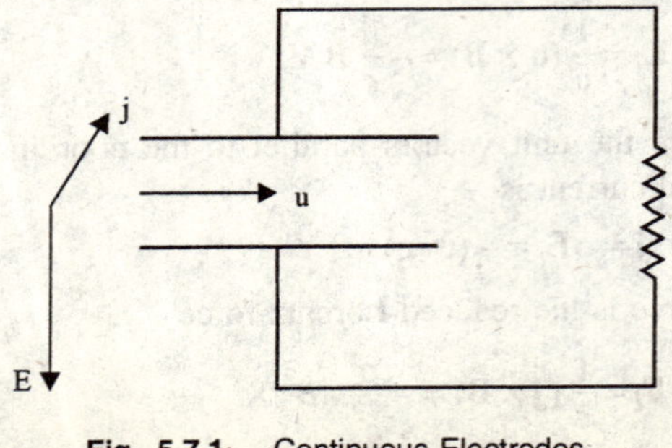

Fig. 5.7.1 Continuous Electrodes.

And, if E_x is eliminated from (5.7.6) and (5.7.7) we obtain

$$j_y = \alpha j_x + \sigma \left(E_y - \frac{u}{c} B \right) \tag{5.7.12}$$

Electrode configurations for extracting the power from MHD generators, are those of either continuous electrodes, Fig. 5.7.1, or segmented electrodes, Fig. 5.7.2, or Hall electrodes, Fig. 5.7.3.

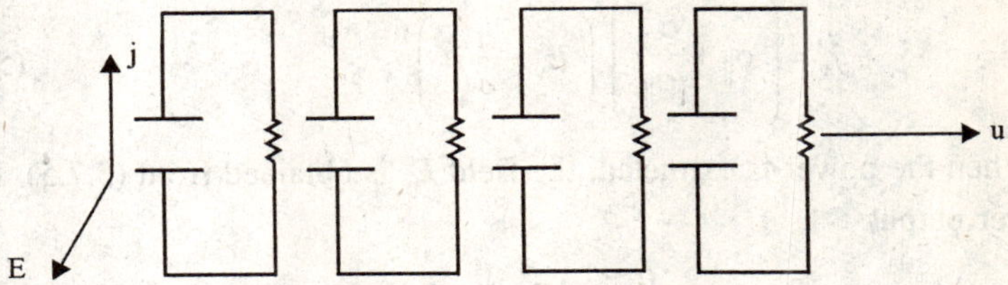

Fig. 5.7.2 Segmented Electrodes.

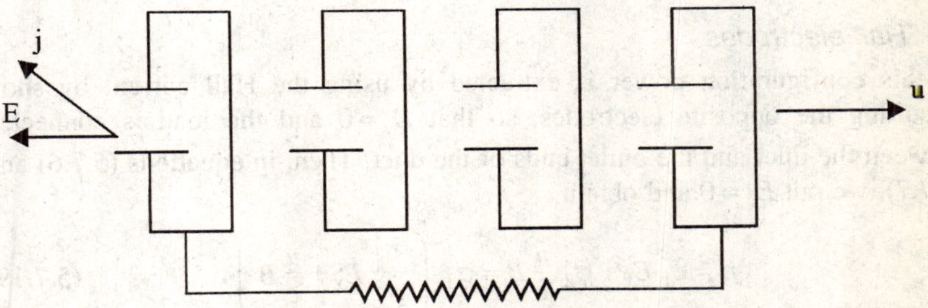

Fig. 5.7.3 Hall Electrodes.

(a) Continuous electrodes

Electrodes being equipotential surfaces, no electric field can appear axially along the duct, so we put $E_x = 0$ in (5.7.6) and (5.7.7), and obtain

$$j_y = \sigma_{\parallel} \left(E_y - \frac{u}{c} B \right) \tag{5.7.13}$$

The open circuit field E_{y0} is obtained by putting $j_y = 0$ in (5.7.13); then

$$E_{y0} = \frac{u}{c} B \tag{5.7.14}$$

However, when power is extracted from the gas j_y is finite and the field is given by (5.7.5) in terms of the load factor K. The power P, produced per unit volume, is

$$P = (\mathbf{E} \cdot \mathbf{j}) = E_y j_y = \sigma_{\parallel} \left(E_y - \frac{u}{c} B \right) E_y = \sigma_{\parallel} K (1 - K) B^2 (u^2/c^2) \tag{5.7.15}$$

(b) Segmented electrodes

In this configuration there is no closed circuit path for the axial current, so $j_x = 0$ and equations (5.7.6) and (5.7.7) give

$$E_x = \frac{\sigma_\perp}{\sigma_\parallel}\left(E_y - \frac{u}{c}B\right) \tag{5.7.16}$$

Hence the current density in the Y direction is

$$j_y = \left(\sigma_\parallel + \frac{\sigma_\perp^2}{\sigma_\parallel}\right)\left(E_y - \frac{u}{c}B\right) \tag{5.7.17}$$

When the power is extracted, the field E_y is obtained from (5.7.5). Hence the power output

$$P = (\mathbf{E}\,.\,\mathbf{j}) = \left(\sigma_\parallel + \frac{\sigma_\perp^2}{\sigma_\parallel}\right) K(1-K)\frac{u^2}{c^2}B^2 \tag{5.7.18}$$

(c) Hall electrodes

In this configuration power is extracted by using the Hall current by short circuiting the opposite electrodes, so that $E_y = 0$ and the load is connected between the inlet and the outlet ends of the duct. Then, in equations (5.7.6) and (5.7.7), we put $E_y = 0$ and obtain

$$j_x = \sigma_\parallel E_x + \sigma_\perp \frac{u}{c}B = \sigma_\perp\left(\frac{\sigma_\parallel}{\sigma_\perp}E_x + \frac{u}{c}B\right) \tag{5.7.19}$$

The open circuit field E_{x0}, obtained by putting $j_x = 0$ in this relation, is

$$E_{x0} = -\frac{\sigma_\perp}{\sigma_\parallel}\frac{u}{c}B \tag{5.7.20}$$

When power is extracted the field is given by

$$E_x = -\frac{\sigma_\perp}{\sigma_\parallel}K\frac{u}{c}B \tag{5.7.21}$$

and

$$P = (\mathbf{E}\,.\,\mathbf{j}) = j_x E_x = \frac{\sigma_\perp^2}{\sigma_\parallel}K(1-K)\frac{u^2}{c^2}B^2 \tag{5.7.22}$$

The relations (5.7.15), (5.7.18) and (5.7.22), for power density, can therefore be written as

$$P = \sigma_{eff} K(1-K)\frac{u^2}{c^2}B^2$$

where $\sigma_{eff} = \sigma_\parallel$ for continuous electrodes

$$= \sigma_{\|} + \frac{\sigma_{\perp}^2}{\sigma_{\|}} \text{ for segmented electrodes}$$

$$= \frac{\sigma_{\perp}^2}{\sigma_{\|}} \text{ for Hall electrodes} \tag{5.7.23}$$

(d) Faraday type generators

In Faraday type of MHD generators loads are connected so that the primary current flow in the gas is in the direction of $\mathbf{U} \times \mathbf{B}$. Consequently, no current remains in the direction of flow. So,

$$j_x = 0 \tag{5.7.24}$$

Then, using (5.7.5), equations (5.7.12) and (5.7.11) become

$$j_y = \sigma \left(E_y - \frac{u}{c} B \right) = -\frac{\sigma}{c} u\, B\, (1 - K) \tag{5.7.25}$$

$$E_x = \frac{\alpha}{\sigma} j_y = \alpha \left(E_y - \frac{u}{c} B \right) = -\frac{\alpha}{c} uB\, (1 - K) \tag{5.7.26}$$

To show that the generator loading coefficient β, defined by (5.7.4), is identical to the load factor k of (5.7.5), we first determine the numerator and denominator of (5.7.4)

$$(\mathbf{E} \cdot \mathbf{j}) = E_y\, j_y = -\frac{\sigma}{c^2} u^2 B^2 (1 - K)\, K, \tag{5.7.27}$$

$$\frac{1}{c} j^2 = \sigma \left(\frac{u^2 B^2}{c^2} \right) (1 - K)^2 \tag{5.7.28}$$

$$(\mathbf{j} \times \mathbf{B})_x = j_y\, B = -\frac{\sigma}{c} uB^2 (1 - \beta), \tag{5.7.29}$$

$$\frac{1}{c} u\, (\mathbf{j} \times \mathbf{B})_x = -\frac{\sigma}{c^2} u^2 B^2 (1 - \beta) \tag{5.7.30}$$

Equations (5.7.27) and (5.7.30), with the help of (5.7.4), show that $\beta = K$. Also it is seen that

$$\mathbf{u} \cdot (\mathbf{j} \times \mathbf{B})_x + \frac{j^2}{\sigma} = (\mathbf{E} \cdot \mathbf{j})$$

Hence, sum of the work done by the gas against the field and internal dissipation equals the output.

(e) Generators in which the Hall current dominates

When the Hall current flow j_x dominates, we put $E_x = 0$ in (5.7.7) and obtain

$$j_y = \frac{\sigma}{1+\alpha^2}\left(E_y - \frac{u}{c}B\right) \tag{5.7.32}$$

Then the electrical conductivity is reduced by the factor $1+\alpha^2$, and the short circuit of the Hall potential is prevented by segmenting the device in the axial direction, giving

$$j_y = -\frac{\sigma u B}{c\,(1+\alpha^2)}(1-K), \tag{5.7.33}$$

$$j_x = -\frac{\sigma \alpha u B}{c\,(1+\alpha^2)}(1-K), \tag{5.7.34}$$

$$(\mathbf{E}\,.\,\mathbf{j}) = E_y\, j_y = -\frac{\sigma u^2 B^2}{c^2\,(1+\alpha^2)}K\,(1-K), \tag{5.7.35}$$

$$\frac{1}{c}u\,(\mathbf{j}\times\mathbf{B})_x = -\frac{\sigma u^2 B^2}{c^2\,(1+\alpha^2)}(1-K), \tag{5.7.36}$$

so that K of (5.7.5) equals β of (5.7.3).

(f) Hall generators

In some Hall generators any circumferential velocity, induced by the radial current, is neglected. This is equivalent to putting

$$E_y = 0 \tag{5.7.37}$$

in (5.7.12), which yields the results

$$j_y = \alpha\, j_x - \frac{\sigma}{c}uB = -\frac{\sigma}{c}uB\,(1-\eta) \tag{5.7.38}$$

where, now, the loading coefficient η is defined by

$$\eta = \frac{\alpha\, j_x\, c}{\sigma u B} \tag{5.7.39}$$

Consequently, equation (5.7.11) gives

$$E_x = \frac{1}{\sigma}(j_x + \alpha\, j_y) = -\frac{u}{c}B\left\{\alpha\,(1-\eta) - \frac{\eta}{\alpha}\right\}, \tag{5.7.40}$$

$$(\mathbf{E}\,.\,\mathbf{j}) = E_x\, j_x = -\frac{\sigma}{c^2}u^2 B^2\left\{\eta\,(1-\eta) - \frac{\eta^2}{\alpha^2}\right\}, \tag{5.7.41}$$

$$\frac{1}{\sigma}j^2 = \frac{\sigma}{c}u^2 B^2\left\{(1-\eta)^2 + \frac{\eta^2}{\alpha^2}\right\}, \tag{5.7.42}$$

$$\frac{1}{c}(\mathbf{j}\times\mathbf{B})_x = -\frac{\sigma}{c^2}uB^2\,(1-\eta) \tag{5.7.43}$$

Then the efficiency β of (5.7.4) is given by

$$\beta = \eta \left\{ 1 - \frac{\eta}{\alpha^2 (1-\eta)} \right\} \tag{5.7.44}$$

and the relation (5.7.31) holds good. The second term in the brackets of (5.7.44) represents the additional internal dissipation due to the axial current. Dissipation is minimized if the generator operates at a high value of the Hall coefficient.

Since β is a function of η, for a given value of α^2, the maximum value,

$$\beta_{max} = \frac{(\sqrt{1+\alpha^2} - 1)}{\sqrt{1+\alpha^2}} \left\{ 1 - \frac{\sqrt{1+\alpha^2} - 1}{\alpha^2} \right\} \tag{5.7.45}$$

is obtained when

$$\eta = 1 - \frac{1}{\sqrt{1+\alpha^2}} \tag{5.7.46}$$

Hence, writing $\eta\, UA = \dot{m}$ where A is the channel area, $\dot{m}$ is the mean flow of the working fluid, and $I\,(= jA)$ is the total current, we obtain

$$\eta = \frac{c}{\sigma} \left(\frac{P\alpha}{B} \right) \left(\frac{I}{\dot{m}} \right) \tag{5.7.48}$$

In a two-terminal generator both I and $\dot{m}$ are constants. Similarly the ratio $P\alpha/\eta$ is approximately constant. Hence, for the peak efficiency, the electrical conductivity σ and the gas temperature should be constant. The value $\eta = 1$ for the loading coefficient corresponds to the short circuit j_{xs} if α is very high.

In these days of energy crisis, in many countries, MHD power generation possibilities are explored seriously. An open cycle MHD power plant is known to be functioning at Tiruchi in Tamil Nadu, which was built by Bhabha Atomic Research Centre and Bharat Heavy Electricals. Commercial MHD power generators are producing 5 MW to 20 MW power in Moscow. However, the British conclusion, achieved through ten years of search, is that MHD power generation would not be economic in British Isles; so research on this programme has been stopped there.

5.8 LABORATORY PLASMAS

A gas becomes electrically conducting even at room temperature in presence of electric field. Electrons receive energy from the applied field, and get accelerated to energies high enough to ionize other atoms of a gas through collisions. Since the gas molecules and ions are not much affected thermally by this field, these remain at the room temperature level, though electron temperature becomes of the order of 1 ev, because the collisional energy transfer from the heated electrons to heavy particles is generally very slow. The energy source in a gas discharge, thus, preferentially heats the electrons to a rather high temperature.

5.9 THERMONUCLEAR REACTIONS AND FUSION PHYSICS

An aim of scientists, since mid fifties of the twentieth century, is to build a viable miniature man-made sun, or a fusion reactor, which could harness the reserves of deuterium (an isotope of hydrogen) abundantly found on earth. Deuterium is present in water in the ratio of 1 atom to 6000 atoms of ordinary hydrogen. Total stock of deuterium in the water of the oceans exceeds 10^{14} tonnes. Energy available from deuterium in 1 litre of water equals the energy of 350 litre of petrol, the total energy from all the deuterium would be (about 10^{18} Megawatt years) enough to meet the requirement for human use for millions of years.

The basic fuel for the fusion reaction is hydrogen. Initial reactions, when deuterium is used as the fuel, are

$$D + D \longrightarrow T + p + 4.03 \text{ MeV}$$

$$D + D \longrightarrow {}^{3}He + n + 3.25 \text{ MeV}$$

where D is a deuteron, T is a triton (nucleus of tritium, which is another isotope of hydrogen). ^{3}He is a helium-3 nucleus, p is a proton, n is a neutron and MeV is million electron volt energy. These two reactions have an equal probability of occurrence. The tritons and helium-3 nuclei react exothermically with deuterons to yield, finally, helium-4 nuclei (α particles) :

$$D + T \longrightarrow {}^{3}He + n + 17.58 \text{ MeV}$$

$$D + He^{3} \longrightarrow {}^{4}He + p + 18.34 \text{ MeV}$$

A mixture of deuterium and tritium may also be employed as fuel, because (1) The $D-T$ reaction takes place at a lower temperature than necessary for the $D-D$ reaction; (2) The probability of occurrence of the $D-T$ reaction is larger; and the overall energy output is higher. But, tritum is not available in nature. It has to be continuously bred by surrounding the wall of the reactor chamber with a blanket of lithium which, upon bombardment by neutrons, yields tritium. Lithium is plenty on earth, so reactors with the $D-T$ fuel may run for many thousand years.

The heat released due to the created D – and T – ions, would maintain the temperature of the particles for a sufficient time (the time of confinement), for continuation of the fusion reactions, as a chain reaction. The $D-T$ reaction produces thermonuclear energy in hydrogen bombs. This bomb is absolutely of devasting type, because its energy, once released, is totally uncontrolled. Fusion reactors are expected to tame hydrogen bombs.

The temperature for starting thermonuclear reactions, employing deuterium as fuel, is about 400 million degrees celsius, and 40 million degrees when the fuel is a mixture of deuterium-tritium. Other factors also cause loss of energy. So the effective temperature for reactors using the $D-T$ fuel is at least 100 million degrees celsius. This temperature is higher than the solar core temperature, 14,000,000°C. Faraday rotation (section 3.14) measurement with waves beamed through fusion plasma is used to probe it. The number of these reactions per second must be sufficiently large for the energy generated to exceed the energy lost due to radiation by an appreciable amount, so that this reaction becomes useful for mankind. This number also depends on the time of plasma confinement.

For production of thermonuclear energy in excess of the input energy required for plasma heating, together with the usual energy losses, the product NT (where N is the number of nuclei per cm^3 of the plasma, and T is the confinement time in seconds) should exceed a given limiting value. This is the Lawson criterion. It must be satisfied for success of controlled fusion reactors. The minimum value of NT is about 10^{16} cm^{-3} s, if the fuel is deuterium; for the DT fuel it is 10^{14} cm^{-3} s. The amount of energy required to assemble and maintain a collection of energetic light nuclei and the energy released by fusion reaction among these nuclei, as a function of density and temperature, can be estimated from the Lawson condition. The thermonuclear energy production, at a fixed ion temperature, is proportional to the Lawson number *NT*. The specific thermonuclear burn rate is given by the law

$$\dot{a} = k\,\rho\,\langle \sigma\, v \rangle$$

where k is the constant of proportionality, a is the fraction burnt up, ρ is the density and $\langle \sigma v \rangle$ is the Maxwell velocity averaged over the reaction cross-section (section 9.12). Electromagnetic wave fields are employed to compress the plasma.

Hydrogen or deuterium gas is injected into the torus chamber, at the beginning of each pulse, by a controllable gas puffing system, in such a way that the working pressure in the vessel is 10^{-4} torr. The gas is pre-ionized by an electron beam for the discharge current pulse.

High electric field is applied along the magnetic field for stabilization. The plasma current creates a longitudinal magnetic field, that converts the purely toroidal field into a toroidal-poloidal field with a desirable value of the rotational transform; a low-density, low-temperature toroidal plasma is thus pulsed in the toroidal direction by the electric field to drive the large toroidal current necessary for production of the poloidal magnetic field. The current inside heats the plasma to a relatively high temperature. Field constructions satisfy the conditions of pressure balance and static configuration.

High speed plasma particles spreadout in all directions. When these come in contact with the material walls of the container (the reaction chamber), which is maintained at a much smaller temperature, the plasma looses very quickly, a considerable amount of heat. No matter can remain solid at so high temperatures. So, the plasma must be kept away from the chamber walls for a time, so that a sufficient number of fusion reaction takes place for production of energy.

Strong magnetic caging, for confining very hot plasma, is achieved by creating an intense toroidal magnetic field, by passing electric current through a number of coils, surrounding the toroidal vessel. The induction field from the toroidal field coils restrains motion of charged particles in perpendicular directions, so that the particles spiral with small radius around these lines. But such perfect motions, along toroidal and other endless field lines, is possible only if the magnetic field is uniform in space and constant in time. This is not practically possible, because magnetic fields have spatial gradients and time variation. The consequent charge separation, drifts and currents tend to quickly drive the plasma away from the magnetic cage.

A poloidal magnetic field, generated by an electric field along the toroidal direction, checks this drive. The resultant field lines are helical lines, about the circular central line, inside the torus. Each of these field lines winds endlessly on a magnetic surface. The ratio of the toroidal field to the poloidal field, at every point inside the torus, should be above a critical value, depending on the torus dimensions, for ensuring suppression of instabilities.

First Indian tokamak, imported from Japan (built by Toshiba Corporation), was installed in SINP, Calcutta, in July of 1987. Second Indian (India made) tokamak, named Aditya (Sun), has come up in Institute of Plasma Research in Gujarat, at Gandhinagar. Some tokamaks of other countries are: TFTR (Tokamak Fusion Test Reactor), Alcator (at MIT) and D-III in USA; T-10 and T-15 in USSR; JT-60 in Japan; JET (Joint European Torus) in UK.

In TFTR the major radius is 285 em, minor radius 85 cm, toroidal magnetic field 5.2×10^4 gauss (5.2 tesla); maximum plasma current 2.5 million amperes, the temperature achieved is about 200 million degrees celsius; but the Lawson criterion was far from the achieved value. In Alcator experiments the Lawson criterion was satisfied, but the plasma temperature was relatively low; moreover, the introduction of a small D-T pellet in the D-T plasma was found to considerably raise the Lawson number. The Aditya parameters are : Major radius 75 cm, minor radius 25 cm, toroidal magnetic field 1.5 tesla, plasma current 250 KA.

The SINP tokamak is 269 cm in length, and 262 cm in height, and in weight 8 tonnes. The toroidal vessel is made of steel. It has the major radius 30 cm and minor radius 8.5 cm. A few limiters in the torus confine the plasma to a minor radius 7.5 cm. After being baked continuously at a temperature of $150^{o}C$ for 3 to 4 days, with the help of heating coils wound around the torus vessel, to drive out the impurities from the inner walls, the vessel is brought to the base pressure 10^{-8} torr. The plasma current maximum, attained so far in the SINP tokamak, is 68.5 KA, which is little more than 90 percent of the maximum design value.

The plans of some investigation with the SINP tokamak are : (*a*) Instabilities in the tokamak plasma; (*b*) Transport processes in the plasma; (*c*) Interaction between the walls and the plasma; (*d*) Study of plasma particles and radiation which are incident on the inner walls, in spite of the magnetic confinement; (*e*) Study of atomic and molecular processes which are manifested particularly in the boundary regions; (*f*) Generation of flat-top plasma current (of the order of 75 KA), of 50 ms duration, by adding pulses forming networks to the capacitor banks, which are used for energising the magnetic coils; (*g*) Enhancement of temperature of the Joule heated plasma, by injecting powerful radio frequency waves into it; (*h*) Tokamak operation in the steady state, for a reasonable period of time, generating plasma current with the help of radio frequency waves; (*i*) Extension of the diagnostic facilities by developing some other systems, employing millimetre waves, laser beams, neutral beams, etc.

Chapter 6

COLLISION PROCESSES IN PLASMAS

6.1 INTRODUCTION

Collisions are elastic or inelastic. Elastic collisions, also called soft collisions or grazing collisions, do not change the internal energy of the colliding particles, so, these are studied classically. Inelastic collisions change the internal energy of the colliding particles; so these are studied quantum mechanically. Both elastic and inelastic collisions are possible in plasmas.

For collision study particles are assumed to be rigid and spherically symmetric point masses, having no spin; so the colliding particles have only three degrees of freedom due to translational motion. A particle, which experiences scattering by others, is called test particle. If one particle is heavier than others, it is called the scattering centre of the scatterers (or the field particles). The delta function representation of a test particle, moving with velocity **u**, is

$$f_t(\mathbf{v}) = a\delta(\mathbf{v} - \mathbf{u})$$

A relaxation time is the time it takes for collisions to effect a substantial change in a given initial distribution of momentum, or energy. In other words, for particles of one kind it is the time in which this kind changes direction by a large angle, say $\pi/2$ radians, due to presence of others; its inverse is the relaxation frequency of this collision.

The single, short range, individual encounters are binary collision processes; these are different from the long range collective effect of collisions. The binary and other individual encounters transfer particles in and out of the six-dimensional phase space volume element $d\mathbf{r}\,d\mathbf{v}$, at a rate that can be expressed via a collision operator, say of the form $(f-\bar{f})/\tau$, whereas collective effects enter by producing streaming losses (or gains) from $d\mathbf{r}\,d\mathbf{v}$.

Head-on collision is possible only for non-interacting spheres, as between two marble balls, or between a charged particle and an uncharged particle. Collision between two uncharged particles when their distance of separation is of the order of atomic dimensions, as in partially ionized plasmas, is a short range encounter, (or an inelastic collision). When an uncharged particle is so close to a charged particle, other particles cannot be so close to it. But a charge experiences long range Coulomb collisions with all the other charges simultaneously. So, while a charge experiences a short range single encounter it experiences a large number of long range Coulomb collision simultaneously, with many charges; the collective influence of all these determines its motion. So, short range single individual collisions are strictly binary, that is at a time only one such collision can be experienced by a particle. Therefore, for weekly ionized plasmas, the assumption of the existence of binary processes is valid.

Sometimes, as in nuclear physics, only individual short range encounters are useful. At levels of sub-nuclear physics high speed collisions, following from relativistic laws, dominate. The collisions are manifestations of particle behaviour. Study of subnuclear physics also reveals that one should not enquire too closely as to what is meant by a particle or by a collision.

(a) Collision, encounter, interaction

The physical processes of collision, encounter and interaction, cannot be clearly distinguished. A continuous change, occurring in a large interval of time, is an interaction, and an almost instantaneously occurring change is a collision, or an encounter. These three words are often used without discretion for both the processes. Actually, for distant encounters, interaction is continuous; so collision frequency in the sense of discrete events, becomes meaningless, and the effects of a series of distant and continuous encounters cannot be isolated.

The collision frequency ν is the probability of certain events per unit time; these might be due to the appearance of some particles so close to a specific particle that their influence on its trajectory cannot be neglected. The mean value of ν is the product of the number density of the colliding particles, the collision cross-section and the mean value of relative velocities. The average probability is a weighted mean of the collision frequency; it depends on the distribution of velocities of both the colliding particles. The probabilistic definition of collisions is used for study of short range interactions.

The overall effect of collisions is the integrated effect of interactions between particles of all velocities. Collisions explain plasma diffusion, transfer of energy between particles, and electrical conductivity. The momentum and energy of particles (electrons or ions) interacting with other particles change significantly in an average time $\bar{t}$, called the mean free time between successive collisions and over an average distance l, called the mean free path.

In soft collisions interaction is continuous and dependent on the interparticle force. So the particles have extended force fields. In hard collisions, the colliding particles have no extended force fields; they move freely between abrupt collisions. For soft collisions l is the average distance that a sequence of small angle collisions takes to stop the motion of a test particle, in a given direction, that is, to give it a deflection by the angle $\pi/2$, and $\bar{t}$ is the time it takes for this change to happen. For hard collisions l and $\bar{t}$ are evaluated from kinetic theory in terms of the particle diameter, number density and average speed.

Neutral particles and charged particles collide among themselves and with one another. Neutral particles transfer momentum only impulsively. In plasma momentum is transferred continuously via grazing collisions (soft collisions). The collisional interval time $\bar{t}$ is the average time in which a particle at a point

P at time t has gone there along a random walk path that commenced at a point P_1 at a time $\overline{t}$ earlier, when it was moving at the angle $\pi/2$ to its final direction at P at time t. The situation is just as if the impulsive force experienced at a hard collision by a neutral particle was smeared out over the whole collision time interval $\overline{t}$.

The term 'collisionless' does not mean that there are no collisions because collisions are continuous in a plasma. It merely means that the time scale of interest is less than the 'collision interval' $\overline{t}$ for a net deflection by the angle $\pi/2$. So, it is wrong to believe that in a collisionless plasma individual particles do not respond to pressure gradient forces. Actually, a particle experiences a continuous nonrandom force $-(\nabla p)/n$, due to the pressure gradient.

(b) Degrees of Freedom

Molecules of a gas have a variety of degrees of freedom. With each degree of freedom is associated an energy. Near the zero degree Kelvin there are only three translational degrees of freedom. At temperatures near 10°K, the rotational degrees of freedom are excited although the molecular bonds remain quite stiff. At temperatures $\geq 1000^{o}$K the vibrational degrees of freedom begin to act. Dissociation begins when the molecular impact becomes so strong that the molecular bond is stretched to the breaking point. At still higher temperatures ($\geq 5000^{o}$K), the electrons in the orbits around the atomic nucleus become excited to quantum states above the ground state. At still higher temperatures (for example at 10^{4o}K and pressure $\approx$ one atmosphere, electrons leave their orbits, and the gas becomes ionized. In diatomic gases, the electron-molecule collisions can excite the rotational and vibrational degrees of freedom of the molecules. Electrons ionize other atoms through inelastic collisions. They gain energy during elastic collision and lose energy during inelastic collision.

If P_i is the sum of all the lost energies in inelastic collisions we can write

$$P_i = P_{\text{rot}} + P_v + P_r \tag{6.1.1}$$

where P_{rot} is the loss due to rotational degrees of freedom of the heavy particles, P_v is the vibrational loss and P_r is the radiation loss. The energy balance relation, then is

$$(\mathbf{E}\,.\,\mathbf{j}) = P_e + P_i \tag{6.1.2}$$

where P_e is the energy lost by elastic collisions and P_i is the inelastic loss. Electrons do not couple to rotational degrees of freedom because their angular momentums, at a given impact parameter, are very small. But exchange of energy by electrons with vibrational degrees of freedom is highly probable.

(c) Collisions, point mechanics and continuum mechanics

A fluid motion is an average motion of a dense assembly of particles. And the particles are subjected to all the forces of fluid dynamics. Then, forces known to act on the particles, but not on the fluid, upon summation must vanish. Pressure gradients act on fluids but not upon particles. Macroscopic material bodies are continuous pieces of matter in specific domains in position space. So, in a sense, continuous mechanics is fundamental in mecahnics and particle dynamics should follow from it as an idealization, to treat very small bodies. Newtonian laws of motion and their generalization in the form of the relativistic laws are valid for particle motion. This validity ceases at the levels of sub nuclear particles which cannot be studied classically. Thus, particle dynamics is fundamental in mechanics, and deserves to be treated first, particularly for the development of the basics of dynamics of continua. Pressure gradients do not act on particles, and collisions, which are effective so long as the mean free path is much larger than the particle dimensions, are the manifestation of particle behaviour. As the density of particles increases, collisions become too numerous to be considered as usual.

In magnetized rarefied plasmas for strong fields, the particle gyration period is much smaller than the mean free time t between two collisions. Moreover, the circular orbits of a particle, about a 'guiding centre' C, are steady. The guiding centre theory for particle orbits relates the motion of C to the forces acting on the particles. But, gradients in B cause drift of the guiding centre of a species of plasma across the magnetic lines. So, forces responsible for the guiding centre motions in presence of magnetic field gradients do not possibly vanish on summation. This shows that the particle orbit approach may give inconsistent results. Strangely, therefore, studies of subnuclear physics show that one should not enquire too closely as to what is meant by a particle or by a collision.

(d) Long range interactions in plasmas

Many-body interactions of long range determine behaviour of a classically described plasma. These account for the dissipation loss. Inelastic collisions should be treated quantum mechanically. These collisions are not important in classical study.

In binary encounters, only two particles are assumed to collide at a time, and the probability of influence of a third particle at that time is negligible. In short range interactions a particle interacts with other particles only in brief intervals of time, and spends the remaining time freely. In fully ionized plasmas long range encounters between two charged particles (section 6.7) dominate over short range interactions. Short range processes are important in weakly ionized plasmas. Even for more than 1 per cent ionization, long range collisions between charged particles dominate over short range collisions, because of their large effective cross-section (Appendix II). The mechanisms of dissipation are classically studied on the basis of binary collision theory. If the plasma is unstable (not in thermodynamic equilibrium), then dissipation will be anomalous, and the nature of dissipative processes cannot be guessed by the classical theory of binary collisions.

On the trajectory of a star the influence of close binary encounters is negligible when compared with the influence of many distinct stars. Coulomb encounters between many particles in plasmas are binary. The long range, small angle, encounters, occurring inside a Debye sphere (section 3.10); accumulate into a large angle, because there are many particles in it. The short range individual collisions, causing scattering through large angles, are also possible. Comparisons of the magnitudes of the mean free paths, relaxation times, probabilities and cross-sections for the two types of scattering, show that the cumulative interactions of long range dominate over the short range encounters.

(e) Scattering cross-sections

The scattering cross-sections (Appendix A11) are functions of the velocity of the incident beam. The total cross-section σ_1 of (A.11.6) is obtained by integrating the cross-section σ of (A.11.3) over all solid angles. The scattering cross-section for diffusion, σ_2 (momentum transfer cross-section), defined in (A.11.7), is a measure of the averaged forward momentum lost by the particles of one species colliding with particles of another species. The cross-section σ_2 is important in the study of transport properties of plasmas.

Due to the collisional spread of energy of the plasma particles, the total cross-section $\sigma(v)$ cannot be determined by diagnostic measurements; the average $\langle\sigma\rangle$, defined by

$$\langle\sigma\rangle = \frac{\langle v\sigma(v)\rangle}{\langle v\rangle} \tag{6.1.3}$$

is thus estimated. Here $\langle v\,\sigma(v)\rangle$, the rate coefficient, and the average velocity $\langle v\rangle$ are defined by

$$\langle v\,\sigma(v)\rangle = \int_v \sigma(v)\; vf(v)\,dv \tag{6.1.4}$$

$$\langle v\rangle = \int_v vf(v)\,dv \tag{6.1.5}$$

where v is the relative velocity of motion of the two species of colliding particles and $f(v)$ is the velocity distribution function (chapter 9) of the incident particles, relative to that of the plasma particles experiencing the collisions.

The total cross-section $\sigma(v)$ is useful in the determination of temperatures, population densities, life times, etc., when the state of the plasma deviates from local thermodynamic equilibrium. Average total cross-sections, derived from plasma diagnostic measurements, are valid for the very special conditions under which they were measured. $\sigma(v)$ is guessed from measurement of $\langle\sigma\rangle$ for a very large temperature range, for chekcing the theoretically predicted collision coefficients. For the unscreened Coulomb inter-action field, σ_1 and σ_2 have been evaluated in section 6.2 and for the screened Coulomb field in section 6.3.

6.2 TWO-PARTICLE COULOMB INTERACTION

We consider two charges e_i and e_j, separated by the distance r; the potential φ of their interaction is given by

$$\varphi(r) = \frac{e_i e_j}{r} \tag{6.2.1}$$

The forces $\mathbf{F}_{ij}$ and $\mathbf{F}_{ji}$, acting on e_i and e_j, respectively, in terms of their position vectors $\mathbf{r}_i$ and $\mathbf{r}_j$ with respect to a fixed origin O, are

$$\mathbf{F}_{ij} = -|e_i e_j|\,\mathbf{r}/r^3, \quad \mathbf{F}_{ji} = |e_i e_j|\,\mathbf{r}/r^3$$

where $\mathbf{r} = \mathbf{r}_j - \mathbf{r}_i$. The condition $(\mathbf{r} \times \ddot{\mathbf{r}}) = 0$ follows from the relation $(\mathbf{r} \times \mathbf{F}) = 0$ for both $\mathbf{F}_{ij}$ and $\mathbf{F}_{ji}$ and also from (A10.5). Using now (A.10.5) and (A.10.11) to solve (A.10.14) we get

$$\frac{1}{r_m} + \frac{e_i e_j}{m v_0^2 p^2} = \left(\frac{1}{p^2} + \frac{e_i^2 e_j^2}{m^2 p^4 v_0^4} \right)^{1/2} \tag{6.2.2}$$

Moreover, if (6.2.1) is used in (A.10.15), we obtain

$$\theta_m = \frac{\pi}{2} - \sin^{-1} \left\{ \frac{e_i e_j}{(m^2 p^2 v_0^4 + e_i^2 e_j^2)^{1/2}} \right\} \tag{6.2.3}$$

Equation (A.10.9) shows that

$$\cot\left(\frac{\chi}{2}\right) = \frac{m p v_0^2}{|e_i e_j|}; \tag{6.2.4}$$

then, if χ is small we can write

$$\chi \approx \frac{2|e_i e_j|}{m p v_0^2} \tag{6.2.5}$$

Hence χ decreases as p increases. Putting, in (6.2.4), the expression for collision cross-section, from (A.11.4) we obtain

$$\sigma = \left(\frac{e_i e_j}{2m v_0^2} \right)^2 \operatorname{cosec}^4 \left(\frac{\chi}{2} \right) \tag{6.2.6}$$

This is the well-known Rutherford formula for scattering cross-section. Since σ depends on the angle χ, scattering is not isotropic and σ diverges to infinity as $\chi \to 0$. The cross-sections σ_1 of (A.11.6) and σ_2 of (A.11.7) have the values

$$\sigma_1 = -\pi \left(\frac{e_i e_j}{m v_0^2} \right)^2 \operatorname{cosec}^2 \frac{\chi}{2}; \quad \sigma_2 = 4\pi \left(\frac{e_i e_j}{m v_0^2} \right)^2 \log \sin \left(\frac{\chi}{2} \right) \tag{6.2.7}$$

Like σ these also do not exist at $\chi = 0$. The divergence appears in the formula (6.2.7) because it uses only two-particle Coulomb interaction resulting in the same final energies and momentum magnitudes as initially. But, in effect, a particle is also under the simultaneous influence of Coulomb fields of all others situated at finite distances from it. The divergence (6.2.6) can be removed by including the many-body Coulomb interaction on the scattering of charges in a plasma. A short-cut method to be discussed now is to artificially draw the minimum distance of approach at the Debye shielding distance λ_D (section 3.10).

The scattering angle is small for distant encounters and large for close collisions. So the critical value of χ, separating the group of short range interactions from distant encounters, is fixed at $\pi/2$ for respulsion and $3\pi/2$ for attraction between opposite charges. Denoting the values of p and σ at the critical value $\chi = \pi/2$ by p_0 and σ_0, respectively, we find that

$$p_0 = \frac{|e_i\, e_j|}{m v_0^2}, \quad \sigma_0 = p_0^2 = \frac{e_i^2\, e_j^2}{m^2\, v_0^4}, \quad r_m = (1 + \sqrt{2})\, p_0 \tag{6.2.8}$$

Hence, (1) for close collisions $\chi > \pi/2$ or $3\pi/2$ and $0 < p \le p_0$; (2) for distant encounters $\chi < \pi/2$ or $3\pi/2$ and $p > p_0$. Relations of (6.2.8) show that p_0 is much greater than the classical radius of an electron or ion. For example, if the average energy per electron in a plasma is 100 ev, then at $\chi = \pi/2$ $p_0 = 14 \times 10^{-10}$ cm; but the classical electron radius, being 2.8×10^{-13} cm, is a much smaller quantity. So these classical results can be used in collision processes in plasma. For $p_0 << p$, with the help of (6.2.5) and (6.2.8), we have

$$\chi \approx \frac{2p_0}{p} \tag{6.2.9}$$

(a) Coulomb logarithm parameter Δ

The dimensionless quantities

$$\Delta = \frac{\lambda_D}{p_0} \text{ and } \log \Delta = \log \frac{\lambda_D}{p_0} \tag{6.2.10}$$

where λ_D is the Debye shielding distance, are important plasma parameters. The quantity $\log \Delta$ is called the Coulomb logarithm. Taking, from kinetic theory, $v^2 = 3K_B\, T/m$ in (6.2.8), we find that

$$\Delta = \left(\frac{3}{\sqrt{4\pi N}}\right)\left(\frac{K_B\, T}{|e_i\, e_j|}\right)^{3/2} \approx 1.1 \times 10^4 \times \sqrt{T^3/N} \tag{6.2.11}$$

because (section 3.10) $\lambda_D = (K_B\, T/4\pi N/|e_i\, e_j|)^{1/2}$. The number N_D of particles in a Debye sphere of relation (3.10.7) is given by

$$N_D = \left(\frac{4\pi N \lambda_D^3}{3}\right) = \frac{1}{3}\left(\frac{1}{4\pi N}\right)^{1/2}\left(\frac{K_B T}{|e_i e_j|}\right)^{3/2} \tag{6.2.12}$$

Hence

$$\Delta = 9N_D \tag{6.2.13}$$

For $N = 10^{15}$ cm^{-3} and $T = 10^{7o}$K we find that $p_0 = 1.7 \times 10^{-10}$ cm.

$$\lambda_D \approx 6.9 \times 10^{-4} \text{ cm and so } \Delta = 4 \times 10^6, \ \log \Delta = 16 \tag{6.2.14}$$

Hence, generally, $p_0 << \lambda_D$ and the Debye screening potential begins to act long before close collisions between electrons and ions can occur. This analysis gives a measure of the degree of domination of collective effects over the individual particle phenomena (section 6.7).

(b) Landau length and some other parameters

The Landau length l_0 is the distance at which the mutual potential energy e^2/l_0 of two charges equals the kinetic energy $K_B T$ of thermal motion. So

$$l_0 = e^2/K_B T = (1.6 \times 10^{-3}/T) \text{ cm}.$$

The potential energy decreases as the distance r between two charges increases; but the kinetic energy is independent of this distance. So, if $r > l_0$, the thermal effect dominates over the Coulomb effect of binary collision, and if $r < l_0$ the situation is reversed. For the same reason, if $p_0 > l_0$, the thermal motion will prevail over binary collisions, but, if $p_0 < l_0$, the binary collisions are close and not negligible. If l_1 is the average distance between charged particles, then the free zone of a particle is the sphere of volume $4\pi\, l_1^3/3$ round it. So, if N is the number density of charged particles, then $(4\pi\, l_1^3/3)\, N = 1$; consequently $l_1 = (3/4\pi N)^{1/3} \approx N^{-1/3}$. Hence, if $l_0 << l$, close collisions are not important. The cross-section for close collisions is about πl_0^2. If the mean free path l_2 is defined from the relation $\pi l_0^2\, l_2\, N = 1$, then $l_2 = 1/\pi l_0^2\, N$.

In an ideal plasma the energy of Coulomb interaction between a charge and the surrounding charges at the mean distance l_1 $(= N^{-1/3})$ between them should be much less than the mean kinetic energy $K_B T$ of the particle due to thermal motion. So $Ne^6 << (K_B T)^3$. This condition is not generally violated in naturally occuring plasmas.

6.3 INTERACTIONS IN SCREENED POTENTIAL

Denoting the Coulomb potential between the charges e_1 and e_2 by $\varphi_c(r)$, and the Debye screened potential for the same two charges by $\varphi_D(r)$, we have

$$\varphi_c(r) = \frac{e_1 e_2}{r}, \tag{6.3.1}$$

$$\varphi_D(r) = \frac{e_1 e_2}{r} \exp\left(-\frac{r}{\lambda_D}\right) = \varphi_c(r) \exp\left(-\frac{r}{\lambda_D}\right) \tag{6.3.2}$$

But σ_1 and σ_2 of (A.11.6) and (A.11.7), evaluated with the help of the potential φ_D of (6.3.2), do not diverge for $\chi = 0$ and π. Instead of using φ_D and that complicate procedure we use the simpler potential field defined by

$$\begin{aligned} \varphi(r) &= \frac{e_1 e_2}{r}, \quad 0 < r \leq \lambda_D, \\ &= 0, \quad r > \lambda_D \end{aligned} \tag{6.3.3}$$

$\varphi_D(r)$ reduces approximately to these values for $r << \lambda_D$ and $r >> \lambda_D$ respectively. Then χ of (6.2.4) must vary from π to χ_1 where χ_1 is given by

$$\tan\frac{\chi_1}{2} = \frac{e_1 e_2}{m\lambda_D v_0^2} \tag{6.3.4}$$

Consequently, p varies from zero to λ_D when χ varies from π to χ_1. Then

$$\sigma_1 = -\pi\left(\frac{e_1 e_2}{mv^2}\right)^2 \left[\frac{p^2}{p_0^2}\right]_{p=\lambda_D}^{p=0} = \pi\lambda_D^2,$$

$$\sigma_2 = 4\pi\left(\frac{e_1 e_2}{mv^2}\right)^2 \log\left(1 + \frac{\lambda_D^2}{p_0^2}\right)$$

Since $\lambda_D^2 >> p_0^2$ and $(1 + \lambda_D^2/p_0^2)^{1/2} = \Delta$, it is evident that

$$\sigma_2 \approx 4\pi\left(\frac{e_1 e_2}{mv^2}\right)^2 \log\Delta.$$

6.4 DIFFUSION AND SCATTERING

If χ is the scattering angle, v is the velocity much before the collision, p is the collision impact parameter and p_0 is given by (6.2.8), then the diffusive velocity $\delta v_\perp$, perpendicular to the direction of motion, is given by

$$(\delta v_\perp)^2 = v^2 \sin^2\chi = \frac{p^2}{p_0^2}\,\frac{4v^2}{\{1 + p^2/p_0^2\}^2} \tag{6.4.1}$$

Defining $\langle\delta v_\perp^2\rangle$ as the mean of $\delta v_\perp^2$ per second, we have

$$\langle(\delta v_\perp^2)\rangle = \langle(v\sin\chi)^2\rangle\, v = \langle\chi^2\rangle\, v^3 \tag{6.4.2}$$

where $\langle\chi^2\rangle$ is the mean square of angle of deflection of the scattered particle after N_0 collisions. So

$$\langle\chi^2\rangle = \sum_{i=1}^{N_0} \langle\chi_i^2\rangle \tag{6.4.3}$$

where $\langle\chi_i^2\rangle$ is the mean square of angle of deflection in a single distant collision.

If the probability of collisions having impact parameters lying between p and $p + dp$ is denoted by dP, then

$$\langle\chi_i^2\rangle = \int_{p_{min}}^{p_{max}} \chi_i^2 \, dP \tag{6.4.4}$$

Taking $p_{max} = \lambda_D, p_{min} = p_0$ and for χ_i the value obtained in (6.2.5), then for charges of equal magnitude we obtain

$$\langle\chi_i^2\rangle = \int_{p_0}^{\lambda_D} \frac{4e^4 \, dP}{m^2 p^2 v^4} \tag{6.4.5}$$

The probability element dP is simply the ratio of the areas $d(\pi p^2)$ and $\pi p_{max}^2 - \pi p_{min}^2$. So putting $dP = d(\pi p^2)/(\pi p_{max}^2 - \pi p_{min}^2)$, and then integrating (6.4.5) with respect to p, we find that

$$\langle\chi_i^2\rangle = \frac{8e^4 \log \Delta}{m^2 v^4 \lambda_D^2} \tag{6.4.6}$$

For perfectly random scattering

$$\langle\chi^2\rangle = N_0 \langle\chi_i^2\rangle \tag{6.4.7}$$

Hence $\langle\delta v\perp^2\rangle$ has the dimensions of Nv^3. The mean free path l (Appendix 12) is the distance in which cumulative effects of many small angle deflections, due to long range encounters, produce the total angular deflection $\pi/2$. Since N_0 is the number of scatterers in a cylinder of length l and radius p_{max}, then

$$N_0 = \pi\lambda_D^2 \, Nl, \; \langle\chi^2\rangle = 8\pi Nl \, p_0^2 \log \Delta, \; \langle\delta v_\perp^2\rangle = 8\pi Nl \, v^3 \, p_0^2 \log \Delta \tag{6.4.8}$$

Now, by virtue of (A.12.6), replacing, in the third relation of (6.4.8), Nl by $1/\sigma$, we get

$$\sigma \langle\delta v_\perp^2\rangle = 8\pi v^3 \, p_0^2 \log \Delta \tag{6.4.9}$$

When scattering by N scatterers is so much that $\langle\delta v_\perp^2\rangle$ becomes about v^3, the particle has been deflected through a large angle, say $\pi/2$. Than

$$\sigma \approx \frac{8\pi e^4 \log \Delta}{m^2 v^4} \tag{6.4.10}$$

Comparing this σ with σ_0 of (6.2.8) we find that $\sigma > \sigma_0$ because $\log \Delta$ is a large number. These results are applicable to scattering of electrons by electrons or ions but not to scattering of ions by electrons.

6.5 RELAXATION PROCESSES

The result (A.12.6) gives

$$t_{ii} = \frac{1}{N_i \sigma v_i}, \quad t_{ee} = \frac{1}{N_e \sigma v_e} \tag{6.5.1}$$

where t_{ii} and t_{ee} are the relaxation times, v_i and v_e are the thermal sound speeds and N_i and N_e are the number densities of ions and electrons respectively. So $v_i^2 = K_B T/M$, $v_e^2 = K_B T/m$ where M and m are respectively the ionic and electronic masses, T is the kinetic temperature and K_B is the Boltzmann constant. Now, if $N_i \approx N_e$, we find that

$$t_{ii} \approx 60 t_{ee} \tag{6.5.2}$$

Hence electrons approach the equilibrium state before ions. Putting (6.4.10) in (6.5.1) we get

$$t_{ii} = \frac{\sqrt{2M}\, W_i^{3/2}}{4\pi e^4 (\log \Delta) N_i}, \quad t_{ee} = \frac{\sqrt{2m}\, W_e^{3/2}}{4\pi e^4 (\log \Delta) N_e} \tag{6.5.3}$$

where $$W_i = \frac{M v_i^2}{2}, \quad W_e = \frac{m v_e^2}{2} \tag{6.5.4}$$

Another relaxation time t_{ei} is defined as

$$t_{ei} = \frac{1}{N_i \sigma v_e} = \frac{\sqrt{2m}\, W_e^{3/2}}{4\pi e^4 N_i \log \Delta} \tag{6.5.5}$$

So we find that $t_{ei} \approx t_{ee}$ because $N_i \approx N_e$. It is the time to cover a series of distant collisions leading to a loss of momentum by an electron equal to the mean momentum. All the three relaxation frequencies ν_{ii}, ν_{ee}, ν_{ei} are proportional to $T^{-3/2}$, and follow from the general relation

$$\nu_{pq} = \left(\frac{\delta \pi e^4 N_q}{m_p^2} \log \Delta \right) \left(\frac{m}{K_B T} \right)^{3/2} \tag{6.5.6}$$

(a) Elastic collisions between electrons and neutral particles

Denoting by l_e the mean free path for collisions between electrons and neutral particles and by ν_{en} the corresponding frequency we have the relation

$$\nu_{en}\, l_e = v_e \tag{6.5.7}$$

where $v_e^2 = K_B\, T/m$. Putting (A.12.10) in (6.5.7) we obtain

$$\nu_{en} = \pi r^2 N \left(\frac{K_B\, T}{m}\right)^{1/2} \tag{6.5.8}$$

The cross-section $\sigma = (r_1 + r_2)^2/4$, given in (A.11.10), can be used in the study of collisions between an electron and a neutral particle. This σ is much less than the cross-section $\pi\, p_0^2$ of collision between electrons and ions or electrons. Moreover, (6.5.8) shows that ν_{en} is proportional to $\sqrt{T}$ whereas $\nu_{ei}\ (= 1/t_{ei})$, obtained from (6.5.5), is proportional to $T^{-3/2}$.

(b) Phenomenological theory of collisions for weak fields

The kinetic energy is conserved, for elastic collision of charges with neutral particles in a plasma. Inelastic collisions excite the target to an electronic, vibrational or rotational state. To determine the average effect an equilibrium state and an average uniform motion in some direction, due to a constant electric field in that direction, are assumed. Then the phenomenological relation

$$v_D = \mu E \tag{6.5.9}$$

is useful, where v_D is the average drift and μ is the constant mobility.

For both electrons and positive charges there are regions at low fields where (6.5.9) is experimentally valid. The drift velocity v_D is also expressed as a function of the ratio E/p for both electrons and ions where p is the total pressure of the gas. When there is no charge multiplication, equation (6.5.9) implies that current density $Ne\, v_D$ is proportional to the applied voltage which is Ohm's law.

In the absence of collisions equation of motion is

$$m\frac{d\mathbf{v}}{dt} = e\mathbf{E} \tag{6.5.10}$$

Ignoring the initial acceleratory motion and taking time average for the velocity over several collisions, the addition of a frictional force $m\mathbf{v}\nu$ in the left side of (6.5.10) gives the relation (6.5.9). If ν_e and ν_i are the effective collision frequencies for electrons and ions, then these are defined as functions of E/p by the conditions

$$\frac{\nu_e}{p} = \left(\frac{e}{mv_{De}}\right)\left(\frac{E}{p}\right),\ \frac{\nu_i}{p} \equiv \left(\frac{e}{mv_{Di}}\right)\left(\frac{E}{p}\right) \tag{6.5.11}$$

where v_{De} and v_{Di} are the average drifts of electrons and ions respectively and m and M are their masses.

6.6 SCATTERING OF IONS BY ELECTRONS

At each encounter with electrons an ion receives an amount of momentum δp_i which is equal to the change δp_e of the momentum of an electron. Since scattering is through small angles we find that

$$\delta p_e \approx mv\chi, \ \chi = \frac{2e^2}{mpv^2}, \tag{6.6.1}$$

$$\delta p_i \approx \frac{2e^2}{pv} \tag{6.6.2}$$

The number N_0 of electrons passing through unit area per unit time can be expressed in terms of the average velocity v_{av} ($\approx v$) and the number density n_e of electrons; using section 9.9, we therefore find that

$$N_0 = \frac{1}{2}(n_0 \, v) \tag{6.6.3}$$

Hence the number dN of electrons passing through the ring formed by the impact parameters p and $p + dp$, in time δt, is given by

$$dN = (2\pi p \, dp) \frac{1}{2}(N_e \, v\delta t) \tag{6.6.4}$$

Integrating the right hand side with respect to p between limits o and λ_D we get

$$N = (\pi\lambda_D^2) \frac{1}{2}(N_e \, v \, \delta t) \tag{6.6.5}$$

Putting this value of N in (6.4.7) we obtain

$$\langle\chi^2\rangle = \frac{1}{2}(\pi \, \lambda_D^2 \, N_e \, \delta t) \ \langle\chi_i^2\rangle = \frac{\delta e^4 \, N \log \Delta}{m^2 \, v^4 \, \lambda_D^2} \tag{6.6.6}$$

Since $\langle\delta p_i^2\rangle = m^2 \, v^2 \, \langle\chi^2\rangle$ we find that

$$\langle\delta p_i^2\rangle = \frac{4\pi}{\omega} e^4 \, N_e \, \delta t \log \Delta \tag{6.6.7}$$

Writing $\langle\delta p_i^2\rangle = 2M \, \delta W_i$, $v^2 = v_e^2 = KT_e/m$, this formula becomes

$$\frac{\delta W_i}{\delta t} = \frac{2\pi N_e \, e^4}{(mK_B \, T_e)^{1/2}} \frac{m}{M} \log \Delta \tag{6.6.8}$$

It determines the rate of change of the ion energy due to encounters of ions with electrons. Replacing in it W_i by $K_B \, T_i/2$ we obtain

$$\frac{dT_i}{dt} = \left(\frac{4\pi N_0 e^4}{K_B M}\right)\left(\frac{m}{K_B T_e}\right)^{1/2} \log \Delta \tag{6.6.9}$$

Putting $T_e = aT_i$, and integrating with respect to time, we find that the time t available to electrons to share their energies with ions is given by

$$t \approx \left\{ K_B T \left(\frac{aK_B}{m}\right)^{1/2} \right\} \frac{T_i^{3/2}}{8\pi N_e e^4 \log \Delta} \tag{6.6.10}$$

Generally, the energy of the field disturbing equilibrium is taken up by electrons. If the gain by electrons is small compared with their thermal kinetic energy, they cannot effectively transfer that to ions and uncharged particles. Only when the external field can sufficiently accelerate the electrons do they transfer energy through collisions to the heavier particles.

6.7 INTERACTIONS OF LONG RANGE AND SHORT RANGE

Cumulative, small angle collisions of long range, resulting in a large net deviation in the direction of motion of a particle and the short range individual encounters through large angles occur in plasmas (section 6.2). In this section we compare these two processes to determine which of them dominates in plasmas.

Let t_1 and l_1 be respectively the time and length of free motion for the long range collisions through small angles, and let t_2 and l_2 be the corresponding quantities for the short range collisions. Then

$$l_1 = v\, t_1,\ l_2 = v\, t_2 \tag{6.7.1}$$

where v is the velocity of the scattered particle relative to this scatterer. Moreover, for the long range process we have

$$t_1 \langle(\delta v_\perp^2)\rangle = l_1\, v^2 \tag{6.7.2}$$

Putting the value of $\langle \delta v_\perp^2 \rangle$ from the last relation of (6.4.8) in (6.7.2) we obtain

$$t_1 = \frac{1}{8\pi N v\, p_0^2 \log \Delta} \tag{6.7.3}$$

For the large angle, short range, collisions the cross-section is $\pi\, p_0^2$. So, by (A.12.10) and (6.7.1), we get

$$l_2 = \frac{1}{\pi p_0^2 N},\ \frac{l_2}{l_1} = \frac{t_2}{t_1} = 8 \log \Delta \tag{6.7.4}$$

The ratio l_2/l_1 is generally a large number because $\log \Delta$ is so; for example, it is about 128 when the data of section 6.2 are used. Hence the distance travelled or time taken by an electron before it is deflected through the large angle $\pi/2$, for a near collision, is about 8 log Δ times that for an equal deflection by the cumulative effects of distant encounters. If σ_1 and σ_2 are the cross-sections for the two types and if σ_t is the total cross-section, then

$$\sigma_t = \pi\lambda_D^2,\ \sigma_2 = \pi p_0^2,\ \frac{\sigma_1}{\sigma_2} = \frac{\sigma_t - \sigma_2}{\sigma_2} = \frac{K_B T}{8\pi N_e^2}\frac{m^2 v^4}{e^4} - 1 \tag{6.7.5}$$

When $N = 10^{13}/\text{cm}^3$, $T = 10^4\ ^{\circ}\text{K}$, $v^2 = 3K_B T/m$, we find that

$$\frac{\sigma_1}{\sigma_2} = 7.7 \times 10^6 \tag{6.7.6}$$

Therefore, the cross-section for a short range and large angle collision is much less than for the cumulative long range interactions. The probability P is shown in relation (A.12.21) to be proportional to the collision cross-section. Hence the probability of a large angle deviation, due to distant cumulative encounters, is several times that of a single short range collision.

6.8 PLASMA CONDUCTIVITY

Resistance to electron motion in plasma is due to momentum transfer through collisions with ions and is given by the equation

$$\frac{d\mathbf{v}}{dt} = -\frac{e}{m}\mathbf{E} - \nu_{ei}\mathbf{v} \tag{6.8.1}$$

where ν_{ei} is the collision frequency. In the mean free time t_{ei} $(= 1/\nu_{ei})$ electrons lose, on the average, a quantity of momentum equal to the mean momentum due to cumulative small angle long range interactions. So, equating the rate of change of momentum $m\,\mathbf{v}/t_{ei}$ with $e\mathbf{E}$, in the steady state, we obtain

$$v = -\frac{e}{m} E\, t_{ei} \tag{6.8.2}$$

This means that the particle looses all its directed motion that it acquired from the field. Since $\mathbf{j} = -en\mathbf{v}$, the conductivity σ is given by

$$\sigma = \frac{|\mathbf{j}|}{|\mathbf{E}|} = \frac{Ne^2}{m\nu_{ei}} = \left\{ m\left(\frac{K_B T}{m}\right)^{3/2}\right\}\frac{1}{8\pi e^2 \log\Delta} \tag{6.8.3}$$

Hence $\quad \sigma = T^{3/2} \times 8.96 \times 10^{14}/\log\Delta$ per sec $\quad$ (6.8.4)

Putting $T = 10^6\ ^{\circ}\text{K}$ and $\log\Delta = 16$ from (6.2.14) we get

$$\sigma = 5.6 \times 10^{22}/\text{sec}$$

So, at thermonuclear temperatures, plasmas are very good conductors of electricity and can be regarded as infinitely conducting when the penetration of electric fields into them is investigated.

Putting $N_e = N_i = N_0$ and $\nu = \nu_e + \nu_i$ in (3.11.16), and since $M >> m$, we find that

$$\sigma = \frac{N_0 e^2}{m\nu} = \frac{N_0 e^2 t_r}{m} \tag{6.8.5}$$

where $t_r = 1/\nu$, t_r is the relaxation time of (A.12.6). Putting the value of t_r from (A.12.6) in (6.8.5) we obtain

$$\sigma = \frac{N_0 e^2}{mNQ v} \tag{6.8.6}$$

where the collision cross-section of (A.12.6), for momentum transfer, is given by the symbol Q (to distinguish between the symbols for conductivity and cross-section) and N_0 is the number density of atoms or molecules. The total conductivity of a gas is obtained by summing the currents carried by the different types of charged particles. The electrical conductivity σ, due entirely to the diffusion of electrons in weakly ionized plasmas, is valid for all charged particles, and depends inversely on the particle mass; other charges being heavy, their diffusion is not much. And the gas resistance is completely dominated by electron-ion and electron-neutral particle collisions.

In weakly ionized plasmas the total electron collision frequency ν_t is the sum of the electron-neutral (ν_{en}) and electron-ion (ν_{ei}) collision frequencies. So $\nu_t = \nu_{ei} + \nu_{en}$, and the total electrical conductivity σ_t is given by

$$\frac{1}{\sigma_t} = \frac{1}{\sigma_{en}} + \frac{1}{\sigma_{ei}}, \tag{6.8.7}$$

because σ, given by (6.8.5), is inversely proportional to ν; here, σ_{en} is the conductivity determined by close electron-neutral encounters, and σ_{ei} is the same determined by distant electron-ion encounters.

6.9 DIFFUSION OF PLASMA

Charged particles diffuse even in stable configurations containing plasma. Actually, diffusion is a slowly occurring process. It is due to collisional transfer and by pressure gradients, which tend to smooth out any density inhomogeneities. The displacement current, the electrostatic force ρ**E**, due to space charge separation, and temporal fluctuations in velocity, are neglected. Non-linear effects of strong applied fields are also not important in diffusion processes. Diffusion results in gradual dissolution of plasma strcutures.

(a) Thermal diffusion of electrons

Consider the MHD momentum balance equation in the quasi-static limit:

$$0 = -\frac{1}{\rho_0}\nabla p + \frac{e}{me}(\mathbf{v}\times\mathbf{H}) - \nu\mathbf{v} \tag{6.9.1}$$

where ν is the collision frequency, e and m are the charge and mass of an electron, ρ_0 and p are the mass density and partial pressure of electrons and $\mathbf{H}$ is the magnetic field. The continuity condition is

$$\dot{N} + N_0 (\nabla . \mathbf{v}) = 0 \tag{6.9.2}$$

Using the particle flux vector $\Gamma (= N \mathbf{v})$ per unit volume, equations (6.9.1) and (6.9.2) reduce to

$$\Gamma = -\frac{K_B T}{m\nu} (\nabla N) + (\Gamma \times \Omega); \ \dot{N} + (\nabla . \Gamma) = 0 \tag{6.9.3}$$

where $\Omega = e\mathbf{H}/mc$. If magnetic field $\mathbf{H}$ is negligible, then these two equations give the familiar equation to determine the thermal diffusion of electrons:

$$\dot{N} = D\nabla^2 N, \ D = K_B T/m\nu \tag{6.9.4}$$

D is called the electron free-diffusion coefficient.

(b) Diffusion in magnetized plasmas

Taking Ω parallel to the z-axis, we follow the equations (3.11.19)–(3.11.25) and write

$$\Gamma = -D (\nabla N) \tag{6.9.5}$$

where
$$\mathbf{D} = \begin{bmatrix} D_\perp & D_{12} & 0 \\ -D_{12} & D_\perp & 0 \\ 0 & 0 & D \end{bmatrix}$$

with
$$D_\perp = \frac{D\nu^2}{\Omega^2 + \nu^2}, \ D_{12} = \frac{D\nu\Omega}{\Omega^2 + \nu^2}, \ D = \frac{K_B T}{m\nu}$$

Eliminating Γ from (6.9.5), with the help of the continuity condition (6.9.2), we obtain the diffusion equation

$$\dot{N} = D_\perp \left(\frac{\partial^2 N}{\partial x^2} + \frac{\partial^2 N}{\partial y^2} \right) + D \frac{\partial^2 N}{\partial z^2} \tag{6.9.6}$$

in a magnetized plasma. Since motion remains unchanged, parallel to the magnetic field $\mathbf{H}$, diffusion of plasma is not affected by that field in that direction. But diffusion across the magnetic field (Bohm diffusion) depends on $\mathbf{H}$. However, this dependence may be different because experiments indicate much higher values of diffusion than theoretical estimates of this type. This diffusion is estimated with the help of the concept of random walk for the molecular theory of dispersion. See Appendix 13 for the random walk results. Assuming that the probability is $P(x, t)\, dx$ for the particle, initially at rest, to be between x and $x + dx$ after time t, from equation (A.13.9) we find that

$$P(x, t)\,dx = \frac{1}{\sqrt{4\pi Dt}} \exp\left[-\frac{(x-x_0)^2}{4D_s t}\right] dx \tag{6.9.7}$$

where D_s is the spatial diffusion coefficient. If ν is the number of steps per unit time and l is the length of each step, then by (A.13.10) we have

$$D_s = \frac{\nu}{2} l^2 \tag{6.9.8}$$

Obviously, ν is also the collisional probability and l is the mean free path length. Taking l of the order of the electron gyro radius (section 3.3), we get

$$l = \rho = \frac{v}{\Omega} \tag{6.9.9}$$

where $\Omega = |eH|/mc$ and $v^2 = KT/m$. Then equation (6.8.8) gives

$$D_s = \left(\frac{K_B T}{m}\right)\left(\frac{\nu}{2\Omega^2}\right) \tag{6.9.10}$$

Replacing, in this equation, ν by ν_{ei}, from (6.5.5), we get

$$D_s = \frac{8\pi}{H^2} Ne^2 c^2 (\log \Lambda)\left(\frac{m}{K_B T}\right)^{1/2} \tag{6.9.11}$$

Hence, for $\nu \ll \Omega$, collisions produce a random transverse jump of the guiding centre by a distance of the order of the gyrating radius ρ, interrupting, thereby, with the circular motion of electrons. This is a random walk of the guiding centre, with frequency ν and step length ρ. Also, equation (6.9.11) shows that the diffusion coefficient is inversely proportional to the square of the magnetic field; so confinement of a plasma would increase with increasing magnetic field.

(c) Ambipolar diffusion

Assume that both electrons and ions are mobile and the neutral particles are immobile in an unmagnetized and collisionally affected plasma. Its momentum transfer equations and mass conservation equations, for quasi-static problems, are

$$0 = -\nabla p_e - m\nu_{ei} N_0 (\mathbf{v}_e - \mathbf{v}_i) - m\,\nu_{en} N_0 (\mathbf{v}_e - 0), \tag{6.9.12}$$

$$0 = -\nabla p_i - M\,\nu_{ie} N_0 (\mathbf{v}_i - \mathbf{v}_e) - M\,\nu_{in} N_0 (\mathbf{v}_i - 0), \tag{6.9.13}$$

$$\dot{N}_e + N_0 (\nabla \cdot \mathbf{v}_e) = 0,\ \dot{N}_i + N_0 (\nabla \cdot \mathbf{v}_i) = 0 \tag{6.9.14}$$

where p_e and p_i are, respectively, the partial pressures of electrons and ions, $\mathbf{v}_e$ and $\mathbf{v}_i$ are their average velocities, ν_e and ν_i are their collision frequencies, N_e and N_i are the perturbations in their number densities from their common equilibrium value N_0; ν_{en} and ν_{in} are the respective collision frequencies with neutral particles. We also have the kinetically obtained relations

$$p_e = K_B T_e (N_0 + N_e),\ p_i = K_B T_i (N_0 + N_i) \tag{6.9.15}$$

Moreover, since momentum must be conserved during interparticle collisions, sum of the densitites of forces of collision between electrons and ions must vanish. So, if ρ_e and ρ_i are the mass densities of electrons and ions, then

$$\rho_e \nu_{ei} (\mathbf{v}_e - \mathbf{v}_i) + \rho_i \nu_{ie} (\mathbf{v}_i - \mathbf{v}_e) = 0 \tag{6.9.16}$$

For purely linearized processes, $\rho_e = mN_0$, $\rho_i = MN_0$; and so

$$m\nu_{ei} = M\nu_{ie} \tag{6.9.17}$$

Now taking divergence of the momentum equtions (6.9.12) and (6.9.13), and using, in the results, the continuity conditions of (6.9.14), we obtain

$$0 = -K_B T_e \nabla^2 N_e + m\nu_{ei} (\dot{N}_e - \dot{N}_i) + m\nu_{en} \dot{N}_e, \tag{6.9.18}$$

$$0 = -K_B T_i \nabla^2 N_i + M\nu_{ie} (\dot{N}_i - \dot{N}_e) + M\nu_{in} \dot{N}_i \tag{6.9.19}$$

Adding the two equations, using in the result (6.9.17) to cancel the terms containing $N_e - N_i$, and then putting the charge neutrality conditon $N_e = N_i = N$, we obtain the ambipolar diffusion equation

$$K_B (T_e + T_i) \nabla^2 N = (m\nu_{en} + M\nu_{in}) \dot{N} \tag{6.9.20}$$

It can be written as

$$\dot{N} = D_A \nabla^2 N \tag{6.9.21}$$

where the ambipolar diffusion coefficient

$$D_A = K_B\ (T_e + T_i)/(m\nu_{en} + M\nu_{in}) \tag{6.9.22}$$

Owing to the great difference between the value of m and M the term $m\nu_{en}$ is neglected in the denominator of D_A. Then

$$D_A = \frac{K_B T_i}{M\nu_i}\left(1 + \frac{T_e}{T_i}\right) = D_i\left(1 + \frac{T_e}{T_i}\right) \tag{6.9.23}$$

where, following the terminology for D of (6.9.4), D_i is called the ion free-diffusion coefficient. If there is a significant charge difference, then a large electric field develops and tends to accelerate the ions and slow down the electrons to achieve simultaneous diffusion of both electrons and ions.

6.10 COLLISION PROCESSES IN THE IONOSPHERE

The ionosphere is rarefied and only slightly ionized. So collisions between electrons and neutral particles determine practically the momentum transfer and dissipation loss. The electron mean free path is very large compared with the wavelengths of radio waves; so, kinetic energy of electrons, between two successive collisions, is considerably increased by the waves.

If v is the electron velocity relative to the neutral particle of mass M, and χ is the deflection angle, then in an encounter, the energy lost by an electron is

$$\xi = \frac{m^2 M v^2}{(M+m)^2}(1 - \cos\chi) \tag{6.10.1}$$

The original electron energy being $mv^2/2$ and M being about 10^4 times m, the energy lost in the encounter is only a very small fraction of the original energy. So the mean energy transformed from the electron to the molecule is negligible. Actually, as remarked after (A.10.25), only the direction of motion of the electron undergoes a large deflection due to the encounter. Hence, in the presence of strong electric field, a slowly moving electron gains on average far more energy than it can lose at an encounter with a molecule. The mean electron energy in the steady state, under the influence of the wave, therefore grows until the small fraction lost at collisions balances the energy gained during a free path. Moreover, the energy distribution of electrons in the steady state should considerably differ from the distribution before the wave was applied.

For computation several simple expressions for the collision frequency ν have been assumed by different authors for ionospheric applications, all these give values of the same order of magnitude for most of the common ionospheric problems. One of these formulas is $\nu = \pi a^2 N (8K_B T/\pi m)^{1/2}$ where a is the molecular radius, and $(8K_B T/\pi m)^{1/2}$ is the mean thermal velocity of electrons at T degrees Kelvin of equation (9.8.8). One suggestion is for the relation $\nu = (4\pi/3) Nr^2 (8K_B T/m\pi)^{1/2}$ where πr^2 is the cross-section, or a suitable mean value of that or is simply a suitably chosen constant. For weak fields in the ionosphere

$$\nu_{en} = \frac{4\pi a^2 N}{3}\left(\frac{8K_B T}{m}\right)^{1/2} \tag{6.10.2}$$

When strong fields are applied the temperature T of these relations should be replaced by the effective temperature T_e; then if, ν_{en}^0 and ν_{ei}^0 are the values of ν_{en} and ν_{ei} in the absence of strong fields, we can use

$$\nu_{en} = \nu_{en}^0\left(\frac{T_e}{T}\right)^{1/2}, \quad \frac{\nu_{ei}}{\nu_{ei}^0} = \left(\frac{T_e}{T}\right)^{3/2}\frac{\log \Delta_e}{\log \Delta} \tag{6.10.3}$$

where $$\Delta_e = \frac{3}{\sqrt{4\pi N}}\left(\frac{K_B T_e}{e^2}\right)^{3/2}$$

The mean free path l and the collision frequency ν give the time rate of displacement $v\,(= l\nu)$. Therefore, if ξ is the kinetic energy of the electron, $\xi = ml^2 v^2/2$, and ν_0 is the collision frequency before the field is applied, and

$\xi(t)$ and $\nu(t)$ be their values at time t, in the steady state reached under the influence of the field, then

$$\nu(t) = \nu_0 \left[\frac{\xi(t)}{\xi_0}\right]^{1/2} \tag{6.10.4}$$

This relation is used in the ionospheric cross-modulation studies (chapter 7). The time interval during which the variation of ν is effective is large compared with the small time period of variation of the applied alternating electric field; the increase of ν, due to the applied electric field, has been verified experimentally.

6.11 THOMSON SCATTERING AND RAYLEIGH SCATTERING

A plasma can scatter radiation. Thomson scattering is the radiation by a free electron which is accelerated by an incident monochromatic electromagnetic wave. If the electron energy is non-relativistic, the scattering will be at the frequency of the incident field. The electric field of the incident wave and the electron acceleration are

$$\mathbf{E}(\mathbf{r}, t) = \mathbf{E}_0 \exp\{i(\mathbf{K}\,.\,\mathbf{r}) - i\omega t\}, \tag{6.11.1}$$

$$\dot{\mathbf{v}} = -\frac{e}{m}\mathbf{E}(\mathbf{r}, t) \tag{6.11.2}$$

Denoting the average of $\dot{v}^2$ over the time period $2\pi/\omega$ of oscillation of the wave field by $\langle\dot{v}^2\rangle$, we obtain

$$\langle\dot{v}^2\rangle = \frac{e^2 E^2}{2m^2} \tag{6.11.3}$$

Putting this value of $\langle\dot{v}^2\rangle$ for $\dot{v}^2$ in the formula for $d\rho/d\Omega$ in (1.14.6) gives

$$\frac{d\rho}{d\Omega} = \frac{c}{8\pi}\left(\frac{e^2 E_0}{mc^2}\right)^2 \sin^2\theta \tag{6.11.4}$$

The differential scattering cross-section $d\sigma/d\Omega$ is defined as the ratio of the energy radiated per unit time per unit solid angle and the incident energy flux. So

$$\frac{d\sigma}{d\Omega} = \frac{1}{|\langle \mathbf{S}\rangle|}\frac{dP}{d\Omega} = \left(\frac{e^2}{mc^2}\right)^2 \sin^2\theta \tag{6.11.5}$$

Integrating this formula over all solid angles we obtain the Thomson cross-section,

$$\sigma_T = \frac{8\pi}{3}\left(\frac{e^2}{mc^2}\right)^2 \tag{6.11.6}$$

For electrons σ_T is about 7×10^{-25} cm^2. If the direction cosines of $\mathbf{E}_0$ and $\mathbf{r}$ are ($\cos\psi$, $\sin\psi$, 0) and ($\sin\xi\cos\varphi$, $\sin\xi\sin\varphi$, $\cos\xi$), then

$$\cos\theta = \frac{(\mathbf{E}_0 \cdot \mathbf{r})}{|\mathbf{E}_0|\,|\mathbf{r}|} = \sin\xi\cos(\varphi - \psi), \tag{6.11.7}$$

$$\frac{d\sigma}{d\Omega} = \left(\frac{e^2}{mc^2}\right)^2 \{1 - \sin^2\xi\cos^2(\varphi - \psi)\} \tag{6.11.8}$$

For unpolarized incident fields in any direction perpendicular to the z-axis, averaging over all values of ψ gives

$$\langle\frac{d\sigma}{d\Omega}\rangle = \frac{1}{2}\left(\frac{e^2}{mc^2}\right)^2 (1 + \cos^2\xi) \tag{6.11.9}$$

The scattering of power in different directions due to an incident electromagnetic field on an electron is incoherent. So Thomson scattering is incoherent; it does not include the scattering from fluctuations due to the discrete nature of the electrons; and its theory allows a plane wave to propagate without any deformation.

The Thomson differential scattering cross-section per unit volume, due to scattering by the electrons in the plasma, is obtained by adding the contributions from all electrons. Hence, if N_e is the electron number per unit volume,

$$\frac{d\sigma}{d\Omega} = N_e \left(\frac{e^2}{mc^2}\right)^2 \sin^2\theta \tag{6.11.10}$$

In experiments it is small and is often mixed up with scattering by walls and other stray light. So, experimentally, it is very difficult to detect.

The classical electron radius r_e is defined by

$$\sigma_T = \frac{8\pi}{3} r_e^2, \tag{6.11.11}$$

$$r_e = \frac{e^2}{mc^2} = 2.8 \times 10^{-13} \text{ cm} \tag{6.11.12}$$

For study of Rayleigh scattering, the equation of electron motion is

$$\ddot{\mathbf{r}} = -\omega_0^2\,\mathbf{r} = -\frac{e}{m}\mathbf{E} + T\dddot{\mathbf{r}} \tag{6.11.13}$$

where $T = 2\,r_e/3c$. Numerically $T = 6.3 \times 10^{-24}$ sec. So the time scale of variation in the motion of the electron should be much greater than T. Frequencies of the order of $1/T = (-\,10^{23}$ cycles/sec) and higher cannot, therefore, be treated classically.

Putting in (6.11.13) the value of E from (6.11.1) we get

$$\mathbf{r} = \frac{e\mathbf{E}}{m(\omega^2 - \omega_0^2 + iT\omega^3)} \tag{6.11.14}$$

Then, following (1.14.9), the power radiated by the electron is

$$P = \frac{2e^2}{3c^3}(\ddot{\mathbf{r}})^2 \tag{6.11.15}$$

The average of its real value,

$$\langle P\rangle = \frac{2e^2}{3c^3}\frac{e^2\omega^4 E_0^2}{2m^2\{(\omega^2 - \omega_0^2)^2 + T^2\omega^6\}} \tag{6.11.16}$$

defines the scattering cross-section σ,

$$\sigma = \frac{|\langle P\rangle|}{|\langle \mathbf{S}\rangle|} \tag{6.11.17}$$

where $\langle \mathbf{S}\rangle$ is the Poynting vector averaged over the time period $2\pi/\omega$. We find that

$$\sigma = \frac{2\pi}{3}\frac{\omega^4 r_e^2}{(\omega^2 - \omega_0^2)^2 + T^2\omega^6} \tag{6.11.18}$$

For a free electron we put $\omega_0 = 0$ and recover the result (6.11.6) when $\omega^2 T^2 << 1$. When the electron is strongly bound to the inner core, the conditions $\omega << \omega_0$ and $T^2\omega^6 << \omega_0^4$ hold good because then radiation damping is negligible; in that case equation (6.11.18) reduces to the Rayleigh scattering cross-section formula

$$\sigma = \frac{8\pi}{3}\frac{\omega^4}{\omega_0^4}r_e^2, \tag{6.11.19}$$

σ is inversely proportional to the fourth power of the wavelength λ of light because, in vacuum $c = \lambda\omega$. Hence, short, blue and violet waves are scattered more than the long wavelength red waves. The sky appears blue because air molecules scatter more blue light than red. The setting sun appears red because of Rayleigh scattering. Evidently, $\omega \approx \omega_0$ is the condition for resonant scattering. For example, a bulb containing sodium vapour (which has a natural frequency in the visible region), when illuminated by light at its natural frequency, appears very luminous.

Theory of non-linear classical effects of optics has become important owing to application in the scattering of light by free electrons, which includes the development of higher harmonics and a "white radiation" due to the radiation reaction. High intensity Thomson scattering is possible in laser diagnostics of high temperature plasmas.

6.12 COLLISIONS IN WEAKLY IONIZED PLASMA

Consider the collision between two particles A_i and A_j in a plasma such that the inner state of the scattered particle A_i is altered by the scatterer A_j. Before collision the state of the scattered particle is described by the addition of the subscript i; to indicate the state after collision the subscript f is attached. So the transition from the ith state to the fth state is due to a collision.

The probability $P(t)$ that A_i has not altered its state in time t is given by the equation

$$\frac{dP}{dt} = -\nu_{\text{if}} P(t) \tag{6.12.1}$$

where ν_{if} is the frequency of transition of a particle before collision, in the state i to the state f after collision with A_j. As the frequency of variation of the state of the particle increases, the incident flux Γ of particles also increases; but the quantity $\frac{\nu_{\text{if}}}{\Gamma}$ does not depend on the density N_j of the particles A_j. This ratio has the dimension of area; it determines the nature of inelastic collision between particles, and is known as the cross-section of inelastic collision between two particles. It is also the ratio of probabilities of transition from one state per unit time and the incident flux of particles. If all the A_j particles have the same velocities $\mathbf{v}_j$, then their flux Γ, in the frame of reference in which A_i is at rest, is given by $\Gamma = |\mathbf{v}_i - \mathbf{v}_j| N_j$; where N_j is the density of A_j and $\mathbf{v}_i$ is the velocity of A_i. So, we put $\frac{\nu_{\text{if}}}{\Gamma} = Q_{\text{if}}$ where Q_{if} is the transition cross-section; Q_{if} depends only on the relative velocity of the colliding particles. Then we find that

$$\nu_{\text{if}} = Q_{\text{if}} N_j |\mathbf{v}_i - \mathbf{v}_j| \tag{6.12.2}$$

Instead of $|\mathbf{v}_i - \mathbf{v}_j|$ we can take the average $\langle|\mathbf{v}_i - \mathbf{v}_j|\rangle$ for different distributions of $\mathbf{v}_i$ and $\mathbf{v}_j$ in the velocity space. Then

$$\nu_{\text{if}} = N_j \langle|\mathbf{v}_i - \mathbf{v}_j|\rangle Q_{\text{if}} = N_j \langle K_{\text{if}}\rangle \tag{6.12.3}$$

So K_{if} is the rate constant of the process and is used to study frequency of transition between two states in a plasma. When frequency transition for averages over the velocities of the particles is necessary, the concept of K_{if} finds application.

The variations of A_i due to the appearance of a new particle in the given state i, and transitions of particles from this state into other states, are given by the balance equation

$$\frac{dN_i}{dt} = N_j \left\{ \sum_f K_{\text{if}} N_f - N_i \sum_f K_{\text{if}} \right\} \tag{6.12.4}$$

Here N_i is the number density of A_i particles and N_f is the number density of particles in the state f, obtained from A_i due to collision with A_j. Equation (6.12.4) is also valid for other conversion processes.

In weakly ionized plasmas the electrical properties depend on the formation and decomposition of charges. Collisions between electrons and atoms or molecules produce charges. Single ionization occurs in one collision between an electron and a neutral particle. When electron density is quite high and the excited atom or molecule does not have the time to decay between two successive collisions with electrons, the originally neutral particle passes through a number of excited states due to several collisions. Then ionization occurs from an excited state due to multistage ionization.

The time rate of radiative recombination of charged particles due to collisions is proportional to the product of their concentration; so the balance equation is

$$\frac{dN_i}{dt} = \frac{dN_e}{dt} = -\alpha N_e N_i \tag{6.12.5}$$

where α is the recombination coefficient and N_e and N_i are the densities of electrons and ions. The coefficient ($\alpha = \langle Q\, v \rangle$, where the average is taken over all electron velocities), includes the sum over all the bound states into which an electron can be captured. The measured values of α, in laboratory plasmas, are generally many times larger than the theoretical estimations. This difference is due to the non-inclusion of a recombination process in which a molecular ion captures an electron and dissociates into two neutral atoms.

6.13 NON-EQUILIBRIUM IONIZATION FIELD EFFECTS

Non-equilibrium ionization is mainly a two temperature process in a three element device; the three elements being the negatively charged electrons, positively charged particles and neutral particles. Though the gas, as a whole, is relatively cool the electron temperature T_e is much above the gas temperature T_g due to acceleration by the applied field. Generally the difference $T_e - T_g$ is about 300°K or more; even a uniform increase of electron temperature by 20 per cent is not insignificant for this purpose. Generally it is assumed that both the number of free electrons and the distribution of bound electrons among the excited states are the same as they would be in thermodynamic equilibrium at temperature T_e. The amount of ionization depends on T_e. Non-equilibrium ionization preferentially heats the electrons of the system without appreciably raising the temperature of the working fluid in this effect known as the Kerrebrock effect.

Even if current flow is sufficiently large only a small fraction of the energy acquired by the electrons is exchanged on each collision with a heavy particle because the electronic mass is much smaller than that of the scatterer. The electron energy should not increase at the expense of dissipation of a large part of the work out put of the generator. Laser radiation can possibly provide the desired increase of electron temperature. If electron beams are used then, due to high rates of recombination, these must be distributed over almost the entire volume of the generator.

Internal dissipation due to the current j is (j^2/σ) per unit volume where σ is the electrical conductivity of the gas. If this power dissipation in the gas produced by the induced emf increases the electron temperature, for achieving non-equilibrium ionization, then we can write

$$\frac{j^2}{\sigma} = \sigma (1-K)^2 E^2 = P_e + P_i \tag{6.13.1}$$

where P_e is the power loss due to elastic collision, P_i is the sum of the inelastic losses and E is the field measured in the coordinate system in which the gas is at rest and has a maximum value vB/c (v is the flow velocity, B is the magnetic field in gauss, c is the vacuum speed of light), and K is the load factor of the generator, because the effective field is less than the induced field by the factor $(1-K)^{-1}$.

Non-equilibrium ionization is not uniform. The non-uniformity can lead to internal short circuits due to the inherent nonlinearity existing near the electrodes and in generators employing monatomic gases. In deriving (6.13.1) the losses through diffusion and heat conduction to walls and by radiation from the gas have been ignored. Actually radiation loss is negligible for electron densities at $10^{13}/\text{cm}^3$ and more, and so its exclusion is a valid approximation in some cases. The electron number density varies between 10^{12} and $10^{17}/\text{cm}^3$. At relatively low energies (less than 0.5 ev) and these high values of density of charged particles, encounters among electrons themselves are very frequent. So the electron energy distribution approximates to Maxwellian at some temperature T_e which may be different from the gas temperature T_g. Encounters between electrons and ions are then negligible.

The energy loss P_e, due to elastic collisions, is

$$P_e = \frac{3K_B}{2}(T_e - T_g)\, N_e \sum_i \left(\frac{2me}{m_i}\right) N_i\, Q_{ie}\, c_e \tag{6.13.2}$$

where K_B is the Boltzmann constant, m_i is the mass and N_i is the number density of the ith species, Q_{ie} is the cross-section of elastic energy transfer for electrons and C_e is the mean thermal velocity of electrons at the temperature T_e.

The collision frequency ν_i of an electron with particles of the ith species is given by equation (A.12.5) : $\nu = N_i\, Q_{ie}\, c_e$. Since the mass m_i of heavy particles is much greater than the electronic mass m_e, the average energy lost per collision is $2m_e/m_i$ times the energy difference $3K_B\,(T_e - T_g)/2$. A factor G (section 7.9 and equation (6.14.13)) $= \delta\,(2m_e/m_i)$, is introduced to take account of the inelastic losses over the elastic losses. Then $\delta = 1$ for ideal elastic collisions. The electric field, measured in the coordinate system in which the gas is at rest, being $E = uB/C$, equation (6.1.2) gives

$$\frac{j^2}{\sigma} = \sigma\left(E + \frac{v}{c}B\right)^2 = P_e \tag{6.13.3}$$

Assuming that there is only one species, we write for the D.C. electrical conductivity (from the second relation of (3.11.20))

$$\sigma = \frac{Ne^2}{m_e\,\nu}$$

and find that equations (6.13.2), (6.13.3) and (A.12.5) give

$$\frac{3K_B}{2}(T_e - T_g) = \frac{m_i}{2}\left\{\frac{e}{m_e\,\nu}\left(E + \frac{u}{c}B\right)\right\}^2 \tag{6.13.4}$$

If $e\,(E + (vB/c))/m_e\,\nu$ is replaced by the ordered velocity v_e, acquired by the electron from the applied field, then

$$\frac{3}{2}K_B\,(T_e - T_g) = \frac{m_i}{2}v_e^2 \tag{6.13.5}$$

This relation shows that the excess electron energy equals the kinetic energy of the heavy particles in the coordinate system in which electrons are at rest. It can also be written as

$$\frac{3}{2}K_B\,(T_e - T_g) = \frac{m_i}{2}\left(\frac{e}{m_e}\right)^2 \frac{(K_B\,T_0)^2}{(Q_{ie}\,v_e)^2}\left(\frac{E'}{p_0}\right)^2 \tag{6.13.6}$$

where $E' = E + UB/C$, $T_0 = 273^{\circ}$K and p_0 is the reduced pressure of equation (A.12.18). Equation (6.13.6) expresses electron temperature as a function of E'/p_0 and thus connects work on gas discharge with that of MHD power generation (sections 5.6-5.7) and some other effects, for example, of section 7.7).

6.14 DISSOCIATION AND RECOMBINATION

Simultaneously with ionization also starts dissociation and recombination. Electrical conductivity depends on the recombination rates, and so on diffusion to the duct walls of the ions or electrons. Recombination is possible of the positive ions of the seed material and the electrons; so the main recombination process is given by the thermal ionization reaction

$$A \rightleftarrows A^+ + \text{electron} \tag{6.14.1}$$

If the ion is molecular then the dissociative process is

$$AB^+ + e \rightarrow A + B \tag{6.14.2}$$

In monatomic gases first occurs the recombination process

$$A^+ + e \rightleftarrows A + e\ (\text{hot}) \tag{6.14.3}$$

and then occurs the electron cooling process

$$e\ (\text{hot}) + M \rightleftarrows e + M\ (\text{hot}) \tag{6.14.4}$$

In the two temperature model the process (6.14.3) is very rapid, so it can be neglected compared to the process of (6.14.4). Electron-ion recombination is not a source of loss of energy of the electron gas because the recombination energy is carried by another electron. Only, in the beginning of ionization, and particularly from the first few excited states to the ground state, radiation loss may be significant; at electron densities greater than $10^{13}/\text{cm}^3$, radiation loss is not much. Electrons, through collision, transfer to the heavy particles energy of translation, vibration and rotation.

The total number of atoms, N_A, in the gas is given by the Avogadro number

$$\frac{6.25}{2.24} \times 10^{25} \times \frac{p}{I} \times \frac{273}{T} / m^3 \tag{6.14.5}$$

where p and T are the pressure and temperature respectively of the gas. In terms of the fraction α of the seed atoms in this gas the density of seed atom is

$$\frac{6.25}{2.24} \times 10^{25} \times \frac{p}{I} \times \frac{273}{T} \times \alpha / \text{m}^3 \tag{6.14.6}$$

The n the number N_i of ionization of seed atoms per second is given by

$$N_i = \frac{6.25}{2.24} \times 10^{25} \times \frac{p}{I} \times \alpha F' N_e c_e Q_i / \text{sec} \tag{6.14.7}$$

where $F' N_e$ is the density of electrons with energies greater than ξev, c_e is mean electron thermal velocity at ξev, Q_i is the collision cross-section for the ionization of the seed atom by an electron and F' is given by (6.14.8).

In non-equilibrium ionization, in combustion products, the energy distribution of electron is not Maxwellian. The exact energy of electrons, as a function of the applied field, is difficult to obtain because changes due to all types of collisions must be included in the electron distribution. Denoting by $f(\xi)$ the fraction of energy in the range ξ and $\xi + d\xi$, the total fraction F' of electrons having energies above the ionization potential of the seed material is calculated:

$$F' = \int_{\xi}^{\infty} f(\xi)\, d\xi \tag{6.14.8}$$

where the lower limit ξ of the integral is 4.34 ev for potassium and 3.89 ev for cesium. The rate of recombination of the electrons with the seed atoms is given by

$$-\frac{dN_e}{dt} = \alpha' N_e^2 \tag{6.14.9}$$

where α' is the recombination coefficient. To attain the steady state electron density at given values of the ratio E/N_M (E is the electric field and N_M is the number density of neutral particles) this rate must equal the rate of ionization of the seed atoms. So, equating (6.14.7) and (6.14.9), we obtain

$$\alpha' N_e^2 = \frac{6.25}{2.24} \times 10^{25} \times \frac{p}{I} \times \frac{273}{T} \times \alpha F' N_e c_e Q_i \tag{6.14.10}$$

Rearrangement of some factors gives

$$N_e = \frac{6.25}{2.24} \times 10^{25} \times \frac{p}{I} \times \frac{273}{T} \times \frac{\alpha}{\alpha'} \times F' C_e Q_i \tag{6.14.11}$$

Equation (6.14.9) can be obtained by putting $N_i = N_e$ in (6.12.5). Integration of (6.14.9) gives

$$N_e = \frac{N_0}{1 + \alpha' N_0 t} \tag{6.14.12}$$

where N_0 is the density of electrons or ions when the source of ionization is removed and so N_0 is the initial density for (6.12.5). Knowing α' the rate of decay of ionization with recombination is determined; this rate can be compared with the ratio of MHD duct length and the transit time through it of the gas. α' is about 10^{-7} cm^3/sec. For cesium at 1700°C, α' is about 3.4×10^{-10} cm^3/sec. If $N_0 = 10^{15}/\text{cm}^3$, equation (6.14.12) gives $t = 3$ micro seconds to reach 50 per cent of the initial value and 26 microseconds to reach 10 per cent. This decay rate is evidently small compared to the rate of flow of gas in the duct. So a continuous source of ionization, magnetically induced by the applied electric field, is necessary in the generator.

When an electron collides with a heavy particle, the amount of energy loss per collision is $(3K_B/2)\ (T_e - T_g)\, G$ where T_e is the electron temperature, T_g is the gas temperature, K_B is the Boltzmann constant and the loss_factor (section 7.9)

$$G = \frac{2m}{M}\,\delta \tag{6.14.13}$$

Here the factor δ expresses the excess of losses due to inelastic collisions over the ideal elastic loss factor unity; it gives reasonable values of inelastic losses for internal degrees of freedom with polyatonic species where the ratio of inelastic to elastic collisions is fixed at a given energy. In molecular gases, due to the excitation of vibration and rotation, δ varies from ten to several hundred. But, for monatomic gases, the losses from inelastic collisions should be calculated separately. The number of collisions per second per unit volume between the electrons of density N_e and the heavy particles of density N_i is $N_i N_e Qc_e$ where Q is the collision cross-section and $C_e^2 = K_B T_e/m$. The power input P_e, required to maintain the temperature difference $T_e - T_g$, is given by

$$P_e = N_i N_e QC_e G \left(\frac{3K_B}{2}\right)(T_e - T_g) \tag{6.14.14}$$

Electronegative particles (for example Chlorine, Oxygen, and Hydroxyl radicals) can reduce the electron density by the formation of negative ions. For several species of heavy particles, P_e is the sum of the energy loss rate from all these species. If, therefore, N_i is the density of the ith species and Q_{ie} is the cross-section of collision with that species, we can write

$$P_e = \left(\frac{3K_B}{2}\right)(T_e - T_g)\, N_e \sum_i (G N_i Q_{ie} C_e) \tag{6.14.15}$$

From (A.12.5) we find that $N_i Q_{ie} c_e$ equals the collision frequency ν_i of the ith species. The summation in (6.14.16) should be the average over all velocities and species. In (6.13.1) ignoring P_i and putting the value of P_e from (6.14.16) we obtain

$$\frac{j^2}{\sigma} = \sigma E^2 = \frac{3}{2} K_B (T_e - T_g)\, N_e \sum_i (GN_i Q_{ie} C_e) \tag{6.14.16}$$

The amount of recombination energy per electron is $e\varepsilon_i + (5K_B T_e/2)$ where e is the electronic charge, ε_i is the ionization potential of seed atoms and other quantities have been already explained. The total amount of this energy per second per unit volume is $(e\varepsilon_i + 5K_B T_e/2)\ (dN_e/dt)$. Equating this quantity to the P_e of (6.14.14) we obtain the energy relation

$$\left(e\varepsilon_i + \frac{5}{2} K_B T_e\right) \frac{dl_n N_e}{dt} = -N_i QC_e G \left(\frac{3K_B}{2}\right)(T_e - T_g) \tag{6.14.17}$$

where T_e and T_g are replaced by the Saha equation (section 9.5):

$$\frac{N_e N_i}{N_s} = \frac{(2\pi m K_B T)^{3/2}}{h^3} \frac{2g_i}{g_0} \exp\left(-\frac{e\varepsilon_i}{K_B T}\right);$$

here N_s is the concentration of the neutral seed atoms, g_i is the statistical weight of the ground state of the ion, and g_0 is the same of the ground state of the neutral particles.

In the two-temperature model, for non-equilibrium ionization, the rate of ionization is determined by the net rate at which electron gets energy. If the energy comes exclusively from joule dissipation, then (6.14.17) should be replaced by

$$\left(e\varepsilon_i + \frac{5}{2} K_B T_e\right) \frac{dN_e}{dt} = \frac{j^2}{\sigma} - N_i N_e QC_e \left(\frac{3K_B}{2}\right)(T_e - T_g) \tag{6.14.18}$$

Chapter 7

IONOSPHERIC EFFECTS

7.1 INTRODUCTION

In June 1933 Tellegen reported to a journal that the radio broadcast from Beromunster station of (Switzerland), when received in Eindhoven (Netherlands), was found to be disturbed by the programme broadcast from Luxembourg city. So the conclusion was that one radio broadcast can disturb other broadcasts.

Townsend developed the mean free path method empirically for study of steady motion of electrons in gases at a given pressure p under the influence of an electric field and a transverse magnetic field. His work explained some ionospheric effects detected subsequently. Bailey, and Martyn, applied the ideas of Townsend to explain this effect of interference of radio broadcasts. The Bailey–Martyn theory draw much attention to applications of mean free path method to problems of ionosphere. The atmosphere near the sea level consists of molecular nitrogen (78 per cent) and molecular oxygen (21 per cent), and the total gas density at the pressure of one atmosphere is $2.7 \times 10^{19}/\text{cm}^3$. The ionosphere is the region between heights 80 km and 120 km above earth's surface. It consists of three layers, known as the D, E and the F layer. The D layer is the region between 80 km and 90 km, E is between 90 and 100 km and F is above 100 km. These heights being small compared to earth's diameter (1.3×10^4 km), the earth's surface and these layers can be assumed absolutely flat for our study.

The E layer has kinetic temperature about 300°K, collision frequency 10^5 cycles/sec, electron mean free path 100 cm, electron thermal velocity 10^7 cm/sec electron kinetic energy 6×10^{-14} erg and the gyro frequency 10^6 cycles/sec. At heights between 80 and 95 km the molecular temperature varies between 205°K and 217°K, electron density between 5×10^2 and $5 \times 10^4/\text{cm}^3$, the magnetic field H between 0.45 and 0.35 gauss; so the gyro-frequency Ω is about 7×10^6cycles/sec; the molecular density is about $10^{14}/\text{cm}^3$. The power flux at E layer is about 8×10^{-7} Watts/cm^2; it corresponds to 0.17 volts/m of electric field independent of oscillation frequency. The plasma conductivity σ and frequency ω_p, at about 100 km above earth, are about 10^9/sec and 2×10^8 rad/sec. Earth's magnetic field H cannot be neglected for waves at 50 MC/sec or less. But neglect of H can be tolerated at radar frequencies, 400 MC/sec and more.

To measure the electron density and temperature at various heights the back scattering can be used of a radar beam at frequencies of 50 to 1000 MC/sec. These frequencies are very large as compared with the electron plasma frequency ω_p $(= \sqrt{4\pi N_0 e^2/m})$, of the order of 0.1 to 10 MC/sec.

7.2 STANDARD TERMINOLOGY

The effect was called 'Luxembourg effect' or 'wave interaction'. The name 'ionospheric cross modulation' has become familiar. This is a weak wave interaction process in plasma. The radio signal, which is disturbed, is called the wanted wave from the wanted radio station, named the station *A* in the Fig. 7.2.1, and *C* (the disturbing radio station) transmits the modulated disturbing wave. The receiving station *B* receives the signals from *A* which have already undergone modulation. The modulation acquired by the wanted signal is called transferred modulation; it is imposed on the wanted field by the disturbing wave in the cross modulating region.

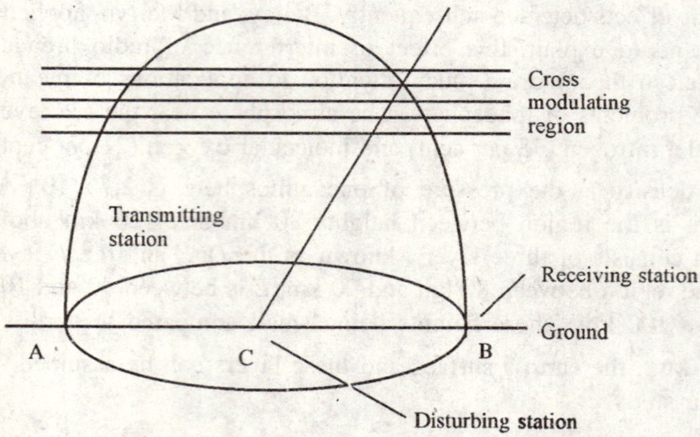

Fig. 7.2.1 The ionospheric cross-modulation effect.

7.3 SOME FACTS

The Beromunster station broadcast was at 650 kilo cycles per sec. and the broadcast of Luxembourg was at 200 kilowatts at 252 kilocycles per sec. The Luxembourg station was found to interfere seriously several different medium wave broadcasts. If the disturbance is from a high power, long wavelength, broadcast, situated near the midpoint and within about 100 to 200 km of the transmission path of the wanted wave then it is well recorded by the latter.

The solar activity exerts influence on this effect, because out of the twelve nights of experiments conducted by a group, the phenomenon was observed only six times, and only on two of these occasions was the effect clearly observed. Cross-modulation occurs generally at the lower *E* layers, particularly in the zone between the heights 80 and 95 km, and appears to decrease at dawn; so experiments are carried out at night. Also the reverse phenomenon of increase of cross-modulation at dawn was reported. The level of cross-modulation is about 2 km below the level of reflection of the wanted wave. And the level at which cross-modulation takes place is found to be independent of the frequency of the wanted wave.

Initially Bailey found that a powerful radio wave *A*, with an output frequency of power 200 Kw, could increase the collision frequency ν by an appreciable amount. To detect this increase a second wave *B* was introduced, which passed through the affected region, because the increase of ν would lead to more absorption of the wave *B*.

Interaction between waves at the local gyro-frequency is maximum between 90 and 95 km above ground. If the wanted wave suffers maximum absorption in this region, then a disturbing wave of only 1 Kw radiated power could impress an observable modulation on the *B* wave. Another consequence. verified by using a 1 Kw power, is that the resonance curve for the impressed modulation is either one-humped or two-humped.

7.4 PHYSICS OF THE PHENOMENA

A powerful broadcast from a radio station can heat up the region of the ionosphere through which it passes. Energy is exchanged between different types of particles and the propagating wave field through collisions, in the cross modulating region. There both the wanted wave and the disturbing wave suffer absorption because energy acquired by electrons from the waves must have come from the energy carried by the wave fields. Energy absorbed in the cross modulating region increases the mean electromagnetic energy, which means increase of the mean relative particle velocities. Since the plasma state is temperature-dependent, temperature increases, and also with it increases the mean collision frequency ν which is dependent on concentration, mean value of relative velocity of the colliding particles and on collision cross-section. Electrical conductivity is affected too by absorption of the waves. Reduction in the field strength of the disturbing wave in any narrow layer in the ionosphere depends on the amount of modulation transferred to that layer, The powerful disturbing modulated wave is also received in a distorted form because it has been energized in the cross-modulation region, due to self-interaction in the ionsophere. The region of maximum powerloss is near the level of reflection and is experimentally found to be about 2 km below that level. Moreover, the height of the regions of cross-modulation for different wanted stations, acted upon by the same disturbing station, are found to be identical.

The collision frequency ν of the medium cannot follow the high frequencies of the wanted or the disturbing field; but ν can easily interact effectively with the small modulation frequency ω_1 of the disturbing wave and can therefore acquire the slower time period $2\pi/\omega_1$ of the modulation. Since ν is small, the collisional heating of the electrons is delayed by a time lapse, and the oscillation of ν at the frequency ω_1 has a phase different from the phase of modulation of the field. The wanted wave catches the oscillation of ν at ω_1 in the form of an acquired modulation at that frequency. The superposed modulation interferes with the wanted wave.

Cross-modulation is influenced by collision frequency, temperature, electron density, the geomagnetic field in the region of cross-modulation, power of the radio frequency, the audio frequency of modulation and polarization of the disturbing wave, and by the polarization and frequency of the wanted wave. Degree of penetration into the ionosphere is determined by the angle of incidence. Cross-modulation in excess of about 10 percent only is useful in research.

The effect is most pronounced when the region of absorption of the wanted wave coincides with the region where the disturbing wave has transferred its modulation. If the two regions do not overlap, which is possible if frequencies of the two waves have certain specified ranges of values, the effect is not appreciable. Transferred modulation is noticeable on long wavelength high power broadcasts (for example of the referred Luxembourg broadcast) but there was no interference from nearby broadcasts at still longer wavelengths. This is due to non-overlapping of the regions of absorption of the disturbing and wanted waves.

7.5 NON-LINEAR NATURE OF THE PROCESS

Non-linearity and dissipative properties of the ionosphere are responsible for the interaction between modulated waves, particles, collisions and other waves. The induced fields change the properties of the region; due to collisional changes and finiteness of the amplitudes of the fields this change acts on all the waves passing through the region.

Linearized field solutions for waves of very small amplitude are useless in the study of intensity induced effects like absorption increasing with increasing power and non-linear mixing of waves to generate new frequency components. If the linearized theory is sufficient the kinetic energy induced on an electron should be small when compared with the random thermal energy; so $(e^2 E^2/m^2 \omega^2) << (K_B T/m)$ where e and m are the electronic charge and mass respectively, ω is the frequency of the applied electric field of amplitude E, K_B is the Boltzmann constant and T is the kinetic temperature. But, this restriction cannot prevent the significant cumulative heating of the electron gas by applied fields. For negligible such heating the condition is $\langle p \rangle << G\nu (\xi - \xi_0)$, where $\langle P \rangle$ is the mean power impressed by the field on the electrons, ξ is the mean energy of random motion of an electron in the steady state in presence of the field, ξ_0 is the mean kinetic energy of a molecule, G is a constant (called the energy loss factor) and ν is the collision frequency in the presence of the field. In unmagnetized plasma, neglecting inelastic collisions, we have

$$\langle P\rangle = \frac{\nu e^2 E^2}{2m\,\omega^2},\ \xi = \frac{3}{2}K_B T,\ \xi_0 = \frac{3}{2}K_B T_0,\ G = \frac{2m}{M}$$

where T_0 is the molecular kinetic temperature and other quantities have been explained already. Hence, for $T_0 << T$, the condition for the linearized approximation to hold good without any effective heating by the field, is

$$\frac{e^2 E^2}{m^2 \omega^2} << \left(\frac{6m}{M}\right)\left(\frac{K_B T}{m}\right)$$

When the non-linear effects are strong, the reverse condition is valid, namely

$$\frac{e^2 E^2}{m^2 \omega^2} >> \left(\frac{6m}{M}\right)\left(\frac{K_B T}{m}\right)$$

7.6 CRITICAL COMMENTS AND CLARIFICATIONS

Temperature variation by the impressed modulation frequency is due to a pressure variation which is an acoustic effect. Parameters depending on electronic temperature, or pressure, vary with the frequency of this acoustic field. Absorption coefficients of the waves are also affected. Since the disturbing acoustic field is localized to the region of transfer of modulation, maximum interaction occurs there. So, the Luxembourg effect is an interaction of acoustic plasma oscillations and electromagnetic waves. The wanted wave is modulated thus in its path by acoustic field, which is converted into electromagnetic wave by static geomagnetic field and inhomogeneities. The acoustic field can be artificially created by a vibrating plate performing simple harmonic motion. Wave propagation is affected in a region where shock waves may exist.

The Bailey-Martyn theory is valid for weak fields having modulation frequencies less than 150 cycles/sec. So, elastic collisions dominate, and simple macroscopic state relations are valid. For modulation frequencies extending to nearly 1500 cycles/sec and high power, and for the dependence of the phase lag of the acoustic disturbance relative to modulation, on the audio frequency, this theory cannot be extrapolated. The kinetic theory of weakly ionized plasma (sections 10.9-10.10), is more general for understanding the ionospheric cross-modulation effects. This theory may be helpful for finding a state relation between the mean thermal energy, and the mean collision frequency ν of electrons, which is valid for all ranges of the modulation frequency.

The wanted wave suffers cross-modulation in one region while ascending, and in another region while descending. These two regions have different thermal and other properties. The absorption coefficient η and the collision frequency ν depend on electron temperature and modulation frequency.

The fluctuating part of the electron distribution in the region of cross-modulation changes appreciably in the time period $2\pi/\omega_1$ (where ω_1 is the modulation frequency). Since this time period is short compared to the thermal relaxation time T, the state parameters depend on ω_1 as well. In the D and E regions the time T is of the order of 100 seconds, and the electron gas is cooled by the effects of inelastic collisions.

In presence of strong electric fields, the mean energy of electrons is large. This energy excites quantum states of internal motion of molecules. So, inelastic collisions become important, and electrons remain free until they combine with positive ions to form molecules. These can also form negative ions by combination with the molecules. Energy required to ionize a molecule is much larger than the mean thermal energy $3S/2$, and the energy of the ionizing agent must exceed this requisite amount by an appreciable fraction; the excess energy may become the kinetic energy of the liberated electrons. Electrons do not make too many collisions with molecules before recombination takes place. So the mean kinetic energy of electrons is likely to be larger than the thermal energy.

Electron collision frequency ν, the electron concentration N and its spatial variation, the local geomagnetic field H, the parameter G (which determine the mean loss of energy by an electron due to collisions with the heavier particles of the neighbourhood) and the heights at which the wave interaction occurs in the E layer, are determined through experiments. By measuring the phase of the transferred modulation, for a series of modulation frequencies, collision frequency at a height is estimated. Measurement of phase and amplitude of transferred modulation gives information on the region where modulation transfer occurs.

In the lower part of the E layer an increase of ion density at night by a factor of 1.7 per km has been observed from the increase in length of path of the wave in the region of absorption and the change in the modulation frequency ω_1, due to curvature of the trajectory. Increase in electron temperature is only about 10 per cent over 520 kilowatts radiated from the ground. So this effect is used for diagnosis of the ionosphere.

Cross-modulation study helps understanding gas discharge. Powerful beams receiving irregularly scattered waves are used for investigating observable wave-interactions at high frequencies. The E layer magnetic field is measured with such precession that the presence of any electric current, responsible for variation of terrestrial magnetic field, can be detected.

Spectroscopic study of the E layer is possible by artificial glows. Experiments, conducted with 40 dipoles connected to a generator producing 500 kw and working at the local gyro frequency, can control the ionsophere. An array of 80 dipoles radiating vertically a beam with mean power 500 kw can produce an observable artificial glow at the gyro frequency at night within 32 square degrees solid angle. A flash of celestial glow, that is about 40 times as bright as the night sky, is produced by the same array of pulses of power 3500 kw for a duration of 0.2 ms. Celestial glow like the moon light would require one million Kw of power.

7.7 THE MEAN FREE PATH METHOD OF TOWNSEND

In this method all electrons are assumed to have the same mean free path l under the influence of an electric field in the time between two collisions. Diffusion velocity through the gas, depends on the mean distance l travelled by an ensemble of electrons per unit time for binary collisions with other electrons and neutral particles. Then motion of particles consists of series of short and long periods of motion. Short periods are times between two collisions. In the long times action of the field dominates, and the short-range encounters become negligible.

The spatial diffusion equations for electrons are $\mathbf{F} = N\mathbf{v} - D\nabla N$ and $(\nabla \cdot \mathbf{F}) = 0$ where **F** is the mass flow velocity, **v** is the mean drift velocity in the direction of electric field **E** and D is the diffusion coefficient. These yield the diffusion equation

$$\nabla^2 N = \frac{v}{D}\frac{\partial N}{\partial x} \tag{7.7.1}$$

when E is constant and parallel to OX. The Townsend relation is

$$\frac{v}{D} = \frac{E}{40.3 K_0} \tag{7.7.2}$$

where ξ and ξ_0 are the mean kinetic energies of an electron and molecule respectively, and $K_0\,\xi_0 = \xi$; K_0 is the Townsend energy factor. Then (7.7.1) becomes

$$\nabla^2 N = \frac{E}{40.3 K_0}\frac{\partial N}{\partial x} \tag{7.7.3}$$

Hence the spatial distribution of electron density depends only on E and K_0 and the boundary conditions. So, measuring the spatial variation of N, the quantity K_0 can be determined. In addition to the mean drift velocity **v** electrons also possess a random velocity which determines the electron kinetic energy. If u is the root mean square of this velocity, then $\xi = mu^2/2$, and so

$$u = \left(\frac{2K_0\,\xi_0}{m}\right)^{1/2} \tag{7.7.4}$$

Equating the mean rate of gain and loss of momentum, due to collisions, we get $v = eE/m\nu$. The expression for the mean free path l then obtains from the relation $\nu = u/l$, and, at any given pressure, ν increases with E, because

$$\frac{\nu}{D} = \frac{E\xi_0}{40.3\,(mu^2/2)} \tag{7.7.5}$$

The mean gain in energy per particle of charge e being $e\,(\mathbf{E}\cdot\mathbf{v})$, and the mean loss due to collisions being $\nu\,(mu^2/2)$, if α is the approximate mean proportion of electron energy lost at a collision with a molecule, we can write

$$\alpha \nu\, mu^2/2 = eE\nu,$$

$$\alpha = 2v^2/u^2 \tag{7.7.6}$$

Thus, from observed values of K_0 and v, the values of u, ν, l, α are determined. If the magnetic field **H** is applied perpendicular to **E**, the electron stream is deflected through angle θ which can be measured to find **v**. But **H** can be adjusted such that the direction of the resultant of forces parallel to **E** and $(\mathbf{v} \times \mathbf{H})/c$ is the same as the direction of the electron beam. Then $\tan \theta = vH/Ec$. In the instrument used for measuring K_0 and **v** the gas pressure p was maintained at a constant value (of the order of few millimeters of mercury) and a constant electric field was applied; so these were found to be functions of the ratio E/p and K_0 is greater than unity. Townsend also extended this theory for the alternating electric field **E**. He showed that ν increases under alternating **E** in times large compared to its period. Bailey used this generalization successfully to explain the Luxembourg effect.

7.8 BASIS OF THE BAILEY-MARTYN THEORY

When a gas in thermal equilibrium at temperature T and pressure p is disturbed by an applied electric field, the random motion of electrons is affected. Electrons then absorb energy from the wave field. As a result their kinetic energy and temperature increase to higher values. Consequently, the root mean square value of random electron velocity and the mean collision frequency of an electron with molecules also increase. The local value of ν decides the extent of absorption of that wave and all other fields passing through the region. Loss of energy, by processes such as scattering and other forms of radiation due to several factors, is negligible in this ionospheric effect.

7.9 THE MEAN-FREE-PATH THEORY FOR ALTERNATING FIELDS

The undisturbed ionosphere has thermal equilibrium; so that collisions occur in such a manner that a mean energy ξ_0 is maintained by both electrons, molecules and ions. Hence, when some particles of one sort gain energy others lose energy to retain thermal equilibrium. In presence of the wave field, which, for simplicity, is assumed to propagate isotropically through the ionosphere, electrons gain energy so that their mean energy ξ is greater than the mean thermal energy of the nearby molecules which is ξ_0. The extra energy is transferred gradually to the surrounding particles through collisions. Thus the indefinite growth of electron thermal energy due to transfer of mean ordered oscillatory energy to the random thermal form, at the end of the free path, is checked. The characteristic time of transfer of the extra energy, through collisions, is much greater than the time period $2\pi/\omega$ of the wave field. Since the collision frequency depends on the relative velocity of the colliding particles, the process of collision is modified by the field. The characteristic time of effective change of the collision frequency is also of the order of time of effective transfer of extra energy through collisions and so is much greater than $2\pi/\omega$.

If ξ_m is the mean energy lost by an electron in a collision, $\langle P\rangle$ is the mean power impressed by the field on the electrons, then the equation for the transfer of energy in electron gas is

$$d(N\xi) + N\nu\, E_m\, dt = \langle P\rangle\, dt \tag{7.9.1}$$

Since the source $\langle P\rangle$ is averaged over the time period $2\pi/\omega$, it affects only the slowly occurring changes (that is, changes in times much greater than $2\pi/\omega$). If ε is the approximate mean proportion of the electron energy lost at a collision, then

$$\xi_m = \varepsilon\xi \tag{7.9.2}$$

For small values of ξ_m, the experimentally satisfactory value of ε is given by

$$\varepsilon = G\left(1 - \frac{1}{K_0}\right) \tag{7.9.3}$$

where $K_0 = \xi/\xi_0$, ξ_0 is the mean kinetic energy of a molecule, and the energy loss factor G is a constant for small changes of energy. For purely elastic collisions $G = 2m/M$ where M is the mass of a molecule and m is that of an electron, as may be evident from (A10.31). For powerful applied fields, due to complications from inelastic collisions, the experimentally more suitable value of G is 2×10^{-3}; so the energy lost is more for inelastic collisions than otherwise, section 6.14. Taking, therefore, $\xi_m = G(\xi - \xi_0)$, for small loss of energy by collisions, the average loss of energy per electron between two consecutive collisions is found proportional to the difference between the actual average kinetic energy ξ and its value ξ_0 under conditions of thermal equilibrium. Hence, power absorbed by an electron, per second, is $G\nu(\xi - \xi_0)$.

Neglecting the variation of the number density N, the rate at which the absorbed energy is transferred to the nearby particles, is given by the transport equation

$$\frac{d\xi}{dt} = \frac{\langle P\rangle}{N} - G\nu(\xi - \xi_0) \tag{7.9.4}$$

If ν_0 is the collision frequency, in the unperturbed state, and $\delta\nu$ is its increase in the presence of the field, then $\nu = \nu_0 + \delta\nu$. When $\xi - \xi_0$ is small we can write $\nu(\xi - \xi_0) = \nu_0(\xi - \xi_0)$. Then the energy balance relation (7.9.4) reduces to

$$\frac{d\xi}{dt} = \langle P_0\rangle - G\nu_0(\xi - \xi_0) \tag{7.9.5}$$

where $\langle P_0 \rangle = \langle P \rangle / N$. If the disturbing field is abruptly switched-off, ξ will ultimately reach the value ξ_1 according to the relation

$$G\nu_0 (\xi_1 - \xi_0) = \langle P_0 \rangle \tag{7.9.6}$$

If electrons were heated up by the wave until their mean energy becomes ξ_2 and then the field is switched-off, the mean agitational energy ξ decreases exponentially to the mean value ξ_0 in time t, given by

$$\xi = \xi_0 + (\xi - \xi_0) \exp(-G\nu t) \tag{7.9.7}$$

In the relaxation time $1/G\nu$ the mean energy loss will decay to $1/e$ times its values at the switch-off. If we take $G = 2m/M = 2.4 \times 10^{-5}$ and $\nu = 8 \times 10^6$ rad/ sec, then $1/G\nu = 4 \times 10^{-3}$ sec. In the D layer and the lower E layer $1/G\nu$ is about 100 sec. Putting $\xi = mu^2/2$ and $\nu = u/l$ in (7.9.1) we obtain the equation

$$ml^2 \frac{d\nu}{dt} + \xi_m = \frac{1}{\nu} \langle P_0 \rangle \tag{7.9.8}$$

for the rate of change of the collision frequency due to the applied field. Here u is the root mean square of the random velocity of electrons, l is their mean free path and time variation of l and N are neglected. Since $\xi_0 = ml^2 \ \nu_0^2/2$ we find that $\xi_m = (ml^2/2) \ G (\nu^2 - \nu_0^2)$ and (7.9.8) reduces to

$$\frac{d\nu}{dt} + \frac{G}{2} (\nu^2 - \nu_0^2) = \frac{\langle P_0 \rangle}{ml^2 \nu} \tag{7.9.9}$$

The extreme values of ν are the positive roots of the equation

$$\frac{G}{2} (\nu^2 - \nu_0^2) = \frac{\langle P_0 \rangle}{ml^2 \nu} \tag{7.9.10}$$

Putting $\nu = \nu_0 + \delta\nu$ in (7.9.9) and neglecting squares and higher powers of $\delta\nu$, we obtain the linearized equation for $\delta\nu$. It is

$$\frac{d}{dt} (\delta\nu) + G\nu_0 (\delta\nu) = \left(\frac{\langle P_0 \rangle}{ml^2 \nu} \right)_{\nu = \nu_0} \tag{7.9.11}$$

Hence the change in collision frequency decays exponentially when the applied electric field is switched off, and the relaxation time is $1/G\nu_0$. Putting $\xi = 3K_B T/2$ and $\xi_0 = 3K_B T_0/2$ in the equation for electron temperature T,

$$\frac{dT}{dt} = \frac{2 \langle P \rangle}{3K_B N} - G \nu (T) \ (T - T_0), \tag{7.9.12}$$

is obtained from (7.9.5).

For collisions between electrons and molecules, following (6.10.4), we write

$$\nu(T) = \nu_0 \left(\frac{T}{T_0} \right)^{1/2} \tag{7.9.13}$$

We assume $\nu = \nu_0$ in the first approximation, and then put that value of T from (7.9.12) in (7.9.13) to find $\nu\ (= \nu_0 + \delta\nu)$. A sufficiently strong field increases the electron temperature and performs work on the plasma at the rate

$$P = (\mathbf{E} \,.\, \mathbf{j}) = -Ne\, E\, v,$$

but there is a loss of energy due to collisions with heavy particles. To find this loss we put $E = E_0 \cos \omega t$ in

$$\frac{dv}{dt} = -\frac{e}{m} E - \nu v \tag{7.9.14}$$

Then the average rate of work done in the plasma by the applied sinusoidal field, over the time period $2\pi/\omega$, is

$$\langle P \rangle = \frac{Ne^2 E_0^2 \nu}{2m(\omega^2 + \nu^2)} \tag{7.9.15}$$

Taking $\nu = \nu_0$ and putting the value of $\langle P \rangle$ in (7.9.12) we obtain

$$\frac{dT}{dt} + G\nu_0 (T - T_0) = \frac{e^2 E_0^2 \nu_0}{3mK(\omega^2 + \nu^2)} \tag{7.9.16}$$

The quasi-stationary solution of this equation is

$$T = T_0 + \frac{e^2 E_0^2}{3m\, GK(\omega^2 + \nu^2)} \tag{7.9.17}$$

Putting $T = T_0 + \delta T$ in this relation, we obtain

$$\delta T = \frac{e^2 E_0^2}{3m\, GK(\omega^2 + \nu_0^2)} \tag{7.9.18}$$

If $E_0 = 0.1$ esu, $G = 10^{-3}$, $\omega = 10^{11}$ rad/sec, $\nu_0 = 10^6$ rad/sec, we find that $\delta T = 6 \times 10^2\ ^{\circ}$K. The heating due to the applied field can therefore be very large, and if sufficient power is fed into the gas, not only the electrons get heated but also further ionization is possible.

7.10 OSCILLATIONS OF COLLISION FREQUENCY

If the disturbing field is modulated at the low frequency ω_1 with modulation index M the field amplitude E can be written as

$$E_2 = E_1 (1 + M \cos \omega_1 t) \tag{7.10.1}$$

where $E_1 = E_0 \cos \omega t$ and E_0 is the constant field amplitude. The quantity M is the modulation depth, ω is the carrier frequency and ω_1 is the modulation frequency. The modulated field E_2 of (7.10.1) can be expressed as

$$E_2 = E_0 \cos \omega t + \frac{E_2}{2} M \cos(\omega t + \omega_1 t) + \frac{E_0}{2} M \cos(\omega t - \omega_1 t) \tag{7.10.2}$$

Hence E_2 is the result of interaction of three waves having the central frequency ω and two frequencies $\omega \pm \omega_1$, which are shifted equally to the right and left.

If P_1 is the power of the E_1 wave field, power of the modulated field is $P_2 = P_1 (1 + M \cos \omega_1 t)^2$ where P_1 is the steady part of P_2 when $M = 0$. So

$$P_2 = P_1 \left(1 + \frac{M^2}{2}\right) + 2\, MP_1 \cos \omega t + \frac{M^2}{2} P_1 \cos 2\, \omega_1 t \tag{7.10.3}$$

Average of P_2 over the time period $2\pi/\omega_1$ gives $\overline{P}_2 = P_1 (1 + M^2/2)$. This $\overline{P}_2$ differs from the value for the unmodulated wave. For energy transport the guiding equation (7.9.5) is

$$\frac{d\xi}{dt} + G\nu_0\, \xi = G\nu_0\, \xi_0 + P_1 \left(1 + \frac{M^2}{2}\right) + 2M\, P_1 \cos \omega_1 t + \frac{M^2}{2} P_1 \cos 2\, \omega_1 t \tag{7.10.4}$$

It determines the forced vibrations in the rhythm of the low frequency ω_1 of the kinetic energy ξ, and consequently of the effective electronic temperature T_e, in the presence of the wave field. The exact solution of this equation, for forced vibrations, is

$$\xi - \xi_0 = \frac{P_1}{G\,\nu_0} \left(1 + \frac{M^2}{2}\right) \left\{ 1 + \frac{4M \cos(\omega_1 t - \alpha)}{(2 + M^2)(G^2 \nu_0^2 + \omega_1^2)^{1/2}} + \frac{M^2 \cos(2\omega_1 t - \beta)}{(2 + M^2)\,(G^2 \nu_0^2 + \omega_1^2)^{1/2}} \right\} \tag{7.10.5}$$

where $\alpha = \tan^{-1}(\omega_1/G\nu_0)$, $\beta = \tan^{-1}(2\omega_1/G\nu_0)$. Hence ξ increases during phases of large field strength. Coefficients of $\cos(\omega_1 t - \alpha)$ and $\cos(2\,\omega_1 t - \beta)$ represent the modulation depths induced by the disturbing field on the electronic energy. Coefficient of $\cos(\omega_1 t - \alpha)$ is the contribution of a wave at the octave frequency of the disturbing modulation. Phase lag is about $\pi/2$ for large values of ω_1. Averaging over the time period $2\pi/\omega_1$ the mean value $\langle\xi\rangle$ is given by

$$\langle\xi\rangle - \xi_0 = \frac{P_1}{G\nu_0}\left(1 + \frac{M^2}{2}\right) \tag{7.10.6}$$

Putting $\xi = ml^2 \nu^2/2$, $\xi_0 = ml^2 \nu_0^2/2$, the modulation of the collision frequency change $\delta \nu$, acquired from the modulation of ξ, is given by

$$\nu (t) = \nu_0 \sqrt{\xi/\xi_0} = \nu_0 \{1 + 2M_1 \cos (\omega_1 t - \alpha) + 2 M_2 \cos (2 \omega_1 t - \beta)\}^{1/2} \tag{7.10.7}$$

$$\text{where } M_1 = \left(\frac{MP_1}{\xi_0 G \nu_0}\right)\left(1 + \frac{\omega_1}{G\nu_0}\right)^{-1/2}, \; M_2 = \left(\frac{M^2 P_1}{4\xi_0 G \nu_0}\right)\left(1 + \frac{2\omega_1}{G \nu_0}\right)^{-1/2} \tag{7.10.8}$$

Since M_1 and M_2 are small compared to unity we can replace (7.10.8) by

$$\nu (t) = \nu_0 \{1 + M_1 \cos (\omega_1 t - \alpha) + M_2 \cos (2\omega_1 t - \beta)\} \tag{7.10.9}$$

Thus the expression for the time varying collision frequency in the presence of the modulated field is obtained approximately. Oscillations of ν affect those of any electric field passing through the region.

(a) Relaxation time

The characteristic time of change of ξ is $2\pi/\omega_1$ and that of the carrier field wave is about $2\pi/\omega$. Since $\omega >> \omega_1$, we find that $T_\omega << T_\omega$, where $T_\omega = 2\pi/\omega$ and $T_{\omega_1} = 2\pi/\omega_1$. Solving the equation (7.9.12) for T and first disregarding the dependence of ν on T, and then putting the solved value of T into the relation (7.9.13), we find ν in terms of time and other parameters. When electric current is switched off, the solution of equation (7.9.12) becomes

$$T - T_0 = A \exp (- G\nu_0 t) \tag{7.10.10}$$

We now define the relaxation time T_r of electronic energy by $T_r = 1/G\nu_0$ When the carrier field varies slowly, $T_\omega >> T_r$ and for high frequency fields $T_\omega << T_r$. In ionosphere, taking $G = 3.4 \times 10^{-5}$, $\nu = 8 \times 10^6$ rad/sec we find that $T_r = 4 \times 10^{-3}$ sec; and since $T_\Omega = 1.4 \times 10^{-7}$ sec for fields at the gyrofrequency, it is clear that $T_\Omega << T_r$.

In ionospheric cross-modulation there are three characteristic times: the time period of the carrier wave T_ω $(= 2\pi/\omega)$, the time period of the modulation T_{ω_1} $(= 2\pi/\omega_1)$ and the relaxation time T_r $(= 1/G\nu_0)$.

7.11 ABSORPTION OF MODULATED FIELDS IN SPACE

We now consider the interaction of electrons with a wave which propagates in the X direction through a slab of thickness δx, containing, on the average, N electrons per cm^3. The slab surface is perpendicular to its direction of propagation; so the line element δx is along the trajectory of a point of the wave front. The wave vector **K** is also along the direction of the x-axis. Number of electrons in unit cross-section of the slab is $N \delta x$. These electrons abstract from the field the power $\langle(\mathbf{E}\,.\,\mathbf{j})\rangle$, which is the average of $(\mathbf{E}\,.\,\mathbf{j})$ over the carrier period $2\pi/\omega$. If $\mathbf{S}$ is the flux of energy, the mean power lost in the thickness δx is $\delta\,|\langle\mathbf{S}\rangle|$. It equals $-(\partial\,|\langle\mathbf{S}\rangle|/\partial x)\,\delta x$ and the mean power absorbed per electron is $-(1/N)\;(\partial\,|\langle\mathbf{S}\rangle|/\partial x)$. The wanted damped electric field being $E = E_0 \exp(-K_i\, x)$ where K_i, the absorption coefficient, is the imaginary part of the complex wave number, we have

$$-\frac{1}{N}\frac{\partial}{\partial x}|\langle\mathbf{S}\rangle| = 2K_i\,|\langle\mathbf{S}\rangle|/N \tag{7.11.1}$$

because $|\langle S\rangle|$ is proportional to the factor $E^2 \exp(-2K_0\, x)$.

Since the time averaged power, $\langle(\mathbf{E}\,.\,\mathbf{j})\rangle$, depends directly on the collision frequency, it equals $N\nu W$ where W is the kinetic energy per electron. Hence $(2K_i\,|\langle\mathbf{S}\rangle|/N) = \nu W$, or

$$K_i = \frac{WN\nu}{2\,|\langle\mathbf{S}\rangle|} \tag{7.11.2}$$

$$E = E_0 \exp\left[-\frac{WNx\nu_0}{2\,|\langle\mathbf{S}\rangle|}\{1 + M_1\cos(\omega_1 t - \alpha) + M_2\cos(2\omega_1 t - \beta)\}\right] \tag{7.11.3}$$

Since coefficients of $\cos(\omega_1 t - \alpha)$ and $\cos(2\,\omega_1 t - \beta)$ are small, the right hand side is expanded to give

$$E = E_0 \exp\left(-\frac{WNx\nu_0}{2\,|\langle\mathbf{S}\rangle|}\right)\left[1 - \frac{WNx\nu_0}{2\,|\langle\mathbf{S}\rangle|}(1 + M_1\cos(\omega_1 t - \alpha) + M_2\cos(2\,\omega_2 t - \beta)\right] \tag{7.11.4}$$

As ν increases during phases of large field strength the interaction tends to decrease the depth of induced modulation of the wanted field E. So, that the collision frequency of the wanted wave (and its absorption) in the region of cross-modulation, varies due to the predominance of sinusoidal modulation of the disturbing field.

7.12 WHEN THE IONOSPHERE IS ASSUMED UNMAGNETIZED

When frequency of applied electric field is much higher than the local gyro-frequency the approximation of unmagnetized ionosphere gives satisfactory results. Since collision frequency is much less than the carrier frequency, many oscillations occur on the average, within the mean free path, and induce an ordered oscillatory velocity to the charges in addition to the random gas kinetic motion. Including the frictional force $m\nu\mathbf{v}$, per electron, the ordered motion is determined from the momentum equation $m\,dv/dt = eE\cos\omega t - m\nu\, v$. Using the initial conditions $v = 0, t = t_1, x = a$ where $v = dx/dt$, the solution is,

$$v = \frac{eE}{m\omega}\left[\frac{\omega}{\omega^2+\nu^2}(\nu\cos\omega t+\omega\sin\omega t)+\left\{\frac{\nu(\nu\sin\omega t_1-\omega\cos\omega t_1)}{\omega^2+\nu^2}-\sin\omega t_1\right\}\exp\{-\nu(t-t_1)\}\right] \quad (7.12.1)$$

Since $\nu >> \omega$, keeping the largest of the exponentially decaying terms only, we obtain

$$v = \frac{eE}{m\omega}\left[\frac{\omega}{\omega^2+\nu^2}(\nu\cos\omega t+\omega\sin\omega t)-\sin\omega t,\ \exp\{-\nu(t-t_1)\}\right] \quad (7.12.2)$$

Denoting the steady part of $mv^2/2$ (the energy gained by the electron) by $\langle\xi\rangle$, we obtain

$$\langle\xi\rangle = \frac{e^2E^2}{2m\omega^2}\left[-\frac{\omega}{4\pi\nu}\left\{\exp\left(-\frac{4\pi\nu}{\omega}\right)-1\right\}\sin^2\omega t,\ \exp(2\nu t_1)+\frac{\omega^2}{\omega^2+\nu^2}\right] \quad (7.12.3)$$

Taking average over all values of t, that is, at all possible times t_1 when the series of electrons had experienced their last collisions, which is equivalent to the average over the time $2\pi/\omega$, we obtain

$$\langle\bar{\xi}\rangle = \frac{e^2E^2}{2m\omega}\left[-\frac{1}{2}\left(\frac{\omega}{4\pi\nu}\right)^2\left\{\exp\left(-\frac{4\pi\nu}{\omega}\right)-1\right\}\left\{\exp\left(\frac{4\pi\nu}{\omega}\right)-1\right\}+\frac{\omega^2}{2(\omega^2+\nu^2)}\right] \quad (7.12.4)$$

If only terms up to ν^2/ω^2 are retained, the relation (7.12.4) simplifies to

$$\langle\bar{\xi}\rangle = \frac{e^2E^2}{2m(\omega^2+\nu^2)}\left(1+\frac{\nu^2}{2\omega^2}\right) \quad (7.12.5)$$

Ignoring ν^2/ω^2 in (7.12.5), we get

$$\langle\bar{\xi}\rangle = \frac{e^2E^2}{2m\omega^2} \quad (7.12.6)$$

The work done, P, by the field on the electron is eEv. Its steady part is

$$\langle P\rangle = \frac{e^2E^2\nu}{2m(\omega^2+\nu^2)} = \frac{\nu\langle\xi\rangle}{(1+\nu^2/\omega^2)}; \quad (7.12.7)$$

so $\langle P\rangle = \nu\,\langle\bar{\xi}\rangle$ for $\nu << \omega$. At the second collision the ordered energy $\langle\bar{\xi}\rangle$ is converted into disordered random thermal energy. As a result the gas kinetic speed of electrons to the next free journey is increased. If φ is the angle between the ordered velocity **v** and the thermal velocity $\mathbf{v}_0$, the total kinetic energy available at the second collision is

$$\frac{m}{2}\,V^2 = \frac{m}{2}\,(v^2 + v_0^2 + 2v\,v_0 \cos\varphi) \tag{7.12.8}$$

Since v_0 is randomly distributed, the average kinetic energy, for all values of φ, is

$$\langle \frac{m}{2}\,V^2 \rangle = \frac{m}{2}\,(v^2 + v_0^2) \tag{7.12.9}$$

Hence energies of both ordered and random motions are summed up, for obtaining the total kinetic energy at the second collision.

For $\omega >> \nu$, ignoring collisional loss and writing x_0 and v_0 for the maximum displacement and maximum velocity, respectively, of an electron, under the influence of an applied electric field, oscillating at frequency ω, we have

$$x = \frac{eE}{m\omega^2},\quad v_0 = \frac{eE}{m\omega},\quad \langle\xi\rangle = \frac{e^2E^2}{4m\omega^2}$$

because $\xi = (mv^2/2)$ and $v = (eE/m\omega)\sin\omega t$.

In the E layer we take T = 300°K, ν = 10^5 rad/sec, mean free path $l = 100$ cm, $\xi_0 = 6\times10^{-14}$ erg, $G = 1.3\times10^{-3}$ and gyrofrequency $\Omega = 10^6$ rad/sec. For ω = 10^7 rad/sec we get Ω/ω = 0.1, $x_0 = 8\times10^{-5}$ cm, gyro radius $\rho = 8\times10^{-5}$ cm, v_0 = 5×10^3 cm/sec, $\langle\xi\rangle$ = 5×10^{-19} erg. When the steady state is reached, the ordered thermal energy $\langle\xi\rangle$, gained in each free path, equals the average energy lost ($=\xi_m$) of (7.9.1) at each encounter. Hence

$$\xi - \xi_0 = \xi_m/G = \langle\xi\rangle/G = 5\times10^{-19}/(1.3\times10^{-3})$$

Since $\xi_0 = 6\times10^{14}$ erg, we find that $(\xi-\xi_0)/\xi_0 = 6.3\times10^{-13} = 0.6$ per cent, and the mean energy of electrons is in excess of about 0.6 per cent compared with the mean energy of the heavier molecules in the presence of applied field. Electron temperature is also raised to this amount. This heating accumulates until the excess of electronic energy is such that the small fraction, which is lost, is sufficient to balance what it receives from the field.

(a) Wave Absorption

Average of the Poynting flux **S**, over the time $2\pi/\omega$, is

$$|\langle\mathbf{S}\rangle| = \frac{c}{4\pi}\,|\langle(\mathbf{E}\times\mathbf{H})\rangle| = \frac{ncE^2}{8\pi} \tag{7.12.10}$$

where n is the refractive index KC/ω, K is the real part of the wave number and C is the vacuum speed of light. From (7.11.1) we get

$$\delta\,|\langle \mathbf{S}\rangle| = \left(\frac{nc}{8\pi}\right) 2k_i\, e^2\, \delta x \qquad (7.12.11)$$

Therefore, mean power absorbed per electron is

$$\frac{1}{N}\frac{\partial}{\partial x}|\langle \mathbf{S}\rangle| = \frac{nc}{8\pi N} 2K_i\, e^2 \qquad (7.12.12)$$

Since this power is equal to the work done on the electron we must equate it to $\langle P\rangle$ of (7.12.7), and get

$$\frac{e^2 E^2 \nu}{2m(\omega^2+\nu^2)} = \frac{nc}{8\pi N} 2K_i\, E^2 \qquad (7.12.13)$$

Hence $K_i = 2\pi N e^2 \nu / mcnE^2(\omega^2 - \nu^2)$; approximately giving

$$K_i = \frac{\nu_0\, \omega_\rho^2}{2nc\omega^2} \qquad (7.12.14)$$

where $\omega_\rho^2 = 4\pi N e^2/m$; ω_p is the characteristic plasma frequency. Then the wanted electric field E of (7.11.4) can be written as

$$E = E_0 \exp\left(-\frac{\nu_0\,\omega_\rho^2}{2nc\omega^2}\right)\left[1 - \frac{\nu_0\,\omega_\rho^2}{2nc\omega^2}\{1 + M_1\cos(\omega_1 t - \alpha) + M_2 \cos(2\omega t_1 - \beta)\}\right] \qquad (7.12.15)$$

where we have first used (7.11.2) to replace $W/|\langle \mathbf{S}\rangle|$ by K_i and then replaced K_i by its value given in (7.12.14).

7.13 WHEN THE IONOSPHERE IS ASSUMED MAGNETIZED

In the presence of geomagnetic field $\mathbf{H}_0$ an applied alternating electric field splits up into an extraordinary and an ordinary modes (section 3.13). In a free path between collisions the component of electron motion parallel to $\mathbf{H}_0$ is unaltered, but across the field electrons are constrained to move in circles. If v is the component of random thermal velocity of electrons, perpendicular to $\mathbf{H}_0$, and ρ is the radius of the corresponding thermal gyration round $\mathbf{H}_0$, then $\rho\Omega = v$. Taking the average of this velocity as $v/2$ and $\Omega/2\pi = 10^4$, we get $\rho = 0.6$ cm. In the E layer the mean free path l is about 100 cm.; so, between two consecutive collisions, on the average, an electron completes 10 revolutions in a helix of radius 0.6 cm and length 100 cm. For the ordinary wave field this rotation is in the sense opposite to that of the free thermal gyration, and the extraordinary field has that in the same sense. Hence, orbits are a combination of the forced circular motion, at the frequency ω, and free circular thermal motion at the gyro frequency Ω. When $\omega^2 = \Omega^2$, one can consider only the extraordinary wave because it has the same sense of rotation as the thermal gyration of electrons round $\mathbf{H}_0$. This wave is heavily absorbed at the gyro frequency. A wave at the gyro frequency is called a gyro wave. Hence, the cross modulation is enhanced when the disturbing field is a gyro wave. But experiments have not always supported the prediction that an artificial glow discharge can be produced in the ionosphere by the field of a properly arranged transmitter operating at the gyro frequency if it is sufficiently strong.

In the presence of an oscillating electric field $\mathbf{E}$, and the static geomagnetic field $\mathbf{H}_0$, the electron equation of motion in the mean free path is

$$\frac{\partial \mathbf{u}}{\partial t} = \frac{e}{m}\mathbf{E} + (\mathbf{u} \times \boldsymbol{\Omega}) - \nu \mathbf{u} \tag{7.13.1}$$

where $\Omega = e\mathbf{H}_0/mc$. It can be written as

$$\partial \mathbf{u} = (e\mathbf{E}/m) + (\mathbf{u} \times \boldsymbol{\Omega}) \tag{7.13.2}$$

where
$$\partial = (d/dt) + \nu = D + \nu \tag{7.13.3}$$

Using equation (11.6.9), the power absorbed from the applied electric field by electrons per unit volume, averaged over the time period $2\pi/\omega$, is

$$\langle P \rangle = \frac{\omega_\rho^2}{4\pi}\left\{\frac{\nu\,(R_e\,E_x')^2}{(\omega+\Omega)^2+\nu^2} + \frac{\nu\,(R_e\,E_y')^2}{(\omega-\Omega)^2+\nu^2} + \frac{\nu E_z^2}{\omega^2+\nu^2}\right\} \tag{7.13.4}$$

where $E_x' = (E_x + iE_y)/\sqrt{2},\ E_y' = (E_x - iE_y)/\sqrt{2}$. Then writing

$(|E_x|^2 + |E_y|^2)^{1/2} = E_0 \sin\psi,\ E_z = E_0 \cos\psi$, so that if ψ is the angle between **E** and **H**, and if E_0 is the amplitude (which is $1/\sqrt{2}$ times the local amplitude of the circularly polarized electric field, having the same power), we find that

$$\langle P \rangle = \frac{\nu\omega_P^2}{8\pi}E_0^2\left\{\frac{\sin^2\psi}{(\omega+\Omega)^2+\nu^2} + \frac{\sin^2\psi}{(\omega-\Omega)^2+\nu^2} + \frac{2\cos^2\psi}{\omega^2+\nu^2}\right\} \tag{7.13.5}$$

The ordered energy per electron, averaged over the time period $2\pi/\omega$, and denoted by $\langle\xi\rangle$, is absorbed from the wave field between two consecutive collisions. so

$$\langle\xi\rangle = \frac{\langle P \rangle}{N\nu} = \frac{e^2 E_0^2 A}{2m} \tag{7.13.6}$$

where
$$A = \frac{\sin^2\psi}{(\omega+\Omega)^2+\nu^2} + \frac{\sin^2\psi}{(\omega-\Omega)^2+\nu^2} + \frac{2\cos^2\psi}{\omega^2+\nu^2} \tag{7.13.7}$$

Replacing $WN\nu$ of (7.11.2) by $\langle\xi\rangle N\nu$ of (7.13.6), we find from (7.11.2) that the absorption coefficient K_i is given by

$$K_i = \frac{N\nu\langle\xi\rangle}{2nc\,(E_0^2/8\pi)} = \frac{2\pi Ne^2\nu}{mcn}\left[\frac{\sin^2\psi}{(\omega+\Omega)^2+\nu^2}+\frac{\sin^2\psi}{(\omega-\Omega)^2+\nu^2}+\frac{2\cos^2\psi}{\omega^2+\nu^2}\right] \quad (7.13.8)$$

where n is the refractive index KC/ω of the wave in the medium. Putting $\omega = \Omega$ in the middle term in the right hand side of (6.13.8), we find that for the extraordinary wave absorption is very much enhanced there. When ω is within 20 per cent of the local Ω, we can take

$$K_i = \frac{2\pi Ne^2\nu}{mcn}\,\frac{\sin^2\psi}{(\omega-\Omega)^2+\nu^2} \quad (7.13.9)$$

Putting the value of $\langle P\rangle$ from (7.13.5) in equation (7.9.12), for the variation δT of temperature, in the quasi-stationary state, we find that

$$\delta T = \frac{e^2E_0^2}{3mGK}\left[\frac{\sin^2\psi}{(\omega+\Omega)^2+\nu^2}+\frac{\sin^2\psi}{(\omega-\Omega)^2+\nu^2}+\frac{2\cos^2\psi}{\omega^2+\nu^2}\right] \quad (7.13.10)$$

At the resonance $\omega = \Omega$ the increase of the temperature δT has an extreme value.

Chapter 8

KINETIC THEORY

8.1 INTRODUCTION

Kinetic theory explains the behaviour of gases, including plasmas, in terms of the particle motions. Since a finite volume contains large number of particles, this description has to be statistical. So, in the study of the behaviour of systems by kinetic theory propabilistic or statistical arguments and the concepts of atomic division of particles in phase space are important. These establish relations between observed macroscopic properties of large systems and properties of particles and their interactions. The phenomenological laws of thermodynamics and other subjects are thus given an atomistic foundation. For simplicity, particles are assumed to be hard elastic spheres in motion, each specified by a radius and mass, and in plasmas, by a charge. Complicated molecular structures and particle spin and other quantum mechanical effects are ignored. The large number of independent variables is condensed by some averaging procedure to obtain the macroscopic properties of a system. Statistical mechanics takes into account more degrees of freedom than fluid mechanics, for extracting from an exact and detailed description of a complex system, those results which are useful in specifying its gross behaviour. So macroscopic averages are taken after finding out the field solutions in phase space. The fluid continum description, on the other hand, deals with the variations of the macroscopic averages in the continuum of position space and time.

Atomic volumes are about 10^{-24} cm^3. Interatomic electric field is 10^8 volts/cm and macroscopic averages of electric fields are taken over regions of volume 10^{-6} cm^3, or over distance 10^{-2} cm and areas 10^{-4} cm^2. About $10^{23 \pm 3}$ particles exist in volumes of macroscopic interest. A volume containing 10^{20} particles will have 3×10^{20} space coordinates if they are without spin, and 3×10^{20} moments. If these elements could be known at time t_0 then their values at any other instant t are known from the solution of 3×10^{20} number of equations of particle motion. But, it is not possible to handle such a large system of equations; and, practically speaking, this is also not necessary, because only values of small number of macroscopic quantities, like pressure, temperature, average velocity, etc., are generally necessary. Average forces, exerted on the walls of the container and taken over small space and time intervals, by the particles in random motion, are considered and not the very rapidly varying force of impingement by the particles on different points of the walls. Sometimes empirical state relations, such as the dependence of pressure on density and temperature, determine quantitatively, the caloric, dielectric or other electromagnetic properties. In fluid description an element around a point is supposed to move with the average velocity $\mathbf{u}$, which is also called the fluid velocity or the gross macroscopic velocity. The momentum density, $\rho(\mathbf{r})\,\mathbf{u}(\mathbf{r})$, where $\rho(\mathbf{r})$ is the mass per unit volume and $\mathbf{u}(\mathbf{r})$ is the velocity of the fluid element, can be expressed as

$$\rho(\mathbf{r})\,\mathbf{u}(\mathbf{r}) = \sum_{i=1}^{N} m_i(\mathbf{r}+\mathbf{r}_i)\,\mathbf{u}_i(\mathbf{r}+\mathbf{r}_i)$$

where $\mathbf{r}_i$ and $\mathbf{r}$ are points inside the unit volume, $\mathbf{r}_i$ is the current point inside the element and the sum over i includes all the N positions around the point $\mathbf{r}$. Difference between $\mathbf{u}_i(\mathbf{r}+\mathbf{r}_i)$ and $\mathbf{u}(\mathbf{r})$ decreases as $\rho(\mathbf{r})$ increases, and increases as the medium becomes rarefied. The fluid velocity $\mathbf{u}(\mathbf{r})$ is not necessarily the velocity of any particle in the neighbourhood of $\mathbf{r}$ at the given time. Each particle in the element has its own mass and a velocity due to internal thermal motion. The thermal velocity $\mathbf{u}_i(\mathbf{r}+\mathbf{r}_i)$ of a microelement is different from the apparent macroscopic velocity $\mathbf{u}(\mathbf{r})$ of a macroelement. In diffuse plasmas, study of motion of microelements, carrying charges, is necessary and a single fluid theory is not always sufficient. Kinetic theory describes well the properties of dilute gases in which the influence of short range forces of interaction cannot be ignored.

8.2 PHASE SPACE FOR MANY-PARTICLE MOTION

The complete mechanical state of a volume of plasma can be specified in terms of ν generalized position coordinates and ν generalised momentum coordinates and so by a total of 2ν coordinates. Mathematically, these 2ν numbers define a point in a space of 2ν dimensions, and this generalized space is called the phase space of position coordinates and momentum coordinates of the system or of the Γ space.

In a container of volume τ a system of n particles (point masses), all free of internal spin, has $3n$ position coordinates and $3n$ components of momentum; all these are functions of time only. So $\nu = 3n$ and the Γ space has $6n$ dimensions, and also the same number of independent coordinate variables.

Denoting the position of the jth particle at time t by $\mathbf{r}_i$ and its three components by r_j^i, with $i = 1, 2, 3;\ = j = 1, 2, n$, we find that r_j^i is the ith position coordinate of the jth particle, and the position vectors of n particles are $\mathbf{r}_1, \mathbf{r}_2, \mathbf{r}_n$; the n momentum vectors are $\mathbf{p}_1, \mathbf{p}_2, \mathbf{p}_n$; and p_j^i is the ith component of the momentum of the jth particle. The $6n$ dimensional phase space Γ has, therefore, the representative points $(\mathbf{r}_1, \mathbf{r}_2, ..., \mathbf{r}_n ; \mathbf{p}_1, \mathbf{p}_2, ..., \mathbf{p}_n) \equiv \mathbf{Z}$. In this scheme $Z_j \equiv (\mathbf{r}_j,\ \mathbf{p}_j)$ is the phase of the jth particle at position $\mathbf{r}_j$, and momentum $\mathbf{p}_j$. Hence

$$\mathbf{Z} \equiv (\mathbf{z}_1,\ \mathbf{z}_2,\ \mathbf{z}_n)$$

Therefore, the $6n$-dimensional Γ space consists of n sub spaces, each of which has 6 dimensions. Each of the n particles has one such subspace and such a subspace having six dimensions is called a μ space. Specifically, the six-dimensional subspace μ_j refers to the phase space $(\mathbf{r}_j, \mathbf{p}_j) \equiv \mathbf{z}_j$ of the jth particle. The particle ensemble is, therefore, characterized by n points in the μ space and one point in the Γ space at each instant of time. If all the n particles are identical the state of the system in the μ space does not change for different permutations of the n particles in fixed positions. If n is large the distribution of particles in the μ space can be described by a smooth function $f(\mathbf{r}, \mathbf{v}, \mathbf{t})$ of the phase point $\mathbf{r}, \mathbf{v}$. When there are three identical particles at three points A, B, C there will be $6\ (= 3\ !)$ ways of distributing these particles in the μ space. Only one phase exists in the Γ space of 18 dimensions, corresponding to the 6 phases in the μ space. For n identical particles the (factorial $n\ (= n\ !)$ phases $n\ !$ phases in the μ space correspond to only one phase in the Γ space. In other words, one and the same specific phase in the Γ space corresponds to n generic phases in the μ space. So, in kinetic theory the following infinitesimal phase space elements are available :

(*a*) $d\mathbf{r}$, three-dimensional, in position space; (8.2.1)

(*b*) $d\mathbf{p}$, three-dimensional, in momentum space; (8.2.2)

(*c*) $d\mathbf{z}_j$, six-dimensional, in the μ space; (8.2.3)

(*d*) $d\mathbf{Z}$, $6n$-dimensional, in the Γ space; (8.2.4)

Integrations are, therefore, three-fold, three-fold, six-fold and $6n$-fold, and

$$d\mathbf{Z} = d\mathbf{z}_1,\ d\mathbf{z}_2 \,.\, d\mathbf{z}_n = \prod_{\substack{i=1,2,3 \\ j=1,2\ldots n}} dr_j^i\ dp_j^i \tag{8.2.5}$$

The number of coordinates in Γ space depends actually on the degrees of freedom possessed by the individual particles or of the system as a whole. If each particle possesses l degrees of freedom, then the number of coordinates in the phase space of n particles is $2ln$. If all the particles are like rigid bodies, then the position of a point on any one of them is fixed by three position coordinates of another fixed point in the body and three angular coordinates of rotation about that point. In that case each particle possesses six degrees of freedom and the number of coordinates in Γ space is $12n$. The subscripts $j\ (= 1, 2, \ldots n)$ are actually arbitrarily assigned to the coordinates of the n particles without any special preference for any of these numbers on particular particles.

8.3 MANY-BODY PARTICLE DYNAMICS

The equations of motion for n particles are

$$\frac{d\mathbf{p}_j}{dt} = m_j \frac{d^2\mathbf{r}_j}{dt^2} = \mathbf{X}_j(\mathbf{Z}),\ \ \mathbf{p}_j = m_j \frac{d\mathbf{r}_j}{dt}, \tag{8.3.1}$$

where
$$\mathbf{X}_j = \sum_{k=1}^{n} \mathbf{X}_{jk}(\mathbf{z}_j, \mathbf{z}_k) + \mathbf{X}_j(\mathbf{Z}), \tag{8.3.2}$$

$$\mathbf{X}_{jk}\,(\mathbf{z}_j,\ \mathbf{z}_k) = \left(\frac{\mathbf{r}_{jk}}{|\mathbf{r}_{jk}|}\,\varphi'\,(\mathbf{r}_{jk}) \right); \tag{8.3.3}$$

$$\mathbf{r}_{jk} = \mathbf{r}_k - \mathbf{r}_j \tag{8.3.4}$$

Here $\varphi\,(\mathbf{r}_{jk})$ is the two-particle potential function of the conservative force of interaction, $\varphi'\,(\mathbf{r}_{jk})$ is the derivative of $\varphi\,(\mathbf{r}_{jk})$ with respect to the length element parallel to $\mathbf{r}_{jk}$ at $\mathbf{r}_j$, and $\mathbf{X}_j$ is the external force acting on the jth particle. In (8.3.1), for simplicity, the non-relativistic equations of motion have been used. Equation (8.3.1) defines a flow in the phase space Γ; the succession of states **Z** is a function of time and can be represented by a curve in Γ space.

If the system is given at time $t = 0$ by the $6n$ coordinates

$$\mathbf{r}_1\,(0),\ \mathbf{r}_2\,(0), \ldots\ \mathbf{r}_n\,(0);\ \mathbf{p}_1\,(0), \mathbf{p}_2\,(0),\ \mathbf{p}_n\,(0),$$

then subsequent evolution, at any time t, is given by the vectors

$\mathbf{r}_1\,(t), \mathbf{r}_2\,(t), \ldots\ \mathbf{r}_n\,(t);\ \mathbf{p}_1\,(t), \mathbf{p}_2\,(t), \ldots\ \mathbf{p}_n\,(t)$ where

$$\mathbf{r}_j\,(t) = \mathbf{r}_j\,(0) + \int_0^t \frac{1}{m_j}\,\mathbf{p}_j\,(t')\,dt', \tag{8.3.4 (a)}$$

$$\mathbf{p}_j\,(t) = \mathbf{p}_j\,(0) + \int_0^t \mathbf{X}_j\,\{\mathbf{r}_j\,(t'),\ \mathbf{p}_j\,(t');\ \Sigma\,\mathbf{r}_i\,(t'),\ \Sigma\,\mathbf{p}_i\,(t')\}\,dt' \tag{8.3.5}$$

and the force $\mathbf{X}_j$ is given by (8.3.1). To study the evolution of the system in time by Hamilton's equations of mechanics, the Hamiltonian H_n of the n-particle system is defined by

$$H_n = \sum_{j=1}^{n} \frac{p_j^2}{2m_j} + \sum_{j}^{n} \sum_{<k}^{n} \varphi\,(r_{jk}) + \sum_{j=1}^{n} \varphi\,(r_j) \tag{8.3.6}$$

where $\varphi\,(r_j)$ is the potential of the sum of external forces acting on the jth particle. The Hamiltonian canonical equations for motion in the ith direction of the jth particle are

$$\dot{p}_j^i = -\frac{\partial H_n}{\partial r_j^i},\quad \dot{r}_j^i = \frac{\partial H_n}{\partial p_j^i}\ (i = 1, 2, 3;\ j = 1, 2, \ldots n) \tag{8.3.7}$$

When every p_j^i and r_j^i are given, equations (8.3.7) determine uniquely the time rate of change of any coordinate in the Γ space.

Motion of n particles can be considered as the motion of a phase point in the $6n$-dimensional Γ space. At time t this space is filled up with a large number of points, of the μ space, each of which has evolved from a certain initial state which is a point in the Γ space at $t = 0$. Since the number of particles is fixed, no phase point is created or destroyed and so the $6n$-dimensional phase space is without any source or sink.

8.4 DISTRIBUTION FUNCTIONS

The aggregate of particles in the Γ space is called an ensemble. Geometrically the ensemble is the continuous distribution of phase points in the Γ space. The phase points can be plotted in the phase space in which they are distributed with a certain density D_n, which is, in general, different in different parts of the Γ space and change in time. So, we write

$$D_n \equiv D_n(\mathbf{Z}) \tag{8.4.1}$$

where the $6n$ components of $\mathbf{Z}$ are functions of time. The number of phases contained in the element $d\mathbf{Z}$, around the phase point $\mathbf{Z}$, is $D_n\,d\mathbf{Z}$. The real physical system, that is observed, has the probability of having the configuration $\mathbf{Z}$ at time t, namely,

$$F_n(\mathbf{Z})\,d\mathbf{Z} = \frac{D_n(\mathbf{Z})\,d\mathbf{Z}}{\int D_n(\mathbf{Z})\,d\mathbf{Z}} \tag{8.4.2}$$

where integration of the integral in the denominator is taken with respect to the $6n$-dimensional variables of $d\mathbf{Z}$ in the entire Γ space. Hence, whatever be the functional form of D_n, we must have

$$\int D_n(\mathbf{Z})\,d\mathbf{Z} = n \tag{8.4.3}$$

at every instant of time. The probability of a phase in Γ space is

$$F_n(\mathbf{Z}) = \frac{1}{n} D_n(\mathbf{Z}) \quad F_n(\mathbf{Z}) > 0,$$

and

$$\int F_n(\mathbf{Z})\,d\mathbf{Z} = 1 \tag{8.4.5}$$

$F_n(\mathbf{Z})$ is symmetric with respect to the $6n$ coordinates of Γ space; so is the specific distribution of section 8.2. If, therefore, $\varphi(\mathbf{Z})$ is a function in Γ space, its average, $\overline{\varphi}(t)$, is given by

$$\overline{\varphi}(t) = \int \varphi(\mathbf{Z})\,F_n(\mathbf{Z})\,d\mathbf{Z} = \frac{\int \varphi(\mathbf{Z})\,D_n(\mathbf{Z}, t)\,d\mathbf{Z}}{\int D_n(\mathbf{Z}, t)\,d\mathbf{Z}} \tag{8.4.6}$$

The measurable macroscopic quantities correspond to the ensemble averages in (8.4.6) when taken over a suitably chosen ensemble. The initial distribution $F_n(t=0)$ can be written as

$$F_n(0) = \prod_{j=1}^{n} \delta(\mathbf{R}_j - \mathbf{r}_j(0))\;\delta(\mathbf{P}_j - \mathbf{p}_j(0)) \tag{8.4.7}$$

where the right hand side is the product of delta functions and the total number of these factors is n^2. The subsequent evolution of F_n in time t is given by

$$F_n(t) = \prod_{j=1}^{n} \delta(\mathbf{R}_j - \mathbf{r}_j(t))\, \delta(\mathbf{P}_j - \mathbf{p}_j(t)) \tag{8.4.8}$$

$\mathbf{R}_j$ and $\mathbf{P}_j$ are the vectorial current coordinates in the spaces of position vectors and momentum vectors respectively; the equations (8.3.4) and (8.3.5) are solved for $\mathbf{r}_j(t)$ and $\mathbf{p}_j(t)$.

In kinetic theory a distribution function $f(\mathbf{r}, \mathbf{v}, t)$ is defined in the six-dimensional μ space of position $\mathbf{r}$ and velocity $\mathbf{v}$ for n identical particles, such that $f(\mathbf{r}, \mathbf{v}, t)\, d\mathbf{r}\, d\mathbf{v}$ is the average number of particles at time t in position space in the volume element $d\mathbf{r}$ around the point $\mathbf{r}$, and in velocity space in the volume element $d\mathbf{v}$ around the point $\mathbf{v}$, that is, in the volume element $d\mathbf{Z}$ (= $d\mathbf{r}\, d\mathbf{v}$) about $\mathbf{Z}$ in μ space. If $f_n(\mathbf{Z})$ is the generic distribution of section 8.2, the probability of finding the n particles in μ space in the n volume elements $d\mathbf{r}_1, d\mathbf{p}_1, d\mathbf{r}_2 \ldots d\mathbf{p}_2, d\mathbf{r}_n\, d\mathbf{p}_n$, around the points $\mathbf{r}_1\, \mathbf{p}_1, \mathbf{r}_2\, \mathbf{p}_2, \ldots \mathbf{r}_n\, \mathbf{p}_n$, is

$$f_n(\mathbf{Z})\, d\mathbf{z}_1\, d\mathbf{z}_2 \ldots d\mathbf{z}_n = n!\, F_n(\mathbf{Z})\, d\mathbf{Z} \tag{8.4.9}$$

Consider two states of an ensembles; in one the ith particle is at the point $\mathbf{r}_i$ and the jth particle at $\mathbf{r}_j$, and in the other the ith particle is at $\mathbf{r}_j$ and the jth particle at $\mathbf{r}_i$. These two states give different points in the Γ space; so these are different states in that space. But, macroscopically, the two states are indistinguishable and so identical. Hence, a macroscopic state of a gas may be any one of the infinitely large number of states of the ensemble in the Γ space.

8.5 REDUCED DISTRIBUTION FUNCTIONS

The probability that a small number j of particles have prescribed positions and momentums, whatever be the position and momentum of each of the remaining $n - j$ particles, is called the reduced or marginal or lower order distribution function, and is denoted by F_j. If the six-dimensional coordinates of the marked out j particles are $\mathbf{z}_1, \mathbf{z}_2, \ldots, \mathbf{z}_j$, then

$$F_j \equiv F_j(\mathbf{z}_1, \mathbf{z}_2, \ldots, \mathbf{z}_j)$$

The function F_j is obtained by multiplying F_n with τ^j and integrating over all possible positions and momentums of the remaining $n - j$ particles where τ is the volume of the container of the n particles. Then

$$F_j(\mathbf{z}_1, \mathbf{z}_2, \ldots, \mathbf{z}_j) = \tau^j \int_{6n-6j} F_n(\mathbf{Z}) \prod_{i=j+1}^{n} d\mathbf{z}_i \tag{8.5.1}$$

Hence $$F_j(\mathbf{z}_j) = \tau \int_{6n-6} F_n(\mathbf{Z})\, d\mathbf{z}_1\, d\mathbf{z}_2 \,.\, d\mathbf{z}_{j-1}\, d\mathbf{z}_{j+1} \,.\, d\mathbf{z}_n \tag{8.5.2}$$

The function F_j is independent of any particular choice of j particles and satisfies the relation

$$\int F_j(\mathbf{z}_1, \mathbf{z}_2, ..., \mathbf{z}_j)\, d\mathbf{z}_1\, d\mathbf{z}_2 \,.\, d\mathbf{z}_j = \tau^n \tag{8.5.3}$$

Clearly $F_1(\mathbf{z}_1)\, d\mathbf{z}_1/\tau$ is the probability that at time t the particle 1 is lying in the 6-dimensional phase space volume $d\mathbf{z}_1$, around the phase point $\mathbf{z}_1$. Similarly, the two-particle distribution function $F_2(\mathbf{z}_1, \mathbf{z}_2)\, d\mathbf{z}_1\, d\mathbf{z}_2/\tau^2$ is the joint probability that at time t the particle 1 is lying in $d\mathbf{z}_1$, around the point $\mathbf{z}_1$, and the particle 2 in $d\mathbf{z}_2$ around $\mathbf{z}_2$. Also $n F_1(\mathbf{Z})\, d\mathbf{Z}/\tau$ is the expected number of particles in μ space in the volume element $d\mathbf{Z}$ around Z. If $\rho(\mathbf{Z})$ is the mass density in μ space, then

$$\rho(\mathbf{Z}) = \frac{mn}{\tau} F_1(\mathbf{Z}) \tag{8.5.4}$$

The probability of finding the n number of particles in n given space points $\mathbf{r}_1, \mathbf{r}_2, ..., \mathbf{r}_n$, whatever be their momentum, therefore, is

$$F_n(\mathbf{r}_1, \mathbf{r}_2, ..., \mathbf{r}_n) = \int_{3n} F_n(\mathbf{Z})\, d\mathbf{p}_1\, d\mathbf{p}_2 \ldots d\mathbf{p}_n \tag{8.5.5}$$

Similarly, the probability of finding the j number of particles in the j specific points $\mathbf{r}_1, \mathbf{r}_2, ..., \mathbf{r}_j$, without specifying their momentum and the momentum and position of the remaining $n - j$ particles, is

$$F_j(\mathbf{r}_1, \mathbf{r}_2, ..., \mathbf{r}_j) = \int_{3j} F_j(\mathbf{z}_1, \mathbf{z}_2, ..., \mathbf{z}_j)\, d\mathbf{p}_1\, d\mathbf{p}_2 \ldots d\mathbf{p}_j$$

$$= \tau^j \int_{3n-3j} F_n(\mathbf{Z})\, d\mathbf{p}_1\, d\mathbf{p}_2 \ldots d\mathbf{p}_j\, d\mathbf{z}_{j+1}\, d\mathbf{z}_{j+2} \ldots d\mathbf{z}_n \tag{8.5.6}$$

Probability of the n particles having specified momentums $\mathbf{p}_1, \mathbf{p}_2, ..., \mathbf{p}_n$, whatever be their positions, is given by

$$F_n(\mathbf{p}_1, \mathbf{p}_2, ..., \mathbf{p}_n) = \tau^n \int_{3n} F_n(\mathbf{Z})\, d\mathbf{r}_1\, d\mathbf{r}_2 \ldots d\mathbf{r}_n \tag{8.5.7}$$

Similarly

$$F_j(\mathbf{p}_1, \mathbf{p}_2 ..., \mathbf{p}_j) = \tau^n \int_{6n-3j} F_n(\mathbf{Z})\, d\mathbf{r}_1\, d\mathbf{r}_2 \ldots d\mathbf{r}_j\, d\mathbf{z}_{j+1}\, d\mathbf{z}_{j+2}\, d\mathbf{z}_n \tag{8.5.8}$$

In the 6-dimensional μ space the probability of finding the j specified particles in the j six-dimensional volume elements $d\mathbf{r}_1\, d\mathbf{p}_1, d\mathbf{r}_2\, d\mathbf{p}_2, ..., d\mathbf{r}_j\, d\mathbf{p}_j$, around the j number of μ-space points $\mathbf{r}_1\, \mathbf{p}_1, \mathbf{r}_2\, \mathbf{p}_2, ..., \mathbf{r}_j\, \mathbf{p}_j$, is the total number of ways of selecting the j particles from the total n. Therefore, since $F_j(\mathbf{z}_1, \mathbf{z}_2, ..., \mathbf{z}_j)/\tau^j$ is the probability of finding the particular j particles in the volume element $d\mathbf{r}_1\, d\mathbf{r}_2 \ldots, d\mathbf{r}_j\, d\mathbf{p}_1\, d\mathbf{p}_2 \ldots d\mathbf{p}_j$, in the $6j$-dimensional Γ space, we can write

$$f_j(\mathbf{z}_1, \mathbf{z}_2, \ldots, \mathbf{z}_j) = \frac{n\,!\,F_j(\mathbf{z}_1, \mathbf{z}_2, \ldots, \mathbf{z}_j)}{(n-\mathbf{j})\,!\,\tau^j} \tag{8.5.9}$$

For $j = 1$, we obtain

$$f_1(\mathbf{r}_1, \mathbf{p}_1) = \frac{n\,!\,F_1(\mathbf{r}_1, \mathbf{p}_1)}{(n-1)\,!\,\tau} = \frac{n\,F_1(\mathbf{r}_1, \mathbf{p}_1)}{\tau} \tag{8.5.10}$$

Putting now the value of F_j from (8.5.9) in (8.5.6) we find that

$$(n-j)\,!\,f_j(\mathbf{z}_1, \mathbf{z}_2, \ldots, \mathbf{z}_j) = \int_{6n-6j} f_n(\mathbf{Z})\, d\mathbf{z}_{j+1}\, d\mathbf{z}_{j+2} \ldots d\mathbf{z}_n \tag{8.5.11}$$

If now $n_j(\mathbf{r}_1, \mathbf{r}_2, \ldots, \mathbf{r}_j)$ is the generic distribution, in the configuration space, then

$$n_j(\mathbf{r}_1, \mathbf{r}_2, \ldots, \mathbf{r}_j) = \frac{n\,!\,F_j(\mathbf{r}_1, \mathbf{r}_2, \ldots, \mathbf{r}_j)}{(n-\mathbf{j})\,!\,\tau^j}$$
$$= \frac{1}{(n-\mathbf{j})\,!} \int n_n(\mathbf{r}_1, \mathbf{r}_2, \ldots, \mathbf{r}_n)\, d\mathbf{r}_{j+1}\, d\mathbf{r}_{j+2} \ldots d\mathbf{r}_n \tag{8.5.12}$$

The distribution function $f(\mathbf{r}, \mathbf{v}, t)$ in μ space is directly related to the one-particle distribution function $F_j(\mathbf{r}_j, \mathbf{v}_j, t)/\tau$ in the $\mathbf{z}$ space of position and velocity. Since $f(\mathbf{r}, \mathbf{v}, t)\, d\mathbf{r}\, d\mathbf{v}$ is the number of particles in the volume element $d\mathbf{r}\, d\mathbf{v}$, in μ space, for simplicity, we consider the case of n identical particles. It is obvious that there is no way to distinguish between different combinations of them in fixed positions. For example, no distinction can be made between ensembles in which positions of two specific particles are interchanged. Hence, the number of particles having position vector $\mathbf{r}$ and velocity $\mathbf{v}$ is $nF_1(\mathbf{r}, \mathbf{v}, t)\, d\mathbf{r}\, d\mathbf{v}/\tau$, and nF_j/τ is the average value of f. Equation for F_1/τ can, therefore, be transformed to the corresponding equation for $f(\mathbf{r}, \mathbf{v}, t)$.

8.6 TRUNCATED DISTRIBUTION FUNCTIONS

The probability density of a function of the total number of particles, within or outside a specified region of the infinite space, is called a truncated distribution. For finding such distributions consider the point $\mathbf{r}_1$, where, at time t, a particle with momentum $\mathbf{p}_1$ is situated. Also consider a sphere of given radius ε with centre at $\mathbf{r}_1$. Particles outside it satisfy the condition

$$|\mathbf{r}_j - \mathbf{r}_i| >> \varepsilon \tag{8.6.1}$$

Also, by D, we denote the $6n - 6$-dimensional domain outside this sphere. The region under D contains all states of the $n - 1$ particles, other than the one at $\mathbf{r}_1$. Then the truncated distribution $F_1^{\varepsilon}(\mathbf{z}_1)$ is defined as

$$F_1^{\varepsilon}(\mathbf{z}_1) = \tau \int_D F_n(\mathbf{Z})\, d\mathbf{z}_2\, d\mathbf{z}_3 \ldots d\mathbf{z}_n \qquad (8.6.2)$$

Also $F_1^{\varepsilon}(\mathbf{z}_1)$ can be written as

$$F_1^{\varepsilon}(\mathbf{z}_1) = F_1(\mathbf{z}_1) - (n-1) \int_{D_2} F_2(\mathbf{z}_1, \mathbf{z}_2)\, d\mathbf{z}_2 +$$

$$+\frac{1}{2}(n-1)(n-2) \int_{\overline{D}_2 \times \overline{D}_3} F_3(\mathbf{z}_1, \mathbf{z}_3)\, d\mathbf{z}_2\, d\mathbf{z}_3 + \ldots \qquad (8.6.3)$$

where $\overline{D}_j$ is the domain which includes those $\mathbf{Z}_j$ which satisfy the inequality $|\mathbf{r}_j - \mathbf{r}_1| << \varepsilon$. When a specified particle is at the point $\mathbf{r}_1$, and has the momentum $\mathbf{p}_1$, the function $F_1^{\varepsilon}(\mathbf{z}_1)$ is the expectation of finding no other particle within the sphere of radius ε with centre at $\mathbf{r}_1$. If the occupied volume is small, then $n\varepsilon^3$ is small in comparison with the volume τ of the container and the difference between F_1^{ε} and F_1 is also small. The quantity $nF_1^{\varepsilon}(\mathbf{Z})\, d\mathbf{Z}$ is the expected number of particles in the volume element $d\mathbf{Z}$ in μ space; the particles of $d\mathbf{Z}$ have no neighbour closer than at the distance ε. The mass densities in μ space of those particles, which have no close neighbours, is

$$\rho^{\varepsilon}(\mathbf{Z}) = nmF_1^{\varepsilon}(\mathbf{Z}) \qquad (8.6.4)$$

Another truncated distribution is

$$F_2^{\varepsilon}(\mathbf{z}_1, \mathbf{z}_2) \equiv \tau^2 \int_{D^i} F_n(\mathbf{Z})\, d\mathbf{z}_3\, d\mathbf{z}_4 \ldots d\mathbf{z}_n \qquad (8.6.5)$$

where $D^1 = D_2 \times D_3 \times \ldots \times D_n$, and $D_j\ (j = 2, 3, \ldots n)$ is the domain containing all states of the particles j with distances from $\mathbf{r}_1$ greater than ε. Since the domain of integration involves $\mathbf{r}_1$, but not $\mathbf{r}_2$, the function $F_2^{\varepsilon}(\mathbf{z}_1, \mathbf{z}_2)$ is not symmetric in $\mathbf{z}_1$ and $\mathbf{z}_2$.

8.7 LIOUVILLE'S THEOREM

The condition of continuity or conservation of a phase point in the $6n$-dimensional Γ space is the theorem of Liouville. For computation no distinction is necessary between any of the $3n$ position coordinates or momentum coordinates. So, for convenience, the $3n$ position coordinates r_j^i are denoted by q_1, $q_2, \ldots, q_{3n}$, and the $3n$ momentums by $p_1, p_2 \ldots, p_{3n}$. Then, in Γ space, we have

$$\mathbf{Z} \equiv (\mathbf{z}_1, \mathbf{z}_2, \ldots, \mathbf{z}_n) \equiv (q_1, q_2, \ldots, q_{3n};\ p_1, p_2, \ldots, p_{3n}) \qquad (8.7.1)$$

Denoting by P and Q the points in Γ space at times t and $t + dt$, respectively, for the same system of particles, we can write

$$P \equiv P(\mathbf{Z}(t)),\ Q \equiv Q(\mathbf{Z}(t+dt)) = Q(\mathbf{Z}+d\mathbf{Z}); \tag{8.7.2}$$

where $\mathbf{Z}+d\mathbf{Z} \equiv q_1 + dq_1, q_2 + dq_2, ..., p_{3n} + dp_{3n}$. It is clear that $d\mathbf{Z}$, appearing in $\mathbf{Z}+d\mathbf{Z}$, has a different interpretation than the volume element $d\mathbf{Z}$ of (8.2.4) in Γ space, and it is here given by

$$d\mathbf{Z} \equiv dq_1\, dq_2 \,....\, dq_{3n}\, dp_1\, dp_2 \ldots dp_{3n} \tag{8.7.3}$$

If dn is the number of phase points in this volume element, then the density $F_n(\mathbf{Z})$ of phase points is defined by

$$F_n(\mathbf{Z}) = \lim_{Q \to P} \frac{dn}{d\mathbf{Z}} \tag{8.7.4}$$

The particle will flow at the point P in the positive direction of the q_1-axis at the rate $= (F_n \dot{q}_1)_\rho\, dq_2 \,....\, dp_{3n}$. If PP_1 be the displacement along the q_1-axis, in time $t + \delta t$, after time t, then $P_1 \equiv P_1(q_1 + dq_1, q_2, ..., p_{3n})$. Therefore, the flow in the positive direction of the q_1-axis at P_1 from the volume element

$$= [(F_n \dot{q}_1)_{P_1} - (F_n \dot{q}_1)_P]\, dq_2 \ldots dp_{3n}$$

$$= [(F_n \dot{q}_1)_{q_1 + dq_1} - (F_n \dot{q}_1)_{q_1}]\, dq_2 \ldots dp_{3n}.$$

Retaining terms up to the first power of dq_1, this expression equals $\{\partial(F_n \dot{q}_1)/\partial q_1\}\, d\mathbf{Z}$. Similarly, the outward flow from $d\mathbf{Z}$, along the remaining directions in the Γ space, are obtained. If all these outward flows, through the walls of the $6n$-dimensional box, with edges $dq_1, dq_2, ..., dp_{3n}$, are summed up, and equated to the rate of loss of particles from the volume, then, for conservation of phase points, we have

$$-\frac{\partial}{\partial t}(F_n\, d\mathbf{Z}) = \sum_{i=1}^{3n} \left[\frac{\partial}{\partial q_1}(F_n \dot{q}_i) + \frac{\partial}{\partial p_i}(F_n \dot{p}_n) \right] d\mathbf{Z} + ...,$$

where terms not mentioned in the right hand side are of higher orders. In the limit $Q \to P$, the differential equation for the Liouville theorem is

$$\frac{\partial F_n}{\partial t} + \sum_{i=1}^{3n} \left[\frac{\partial}{\partial q_i}(F_n \dot{q}_i) + \frac{\partial}{\partial p_i}(F_n \dot{p}_i) \right] = 0 \tag{8.7.5}$$

This is also the law of conservation of n particles in Γ space.

Using the Hamiltonian equations

$$\frac{dq_i}{dt} = \frac{\partial H_n}{\partial p_i},\ \frac{dp_i}{dt} = -\frac{\partial H_n}{\partial q_i} \tag{8.7.6}$$

we obtain $$\frac{\partial}{\partial q_i}(\dot{q}_i) = \frac{\partial^2 H_n}{\partial p_i\, \partial q_i} = -\frac{\partial}{\partial p_i}(\dot{p}_i); \tag{8.7.7}$$

so the equation (8.7.5) reduces to

$$\frac{\partial F_n}{\partial t} + \sum_{i=1}^{3n} \left(\dot{q}_i \frac{\partial F_n}{\partial q_i} + \dot{p}_i \frac{\partial F_n}{\partial p_i} \right) = 0 \tag{8.7.8}$$

Replacing, in this equation, $\dot{q}_i$ by $(\partial H_n/\partial p_i)$ and $\dot{p}_i$ by $-(\partial H_n/\partial q_i)$, we find that

$$\frac{\partial F_n}{\partial t} + \sum_{i=1}^{3n} \left(\frac{\partial H_n}{\partial p_i} \frac{\partial F_n}{\partial q_i} - \frac{\partial H_n}{\partial q_i} \frac{\partial F_n}{\partial p_i} \right) = 0 \tag{8.7.9}$$

A familiar form of this equation is

$$\frac{\partial F_n}{\partial t} + [F_n, H_n] = 0 \tag{8.7.10}$$

where $[F_n, H_n]$ is the well-known Poisson bracket. We introduce the generalized flow vector **V**, with the *i*th component

$$V_i = \frac{dq_i}{dt} + \frac{dp_i}{dt} \tag{8.7.11}$$

The divergence of **V** in Γ space is

$$\text{div}\,\mathbf{V} = \sum_{i=1}^{3n} \left[\frac{\partial}{\partial q_i}\left(\frac{dq_i}{dt}\right) + \frac{\partial}{\partial p_i}\left(\frac{dp_i}{dt}\right) \right]$$

Using, in this relation, (8.7.6), we get

$$\text{div}\,\mathbf{V} = \sum_{i=1}^{3n} \left(\frac{\partial^2 H_n}{\partial q_i\, \partial p_i} - \frac{\partial^2 H_n}{\partial p_i\, \partial q_i} \right) = 0 \tag{8.7.12}$$

So, flow in Γ space is solenoidal (free from sources and sinks). Another form of the Liouville equation is

$$\frac{\partial F_n}{\partial t} + \text{div}\,(\mathbf{V}\, F_n) = 0 \quad \text{or} \quad \frac{\partial F_n}{\partial t} + (\mathbf{V}\,.\,\text{grad})\, F_n = 0 \tag{8.7.13}$$

where the generalized gradient operator ∇ is such that

$$(\mathbf{V}\,.\,\nabla) = \sum_{i=1}^{3n} \left(\frac{dq_i}{dt} \frac{\partial}{\partial q_i} + \frac{dp_i}{dt} \frac{\partial}{\partial p_i} \right) \tag{8.7.14}$$

The generalized time derivative operator,

$$\frac{D}{Dt} \equiv \frac{\partial}{\partial t} + \dot{q}_i \frac{\partial}{\partial q_i} + \dot{p}_i \frac{\partial}{\partial p_i}, \tag{8.7.15}$$

can be written as

$$\frac{D}{Dt} \equiv \frac{\partial}{\partial t} + (\mathbf{v} \cdot \nabla_{\mathbf{r}}) + (\mathbf{F} \cdot \nabla_{\mathbf{p}}), \tag{8.7.16}$$

where
$$v_i = \frac{dq_i}{dt}, \quad F_i = \frac{dp_i}{dt}, \tag{8.7.17}$$

$$\nabla_{\mathbf{r}} \equiv \left(\frac{\partial}{\partial q_i}, \frac{\partial}{\partial q_2}, \ldots, \frac{\partial}{\partial q_{3n}} \right) \tag{8.7.18}$$

$$\nabla_{\mathbf{p}} \equiv \left(\frac{\partial}{\partial p_1}, \frac{\partial}{\partial p_2}, \ldots, \frac{\partial}{\partial p_{3n}} \right) \tag{8.7.19}$$

The Liouville equation (8.7.5) now takes the form

$$\frac{DF_n}{Dt} = 0 \tag{8.7.20}$$

In Γ space the quantity DF_n/Dt is the mobile derivative of F_n, or the derivative following the motion of a phase point; equation (8.7.20) is the condition of incompressibility of the phase point. So, integrating, we can write

$$F_n(\mathbf{Z}, t) = F_n(\mathbf{Z}_0, t) \tag{8.7.21}$$

where $\mathbf{Z}_0$ is the position of the phase point at time t_0. Hence, the moving phase points, in a given volume in Γ space, occupy, at a later time, an equal volume having any shape, and any given function of $F_n(\mathbf{Z}, t)$ also remains constant. For example,

$$h_n(F_n) = F_n \log F_n, \tag{8.7.22}$$

is constant along the trajectory. From the incompressibility of flow in Γ space, we conclude that $\int h \, d\mathbf{Z}$ is independent of time. If, therefore,

$$H_\Gamma(t) = \int F_n(\mathbf{Z}, t) \log F_n(\mathbf{Z}, t) \, d\mathbf{Z}, \tag{8.7.23}$$

then
$$H_\Gamma = \text{constant} \tag{8.7.24}$$

If the phase point $\mathbf{Z}$ is known to be somewhere within the domain D_0 at time $t = 0$, then the point $\mathbf{Z}(t)$, at time t, can be found out in the domain D_t where D_t is the image of D_0 at time t. This leads to the postulation of conservation of probability, in integral form, of the Liouville theorem:

$$\frac{d}{dt} \int_{D_t} F_n(\mathbf{Z}, t) \, d\mathbf{Z} = 0 \tag{8.7.25}$$

In the Liouville equation (8.7.8) the first term on the left hand side is the time rate of change of F_n at a fixed point in Γ space, and is referred to as the local rate of change; the second term is the rate of change of F_n due to the flow velocity in position space; the third term gives the change of F_n in momentum space, owing to acceleration in the coordinate space.

8.8 THE BBGKY HIERARCHY OF EQUATIONS

Evolution of the n particle distribution function F_n in time requires the solution of the Liouville equation in $6n$-dimensional Γ space. The necessary macroscopic properties can be expressed in terms of the one-particle and two-particle distribution functions F_1 and F_2 ; so this solution is not necessary. To obtain a partial differential equation for F_1 we need only integrate each term of (8.7.8), over the position and momentum coordinates of the remaining $(n-1)$ particles, where the phase space $\mathbf{Z}_j$ of the particle at $\mathbf{r}_j$, with momentum $\mathbf{p}_j$, has been untouched. For simplicity of calculations, we exclude the particle with position $\mathbf{r}_1$ and momentum $\mathbf{p}_1$ at time t. Since the n particles are identical, $\mathbf{p}_j$ is proportional to $\mathbf{v}_j$; so the change from $\mathbf{p}$ to $\mathbf{v}$ is straight forward. Using the notations of section 8.2 we write the Liouville equation as

$$\frac{\partial F_n}{\partial t}+\sum_{j=1}^{n}\sum_{i=1}^{3}\left[v_j^i\frac{\partial F_n}{\partial r_j^i}+a_j^i\frac{\partial F_n}{\partial v_j^i}\right]=0 \tag{8.8.1}$$

where $a_j^i\,(=dv_j^i/dt)$ is the acceleration along the ith direction of the jth particle. We now denote by B_1, B_2, B_3 the integrals of the three terms of the left hand side of this equation in the $6n-6$-dimensional phase space of $\mathbf{z}_2, \mathbf{z}_3, \ldots, \mathbf{z}_n$. Then

$$B_i=\int\left(\frac{\partial F_n}{\partial t}\right)\prod_{j=2}^{n}d\mathbf{z}_j=\frac{1}{\tau}\,\frac{\partial F_i}{\partial t} \tag{8.8.2}$$

The operator $\partial/\partial t$ is taken out of the integral because limits are independent of time. The second term is

$$B_2=\sum_{j=1}^{n}\sum_{i=1}^{3}\int v_j^i\frac{\partial F_n}{\partial r_j^i}\prod_{k=2}^{n}\prod_{i=1}^{3}dr_k^i\,dv_k^i$$

$$=\sum_{i=1}^{3}\int v_1^i\frac{\partial F_n}{\partial r_1^t}\prod_{k=2}^{n}\prod_{l=1}^{3}dr_k^l\,dv_k^l+\sum_{j=2}^{n}\sum_{i=1}^{3}\int v_j^i\frac{\partial F_n}{\partial r_j^i}\prod_{k=2}^{n}\prod_{l=1}^{3}dr_k^l\,dv_k^l$$

$$=\sum_{j=2}^{n}\sum_{i=1}^{3}\int\left[\int_{-\infty}^{\infty}\frac{\partial F_n}{\partial r_j^i}dr_j^i\right]v_j^i\,dv_j^i\prod_{\substack{k=2\\k\neq j}}^{n}\prod_{\substack{l=1\\l\neq i}}^{3}dr_k^l\,dv_k^l+\frac{1}{\tau}\sum_{i=1}^{3}\left(v_1^i\frac{\partial F_1}{\partial r_j^i}\right)$$

$$= \sum_{j=2}^{n} \sum_{i=1}^{3} \int [F_n]_{r_j^i=-\infty}^{r_j^i=\infty} v_j^i \, dv_j^i \prod_{\substack{k=2 \\ k \neq j}}^{n} \prod_{\substack{l=1 \\ l \neq i}}^{3} dr_k^l \, dv_k^l + \frac{1}{\tau} \sum_{i=1}^{3} v_1^i \frac{\partial F_1}{\partial r_1^i}$$

$$= \frac{1}{\tau} v_1 \frac{\partial F_1}{\partial r_1^i} \tag{8.8.3}$$

where, in the second term in the last step, and also subsequently, the adoption of repeated index means summation over that index. Since the integral $\int F_n \, dr_j^i$ must be bounded, we get

$$\lim_{r_j^i \to \pm\infty} F_n = 0 \tag{8.8.4}$$

For evaluating B_3 we remember that A_j^i is the ith component of acceleration of the jth particle, due to the external forces, and a_{jk}^i is the same due to its interaction with the α th particle; so, we can write

$$a_j^i = A_j^i + \sum_{\alpha=1}^{n} a_{j\alpha}^i \tag{8.8.5}$$

The number $\alpha = j$ has to be excluded while summing over all αs, because a particle cannot interact with itself. Assuming that B_3' and B_3'' are parts of B_3, corresponding to $a_j^i = A_j^i$ and $a_j^i = \sum_{\substack{\alpha=1 \\ \alpha \neq j}}^{n}{}' a_{j\alpha}^1$, we find that

$$B_3' = \sum_{j=1}^{n} \sum_{i=1}^{3} \int A_j^i \frac{\partial F_n}{\partial v_j{}^i} \prod_{k=2}^{n} \prod_{l=1}^{3} dr_k^l \, dv_k^l$$

$$= \sum_{i=1}^{3} \int A_1^i \frac{\partial F_n}{\partial v_1^i} \prod_{K=2}^{n} \prod_{l=1}^{3} dr_k^l \, dv_k^l + \sum_{j=2}^{n} \sum_{i=1}^{3} \int A_j^i \frac{\partial F_n}{\partial v_j^i} \prod_{k=2} \prod_{l=1}^{3} dr_k^l \, dv_k^l \tag{8.8.6}$$

If A_1^i is independent of v_1^i, then we can write

$$\sum_{i=1}^{3} \int A_1^i \frac{\partial F_n}{\partial v_1^i} \prod_{k=2}^{n} \prod_{l=1}^{3} dr_k^l \, dv_k^l = \frac{A_1^i}{\tau} \frac{\partial F_1}{\partial v_1^i} \tag{8.8.7}$$

A_1^i can depend on the component of $\mathbf{v}_1$ along directions which are perpendicular to $\mathbf{A}_1$. So it can depend on v_1^j if $i \neq j$. For example, among the permissible acceleration vectors are those which are proportional to the vector product of $\mathbf{v}_1$ and any other vector. The velocity-dependent Lorentz force $e(\mathbf{v}\times\mathbf{B})/C$, due to a magnetic field $\mathbf{B}$, is one such example. For the second term, on the right hand side of (8.8.6), assume that A_j^i is independent of v_j^i, because, otherwise further calculations become very complicated. Using this assumption, and the condition (8.8.4) for boundedness, we have

$$\sum_{j=2}^{n}\sum_{i=1}^{3}\int A_j^i \frac{\partial F_n}{\partial v_j^i}\prod_{k=3}^{n}\prod_{l=1}^{3} dr_k^l\, dv_k^l$$

$$=\sum_{j=2}^{n}\sum_{i=1}^{3}\int\left[\int_{-\infty}^{\infty}\frac{\partial F_n}{\partial v_j^i}dv_j^i\right]A_j^i\, dr_j^i \prod_{k=2,\,k\neq j}\ \prod_{l=1}^{3} dr_k^l\, dv_k^l$$

$$=\sum_{j=2}^{n}\sum_{i=1}^{3}\int [F_n]_{v_j^i=-\infty}^{\infty} A_j^i\, dr_j^i \prod_{k=2,\,k\neq j}\ \prod_{l=1\, l\neq i}^{3} dr_k^l\, dv_k^l = 0 \quad (8.8.8)$$

Hence

$$B_3' = A_1^i \frac{\partial F_1}{\partial v_1^i} \quad (8.8.9)$$

$$B_3'' = A_1^i \frac{\partial F_1}{\partial v_1^i} \quad (8.8.10)$$

The acceleration component $a_{j\,\alpha}^i$, due to interactions between the jth and the αth particles, depends on the position vector $\mathbf{r}_j - \mathbf{r}_\alpha$, and not on the velocities $\mathbf{v}_j$ and $\mathbf{v}_\alpha$, So, we write

$$B_3'' = \sum_{i=1}^{3}\sum_{\alpha=2}^{n}\int a_{1\alpha}\frac{\partial F_n}{\partial v_1^i}\prod_{k=2}^{n} d\mathbf{z}_k + \sum_{i=1}^{3}\sum_{j=2}^{n}\sum_{\alpha=2}^{n}\int a_{j\alpha}^i \frac{\partial F_n}{\partial v_j^i}\prod_{k=2}^{n} d\mathbf{z}_k$$

$$=\sum_{i=1}^{3}\sum_{\alpha=2}^{n}\int a_{1\alpha}^i \frac{\partial}{\partial v_1^i}\left[\frac{F_2(\mathbf{z}_1,\mathbf{z}_\alpha)}{\tau^2}\right] d\mathbf{z}_\alpha +$$

$$+\sum_{i=1}^{3}\sum_{j=2}^{n}\sum_{\alpha=2,\,\alpha\neq j}^{n}\int a_{j\alpha}\, a_{j\alpha}^i\, [F_n]_{v_j^i=-\infty}^{+\infty} dr_j^i \prod_{k=2\, k\neq j}\ \prod_{l=1,\, l\neq i} dr_k^l\, dv_k^l \quad (8.8.10\ (a))$$

where $F_2(\mathbf{z}_1, \mathbf{z}_2)$ is the two-particle point probability function, defined by (8.5.1). Since $[F_n]_{v_n^i\to\pm\infty} = 0$, the above relation yields

$$B_3'' = \frac{1}{\tau^2}\sum_{\alpha=2}^{m}\frac{\partial}{\partial v_1^i}\int a_{i\,\alpha}^i\, F_2(\mathbf{z}_1,\mathbf{z}_2)\, d\mathbf{z}_\alpha \quad (8.8.11)$$

where the differential operator $\partial/\partial v_1^i$ is taken out of the integration sign, because $a_{1\alpha}^i$ is independent of v_1^i, and the integration variables do not contain v_i^j. Hence, integrating (8.8.1) over $\mathbf{z}_2, \mathbf{z}_3 ..., \mathbf{z}_n$, we get

$$\frac{\partial F_1}{\partial t} + \sum_{i=1}^{3}\left[v_1^i \frac{\partial F_1}{\partial r_1^i} + A_1^i \frac{\partial F_1}{\partial v_1^i}\right] + \frac{1}{\tau}\sum_{\alpha=2}^{n}\sum_{i=1}^{3} \frac{\partial}{\partial v_1^i}\int a_{1\alpha}^i F_2(\mathbf{z}_1, \mathbf{z}_\alpha)\, d\mathbf{z}_\alpha = 0 \tag{8.8.12}$$

Since the distribution F_n is defined for n identical particles, the reduced two-particle distribution $F_2(\mathbf{z}_1\, \mathbf{z}_\alpha)$ will have the same form for all the $n - 1$ particles, obtained by putting $\alpha = 2, 3, ..., n$. Hence, each of the $n - 1$ terms in the summation being equal, (8.8.11) reduces to

$$B_3'' = \frac{(n-1)}{\tau^2}\frac{\partial}{\partial v_1^i}\int a_{12}^i F_2(\mathbf{z}_1,\ \mathbf{z}_2)\, d\,\mathbf{z}_2 \tag{8.8.13}$$

Now we replace $\mathbf{Z}_1$ by $\mathbf{Z}$ and $\mathbf{Z}_2$ by $\mathbf{Z}^1$; then if a_{12}^i is replaced simply by a_i, it is the acceleration of the particle at $\mathbf{Z}$ due to the particle $\mathbf{Z}^1$ in the 6-dimensional phase space; so

$$B_3'' = \frac{(n-1)}{\tau^2}\sum_{i=1}^{3}\frac{\partial}{\partial v_i}\int a_i F_2(\mathbf{z}, \mathbf{z}')\, d\mathbf{z}' \tag{8.8.14}$$

where v_i is the ith component of velocity of the particle at $\mathbf{Z}$. Then (8.8.12) reduces to

$$\frac{\partial F_1}{\partial t} + \sum_{i=1}^{3}\left(v_1 \frac{\partial F_1}{\partial r_i} + A_i \frac{\partial F_1}{\partial v_i}\right) + \sum_{i=1}^{3}\frac{(n-1)}{\tau}\frac{\partial}{\partial v_i}\int a_i F_2(\mathbf{z}_1\, \mathbf{z}')\, d\mathbf{z}' = 0 \tag{8.8.15}$$

where A_i, like a_i, is the acceleration component along the ith direction of the particle at $\mathbf{z}$, induced by the sum of all the external forces. It gives the BBGKY (Bogoliubov, Born, Green, Krikwood, Yvon) hierarchy of equations. Denoting summation by use of repeated index, we get

$$\frac{\partial F_1}{\partial t} + v_i\frac{\partial F_1}{\partial r_i} + A_i\frac{\partial F_1}{\partial v_i} + \frac{(n-1)}{\tau}\frac{\partial}{\partial v_i}\int a_i F_2(\mathbf{z}_1\, \mathbf{z}')\, d\mathbf{z}' = 0 \tag{8.8.16}$$

It is not a closed system of equations, because when it is an equation for F_1, it contains F_2 – a quantity, for which another similar equation, depending on F_3, exists; and like that, one can proceed indefinitely. So F_1 cannot be solved from (8.8.16) without some further simplifying assumptions for the two-particle distribution F_2. It gives the rate of change of F_1, as a sum of (*a*) the streaming term $v_i\, \partial F_1/\partial r_i$, (*b*) the collision term $\{(n-1)/\tau\}\,(\partial/\partial v_i) \int a_i\, F_2\,(\mathbf{z}, \mathbf{z}')\, d\mathbf{z}'$, and (*c*) the force term $A_i\,(\partial F_1/\partial v_i)$. The streaming term gives the change in F_1, due to motion of the particle at $\mathbf{z}$ at time t, in six-dimensional μ space, and is governed by the Hamiltonian H_1 of the particle. The collision term gives the change of F_1 due to interaction with the remaining $n-1$ particles. The force term gives the change of F_1 due to mainly, the external sources, that is, those sources which are influencing all the n particles from outside.

8.9 EQUATION FOR F_j

Following the derivation of equation for $F_1\,(\mathbf{z}_1)$, the equation for F_j is obtained in a straight forward manner. Denoting, again, the four terms of integration of (8.8.1) (where a_j^i is given by the two terms of the right hand side of (8.8.4)) with respect to the variables $\mathbf{z}_{j+1}, \mathbf{z}_{j+2}, ..., \mathbf{z}_n$, by B_1, B_2, B_3', B_3'', we find that

$$B_1 = \frac{\partial}{\partial t} \int F_n\,(\mathbf{Z}) \prod_{i=j+1}^{n} d\mathbf{z}_i = \frac{1}{\tau^j} \frac{\partial F_j}{\partial t},$$

$$B_2 = \sum_{k=1}^{n} \sum_{i=1}^{3} v_k^i \frac{\partial F_n}{\partial r_k^i} \prod_{q=j+1}^{n} d\mathbf{z}_q = \sum_{i=1}^{3} \left[\sum_{k=1}^{j} \int v_k^i \frac{\partial F_n}{\partial r_k^i} + \sum_{k=j+1} \int v_k^i \frac{\partial F_n}{\partial r_k^i} \right] \prod_{q=j+1}^{n} d\mathbf{z}_q$$

$$= \frac{1}{\tau^j} \sum_{k=1}^{j} \sum_{i=1}^{3} v_k^i \frac{\partial F_j}{\partial r_k^i} + 0 \qquad (8.9.1)$$

$$B_3' = \sum_{k=1}^{n} \sum_{i=1}^{3} \int A_k^i \frac{\partial F_n}{\partial v_k^i} \prod_{q=j+1}^{n} d\mathbf{z}_q = \sum_{k=1}^{n} \sum_{i=1}^{3} \int A_k^i \frac{\partial F_n}{\partial v_k^i} \prod_{q=j+1}^{n} d\mathbf{z}_q + 0$$

$$= \sum_{k=1}^{n} \sum_{i=1}^{3} A_k^i \frac{\partial F_j}{\partial v_k^i} \qquad (8.9.2)$$

and B_3'' is given by (8.9.3). The last result for B_2 is obtained by virtue of the fact that $\int [F_n]_{r_k^i=-\infty}^{+\infty} v_k^i\, dv_k^i \prod_{q=j+1,\, q\neq k} d\mathbf{z}_q = 0$ for all K, and the last result for B_3' is obtained by assuming that A_k^i is independent of v_k^i, for $a_{j\alpha}^i$ depends only on the position vector $\mathbf{r}_j - \mathbf{r}_\alpha$. Hence, for B_3'' we obtain

$$B_3'' = \sum_{k=1}^{n} \sum_{i=1}^{3} \sum_{\alpha=1} \sum_{\alpha \neq k} \int a_{k\alpha}^{i} \frac{\partial F_n}{\partial v_k^i} \prod_{q=j+1}^{n} d\mathbf{z}_q$$

$$= \sum_{k=1}^{j} \sum_{i=1}^{3} \sum_{\alpha=1, \alpha \neq k} \int a_{k\alpha}^{i} \frac{\partial F_n}{\partial v_k^i} \prod_{q=j+1}^{n} d\mathbf{z}_q + \sum_{k=1}^{1} \sum_{i=1}^{3} \sum_{\alpha=j+1}^{n} \frac{\partial}{\partial v_k^i} \int a_{k\alpha}^{i}$$

$$F_n \, d\mathbf{Z}_\alpha \prod_{\substack{q=j+1 \\ q \neq \alpha}}^{n} d\mathbf{z}_q$$

$$= \frac{1}{\tau^j} \sum_{k=1}^{j} \sum_{\alpha=1}^{3} \left[\sum_{\alpha=1\, \alpha \neq k} a_{k\alpha}^{i} \frac{\partial F_j}{\partial v_k^i} + \frac{1}{\tau} \sum_{\alpha=j+1} \frac{\partial}{\partial v_k^i} \int a_{k\alpha}^{i} F_{j+1} (\mathbf{z}_1, \mathbf{z}_2, \ldots, \mathbf{z}_j, \mathbf{z}_\alpha) \, d\mathbf{z}_\alpha \right] \quad (8.9.3)$$

Thus calculating, the equation for F_j is found to be

$$\frac{\partial F_j}{\partial t} + \sum_{k=1}^{j} \sum_{i=1}^{3} \left[v_k^i \frac{\partial F_i}{\partial r_k^i} + A_k^i \frac{\partial F_j}{\partial v_k^i} + \sum_{\alpha=1, \alpha \neq R} a_{k\alpha}^{i} \frac{\partial F_i}{\partial v_k^i} \right\}$$

$$+ \frac{1}{\tau} \sum_{k=1}^{j} \sum_{i=1}^{3} \sum_{\alpha=j+1}^{n} \frac{\partial}{\partial v_k^i} \int a_{k\alpha}^{i} F_{j+1} (\mathbf{z}_1, \mathbf{z}_2, \ldots, \mathbf{z}_j, \mathbf{z}_\alpha) \, d\mathbf{z}_\alpha = 0 \quad (8.9.4)$$

In the last term of (8.9.3) and (8.9.4) $j \neq n$; so j takes values ranging from 1 to $n-1$. Equations for $F_1, F_2 \ldots, F_{n+1}$ are those of the *BBGKY* hierarchy. When particles are identical, each one of the $n-j$ terms in the summation over α is equal, and (8.9.4) simplifies to

$$\frac{\partial F_j}{\partial t} + \sum_{k=1}^{j} \sum_{i=1}^{3} \left[v_k^i \frac{\partial F_j}{\partial r_k^i} + A_k^i \frac{\partial F_i}{\partial v_k^i} + \sum_{\alpha=1, \alpha \neq k} a_{k\alpha}^{i} \frac{\partial F_j}{\partial v_k^i} \right.$$

$$\left. + \frac{(n-j)}{\tau} \frac{\partial}{\partial v_k^i} \int a_{k,\, j+1}^{i} F_{j+1} (\mathbf{z}_1, \mathbf{z}_2, \ldots, \mathbf{z}_{j+1}) \, d\mathbf{z}_{j+1} \right] = \quad (8.9.5)$$

Since (8.9.5) contains F_{j+1}, for all values of j, from 1 to $n-1$, the set of *BBGKY* equations for $F_1, F_2, \ldots, F_{n-1}$ are not closed. This hierarchy is studied generally for low values of j; then $\tau \to \infty$, $n \to \infty$ such that $(n/\tau) \to N_0$ where N_0 is the number density, and we get

$$\frac{\partial F_j}{\partial t}+\sum_{k=1}^{j}\left[v_k^i\frac{\partial F_j}{\partial r_k^i}+A_k^i\frac{\partial F_j}{\partial r_k^i}+A_k^i\frac{\partial F_j}{\partial v_k^i}+\sum_{\alpha=1,\alpha\neq k}\left(a_{k\alpha}^i\frac{\partial F_j}{\partial v_k^i}\right)\right.$$
$$\left.+N_0\frac{\partial}{\partial v_k^i}\int a_{k,j+1}^i F_{j+1}(\mathbf{z}_1,\mathbf{z}_2,\ldots,\mathbf{z}_{j+1})\,d\mathbf{z}_{j+1}\right]=0 \quad (8.9.6)$$

These are a set of coupled equations in which F_j is expressed in terms of F_{j+1} for all positive integral values of j, including zero. This set is equivalent to the Liouville equation and possesses properties such as evolutional reversibility of that equation. It is customary to obtain the Boltzmann equation (chapter 9) by using some further assumptions. Two well-known approaches for the transition from the *BBGKY* hierarchy to the Boltzmann equation, are (*a*) Grad's truncated or cut-off distribution method and (*b*) Bogoliubov's method.

(a) Grad's method

In this method, (section 9.6) the length ε of the cut-off distribution F_j^{ε} is such that the interaction between particles can be neglected for interparticle distance greater than ε. Then the properties of the equation for F_1^{ε} are also in the Boltzmann equation. Additional assumptions, like binary collision, random particle distribution and slow variation of F_1, render the term containing F_2^{ε} into the familiar collision term of the Boltzmann equation.

(b) Bogoliubov's method

This method yields a set of uncoupled equations from which F_j is obtained by successive approximation. Solutions are sought for which the functions $F_j\,(j\geq 2)$ depend on time only through the single particle distribution F_1. Actually, by expansion of F_j in powers of the number density, a sequence of approximate equations is obtained; the first of these is the Boltzmann equation. The Boltzmann equation is suitable for application to dilute gases only, in which pressure is not greater than one atmosphere or so. The higher order approximations take into account triple and higher order collisions, in moderately dense gases.

8.10 ENERGY STATISTICS

If ρ is the density in the phase space of a gas in thermodynamic equilibrium, then one form of its expression is

$$\rho=\frac{\exp(-\varepsilon/KT)}{\int\exp(-\varepsilon/KT)\,d\mathbf{Z}} \quad (8.10.1)$$

where $d\mathbf{Z}$ is the volume element in the $6n$-dimensional phase space of position and velocity (or momentum) coordinates of n particles of equation (8.2.5), ε is the total energy of the gas in the given configuration, K is the Boltzmann constant and T is the kinetic temperature. The total energy being the sum of energies in kinetic and potential forms, we write

$$\varepsilon = \varepsilon_k + \varepsilon_p \tag{8.10.2}$$

where
$$\varepsilon_k = \sum_i \frac{m}{2} v_i^2, \tag{8.10.3}$$

$$\varepsilon_p = \sum_i \varphi_i + \sum_i \sum_{<j} \varphi_{ij} \tag{8.10.4}$$

ε_k is the kinetic energy of the particles and ε_p is the potential energy, both due to external fields and particle interaction. ε_k depends on the velocity coordinates of the particles, and ε_p depends only on their position coordinates.

The density ρ of (8.10.1) can be written as the product

$$\rho = \rho_k \, \rho_r \tag{8.10.5}$$

with
$$\rho_k = \frac{\exp(-\varepsilon_k/KT)}{\int \exp(-\varepsilon_k/KT)\, d\mathbf{z}_\mathbf{v}}, \tag{8.10.6}$$

$$\rho_r = \frac{\exp(-\varepsilon_p/kT)}{\int \exp(-\varepsilon_p/KT)\, d\mathbf{z}_\mathbf{r}} \tag{8.10.7}$$

where $d\mathbf{z}_\mathbf{v}$ is the $3n$-dimensional volume element in the velocity space of the n particles and $d\mathbf{z}_r$ is the $3n$-dimensional volume element in the position space. So ρ_r describes the statistics of position for the potential energy and ρ_k describes that of velocities for the kinetic energies. Hence, using equation (8.10.3), we can write

$$\rho_k = \rho_1 \, \rho_2 \cdot \rho_n \tag{8.10.8}$$

with
$$\rho_i = \frac{\exp(-m v_i^2/2KT)}{\int \exp(-m v_i^2/2KT)\, d\mathbf{Z}_\mathbf{v}} \tag{8.10.9}$$

where $d\mathbf{Z}_\mathbf{v}$ is the 3-dimensional volume element in the velocity space of the ith particle. Equation (8.10.9) expresses the absence of velocity correlation in thermodynamic equilibrium. So, ρ_k is the product of n Maxwellian velocity distribution functions of section 9.8.

Statistics in position spaces involves correlation owing to the existence of φ_{ij} in (8.10.4). However, if the forces of interaction between molecules is negligible, the particles in thermodynamic equilibrium do not interact, we can put $\varphi_{ij} = 0$, and ignore position correlation, and have

$$\rho_r = \rho_{r_1} \, \rho_{r_2} \cdots \rho_{rn} \tag{8.10.10}$$

with
$$\rho_{ri} = \frac{\exp(-\phi_i/KT)}{\int \exp(-\phi_i/KT)\, d\mathbf{Z}_\mathbf{ri}} \tag{8.10.11}$$

Chapter 9

DISTRIBUTION FUNCTIONS IN SIX-DIMENSIONAL PHASE SPACE

9.1 INTRODUCTION

In this chapter we study the average behaviour of dilute gases with the help of distribution functions defined in the six-dimensional phase space of position coordinates and velocity coordinates. Individual particle motion is then regarded as distinct from average mass motion, because the mean interparticle distance is much greater than the dimensions of the particles. For example, at NTP, the number of particles per $cm^3 = N = 3 \times 10^{19}\ cm^{-3}$, and the collision cross-section $\sigma = 3 \times 10^{-15}\ cm^2$. Hence the mean free path (equation (A. 12.4)), $l = 10^{-5}$ cm, and the average interparticle distance $r_0 = 1/N^{1/3} = 3 \times 10^{-7}$ cm. Both r_0 and l are much greater than the molecular dimension which is about 0.5×10^{-18} cm. If the average speed is 10^4 cm/sec. the average time of collisions, T, is about 10^{-12} sec. Therefore, in times of this order, the phase point, consisting of about 10^{19} coordinates, abruptly changes position in Γ space. It is not practically possible to follow the rapid variations over such small time intervals.

Practically, only quantities such as density, average rate of displacement, temperature (averaged over small times and space intervals), etc., are necessary. So, we introduce some concepts and assumptions of statistical nature, which do not follow from the consequences of the fundamental dynamical laws of motion; and obtain the Boltzmann equation in the six-dimensional μ space, and the higher order equations from the basic Liouville equation. In the uniform steady state, in the absence of all external forces, the distribution becomes Maxwellian and kinetic equilibrium sets in, owing to equipartition of energy among the three translational degrees of freedom. Then the material is said to have a kinetic temperature. In a mixture of several gases, a large value for kinetic temperature of one or some of the components does not necessarily make the gas mixture very hot. For instance, though electron temperature is about $2 \times 10^4\ ^oK$, inside a fluorescent bulb, it does not appear hot because the electron density is very little compared to gas density at atmospheric pressure and room temperature. So the total number of electrons striking the bulb walls is not much. Actually, when the heat capacity is very little, heat transferred is insignificant.

9.2 SIX-DIMENSIONAL PHASE SPACE AND DISTRIBUTION FUNCTION

We denote by $d^6 n$ the number of particles with velocities between $\mathbf{v}$ and $\mathbf{v} + d\mathbf{v}$, and position coordinates between $\mathbf{r}$ and $\mathbf{r} + d\mathbf{r}$. Denoting by $f(\mathbf{r}, \mathbf{v}, t)$ the number of particles per unit volume, in the six-dimensional μ space of position coordinates and velocity coordinates at the phase point $\mathbf{r}, \mathbf{v}$, we find that

$$d^6 n = f(\mathbf{r}, \mathbf{v}, t)\, d\mathbf{r}\, d\mathbf{v} \tag{9.2.1}$$

Then, whatever be the functional form of $f(\mathbf{r}, \mathbf{v}, t)$, we have

$$\int f(\mathbf{r}, \mathbf{v}, t)\, d\mathbf{r}\, d\mathbf{v} = n \tag{9.2.2}$$

where n is the total number of particles in the container at time t. The volume element $d\mathbf{r}$ or $d\mathbf{v}$ has a finite size, it is large enough to contain a large number of particles and small enough, compared with lengths of physical interest, to be practically regarded as a point. For example, when the particle number density is about $10^{19}/\text{cm}^3$, we can choose the volume element $d\mathbf{r}$ to be $10^{-9}/\text{cm}^3$. The sides of a cube of this size are of length 10^{-3} cm. This volume should contain about 10^{10} particles; this volume element is so small that it can be regarded as a point, and also so large that it contains enormously large number of particles. If a container can accomodate a large number of particles, and if the density of these does not change abruptly, during transition to the neighbourhood element, then $f(\mathbf{r}, \mathbf{v}, t)$, for all practical purposes, is a continuous function of $\mathbf{r}, \mathbf{v}$.

Some of the particles with position and velocity in the range $\mathbf{r} + d\mathbf{r}$, $\mathbf{v} + d\mathbf{v}$ and $\mathbf{r}, \mathbf{v}$, at time t, have acquired this velocity due to the acceleration $\mathbf{a}$ and one set of initial conditions $\mathbf{r}_0\ (t = 0)$, and $\mathbf{v}\ (t = 0)$. The acceleration $\mathbf{a}$ and the initial conditions are not necessarily unique, they are rather arbitrary. So, we can define the more general distribution function $f(\mathbf{r}, \mathbf{v}, \mathbf{a}, t)$, per unit volume, in the phase space of position, velocity and acceleration, at time t, at the nine-dimensional phase point with coordinates $\mathbf{r}, \mathbf{v}, \mathbf{a}$. Then the number of particles with positions in the range $\mathbf{r}, \mathbf{r} + d\mathbf{r}$, velocity in $\mathbf{v}, \mathbf{v} + d\mathbf{v}$ and acceleration in $\mathbf{a}, \mathbf{a} + d\mathbf{a}$ is $d^9 n$; so

$$d^9 n = f(\mathbf{r}, \mathbf{v}, \mathbf{a}, t)\, d\mathbf{r}\, d\mathbf{v}\, d\mathbf{a} \tag{9.2.3}$$

In this way, a multidimensional distribution function $f(\mathbf{x}_1, \mathbf{x}_2, ..., \mathbf{x}_N, t)$, in the space of mutually independent vectors $\mathbf{x}_1, \mathbf{x}_2, \mathbf{x}_N$, is defined such that

$$d_n^{3N} = f(\mathbf{x}_1, \mathbf{x}_2, ..., \mathbf{x}_N)\, d\mathbf{x}_1\, d\mathbf{x}_2 \ldots d\mathbf{x}_N \tag{9.2.4}$$

where

$$\mathbf{x}_{r-1}^{(t)} = \mathbf{x}_{r-1}\ (t = 0) + \int_0^t \mathbf{x}_r\, (t')\, dt' \tag{9.2.5}$$

for $r = 2, 3, ..., N$. The dimensions of $\mathbf{x}_r$ equal the dimensions of $\mathbf{x}_{r-1}/t$. Though the vectors $\mathbf{x}_1, \mathbf{x}_2, ..., \mathbf{x}_N$ are independent of each other, the vector $\mathbf{x}_{N+1}$, defined similarly, should be a given function of $\mathbf{x}_1, \mathbf{x}_2, ..., \mathbf{x}_N$; otherwise, the system, like the *BBGKY* hierarchy, is not closed.

9.3 BOLTZMANN EQUATION

The number of particles dn, in the six-dimensional volume element $d\mathbf{r}\,d\mathbf{v}$, at time t, as in (9.2.1), is

$$dn = f(\mathbf{r}, \mathbf{v}, t)\, d\mathbf{r}\, d\mathbf{v} \tag{9.3.1}$$

At time $t + \delta t$ the coordinates $\mathbf{r}, \mathbf{v}$ change to

$$\mathbf{r}' = \mathbf{r} + \mathbf{v}\,\delta t, \quad \mathbf{v}' = \mathbf{v} + \mathbf{a}\,\delta t \tag{9.3.2}$$

where $\mathbf{a}$ is the particle acceleration. The number dn', at time $t'\,(= t + \delta t)$, is $dn' = f(\mathbf{r}', \mathbf{v}', t')\, d\mathbf{r}'\, d\mathbf{v}'$. Since $\mathbf{v}$ and $\mathbf{r}$ are independent, from (9.3.2) we have

$$d\mathbf{r}' = d\mathbf{r} \tag{9.3.3}$$

But the acceleration $\mathbf{a}$ may depend on velocity. So, if J is the Jacobian of transformation from $\mathbf{v}$ to $\mathbf{v}'$, then

$$d\mathbf{v}' = J d\mathbf{v} \tag{9.3.4}$$

In plasmas a part of the acceleration is provided by the Lorentz force; so, for that part, $\mathbf{a} = (e/m)\,\mathbf{E} + e\,(\mathbf{v} \times \mathbf{H})/mc$. Though this $\mathbf{a}$ is velocity-dependent, the ith component, a_i, does not depend on v_i, and, since $\mathbf{E}$ is independent of $\mathbf{v}$, we find that

$$J = \begin{vmatrix} 1 & (eH_z\,\delta t/mc) & -(eH_y\,\delta t/mc) \\ -(eH_z\,\delta t/mc) & 1 & (eH_x\,\delta t/mc) \\ (eH_y\,\delta t/mc) & -(eH_X\,\delta t/mc) & 1 \end{vmatrix}$$

As $\delta t \to 0$ the Jacolian $J \to 1$. Hence, in plasmas, taking into account only the Lorentz force, we have

$$d\mathbf{v}' = d\mathbf{v} \tag{9.3.5}$$

If the time element δt is much greater than the duration of a short-range collision, then several collisions will occur in time δt. Some of these collisions scatter particles into the velocity element $d\mathbf{v}$, and some out of $d\mathbf{v}$. Hence, in the phase space volume element $d\mathbf{r}'\,d\mathbf{v}'$, at time $t'\,(= t + \delta t)$ those particles of the volume element $d\mathbf{r}\,d\mathbf{v}$ at time t will remain which have not been scattered out by collisions; moreover, some new particles come from outside, that is, from other phase space elements, by collisions, occurring outside $d\mathbf{r}\,d\mathbf{v}$. If, by $(\partial f/\partial t)_{\text{coll}}$ is denoted the density of the rate at which collisions change the distribution function f, in the μ space, then during time δt, after t, the difference between dn' and dn must arise due to the short range collisions; so

$$dn' - dn = \left(\frac{\partial f}{\partial t}\right)_{\text{coll}} d\mathbf{r}\, d\mathbf{v}\, \delta t \tag{9.3.6}$$

Expanding dn', using (9.3.1)-(9.3.3), dividing by $d\mathbf{r}\,d\mathbf{v}\,\delta t$, as $\delta t \to 0$, we get the Boltzmann equation

$$\frac{\partial f}{\partial t}+\left(\mathbf{v}\cdot\frac{\partial f}{\partial \mathbf{r}}\right)+\left(\mathbf{a}\cdot\frac{\partial f}{\partial \mathbf{v}}\right)=\left(\frac{\partial f}{\partial t}\right)_{\text{coll}} \tag{9.3.7}$$

Comparison with equaton (8.8.16) shows that the collision term represents the contribution from the higher order reduced distribution functions in the evolution of F_1. Using the mobile derivative operator,

$$\frac{D}{Dt}=\frac{\partial}{\partial t}+(\mathbf{v}\,.\,\nabla_{\mathbf{r}})+(\mathbf{a}\,.\,\nabla_{\mathbf{v}}), \tag{9.3.8}$$

the Boltzmann equation (9.3.7) is written as

$$\frac{Df}{Dt}=\left(\frac{\partial f}{\partial t}\right)_{\text{coll}} \tag{9.3.9}$$

Denoting the six coordinates **r**, **v** by x_i with $i = 1, 2, ..., 6$, the components of a six-dimensional vector **R**, satisfying the orthogonality condition $\Sigma\, x_i x_j = 0$, if $i \neq j$, are obtained. Since a volume in the six-dimensional μ space is bound by a five-dimensional hyper surface, the current of particles is $\mathbf{R}f$ and the rate of flow across a closed surface into the volume element is $-\int (\mathbf{R}\,.\,d\,\Sigma)\,f$ where $d\Sigma$ is such a surface element. By the generalized divergence theorem, in μ space, the flow of particles into the volume element $d\,\mathbf{r}\,d\mathbf{v}$ is

$$[\nabla_{\mathbf{r}}\,(\mathbf{v}f)+\nabla_{\mathbf{v}}\,(\mathbf{a}f)]\,d\mathbf{r}\,d\mathbf{v} \tag{9.3.10}$$

Since the number of particles is conserved, the rate at which particles flow in through the surface of $d\mathbf{r}\,d\mathbf{v}$, plus those redistributed inside, per unit time, equal the rate at which the particle number changes in time due to collisions. Hence, in the limit, we get

$$\frac{\partial f}{\partial t}+\text{div}_{\mathbf{r}}\,(\mathbf{v}f)+\text{div}_{\mathbf{v}}\,(\mathbf{a}f)=\left(\frac{\partial f}{\partial t}\right)_{\text{coll}} \tag{9.3.11}$$

In this form the Boltzmann equation is relativistically correct. When $\text{div}_{\mathbf{v}}\,\mathbf{a} = 0$, the non-relativistic form (9.3.7) is obtained. The interpretation of the various terms of (9.3.11) are : (*i*) the term $\text{div}_{\mathbf{r}}\,(\mathbf{v}f)$ is the spatial gradient and expresses the influence of diffusion in position space due to motion of particles from denser to rarer regions. (*ii*) The term $\text{div}_{\mathbf{v}}\,(\mathbf{a}f)$ expresses, nonexplicitly, the influence of large angle changes in motion of particles due to short-range encounters with the neighbours. When several spacies of particles are present in the volume, a distribution function and a Boltzmann equation for each species should be used. For instance, if there are two species, the distribution functoins are $f_1\,(\mathbf{r}, \mathbf{v}, t)$ and $f_2\,(\mathbf{r}, \mathbf{v}, t)$ and the two Boltzmann equations are

$$\frac{\partial f_1}{\partial t}+\text{div}_{\mathbf{r}}\,(\mathbf{v}f_1)+\text{div}_{\mathbf{v}}\,(\mathbf{a}_1 f_1)=\left(\frac{\partial f}{\partial t}\right)^{11}_{\text{coll}}+\left(\frac{\partial f}{\partial t}\right)^{12}_{\text{coll}}, \tag{9.3.12}$$

$$\frac{\partial f_2}{\partial t} + \text{div}_{\mathbf{r}}\,(\mathbf{v} f_2) + \text{div}_{\mathbf{v}}\,(\mathbf{a}_2 f_2) = \left(\frac{\partial f}{\partial t}\right)^{21}_{\text{coll}} + \left(\frac{\partial f}{\partial t}\right)^{22}_{\text{coll}} \tag{9.3.13}$$

So, for the *i*th species, we can write

$$\frac{\partial f_i}{\partial t} + \text{div}_{\mathbf{r}}\,(\mathbf{v} f_i) + \text{div}_{\mathbf{v}}\,(\mathbf{a}_i f_i) = \sum_j \left(\frac{\partial f}{\partial t}\right)^{ij}_{\text{coll}} \tag{9.3.14}$$

In fully ionized plasmas only the Lorentz force acts, and the short range encounters, responsible for the collision terms, are negligible. Equation (9.3.11) is then called the Vlasov equation or the Boltzmann-Vlasov equation. It is treated in Chapter 10.

9.4 COLLISION TERMS, QUASI-STATIC SOLUTIONS AND BOLTZMANN STATISTICS

(a) Collision terms

To neglect correlation betwen distributions, in the collision term $(\partial f/\partial t)_{\text{coll}}$, we assume (*i*) spherical symmetry of particles, (*ii*) the prevalence of binary collisions under given force fields and (*iii*) the existence of perfect chaos or randomness, in the particles.

We denote by N^+_{ij} the number of particles of type *i*, having velocity $\mathbf{v}_i$, which have entered the unit volume in μ space at the point **r**, as a consequence of collisions with particles of *j*th type, having velocity $\mathbf{v}_j$; also N^-_{ij} denotes the number of particles of type *i* which leave this volume, owing to collisions with particles of *j* type in time δt. Then, adding contributions from all particles of the *j* th category, we write

$$\left(\frac{\partial f}{\partial t}\right)_{\text{coll}} = \sum_j (N^+_{ij} - N^-_{ij}) \tag{9.4.1}$$

To find N^-_{ij} consider collision of a particle of *i* type at the origin of Fig. 9.4.1. In Fig. 9.4.2 the Coulomb collision is shown between two identical particles in the laboratory frame. The projected initial path of a *j* type molecule must intersect the plane containing the differential area $p\,dp\,d\varepsilon$; and the numbers of *j* type particles having velocities between $\mathbf{v}_j$ and $\mathbf{v}_j + d\mathbf{v}_j$, colliding in time dt with a single particle of the *i* type, is $f(\mathbf{r}, \mathbf{v}_j, t)\,|g|\,d\mathbf{v}_j\,p\,dp\,d\varepsilon\,dt$, where **g** is the relative velocity $\mathbf{v}_i - \mathbf{v}_j$. So, the number of molecules of *i*th type, in the volume $d\mathbf{r}$, and with velocities in the range $\mathbf{v}_i$ and $\mathbf{v}_i + d\mathbf{v}_i$, is

$$f(\mathbf{r}, \mathbf{v}_i, t)\,d\mathbf{r}\,d\mathbf{v}_i\,f(\mathbf{r}, \mathbf{v}_j, t)\,d\mathbf{v}_j\,|g|\,p\,dp\,d\varepsilon\,dt$$

Integrating this number with respect to $\mathbf{v}_j$, **p** and ε, we obtain the number of particles of *i* type which are removed from the velocity box $d\mathbf{v}_j$, by collision.

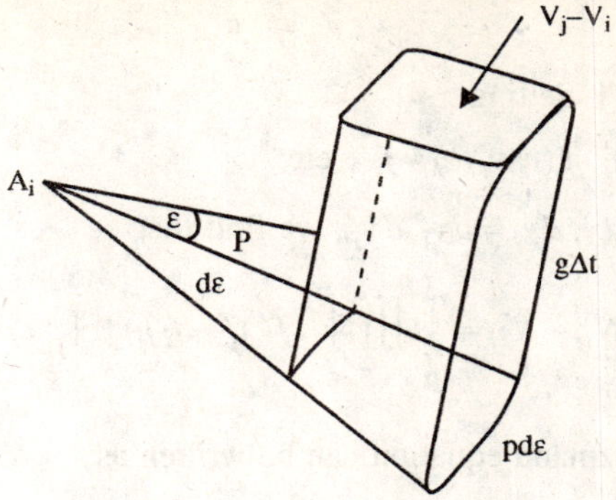

Fig. 9.4.1 Collision geometry to evaluate N_{ij}.

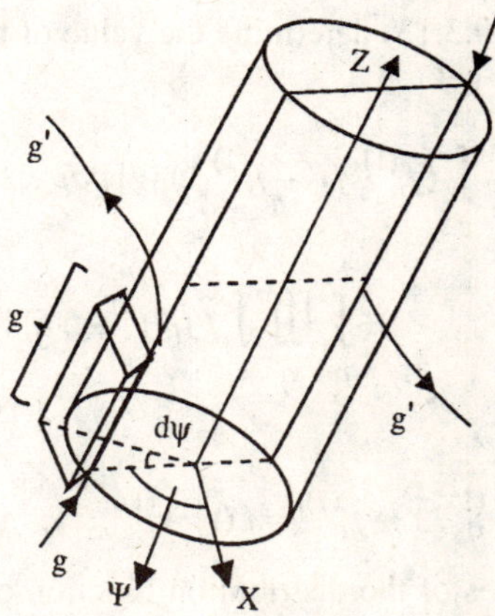

Fig. 9.4.2 Coulomb collision between two identical particles.

Hence the number of collisions per unit time ($dt = 1$), for uncorrelated encounters, is given by

$$N_{ij}^{-}\, d\mathbf{r}\, d\mathbf{v}_j = f(\mathbf{r}, \mathbf{v}_i, t)\, d\mathbf{r}\;\; d\mathbf{v}_i \int_p \iiint_{\mathbf{v}_j} \int_\varepsilon f(\mathbf{r}, \mathbf{v}_j, t)\, |g|\, dp\, d\mathbf{v}_j \tag{9.4.2}$$

When the collision parameters p and χ are not changed, and colliding particles, before collision, move with velocities $\mathbf{v}_i'$ and $\mathbf{v}_j'$ (these are final velocities of direct collision), the final velocities will be $\mathbf{v}_i$ and $\mathbf{v}_j$ respectively, see Appendix 10(*d*). Hence, if arguments to find N_{ij}^{-} are used, also for N_{ij}^{+}, we obtain

$$N_{ij}^{+}\, d\mathbf{r}\, d\mathbf{v}' = f(\mathbf{r}, \mathbf{v}_i', t)\, d\mathbf{r}\, d\mathbf{v}_i' \int_p \iiint_{\mathbf{v}_j'} \int_\varepsilon f(\mathbf{r}, \mathbf{v}_j, t)\, |g|\, p\, dp\, d\varepsilon\, d\mathbf{v}_j \tag{9.4.3}$$

where $p = p'$, $|\mathbf{g}| = |\mathbf{g}'|$. Setting

$$f(\mathbf{r}, \mathbf{v}_i, t) = f_i,\; f(\mathbf{r}, \mathbf{v}_i'\, t) = f_i', \text{ etc.} \tag{9.4.4}$$

and using the result $d\mathbf{v}_i\, d\mathbf{v}_j = d\mathbf{v}_i'\, d\mathbf{v}_j'$, we find that

$$\left(\frac{\partial f}{\partial t}\right)_{\text{coll}} = N_{ij}^{+} - N_{ij}^{-} = \int_p \iiint_{\mathbf{v}_j} \int_\varepsilon (f_i' f_j' - f_i f_j)\, |g|\, p\, dp\, d\varepsilon\, d\mathbf{v}_j \tag{9.4.5}$$

Hence the Boltzmann equation can be written as

$$\frac{Df_i}{Dt} = \int_p \iiint_{\mathbf{v}_j} \int_\varepsilon (f_i' f_j' - f_i f_j)\, |g|\, p\, dp\, d\varepsilon\, d\mathbf{v}_j \tag{9.4.6}$$

Encounters between particles of same type as well as of different types, as indicated by (9.3.13) and (9.3.13) determine the value of the collision terms. So, using the form (9.3.9), we get

$$\frac{Df}{Dt} = \int_p \iiint_{\mathbf{v}_j^{(1)}} \int_\varepsilon (f_i^{(1)\prime} fi' - f_i^{(1)} f_i)\, |g|\, pdp\, d\varepsilon\, d\mathbf{v}_i^{(1)}$$

$$+ \int_p \iiint_{\mathbf{v}_j} \int_\varepsilon (f_i' f_j' - f_i f_j)\, |g|\, pdp\, d\varepsilon\, d\mathbf{v}_j \tag{9.4.7}$$

where

$$f_i^{(1)} = f(\mathbf{r},\, \mathbf{v}_i^{(1)},\, t),\; f_i^{(1)\prime} = f(\mathbf{r},\, \mathbf{v}_i^{(1)\prime},\, t), \tag{9.4.8}$$

because these are the values of the distribution function of the same species for different velocities. We can also write (9.4.7) as

$$\frac{Df}{Dt} = \int_p \iiint_{\mathbf{v}_i^{(1)}} \int_\varepsilon f_i^{(1)\prime} f_i' - f_i^{(1)} f_i)\, |g|\, \sigma \sin\chi\, d\varepsilon\, d\mathbf{v}_j^{(1)}\, d\chi +$$

$$+ \int_p \iiint_{\mathbf{v}_j} \int_\varepsilon (f_i' f_j' - f_i f_j)\, |g|\, \sigma \sin\chi\, d\varepsilon\, d\mathbf{v}_j\, d\chi \tag{9.4.9}$$

where σ is the differential scattering cross-section and χ is the angle of scattering. Negative sign appears in equation (A. 11.4), but is not used here because the limits of integration in (9.4.9) cover all values of angle χ.

Since the collision terms are derived for binary collisions, the Boltzmann equation (9.4.6) holds good only for dilute gases, in which the mean free path is much greater than the dimensions of the constituent particles.

(b) Collision-free quasistatic solutions

The Boltzmann equation (9.3.11), without the collision term, in the steady state, reduces to

$$v_i \frac{\partial f}{\partial x_i} + a_i \frac{\partial f}{\partial v_i} = 0 \tag{9.4.10}$$

where repeated index means summation over that index. When the force $m\,\mathbf{a}$ is conservative we put $m\,\mathbf{a} = \nabla\Phi$; then the substitution

$$f(\mathbf{r}, \mathbf{v}) = N_0 \left(\frac{m}{2\pi\kappa T}\right)^{3/2} \exp\left[-\frac{mv_i\,v_i}{2\kappa T} + \frac{\Phi(\mathbf{r})}{\kappa T}\right] \tag{9.4.11}$$

satisfies (9.4.10), where κ is the Boltzmann cosntant and T is the kinetic temperature. The quantity $\exp[\Phi(r)/\kappa T]$ is called the Boltzmann factor, and

$$N_0 \left(\frac{m}{2\pi\kappa T}\right)^{3/2} \exp\left[-\frac{mv_i\,v_i}{2\kappa T}\right] = f_0(v^2) \tag{9.4.12}$$

is the Maxwell velocity distribution. We find that

$$\iiint_{-\infty}^{\infty} f(\mathbf{r}, \mathbf{v})\, d\mathbf{v} = N_0 \left(\frac{m}{2\pi KT}\right)^{3/2} \exp(\Phi/KT) \iiint_{-\infty}^{\infty} \exp$$

$$\left\{-\frac{m}{2KT}(v_x^2 + v_y^2 + v_z^2)\right\} dv_x\, dv_y\, dv_z = N_0 \exp\left[\frac{\Phi}{KT}\right] \tag{9.4.13}$$

For an electron we write $m\mathbf{a} = \nabla\Phi = -e\mathbf{E}$ and for an ion $M\,\mathbf{a} = \nabla\Phi = e\mathbf{E}$ where φ is the electrostatic potential, and find N_e and N_i of section 3.10 (*b*).

Exercise 9.4.1. Find the average velocity and average speed when Φ is the gravitational potential energy mg**Z**.

(c) Distributions of particles of different states

Particles in an ensemble can exist in one of several states, depending on the associated quantum number i, in a given vibration-rotational state. If N_i is the number of particles in the ith state, then

$$N = \sum_i N_i \tag{9.4.14}$$

where N is the constant number of particles of the system. The system of particles in thermodynamic equilibrium is closed, so there is no exchange of energy with the outside world, and energy of the system is conserved, giving

$$\varepsilon = \sum_i \varepsilon_i N_i \tag{9.4.15}$$

where ε is the total energy of the system and ε_i is the energy per particle in the ith state. The states of the particles change by collision; so the number of particles in a given state changes continuously, and the probability that a given number of particles is in a given state is proportional to the number of possible realisations of this distribution.

We denote by $P(N_1, N_2, ..., N_i, ...)$ the probability that N_1 of the particles are in the first state, N_2 in the second system, etc. To find the number of possible realisations of this distribution we first determine the quantities $P_{N_1}^{N}$, $P_{N_2}^{N-N_1}$, etc., where N is the constant total number of particles and $P_{N_1}^{N}$ is the number of ways of selecting N_1 particles (which are in the first excited state) from N, $P_{N_2}^{N-N_1}$ is the number of ways of selecting N_2 particles (which are in the second excited state) from the remaining $N - N_1$ particles and like that. Thus,

$$P_{N_1}^{N} = \frac{N\,!}{(N-N_1)\,!\,N_1\,!},\quad P_{N_2}^{N-N_1} = \frac{(N-N_1)\,!}{(N-N_1-N_2)\,!\,N_2\,!} \tag{9.4.16}$$

$$P(N_1, N_2, ..., N_i, ...) = \frac{AN\,!}{N_1\,!\,N_2\,!\,.\,N_i\,!} = \frac{AN\,!}{\prod_i (N_i\,!)} \tag{9.4.17}$$

where Π denotes the product and A is the constant of normalization. When N_i is a large number we can write

$$l_n N_i\,! = l_n \prod_{r=1}^{N_i} (r) = \sum_{r=1}^{N_i} (l_n r) = \int_0^{N_i} (l_n x)\, dx \tag{9.4.18}$$

and obtain

$$\frac{d}{dN_i}(l_n N_i\,!) = l_n N_i,\quad \frac{d^2}{dN_i^2}(l_n N_i\,!) = \frac{1}{N_i}, \text{ etc.} \tag{9.4.19}$$

Hence, we find that

$$\left(\frac{d}{dN_i}(l_n N_i\,!)\right)_{N_i=\overline{N}_i} = l_n \overline{N}_i,\quad \left(\frac{d^2}{dN_i^2}(l_n N_i\,!)\right)_{N_i=\overline{N}_i} = \frac{1}{\overline{N}_i} \tag{9.4.20}$$

(d) The Boltzmann distribution

To determine the most probable number of particles N_i in the ith state we have $\overline{N}_i >> 1$, and the probability p of distribution of particles according to their states and the logarithm of this probability have extreme values at $N_i = \overline{N}_i$. If $dN_i = N_i - \overline{N}_i$, with $N_i >> dN_i >> 1$, then we assume that $N_i >> 1$. Next, expand $l_n P$ at the extreme values; then using (9.4.20), we find that

$$l_n P(N_1, N_2, ..., N_i, ...) = l_n P(\overline{N}_1, \overline{N}_2, ..., \overline{N}_i, ...) - \sum_i (dl_n \overline{N}_i)\, dN_i - \frac{1}{2} \sum_i \frac{dN_i^2}{d\overline{N}_i} \quad (9.4.21)$$

The condition for extreme value is that the first derivative term must vanish. So

$$\sum_i (dl_n \overline{N}_i)\, dN_i = 0 \quad (9.4.22)$$

Moreover, by virtue of (9.4.14) and (9.4.15), we obtain

$$\sum_i (dN_i) = 0, \sum_i \varepsilon_i\, dN_i = 0 \quad (9.4.23)$$

The three equations of (9.4.22) and (9.4.23) are linearly combined into the single equation:

$$\sum_i \left(l_n \overline{N}_i - l_n C + \frac{1}{\kappa T} \varepsilon_i \right) dN_i = 0 \quad (9.4.24)$$

where $-l_n C$ and $1/\kappa T$ are arbitrary parameters of the system. Since this equation is valid for any dN_i, we put

$$l_n \overline{N}_i - l_n C = \frac{1}{\kappa T} \varepsilon_i \quad (9.4.25)$$

and obtain
$$N_i = C \exp\left(-\frac{\varepsilon_i}{\kappa T} \right) \quad (9.4.26)$$

It gives the expression for the most probable number of particles in a given state, and is called the Boltzmann distribution.

The constant C is determined by putting (9.4.26) into (9.4.14), which is also valid at $N_i = \overline{N}_i$. So, evidently,

$$C \sum_i \exp\left(-\frac{\varepsilon_i}{\kappa T} \right) = N \quad (9.4.27)$$

The quantity κT has the dimensions of energy per particle, κ is the Boltzmann constant and T is the temperature in degree Kelvin. Putting (9.4.22) into (9.4.21) we obtain

$$P(N_1, N_2, ..., N_i, ...) = P(\overline{N}_1, \overline{N}_2, ..., \overline{N}_i, ...) \exp\left[-\sum_i \left\{ \frac{(N_i - \overline{N}_i)^2}{2\overline{N}_i} \right\} \right] \quad (9.4.28)$$

(e) Statistical weight of a state

If the ith state is a degenerate state then the statistical weight g_i equals the number of states within the same quantum number; for instance, a rotational state of a neutral particle, with the rotational quantum number j, must equal the number of possible angular momentum projections on a given axis and so equals $2j+1$. Summing up the degenerate states in equation (9.4.27) and using the subscript j to denote a group of states, we find that

$$N_j = N_0 \frac{g_j}{g_0} \exp\left(-\frac{\varepsilon_j}{\kappa T}\right) \tag{9.4.29}$$

If, in the ground state, N_0 is the density of particles and g_0 is the statistical weight, then ε_0 being zero we have $N_0 = cg_0$; so (9.4.29) gives

$$N_j = N_0 \frac{g_j}{g_0} \exp\left(-\frac{\varepsilon_j}{\kappa T}\right) \tag{9.4.30}$$

The wave function of a free particle, with momentum p_x along the x-axis, up to an arbitrary factor, is given by $\exp(ip_x x/\hbar)$ if the particle moves in the positive direction, and by $\exp(-ip_x x/\hbar)$ if the particle moves in the negative direction where $\hbar = h/2\pi$ and $\hbar$ is the Planck constant. If the particle is put into a potential well, with infinitely high walls, it moves freely only in the region $0 \le x \le l$; then the wave function of the particle, in the well, is a combination of these wave functions and must be zero at the walls of the well; since the boundary wall is $x=0$, it must be proportional to $\sin(p_x x/\hbar)$, and since $x = L$ is another boundary, the particle momentum is given by

$$\frac{p_x L}{\hbar} = \pi\alpha \tag{9.4.31}$$

where α is an integer. Hence a particle, with momentum, in the range between p_x and $p_x + dp_x$, can exist in the states $d\alpha = \frac{(l\,dp_x)}{(2\pi\hbar)}$ if we take into account the sign of the momentum; if the particle is in the interval dx, the number of states for a free particle is given by

$$d\alpha = \frac{dp_x\, d_x}{(2\pi\hbar)} \tag{9.4.32}$$

In the six-dimensional phase space of position and momentum co ordinates, therefore,

$$d\alpha = \frac{d\mathbf{p}\, d\mathbf{r}}{(2\pi\hbar)^3} \tag{9.4.33}$$

This $d\alpha$ determines the number of states corresponding to a given range of continuously varying parameters; so the number of states, given by (9.4.33), is the statistical weight for the states with continuous spectrum.

9.5 IONIZATION PROCESSES AND THE SAHA IONIZATION

Partial ionization, occurs in all laboratory experiments, except in some containing hydrogen and helium. The assumption of equilibrium is not always justified, and inclusion of concepts of the specific rates of ionization, recombination and ion loss is necessary to estimate properly the degree of ionization. The average energy, per particle, determines the state in which matter exists. Ionization is possible by electron collisions, by collisions with ions or neutral atoms, by radiation (by laser, for example), by collisions of the second kind which removes the second electron from its orbit in an atom, etc.; by electric field of high frequency, or by electron attachment. Then there are the processes of cumulative ionization, thermal ionization, or surface ionization. The ionization energy (the ionization potential) is the energy required to remove an electron from its orbital motion in the atom, and place it at an infinite distance at rest. An element is ionized when the kinetic energy, per particle of it, W, exceeds the ionizing potential (a few electron volt). If u_n, u_{n+1} are the respective binding energies in the nth and $(n + 1)$ *th* states, then

$$u_n < W_{n+1} < u_{n+1} \tag{9.5.1}$$

For a gas of free nucleons and electrons, the extension of (9.5.1) is

$$2 < W_5 < 200 \text{ (MeV)} \tag{9.5.2}$$

For still higher states of matter, containing free mesons, mutilated nucleus and electron, the condition is

$$0{\cdot}2 < W_6 < 4 \text{ (GeV)} \tag{9.5.3}$$

Both short range and long range forces of interaction are effective in these higher states; so they exhibit an even greater variety of behaviour than plasmas. The processes of ionization and recombination are represented by the relation

$$A^+ + e = A \tag{9.5.4}$$

where A^+ is the ion, e is the electron and A is the neutral atom. The Saha ionization equation relates the amount of ionization in a gas in thermal equilibrium with the densities of electrons, ions and neutral particles, following (9.5.4). If I is the ionization potential of the atom, the number of ergs to remove the outermost electron from an atom, and **p** is the free-electron momentum, then

$$\varepsilon = I + \frac{p^2}{2m} \tag{9.5.5}$$

where ε is the energy needed for removing the electron from the atom, and its kinetic energy is $p^2/2m$. So, putting $\overline{N}_e = N_i$, and assuming $\overline{N}_a$ to be the mean number density of atoms, in the ground state, equation (9.4.30) gives

$$\frac{\overline{N}_i}{\overline{N}_a} = \frac{g_e\, g_i}{g_a} \iiiint\!\!\iint \frac{d\mathbf{p}\, d\mathbf{r}}{(2\pi\hbar)^3} \exp\left(-\frac{I + (p^2/2\mathrm{m})}{\kappa T}\right) \tag{9.5.6}$$

Here g_e is the statistical weight of the electronic states, and g_i and g_a are the corresponding statistical weights of the ion and the atom. Integrating the right hand side, in the momentum space, we get

$$\frac{\overline{N}_i}{\overline{N}_a} = \frac{g_e\, g_i}{g_a} \left(\frac{m\kappa T}{2\pi\hbar^2}\right)^{3/2} \exp\left(-\frac{I}{\kappa T}\right) \iiint d\mathbf{r} \tag{9.5.7}$$

In this derivation the state of the electron system is not changed if the coordinates of two electrons are interchanged. So, to find the number of states, per electron, we must consider only the volume per electron. Hence,

$$\iiint d\mathbf{r} = \frac{\tau}{\overline{N}^e}$$

if τ is the total volume of the system, and $N_e = \overline{N}_e/\tau$, $N_i = \overline{N}_i/\tau$, $N_a = \overline{N}_a/\tau$, where N_e is the electron density, N_i is the ion density and N_a is the atomic density. Then

$$\frac{N_e\, N_i}{N_a} = \frac{g_e\, g_i}{g_a} \left(\frac{m\,\kappa T}{2\,\pi\,\hbar^2}\right)^{3/2} \exp\left(-\frac{I}{\kappa T}\right) \tag{9.5.8}$$

This is the Saha equation or the Saha distribution. It determines the degree of ionization in conditions of thermal equilibrium, and the distribution, at equilibrium between the continuous-spectrum and the discrete-spectrum states. So, the Saha distribution can be written in the form of the Boltzmann distribution (9.4.30):

$$\frac{N_e}{N_a} = \frac{g_s}{g_a} \exp\left(-\frac{I}{\kappa T}\right) \tag{9.5.9}$$

where $g_s = (g_e\, g_i/N_i)\,(m\kappa T/2\pi\hbar^2)^{3/2}$; g_s is the effective statistical weight of the continuous spectrum. For an ideal plasma the weight g_s is rather high; hence the degree of ionization is about unity for $T << I/\kappa$, which is a low temperature, compared to the excitation energy of the atom. At the temperature comparable to the excitation energy, the relative number of excited atoms is small because almost all the atoms dissociate into ions and electrons. In alkali metals, at the temperatures for MHD power generation, $2g_i/g_0 \approx 1$ and $N_e = N_i$. In some situations there may be more than one species that ionizes appreciably at a given temperature. Then, $N_e = N_{i1}, + N_{i2}$, and the Saha equation must be solved simultaneously for N_e of each such species. Though temperature of the particles and temperature of the radiation field are different, the Saha equation assumes thermodynamic equilibrium, given by the Maxwell velocity distribution, and radiation equilibrium, given by the Planck distribution for radiation. In a thermonuclear device, due to the existence of non-equilibrium ionization, the Saha equation can give misleading results.

9.6 MACROSCOPIC (AVERAGED) QUANTITIES

Since relativistically no particle can attain the vacuum velocity of light, it is evident that $\lim_{|v_i| \geq c} f(\mathbf{r}, \mathbf{v}, t) = 0$ where v_i is the ith component of $\mathbf{v}$. Hence, and any divergence and singularities, appearing in the distribution function at velocities greater than C, are neglected. The number of particles, in the volume element $d\mathbf{r}$, having all possible velocities, is

$$\iiint_{\mathbf{v}=-\infty}^{\infty} d^6 N = N(\mathbf{r}, t)\, d\mathbf{r} \tag{9.6.1}$$

where
$$N(\mathbf{r}, t) = \iiint_{-\infty}^{\infty} f(\mathbf{r}, \mathbf{v}, t)\, d\mathbf{v} \tag{9.6.2}$$

Hence $N(\mathbf{r}, t)$ is the number of particles, per unit volume, at the point $\mathbf{r}$, at time t. If $\langle\varphi\rangle$ is the average of $\varphi(\mathbf{r}, \mathbf{v}, t)$, in velocity space, then

$$\langle\varphi(\mathbf{r}, \mathbf{v}, t)\rangle = \frac{\iiint \varphi f\, d\mathbf{v}}{\iiint f\, d\mathbf{v}} = \frac{1}{N}\iiint \varphi f\, d\mathbf{v} \tag{9.6.3}$$

The average velocity $\langle\mathbf{v}\rangle$, denoted by $\mathbf{u}$, is given by

$$\mathbf{u} = \frac{1}{N}\iiint \mathbf{v} f\, d\mathbf{v} \tag{9.6.4}$$

The quantity $\mathbf{c} = \mathbf{v} - \mathbf{u}$ is the deviation of the particle velocity from its mean value $\mathbf{u}$, and is called the random velocity (or peculiar, intrinsic or thermal velocity) at $\mathbf{r}$, at time t. Obviously.

$$\langle\mathbf{c}\rangle = \langle\mathbf{v}\rangle - \mathbf{u} = 0,\ \langle c^2\rangle = \langle v^2\rangle - 2\langle(\mathbf{u} \cdot \mathbf{v})\rangle + u^2 \tag{9.6.5}$$

The Kinetic temperature T is proportional to the average of the random kinetic energy $m\mathbf{c}^2/2$. Taking the constant of proportionality to be $\kappa/2$, in one-dimensional motion, and $3\kappa/2$, in three-dimensional motion, we have accordingly

$$\langle\frac{m\mathbf{c}^2}{2}\rangle = \left(\frac{1}{2} \text{ or } \frac{3}{2}\right)\kappa T \tag{9.6.6}$$

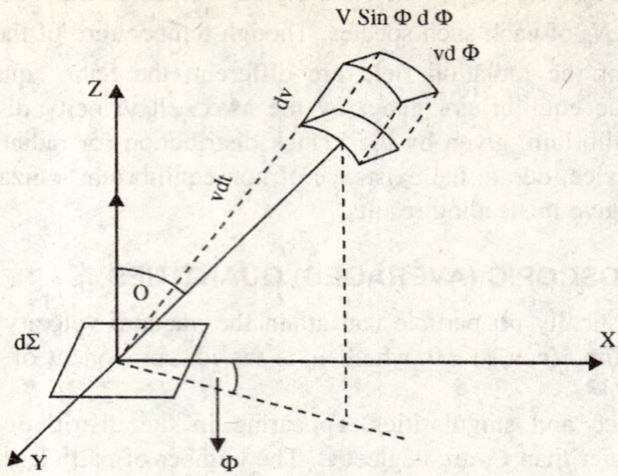

Fig. 9.6.1 Particles striking the surface element $d\Sigma$ in time dt.

where κ is the Boltzmann constant ($= 1{\cdot}38 \times 10^{-16}$ erg/°K) . When we consider the one-dimensional problem of propagation of waves along a fixed direction, we use the value $\kappa T/2$; however, for three translational degrees of freedom, the value is $3\kappa T/2$. For this reason, discrepancies are possible in results of different authors, involving these averages. The pressure tensor p_{ij} and the heat tensor Q_{ijk} are defined as

$$p_{ij} = N \langle mc_i\, c_j \rangle = m \iiint c_i\, c_j f\, d\mathbf{v}, \tag{9.6.7}$$

$$Q_{ijk} = N \langle \frac{m}{2}\, c_i\, c_j\, c_k \rangle = \frac{m}{2} \iiint c_i\, c_j\, c_k f\, d\mathbf{v} \tag{9.6.8}$$

p_{ij} is the flux of the ith component of momentum through a surface, having the normal along the jth space direction. The heat flow vector **Q** is defined as the flux of random thermal energy,

$$\mathbf{Q} = N \langle \frac{m}{2}\, \mathbf{c}^2\, \mathbf{c} \rangle = \frac{m}{2} \iiint \mathbf{c}^2\, \mathbf{c} f\, d\mathbf{v} \tag{9.6.9}$$

Further calculations give

$$\frac{p_{ij}}{m} = \iiint (v_i - u_i)\,(v_j - u_j) f\, d\mathbf{v} = N \langle v_i\, v_j \rangle - N\, u_i\, u_j; \tag{9.6.10}$$

so $$p_{ij} = \rho \langle v_i\, v_j \rangle - \rho\, u_i\, u_j \tag{9.6.11}$$

where $\rho = MN$. Proceeding similarly with Q_{ijk}, we obtain

$$\frac{2}{m} Q_{ijk} = \iiint (v_i - u_i)\,(v_j - u_j)\,(v_k - u_k) f\, d\mathbf{v}$$

$$= \frac{2}{m} q_{ijk} - \left(\frac{p_{jk}}{m} + N u_j u_k \right) u_i - \left(\frac{p_{ik}}{m} + N u_i u_k \right) u_j$$

$$- \left(\frac{p_{ij}}{m} + N u_i u_j \right) u_k + 2N u_i u_j u_k \quad (9.6.12)$$

where $q_{ijk} = \langle m v_i v_j v_k / 2 \rangle$. Hence we have

$$Q_{ijk} = q_{ijk} - \frac{1}{2} (u_i p_{jk} + u_j p_{ki} + u_k p_{ij}) - \frac{\rho}{2} u_i u_j u_k \quad (9.6.13)$$

Denoting, by Γ_k, the number of particles hitting a portion of the *yz* plane, from the side $X < 0$, we get

$$\Gamma_x = \int_{v_x = 0}^{+\infty} \int_{v_y = -\infty}^{+\infty} \int_{v_z = -\infty}^{+\infty} v_x f \, d\mathbf{v} \quad (9.6.14)$$

The quantity $f(v)\, dv\, d\Omega$ is the number of particles having speeds between v and $v + dv$ and direction of motion lying in the solid angle element $d\Omega$, about the direction of $\mathbf{v}$. If $f(v)$ is the distribution of speeds then $f(v)\, dv$ is the number of particles moving with speed between v and $v + dv$, whatever be their directions. Hence

$$f(v) = \int_{\theta = 0}^{\pi} \int_{\varphi = 0}^{2\pi} f(\mathbf{v}) \, d\Omega, \quad (9.6.15)$$

because $d\Omega = \sin\theta \, d\theta \, d\varphi$, and to cover all directions, integration is taken over all solid angles. If there is no preferred direction, for the distribution of particles, the probability of finding $\mathbf{v}$ in the solid angle $d\Omega$ is $d\Omega/4\pi$. Hence the probability that velocity lies between $\mathbf{v}$ and $\mathbf{v} + d\mathbf{v}$ is the product of the probability that the speed lies between v and $v + dv$ and the probability that the direction of motion is within $d\Omega$ around $\mathbf{v}$; so

$$f(\mathbf{v})\, d\mathbf{v}\, d\Omega = f(v)\, dv \frac{d\Omega}{4\pi} \quad (9.6.16)$$

The particle number density N is given by

$$N = \int_0^{\infty} f(v)\, dv = \int_{v=0}^{\infty} \int_{\theta = 0}^{\pi} \int_{\varphi = 0}^{2\pi} f(\mathbf{v})\, d\Omega\, dv \quad (9.6.17)$$

Those particles with velocitites between $\mathbf{v}$ and $\mathbf{v} + d\mathbf{v}$ strike the surface element $d\Sigma$ normal to the *z*-axis in time δt at time t, which lie in the oblique cylinder with base $d\Sigma$, slant side of length $\mathbf{v}dt$, and axis parallel to $\mathbf{v}$, as in Fig. 9.6.1. Here the length vdt, and the time dt, should be much shorter than the mean free path and the mean collision time respectively. The number of particles, with velocitites between $\mathbf{v}$ and $\mathbf{v} + d\mathbf{v}$, striking $d\Sigma$ in time dt, is

$$f(\mathbf{v})\, dv\, d\Omega\, v_z\, dt\, d\Sigma; \tag{9.6.18}$$

and the number of particles striking $d\Sigma$ per unit area per unit time is $f(v)\, v \cos\theta\, dv \sin\theta\, d\theta\, d\varphi/4\pi$, where $f(v)$ is the distribution of speeds. Integrating over all possible directions, with the help of spherical polar coordinates, we obtain

$$dN = \int_{\varphi=0}^{2\pi} \int_{\theta=0}^{\pi/2} f(v) \cdot \frac{v}{4\pi} \cdot dv\, d\varphi \cos\theta \sin\theta\, d\theta = \frac{v}{4} f(v)\, dv \tag{9.6.19}$$

The total number N_1 of particles striking unit area of the wall, in unit time, is

$$N_1 = \int_0^\infty \frac{v}{4} f(v)\, dv = \frac{n\overline{v}}{4} \text{ with } \overline{v} = \frac{\int_0^\infty v f(v)\, dv}{\int_0^\infty f(v)\, dv} = \frac{1}{N} \int_0^\infty f(v)\, v\, dv \tag{9.6.20}$$

If e_i is the charge in the ith particle, then the macroscopic charge density q, and current density $\mathbf{j}$, are

$$q = \sum_i e_i \iiint f_i(\mathbf{r}, \mathbf{v}, t)\, d\mathbf{v}, \mathbf{j} = \sum_i \varepsilon_i \iiint \mathbf{v} f_i(\mathbf{r}, \mathbf{v}, t)\, d\mathbf{v} \tag{9.6.21}$$

Exercise 9.6.1. Find $\overline{v}$ for a Maxwellian velocity distribution and compare with the average speed v_{av} of relation (9.8.8).

9.7 BOLTZMANN'S H-THEOREM

We assume that particles are uniformly distributed in space, and that there is no spatial gradient of the distribution function f. Then, if $f(\mathbf{r}, \mathbf{v}, t)$, at $t = 0$, is every where positive, it is positive at later times also, otherwise f has no meaning as a distribution function. So, at an extreme value of f, in the $\mathbf{r}, \mathbf{v}$ space, we must have

$$\frac{\partial f}{\partial t} \geq 0, \quad \frac{\partial f}{\partial \mathbf{r}} = 0, \quad \frac{\partial f}{\partial \mathbf{v}} = 0 \tag{9.7.1}$$

Then the Boltzmann equation (9.4.6) simlifies to

$$\frac{\partial f_i}{\partial t} = \int_p \iiint_{\mathbf{v}_j} \int_\varepsilon (f_i' f_j' - f_i f_j)\, |g|\, p dp\, d\varepsilon\, d\mathbf{v} \tag{9.7.2}$$

The Boltzmann function of time t, defined by

$$H(t) = \iiint f(\mathbf{v}, t) \log f(\mathbf{v}, t)\, d\mathbf{v} \tag{9.7.3}$$

is independent of the choice of velocity element and depends only on the form of the distribution function f. Differentiating H with respect to time we obtain

$$\frac{\partial H}{\partial t} = \iiint (1 + \log f) \frac{\partial f}{\partial t} \, d\mathbf{v}$$

$$\equiv \iiiint\!\!\!\!\iiiint (1 + \log f_i) \; (f_i' f_j' - f_i f_j) \, |g| \, p dp \, d\varepsilon \, d\mathbf{v}_i \, d\mathbf{v}_j \qquad (9.7.4)$$

Interchanging $\mathbf{v}_i$ and $\mathbf{v}_j$, by simply interchanging the indices i and j, gives

$$\frac{\partial H}{\partial t} = \iiiint\!\!\!\!\iiiint (1 + \log f_i) \, (f_i' f_j' - f_i f_j) \, |g| \, p dp \, d\varepsilon \, d\mathbf{v}_i \, d\mathbf{v}_j \qquad (9.7.5)$$

Adding (9.7.4) and (9.7.5) we obtain

$$\frac{\partial H}{\partial t} = \frac{1}{2} \iiiint\!\!\!\!\iiiint [2 + \log (f_i f_j)] \, (f_i' f_j' - f_i f_j) \, |g| \, p dp \, d\varepsilon \, d\mathbf{v}_i \, d\mathbf{v}_j \qquad (9.7.6)$$

For inverse collisions also this calculation is repeated; so

$$\frac{\partial H}{\partial t} = \frac{1}{2} \iiiint\!\!\!\!\iiiint [2 + \log (f_i' f_j')] \, (f_i f_j - f_i' f_j') \, |g| \, p dp \, d\varepsilon \, d\mathbf{v}_i \, d\mathbf{v}_j \qquad (9.7.7)$$

Adding now (9.7.6) and (9.7.7) we get

$$\frac{\partial H}{\partial t} = \frac{1}{4} \iiiint\!\!\!\!\iiiint \left(\log \frac{f_i f_j}{f_i' f_j'} \right) (f_i' f_j' - f_i f_j) \, |g| \, p dp \, d\varepsilon \, d\mathbf{v}_i \, d\mathbf{v}_j \qquad (9.7.8)$$

Hence, $\log (f_i f_j / f_i' f_j')$ is positive when $f_i f_j > f_i' f_j'$, and negative when $f_i f_j < f_i' f_j'$. In both cases the product $\log (f_i f_j / f_i' f_j') \; (f_i' f_j' - f_i f_j)$ is negative. As other factors in (9.7.8) are positive, we obtain the Boltzmann H-theorem, H can never grow, that is

$$\frac{\partial H}{\partial t} \le 0 \qquad (9.7.9)$$

So, the Boltzmann equation is irreversible, and there is a preferred direction in time such that (9.7.9) is satisfied. Thus H corresponds to an entropy of opposite sign. Since H possesses a lower limit, it tends to a limit as $t \to \infty$. In fact, only if the integral in (9.7.3) diverges can $H \to -\infty$. Now, as $v \to \infty$ and $f \to 0$, we have $\log f \to -\infty$. But the kinetic energy integral $\frac{1}{2} \iiint m v^2 f \, d\mathbf{v}$ must be convergent. So (9.7.3) diverges only when $\log f \to \infty$ faster than v^2. This means that $f \to 0$ more rapidly than $\exp(-v^2)$; then (9.7.3) definitely converges. So H cannot tend to $-\infty$. For extreme values of H we must have

$$\frac{\partial H}{\partial t} = 0 \qquad (9.7.10)$$

For all values of $\mathbf{v}_i$ and $\mathbf{v}_j$ this condition is valid only if

$$f_i f_j = f_i' f_j' \tag{9.7.11}$$

Then $$\log f_i + \log f_j = \log f_i' + \log f_j' \tag{9.7.12}$$

Hence, as defined in Appendix 10 (e), $\log f$ is a collisional invariant; so $\log f$ must be a linear combination of the basic five collisional invariants m, $m\mathbf{v}$ and $mv^2/2$. This means that we can write

$$\log f = \left[a + m\,(\mathbf{b}\,.\,\mathbf{v}) - c\frac{mv^2}{2} \right] \tag{9.7.13}$$

where a, $\mathbf{b}$ and c are constants and the negative sign in the last term is taken for convenience. This distribution function can be written as

$$f = A\,\exp\left[-\frac{m}{2} c \left\{ \left(v_x - \frac{b_x}{c} \right)^2 + \left(v_y - \frac{b_y}{c} \right)^2 + \left(v_z - \frac{b_z}{c} \right)^2 \right\} \right] \tag{9.7.14}$$

where A is another constant. This is a Maxwellian velocity distribution. The five constants $a, \mathbf{b}, c$ are determined by putting this expression for f in the five relations

$$\iiint f\,d\mathbf{v} = N, \iiint \mathbf{v}\,f\,d\mathbf{v} = N\,\mathbf{u}, \iiint \frac{m}{2}\,v^2 f\,d\mathbf{v} = N\left(\frac{3}{2}\,\kappa T + \frac{m}{2}\,u^2 \right) \tag{9.7.15}$$

We then obtain the standard form of the Maxwellian velocity distribution:

$$f(\mathbf{v}) = N\left(\frac{m}{2\pi\,\kappa T} \right)^{3/2} \exp\left[-\frac{m}{2\kappa T}\,(\mathbf{v} - \mathbf{u})^2 \right] \tag{9.7.16}$$

Thus, H increases so long as f attains a value which satisfies (9.7.13) and makes H a minimum (the extreme value is assumed to be minimum). Hence, in the absence of external forces, all distributions approach the Maxwellian as t tends to infinity. The H - theorem is, therefore, similar to the law of increase of entropy.

9.8 ANALYSIS OF THE MAXWELL VELOCITY DISTRIBUTION

The Maxwell distribution in velocity space, also called the normal distribution function, is isotropic; its normalized form is

$$f(v^2) = N\left(\frac{m}{2\pi\,\kappa T} \right)^{3/2} \exp\left\{ -\frac{m}{2\,\kappa T}\,(v_1^2 + v_2^2 + v_3^2) \right\} \tag{9.8.1}$$

where κ is the Boltzmann constant, T is the temperature and v_1, v_2, v_3 are the components of $\mathbf{v}$ in three mutually perpendicular directions. For any positive value of n,

$$\lim_{\mathbf{v}_i \to \pm\infty} v_i^n f(v_1^2 + v_2^2 + v_3^2) = 0 \; (i = 1, 2, 3) \tag{9.8.2}$$

The dependence of f on the components of $\mathbf{v}$ is of the type $\exp(-x^2)$ with x varying between $-\infty$ and $+\infty$. When r is an even positive integer,

$$\int_0^{\infty} \exp(-\alpha x^2)\, x^r\, dx = \frac{\sqrt{\pi}}{2}\, \frac{1}{2} \cdot \frac{3}{2} \cdot \frac{5}{2} \cdot \frac{(r-1)}{2} \cdot \alpha^{-(r+1)/2}, \tag{9.8.3}$$

and when r is an odd positive integer.

$$\int_0^{\infty} \exp(-\alpha x^2)\, x^r\, dx = \frac{1}{2}\, \alpha^{-(r+1)/2} \cdot \frac{(r-1)\,!}{2} \tag{9.8.4}$$

Some of the useful higher order moments of the normal distribution function are

$$\int_{-\infty}^{\infty} \exp(-x^2)\, dx = 2\int_0^{\infty} \exp(-x^2)\, dx = \sqrt{\pi},\ \int_{-\infty}^{\infty} x \exp(-x^2)\, dx = 0,$$

$$\int_0^{\infty} x \exp(-x^2)\, dx = \frac{1}{2},\ \int_{-\infty}^{\infty} x^2 \exp(-x^2)\, dx = 2\int_0^{\infty} x^2 \exp(-x^2)\, dx = \frac{\sqrt{\pi}}{2} \tag{9.8.5}$$

If r is a positive integer,

$$\int_{-\infty}^{+\infty} x^{2r+1} \exp(-x^2)\, dx = 0$$

$$\int_{-\infty}^{+\infty} x^{2r} \exp(-x^2)\, dx = 2\int_0^{\infty} x^{2r} \exp(-x^2)\, dx = \frac{\sqrt{\pi}}{2^r}\, 1 \cdot 3 \cdot 5 \ldots (2r-1)$$

Hence

$$\left.\begin{aligned} &\iiint_{-\infty}^{+\infty} f\, d\mathbf{v} = N,\ \iiint_{-\infty}^{+\infty} \mathbf{v} f\, d\mathbf{v} = 0\ \iiint_{-\infty}^{+\infty} v_i^2 f\, d\mathbf{v} = N\frac{K_B T}{m}, \\ &\langle v_i^2 \rangle = \frac{K_B T}{m},\ \langle v^2 \rangle = \langle v_1^2 + v_2^2 + v_3^2 \rangle = \frac{3K_B T}{m} \end{aligned}\right\} \tag{9.8.6}$$

If the number of particles, having speed v, is denoted by N_1, then

$$N_1 = \int_{\theta=0}^{\pi} \int_{\varphi=0}^{2\pi} f(v^2)\, v^2\, d\theta \sin\theta\, d\varphi = 4\pi v^2 f(v^2) \tag{9.8.7}$$

The maximum of N_1, obtained from solution of the equation $dN_1/dv = 0$, is $v_{\max} = (2K_B T/m)^{1/2}$. It is also the most probable speed. The average speed, v_{av}, is defined as

$$v_{av} = \frac{\int_0^\infty N_1 v\, dv}{\int_0^\infty N_1\, dv} = \left(\frac{8K_B T}{\pi m} \right)^{1/2} \tag{9.8.8}$$

At this speed the kinetic energy is $4K_B T/\pi$. If ε is the kinetic energy of a particle moving with speed v then $\varepsilon = mv^2/2$, and the number of particles having speeds between v and $v + dv$, or kinetic energy lying between the values ε and $\varepsilon = d\varepsilon$, is $N_1\, v$ where

$$N_1 = \frac{2N\sqrt{\varepsilon}}{(K_B T)^{3/2}\sqrt{\pi}} \exp\left(-\frac{\varepsilon}{K_B T} \right) d\varepsilon \tag{9.8.9}$$

The pressure tensor p_{ij} is a scalar quantity, because

$$p_{ij} = \iiint_{-\infty}^{\infty} mc_i\, c_j f\, d\mathbf{v} = m \iiint_{-\infty}^{\infty} v_i\, v_j f\, d\mathbf{v} = 0 \text{ if } i \neq j$$

$$= NK_B T \text{ if } i = j; \tag{9.8.10}$$

so $$p_{xx} = p_{yy} = p_{zz} = K_B TN \text{ and } p_{ij} = 0 \text{ if } i \neq j$$

Electrons, ions and neutral particles have their own equilibrium Maxwellian velocity distributions at temperatures T_e, T_i and T_n respectively; collision rates among electrons or among ions is larger than the rate between electrons and ions. So, each species can have its own state of thermal equilibrium; and the plasma may not last sufficiently long due to these differences. In a magnetized plasma the anisotropy (the CGL approximation) due to one temperature, for the distribution of velocities parallel and another perpendicular to the magnetic field **H**, has been discussed in section 4.12.

The relativistically correct Maxwellian equilibrium distribution of velocities is

$$f(v^2) = \frac{(mc^2/K_B T)}{4\pi c^2 K_2 (mc^2/K_B T)} \exp\left[-\frac{mc^2}{K_B T} \left(1 - \frac{v^2}{c^2} \right)^{1/2} \right] \tag{9.8.11}$$

where $K_2(x)$ is the modified Bessel function of second kind. Since $mc^2/K_B >> 1$ this $f(v^2)$ reduces to (9.8.1) for non-relativistic velocity distributions. See section (12.5) for further discussion.

Exercise 9.8.1. Find the number of Maxwellian particles in a system of N particles which have speeds greater than the average speed of the system.

9.9 PARTICLES STRIKING A WALL

We consider a small plane surface $d\Sigma$ at an arbitrary point within the gas and examine an element of volume $d\tau$, located arbitrarily with respect to $d\Sigma$, as in Fig. 9.6.1. The number dN of particles, with speed between v and $v + dv$, for isotropic velocity distributions (the Maxwell velocity distribution), is

$$dN = 4\pi v^2 f(v^2)\, dv \,.\, v \,.\, d\theta \,.\, v \sin\theta \, d\varphi; \tag{9.9.1}$$

and the number of particles leaving $d\tau$ per second is

$$\nu\, dN = \frac{1}{l} v_{av}\, dN \tag{9.9.2}$$

where ν is the frequency, v_{av} is the mean speed of equation (9.8.8) and l is the mean free path. Those particles of dN, which are scattered within the solid angle subtended at $d\tau$ by $d\Sigma$, will be headed towards $d\Sigma$ and their number is $d\Sigma \cos\theta / 4\pi r^2$ times dN. Of this group only the fraction $\exp(-r/l)$ will reach $d\Sigma$ and the remaining part will be further scattered. Thus, the number leaving $d\tau$, per second, with speeds between v and $v + dv$, and passing directly through $d\Sigma$, is

$$v_{av} \frac{dN}{l} \frac{d\Sigma \cos\theta}{4\pi r^2} \exp\left(-\frac{r}{l}\right) \tag{9.9.3}$$

Therefore, the total number N_1 of particles passing through $d\Sigma$, per second, from the half space containing within it $d\tau$, is given by

$$N_1 = \int_{r=0}^{\infty} \int_{\theta=0}^{\pi/2} \int_{\varphi=0}^{2\pi} \int_{v=0}^{\infty} \frac{v_{av}}{N} dN \frac{d\Sigma \cos\theta}{4\pi r^2} \exp\left(-\frac{r}{l}\right) = \frac{N}{4} v_{av}\, d\Sigma \tag{9.9.4}$$

An equal number will cross $d\Sigma$ per second from the other side also. Hence, the total number of particles, passing through unit area of $d\Sigma$, per second, is

$$N_2 = \frac{1}{2} N v_{av} \tag{9.9.5}$$

This relation is valid only when the hole is very small compared with the mean free path, because then only the distribution of velocities can be disturbed very little by the effusion of gas. Hence, if pressure outside is zero, the rate of effusion from a vessel through a very small hole is $(N v_{av})/4$. If the size of the hole is large, then the problem reduces to the hydrodynamic streaming of fluids.

9.10 EQUATIONS OF MACROSCOPIC MOTION

Multiplying all the terms of the Boltzmann equation, successively, by 1, **v**, $(v_i \; v_j)$, and then integrating the resultant relations in velocity space, we obtain the continuity equation, the momentum transfer relations and the energy transfer relations, respectively.

(a) Conservation of mass, momemtum and energy

For binary collisions m, $m\,\mathbf{v}$, mv^2 are the collisional invariants, Appendix 10 (e); so, when particles have only translational degrees of freedom, the integrals $\iiint (m, m\mathbf{v}, mv^2)\left(\frac{\partial f}{\partial t}\right)_{\text{coll}} d\mathbf{v}$ vanish. To obtain the equations of continuum mechanics only these moments are necessary. The Boltzmann equation, without the collision term, is

$$\frac{\partial f}{\partial t} + v_i \frac{\partial f}{\partial x_i} + a_i \frac{\partial f}{\partial v_i} = 0 \tag{9.10.1}$$

Integrating all the terms of this equation in velocity space, we obtain

$$\frac{\partial}{\partial t} \iiint f\, d\mathbf{v} + \iiint v_i \frac{\partial f}{\partial x_i} d\mathbf{v} + \iiint a_i \frac{\partial f}{\partial v_i} d\mathbf{v} = 0 \tag{9.10.2}$$

The first term $= \partial N/\partial t$, the second term $= (\partial/\partial x_i) \iiint v_i f\, d\mathbf{v}$

$= (\partial/\partial x_i)\,(Nu_i) = \text{div}\,(N\,\mathbf{u})$, and the third term $= \iiint a_i \frac{\partial f}{\partial v_i} d\mathbf{v}$. When a_i is independent of v_i, the third term

$$= \iiint \frac{\partial}{\partial\, v_i}(a_i f)\, dv_i\, dv_j\, dv_k = \iint [a_i f]_{v_i=-\infty}^{+\infty} dv_j\, dv_k = 0$$

Hence, equation (9.10.2) gives the continuity condition

$$\frac{dN}{dt} + N\,\text{div}\,\mathbf{u} = 0 \quad \text{or} \quad \frac{\partial N}{\partial t} + \text{div}\,(N\,\mathbf{u}) = 0 \tag{9.10.3}$$

Since the ith component of the Lorentz acceleration $e\,(\mathbf{v} \times \mathbf{H})/mc$ is independent of v_i, the continuity condition (9.10.3) holds good in plasmas. Multiplying next all the terms of (9.10.1) by v_j, and then integrating in the velocity space, we obtain

$$\frac{\partial}{\partial t} \iiint v_j f\, d\mathbf{v} + \iiint v_i\, v_j \frac{\partial f}{\partial x_i} d\mathbf{v} + \iiint a_i\, v_j \frac{\partial f}{\partial v_i} d\mathbf{v} = 0 \tag{9.10.4}$$

Again the first term $= (\partial/\partial t)\,(Nu_j)$; the second term

$$= \frac{\partial}{\partial x_i} \iiint v_i\, v_j f\, d\mathbf{v}\, \frac{\partial}{\partial x_i}(N\langle v_i\, v_j\rangle) = \frac{\partial}{\partial x_i}\left(\frac{p_{ij}}{m} + Nu_i\, u_j\right)$$

$$= \frac{1}{m}\,\frac{\partial p_{ij}}{\partial x_i} + \text{div}\,(N\,\mathbf{u}\,u_j)\text{ ; the third term} = \iiint a_i\, v_j \frac{\partial f}{\partial v_i}\, d\mathbf{v}$$

$$= \iiint \left[\frac{\partial}{\partial v_i}(a_i\, v_j f) - a_j f\right] d\mathbf{v} = 0 - N\langle a_j\rangle = -\frac{1}{m} F_j$$

where **F** is the average force per unit volume and $\lim_{\mathbf{v}\to\pm\infty}(a_i \mathbf{v} f) = 0$. Therefore, the macroscopic equation of momentum transfer is

$$\frac{\partial}{\partial t}(Nu_j) + \frac{1}{m}\frac{\partial p_{ij}}{\partial x_i} + \text{div}\,(N\mathbf{u}\,u_j) - \frac{1}{m}F_j = 0 \tag{9.10.5}$$

By readjustment it becomes

$$\frac{\partial}{\partial t}(\rho u_j) + \text{div}\,(\rho\mathbf{u}\,u_j) = -\frac{\partial}{\partial x_i}p_{ij} + F_j$$

Using the continuity condition (9.10.3) this equation reduces to

$$\rho\frac{du_j}{dt} = -\frac{\partial p_{ij}}{\partial x_i} + F_j \tag{9.10.6}$$

Next, multiplying all terms of the Boltzmann equation by $v_i\,v_j$ and then integrating them in velocity space, we obtain

$$\frac{\partial}{\partial t}\iiint v_i\,v_j f\,d\mathbf{v} + \iiint v_i\,v_j\,v_k\frac{\partial f}{\partial x_k}d\mathbf{v} + \iiint v_i\,v_j\,a_k\frac{\partial f}{\partial v_k}d\mathbf{v}$$

$$= \iiint\left(\frac{\partial f}{\partial t}\right)_{\text{coll}} v_i\,v_j\,d\mathbf{v} \tag{9.10.7}$$

On the left hand side, the first term $= (\partial/\partial t)\,((p_{ij}/m) + Nu_i\,u_i)$; the second term $= (1/m)\;(\partial/\partial x_k)\;[2Q_{ijk} + \rho u_i\,u_j\,u_k + u_i\,p_{jk}]$; the third term

$$= \iiint\left[\frac{\partial}{\partial v_k}(a_k\,v_i\,v_j f) - a_i\,v_j - a_j\,v_i\right] f\,d\mathbf{v} = 0 - N\langle a_i\,v_j + a_j\,v_i\rangle$$

Using these results, (9.10.7) becomes

$$\frac{\partial}{\partial t}(p_{ij} + |\,\rho u_i\,u_j) + \frac{\partial}{\partial x_k}[2Q_{ijk} + \rho\,u_i\,u_j\,u_k + u_i\,p_{jk}] - \rho\,\langle a_i\,v_j + a_j\,v_i\rangle$$

$$= \iiint\left(\frac{\partial f}{\partial t}\right)_{\text{coll}} v_i\,v_j\,d\mathbf{v} \tag{9.10.8}$$

The diagonal components of the pressure tensor give

$$\frac{\partial}{\partial t}(p_{ij} + \rho u_i^2) + \frac{\partial}{\partial x_l}[2Q_{iil} + 2u_i\,p_{il} + u_l\,p_{ii} + \rho\,u_i^2\,u_l] - 2\,\langle\rho\,a_i\,v_i\rangle$$

$$= \iiint\left(\frac{\partial f}{\partial t}\right)_{\text{coll}} v_i^2\,d\mathbf{v}, \tag{9.10.9}$$

$$\frac{\partial}{\partial t}(p_{ij} + \rho\,u_j^2) + \frac{\partial}{\partial x_l}[2Q_{jjl} + 2u_j\,p_{jl} + u_l\,p_{ji} + \rho\,u_j^2\,u_l] - 2\,\langle\rho\,a_j\,v_j\rangle$$

$$= \iiint \left(\frac{\partial f}{\partial t}\right)_{\text{coll}} v_j^2 \, d\mathbf{v}, \tag{9.10.10}$$

$$\frac{\partial}{\partial t}(p_{kk} + \rho u_k^2) + \frac{\partial}{\partial x_l}[2Q_{kkl} + 2\, u_k\, p_{kl} + u_l\, p_{kk} + \rho\, u_k^2\, u_l] - 2\,\langle \rho\, a_k\, v_k \rangle$$

$$= \iiint \left(\frac{\partial f}{\partial t}\right)_{\text{coll}} v_k^2 \, d\mathbf{v} \tag{9.10.11}$$

Adding these three relations, and since the kinetic energy ξ is a collisional invariant (Appendix 10 (*e*)), so that $\iiint \left(\frac{\partial f}{\partial t}\right)_{\text{coll}} \xi d\mathbf{v} = 0,$ we obtain

$$\frac{\partial \xi}{\partial t} + \frac{\partial}{\partial x_l}[Q_l + u_i\, p_{il} + u_j\, p_{jl} + u_k\, p_{kl} + \xi\, u_l] - \langle \rho\, (\mathbf{a}\,.\,\mathbf{v}) \rangle = 0 \tag{9.10.12}$$

where $Q_l = Q_{iil} + Q_{jjl} + Q_{kkl}$ and

$$\xi = \frac{1}{2}\, m \iiint (v_i^2 + v_j^2 + v_k^2)\, f\, d\mathbf{v} = \frac{1}{2}[p_{ii} + p_{jj} + p_{kk} + \rho\,(u_i^2 + u_j^2 + u_k^2]$$

Simplifying further we get

$$\frac{d\xi}{\partial t} + \text{div}\,(\xi \mathbf{u}) + \frac{\partial}{\partial x_l} Q_l + \frac{\partial}{\partial x_l}(u_i\, p_{il}) = \langle \rho\, (\mathbf{a}\,.\,\mathbf{v}) \rangle \tag{9.10.13}$$

If $\rho\, \mathbf{a}$ is the Lorentz force density $(= Ne\, \mathbf{E} + Ne\, (\mathbf{v} \times \mathbf{H})/c)$, then since $\mathbf{E}$ is independent of $\mathbf{v}$, this equation reduces to

$$\frac{\partial \xi}{\partial t} + \frac{\partial}{\partial x_l}(\xi\, u_l) + \frac{\partial}{\partial x_l} Q_l + \frac{\partial}{\partial x_l}(u_i\, p_{il}) = (\mathbf{E}\,.\,\mathbf{j}) \tag{9.10.14}$$

because $\langle \rho\, (\mathbf{a}\,.\,\mathbf{v}) \rangle = \langle eN\, (\mathbf{E}\,.\,\mathbf{v}) \rangle = (\mathbf{E}\,.\,\mathbf{j})$.

The moment equations, namely (9.10.3), (9.10.6) and (9.10.14), are not closed because the equation to determine the integral $\iiint f\, d\mathbf{v}$, contains the integral $\iiint \mathbf{v}\, f\, d\mathbf{v}$, the equation for $\iiint v_i f\, d\mathbf{v}$ contains a term proportional to $\iiint v_i\, v_j f\, d\mathbf{v}$ and so on. It is, therefore, not possible to determine any of these integrals without further assumptions to close the system, which give rise to some empirical state relations. For instance, the adiabatic thermal law $\nabla p = c_s^2\, \nabla \rho$ in terms of the kinetic theory, is equivalent to the assumptions

$$p_{ij} = 0 \text{ if } i \neq j,\;\; p_{ii} = p_{jj} = p_{kk} = p = m \iiint c_i^2\, f\, d\mathbf{v} = mc_s^2 \iiint f\, d\mathbf{v} \tag{9.10.15}$$

When the pressure is a scalar quantity, equation (9.10.14) can be reduced to another standard form, with the help of the continuity condition (9.10.3). For this purpose we write

$$p_{ij} = p\delta_{ij} \tag{9.10.15 (a)}$$

where δ_{ij} is the Kronecker delta. Then equation (9.10.14) can be expressed as

$$\frac{\partial \xi}{\partial t} + \text{div}\,(\xi \mathbf{u}) + \text{div}\,(\mathbf{u}\,p) = (\mathbf{E}\,.\,\mathbf{j}) - \text{div}\,\mathbf{Q} \tag{9.10.15 (b)}$$

or as $\dfrac{d\xi}{dt} + \xi\,\text{div}\,\mathbf{u} + p\,\text{div}\,\mathbf{u} + (\mathbf{u}\,.\,\nabla)\,p = (\mathbf{E}\,.\,\mathbf{j}) - \text{div}\,\mathbf{Q}$ (9.10.15 (c))

With the help of (9.10.3), equation (9.10.15 (c)) reduces to

$$\frac{d\xi}{dt} - \frac{\xi}{\rho}\frac{d\rho}{dt} - \frac{p}{\rho}\frac{d\rho}{dt} = (\mathbf{E}\,.\,\mathbf{j}) - \text{div}\,\mathbf{Q} - (\mathbf{u}\,.\,\nabla)\,p \tag{9.10.15 (d)}$$

Multiplying both sides of (9.10.15 (d)) by $1/\rho$ it transforms to a standard form:

$$\frac{d}{dt}\left(\frac{\xi}{\rho}\right) - \frac{p}{\rho^2}\frac{d\rho}{dt} = \frac{1}{\rho}(\mathbf{E}\,.\,\mathbf{j}) - \frac{1}{\rho}\text{div}\,\mathbf{Q} - \frac{1}{\rho}(\mathbf{u}\,.\,\nabla)\,p \tag{9.10.15 (e)}$$

The macroscopic energy transport relation, for the one fluid theory of MHD approximation, is considered in section (4.4 (a)). In plasmas, regarded as mixtures of fluids of positive charges, negative charges and neutral particles, the energy transport phenomenon is considered in section 3.7; and for small amplitude waves in the same plasma in section 3.8.

(b) Grad's thirteen-moment equations

If $\varphi\,(\mathbf{v})$ is an arbitrary function of $\mathbf{v}$, the more general moment equation is

$$\iiint \varphi \frac{\partial f}{\partial t}\,d\mathbf{v} + \iiint \varphi\,(\mathbf{v}\,.\,\nabla_r)\,f d\mathbf{v} + \iiint \varphi\,(\mathbf{a}\,.\,\nabla_v)\,f d\mathbf{v}$$

$$= \iiint \varphi \left(\frac{\partial f}{\partial t}\right)_{\text{coll}} d\mathbf{v} \tag{9.10.16}$$

The first and second terms on the left hand side are of the type $\iiint \varphi \dfrac{\partial f}{\partial \psi}$ $d\mathbf{v}$, where ψ is either t or $\mathbf{r}$. Since ψ is independent of $\mathbf{v}$, we can take $\partial/\partial\psi$ outside the integration, and obtain

$$\iiint \varphi \frac{\partial f}{\partial \psi}\,d\mathbf{v} = \frac{\partial}{\partial \psi}(N\langle\phi\rangle) - N\langle\frac{\partial \varphi}{\partial \psi}\rangle \text{ with } \langle\phi\rangle = \frac{\iiint \varphi f\,d\mathbf{v}}{\iiint f\,d\mathbf{v}}$$

The third term on the left hand side

$$= \iiint [\nabla_v\,(\varphi \mathbf{a} f) - f\nabla_v\,(\mathbf{a}\varphi)]\,d\mathbf{v} = 0 - \iiint f\nabla_v\,(a\varphi)\,d\mathbf{v}$$

because $\lim\limits_{\mathbf{v}\to\pm\infty} (\varphi \mathbf{a} f) \to 0$. Hence, the moment equation is

$$\frac{\partial}{\partial t}(N\langle\phi\rangle)+\nabla_{\mathbf{r}}(N\langle\varphi\mathbf{v}\rangle)-N\left[\langle\frac{\partial\varphi}{\partial t}\rangle+\langle(\mathbf{v}\,.\,\nabla_{\mathbf{r}})\,\varphi\rangle+\langle\mathrm{div}_{\mathbf{v}}\,(\mathbf{a}\varphi)\rangle\right]$$

$$=\iiint\varphi\left(\frac{\partial f}{\partial t}\right)_{\text{coll}}d\mathbf{v} \tag{9.10.17}$$

When φ is independent of t and $\mathbf{r}$, this equation reduces to

$$\frac{\partial}{\partial t}(N\langle\varphi\rangle)+\mathrm{div}_{\mathbf{r}}\,(N\langle\varphi\mathbf{v}\rangle)-N\langle\mathrm{div}\;\mathbf{v}\,(\mathbf{a}\varphi)\rangle=\iiint\varphi\left(\frac{\partial f}{\partial t}\right)_{\text{coll}}d\mathbf{v} \tag{9.10.18}$$

Mass continuity, momentum transfer and energy transfer conditions, namely (9.10.3), (9.10.6) and (9.10.14), respectively, are obtained from (9.10.18) by putting, in turn, $\varphi = m,\ m\mathbf{v},\ mv^2/2$ in that relation, and then doing some integration and algebraic work. Putting $\varphi=(v_i-u_i)\ (v_j-u_j)$ is (9.10.16) we obtain

$$\iiint(v_i-u_i)\,(v_j-u_j)\frac{\partial f}{\partial t}d\mathbf{v}+\iiint(v_i-u_i)\,(v_j-u_j)\,v_k\frac{\partial f}{\partial x_k}d\mathbf{v}+$$

$$+\iiint(v_i-u_i)\,(v_j-u_j)\,a_k\frac{\partial f}{\partial v_k}d\mathbf{v}=\iiint\left(\frac{\partial f}{\partial t}\right)_{\text{coll}}(v_i-u_i)\,(v_j-u_j)\,d\mathbf{v} \tag{9.10.19}$$

In the left hand side, the first term $=(1/m)\ \partial p_{ij}/\partial t$, (9.10.20)

the second term

$$=\iiint\{(v_i-u_i)\,(v_j-u_j)\,(v_k-u_k)\}\,\{+u_k\,(v_i-u_i)\,(v_j-u_j)\}\frac{\partial f}{\partial x_k}d\mathbf{v}$$

$$=\frac{2}{m}\,\frac{\partial}{\partial x_k}Q_{ijk}+\frac{\partial u_j}{\partial x_k}\frac{p_{kl}}{m}++\frac{\partial u_i}{\partial x_k}\frac{p_{jk}}{m}+(\mathbf{u}\,.\,\nabla)\left(\frac{p_{ij}}{m}\right)-\frac{1}{N}\frac{dN}{dt}\frac{p_{ij}}{m}, \tag{9.10.21}$$

$$\text{the third term}=-\frac{e}{mc}\iiint\{(\mathbf{v}\times\mathbf{B})_i\,(v_j-u_j)+(\mathbf{v}\times\mathbf{B})_j\,(v_i-u_i)\}\,f\,d\mathbf{v}$$

$$=-\frac{eB_k}{mc}(\varepsilon_{ink}\,p_{nj}+\varepsilon_{jnk}\,p_{ni})=-F_{ij} \tag{9.10.22}$$

where ε_{ink} is the alternate unit tensor (Appendix 4). The last term in (9.10.21) is obtained by using the continuity condition (9.10.3). The term containing $\mathbf{E}$, being independent of $\mathbf{v}$, vanishes. Putting, therefore, (9.10.20), (9.10.21) and (9.10.22) in (9.10.19), we obtain

$$\frac{\partial p_{ij}}{\partial t}+(\mathbf{u}\,.\,\nabla)\,p_{ij}=\left(\frac{p_{ij}}{N}\right)\frac{dN}{dt}+p_{ki}\frac{\partial u_j}{\partial x_k}+\frac{\partial u_i}{\partial x_k}p_{jk}+2\frac{\partial Q_{ijk}}{\partial x_k}$$

$$=\frac{eB_k}{c}(\varepsilon_{ink}\,p_{nj}+\varepsilon_{jnk}\,p_{ni})+L_{ij} \tag{9.10.23}$$

where $$L_{ij} = m \iiint \left(\frac{\partial f}{\partial t}\right)_{coll} (v_i - u_i)\ (v_j - u_j)\ d\mathbf{v} \tag{9.10.24}$$

This equation can be written as

$$\frac{d}{dt}\left(\frac{p_{ij}}{N}\right) + \frac{p_{ki}}{N}\frac{\partial u_j}{\partial x_k} + \frac{p_{jk}}{N}\frac{\partial u_i}{\partial x_k} + \frac{2}{N}\frac{\partial}{\partial x_k} Q_{ijk} = \frac{1}{N}(F_{ij} + L_{ij}) \tag{9.10.25}$$

Hence we find that

$$\frac{d}{dt}\left(\frac{p_{xx}}{N}\right) + \frac{2p_{lx}}{N}\frac{\partial u_x}{\partial x_l} + \frac{2}{N}\frac{\partial}{\partial x_l} Q_{xxl} = \frac{2e}{Nmc}(B_z p_{xy} - B_y p_{zx}) + \frac{1}{N}L_{xx}, \tag{9.10.26}$$

$$\frac{d}{dt}\left(\frac{p_{xy}}{N}\right) + \frac{p_{xy}}{N}\left(\frac{\partial u_x}{\partial x} + \frac{\partial u_y}{\partial y}\right) + \frac{1}{N}p_{xx}\frac{\partial u_y}{\partial x} + \frac{p_{yy}}{N}\frac{\partial u_x}{\partial y} + \frac{p_{yz}}{N}\frac{\partial u_x}{\partial z}$$

$$+ \frac{p_{zx}}{N}\frac{\partial u_y}{\partial z} + \frac{2}{N}\frac{\partial Q_{xyl}}{\partial x_l} \tag{9.10.27}$$

$$= \frac{e}{Nmc}[-B_z(p_{xx} - p_{yy}) + B_x p_{zx} - B_y p_{yz}] + \frac{L_{xy}}{N}$$

Following the simple rule of cyclic rotation of indices x, y, z or i, j, k, we can write down the equations for p_{yy} and p_{zz} from (9.10.26) and for p_{yz} and p_{zx} from (9.10.27). Putting, in (9.10.18), $\varphi(\mathbf{v}) = 1$, v_x, v_y, v_z, $v_x v_x$, $v_x v_y$, $v_x v_z$, $v_y v_y$, $v_y v_z$, $v_z v_z$, $(mv^2 v_x)/2$, $(mv^2 v_y)/2$, $(mv^2 v_z)/2$, we obtain a series of thirteen moment equations, known as Grad's thirteen moment equations. For $\varphi(v) = v_x v_x$ and $v_x v_y$ we obtain the equations (9.10.26) and (9.10.27).

(c) Double-adiabatic state relations

For finding the state relations of the double adiabatic theory from axially symmetric anisotropic distributions, we neglect the divergence of heat flow tensor Q_{ijk}, and energy exchange by collision. Assuming that the distribution function is symmetric about the z-axis we put in (9.10.26)

$$p_{xx} = p_{yy} = p_\perp,\ p_{zz} = p_\parallel,\ p_{ij} = 0 \text{ if } i \neq j \text{ and } \partial/\partial x = 0 = \partial/\partial y, \tag{9.10.28}$$

and find that

$$\frac{d}{dt}\left(\frac{p_\perp}{N}\right) = 0,\ \frac{d}{dt}\left(\frac{p_\parallel}{N}\right) + \frac{2}{N}p_\parallel\frac{\partial u_z}{\partial z} = 0 \tag{9.10.29}$$

With the help of the continuity condition (9.10.3) we eliminate $\partial u_z/\partial z$; then integrating we obtain

$$\frac{p_\perp}{N} = \text{constant}, \ \frac{p_\parallel}{N^3} = \text{constant} \tag{9.10.30}$$

Defining the kinetic temperatures $T_\parallel$ and $T_\perp$ by

$$p_\parallel = NK_B T_\parallel, \ p_\perp = NK_B T_\perp, \tag{9.10.31}$$

where K_B is the Boltzmann constant, we find that

$$T_\perp = \text{constant}, \ \frac{T_\parallel}{N^2} = \text{constant} \tag{9.10.32}$$

9.11 EXPANSION OF DISTRIBUTION FUNCTIONS AND MACROSCOPIC VARIABLES

Field variables are often expanded for obtaining approximate solutions of field equations. Macroscopically averaged field variables are functions of position coordinates and time; and the distribution function is described in the phase space. The relations between the macroscopic field variables and the distribution functions have not been exhaustively discussed in literature; so the validity of some relations is doubted, as may be seen from the following remarks of Tidman and Stainer:[1] "... We make use of the first three moment equations for collisionless electron plasma since it is not clear that the same methods (namely, the method of Bogoliubov, Krilov and Mitropolskii for the secular-free solution of certain types of non-linear equations by power series expansion) are applicable to the Vlasov equation''. Actually it is not clear whether the asymptotic methods of solution for the non-linear problems are valid at the singularity which arises when the phase velocity of a wave equals the particle velocity, as in Landau damping (section 10.3). Another difficulty will be discussed now.

The study of relations between macroscopic and microscopic variables (and, more generally, between the solutions of equations of kinetic theory in phase space and macroscopic equations for the fluid continuum) led Whitham (1965)[2] to remark "... the detailed microscopic structure of (a field variable) φ is not necessarily governed by hyperbolic equations, but the equation for the 'microscopic' variables, such as amplitude, wave number etc. are hyperbolic ... the Boltzmann equation for a gas is not hyperbolic ... and sound waves appear from appropriate averaged equations... The Maxwell distribution describes a uniform state in terms of density ρ, temperature T, mean velocity $\mathbf{u}$ ... This is a solution for constant values of ρ, T, $\mathbf{u}$, but it is used to derive continuum equations for ρ, T, $\mathbf{u}$ when they are slowly varying functions relative to the mean free path and collision time.''

To show the distinction between the expansion of the distribution function $f(\mathbf{r}, \mathbf{v}, t)$ and the expansion of the macroscopic field variables which depend on its velocity moments, we write the expansions

---

1 D.A. Tidman and H.M. Stainer, *Physics of Fluids*, **8**, 345 (1965).

2 G.B. Whitham, *Prob. Roy, Solu.*, London A 283, 238-261, (1945).

$$f(\mathbf{r}, \mathbf{v}\, t) = Nf_0 + \varepsilon f_1 + \varepsilon^2 f_2 +, \ \langle \mathbf{v} \rangle = \mathbf{u}_0 + \varepsilon \mathbf{u}_1 + \varepsilon^2\, \mathbf{u}_2 + \ldots$$

The relation

$$\langle \mathbf{v} \rangle \iiint f\, d\mathbf{v} = \iiint \mathbf{v} f\, d\mathbf{v},$$

shows that

$$(\mathbf{u}_0 + \varepsilon \mathbf{u}_1 + \varepsilon^2\, \mathbf{u}_2 + \ldots) \iiint (N_0 f_0 + \varepsilon f_1 + \varepsilon^2 f_2 + \ldots)\, d\mathbf{v}$$

$$= \iiint \mathbf{v}\,(N_0 f_0 + \varepsilon f_1 + \varepsilon^2 f_2 + \ldots)\, d\mathbf{v} \qquad (9.11.1)$$

where (9.6.9) and (9.6.2) have been used. When the coefficients of ε^1, ε^2, ε^3, ε^4 from both sides, are equated, a series of equations is obtained. The first three of these are

$$\mathbf{u}_0 = \frac{\iiint \mathbf{v} f_0\, d\mathbf{V}}{\iiint f_0\, d\mathbf{V}}, \ \mathbf{u}_1 = \frac{\iiint \mathbf{v} f_1\, d\mathbf{V}}{N_0 \iiint f_0\, d\mathbf{V}} - \frac{(\iiint \mathbf{v} f_0\, d\mathbf{v})\, (\iiint f_1\, d\mathbf{V})}{N_0\, (\iiint f_0\, d\mathbf{V})^2}$$

$$\mathbf{u}_2 = \frac{\iiint \mathbf{v} f_0\, d\mathbf{V}}{N_0 \iiint f_0\, d\mathbf{V}} - \frac{(\iiint \mathbf{v} f_0\, d\mathbf{V})\, (\iiint f_2\, d\mathbf{V})}{N_0\, (\iiint f_0\, d\mathbf{V})^2} - \frac{(\iiint f_1,\, d\mathbf{V})\, (\iiint \mathbf{v} f_1\, d\mathbf{V})}{N_0^2\, (\iiint f_0\, d\mathbf{V})^2} +$$

$$+ \frac{(\iiint \mathbf{v} f_0\, d\mathbf{V})\, (\iiint f_1\, d\mathbf{V})^2}{N_0^2\, (\iiint f_0\, d\mathbf{V})^3}, \text{ etc.} \qquad (9.11.2)$$

Moreover, expansion of the number density N gives

$$N = N_0 + \varepsilon N_1 + \varepsilon^2\, N_2 + \varepsilon^3\, N_3 + \ldots = \iiint f\, d\mathbf{V}$$

$$= N_0 \iiint f_0\, d\mathbf{V} + \varepsilon \iiint f_1\, d\mathbf{V} + \varepsilon^2 \iiint f_2\, d\mathbf{V} + \ldots \qquad (9.11.13)$$

Equating terms of same powers of ε, from both sides, we obtain

$$\mathbf{u}_0 = \iiint \mathbf{v} f_0\, d\mathbf{V}, \ 1 = \iiint f_0\, d\mathbf{V}, \ N_0\, \mathbf{u}_1 + N_1\, \mathbf{u}_0 = \iiint \mathbf{v} f_1\, d\mathbf{V},$$

$$N_0\, \mathbf{u}_2 + N_1\, \mathbf{u}_1 + N_2\, \mathbf{u}_0 = \iiint \mathbf{v} f_2\, d\mathbf{V}, \text{etc.} \qquad (9.11.4)$$

These relations show that there is no one-to-one correspondence between the integrals of f_i and the velocity averages $\mathbf{u}_i$ of different orders of magnitude. Expansion of $\mathbf{E}$ and $\iiint \mathbf{v} f\, d\mathbf{V}$ shows that the same absence of one-to-one correspondence exists between $\mathbf{E}_1$ and $\iiint \mathbf{v} f_1\, d\mathbf{V}$, and between $\mathbf{E}_2$ and $\iiint \mathbf{v} f_2\, d\mathbf{V}$, etc. To further illustrate this point clearly, we take up the Maxwell equation curl $\mathbf{H}$ = $(\partial \mathbf{E}/(\partial t)$ – $(4\pi e N\, \mathbf{u}/c)$ of the hydrodynamic model, and the corresponding equation of Vlasov plasma, curl $\mathbf{H}$ = $(\partial \mathbf{E}/(\partial t)$ – $(4\pi e/c)$ $\iiint \mathbf{v}\,(N f_0 - f)\, d\mathbf{V}$. Using the expansions $\mathbf{u} = \mathbf{u}_0 + \varepsilon \mathbf{u}, + \varepsilon^2\, \mathbf{u}_2 + \ldots$, $f = N_0 f_0 +$

$\varepsilon f_1 + \varepsilon^2 f_2 + ...$, $\mathbf{E} = \mathbf{E}_0 + \varepsilon \mathbf{E}_1 + \varepsilon^2 \mathbf{E}_2 + ...$, and equating coefficients of ε^0, ε^1, ε^2, from both sides of the Maxwell equations we get the equations for the quantities of different orders of magnitude. For example, equating coefficients of ε^3, we obtain

$$\nabla \times \mathbf{H}_3 = \frac{1}{c}\frac{\partial \mathbf{E}_3}{\partial t} - \frac{4\pi l}{c}(N_0\,\mathbf{u}_3 + N_1\,\mathbf{u}_2 + N_2\,\mathbf{u}_1 + N_3\,\mathbf{u}_0) \qquad (9.11.5)$$

and

$$\nabla \times \mathbf{H}_3 = (\partial \mathbf{E}_3/(\partial t) - (4\pi e/c) \iiint \mathbf{v} f_3\, d\mathbf{v} \qquad (9.11.6)$$

The non-linear terms in the right hand side of (9.11.5) are combinations of lower order interactions, but, in (9.11.6), there is no such non-linear term depending on the distributions f_0, f_1, f_2, etc. Hence, it will be improper to identify $\mathbf{H}_3$ of (9.11.5) with the $\mathbf{H}_3$ of (9.11.6), though both are of the same order of magnitude; this distinction gives rise to some qualitatively non-identical results of higher order non-linear corrections. For instance, the non-linear corrections due to superluminuous electromagnetic waves in relativistic Vlasov plasmas, obtained by Winkles and Eldridge (1972)[3], should not necessarily coincide, in the cold plasma limit, with the results of Slujter and Montgomery (1965)[4]; or the results of their section 2, in the same limit, with the results of section 3 of Arons and Max (1975).[5]

9.12 REACTION RATE

The two distributions $f_1(\mathbf{v}_1)$ and $f_2(\mathbf{v}_2)$ are unlike in the velocity space because, per unit volume, their numbers are N_1 and N_2. If v is the relative speed of the interacting particles, then $v = |\mathbf{v}_1 - \mathbf{v}_2|$, and if t is the average confinement time, then vt is the reaction length. The volume swept away by the test particle, during the confinement, is that of the cylinder, σvt, where σ is the cross-section of a reaction; the number of particles in the cylinder is $N\sigma vt$ if N is their number per unit volume. Any other particle in this volume can experience a fusion reaction with the test particle. The probability per unit time for a particle to have a reaction is $N\sigma v$.

The particles of distribution l with the number $f_1(\mathbf{v}_1)\, d\mathbf{v}_1$, in a velocity volume element $d\mathbf{v}_1$, react with particles of distribution 2 with the number $f_2(\mathbf{v}_2)\, d\mathbf{v}_2$ in velocity volume element $d\mathbf{v}_2$, at the rate dr, given by

$$dr = f_1(\mathbf{v}_1) f_2(\mathbf{v}_2)\, |\mathbf{v}_1 - \mathbf{v}_2|\, \sigma(|\mathbf{v}_1 - \mathbf{v}_2|)\, d\mathbf{v}_1\, d\mathbf{v}_2 \qquad (9.12.1)$$

because dr must be equal to product of these two reacting elements. The rate r at which reaction proceeds in a uniform gas mixture is the product of the number

---

3 B. Winkles and O. Eldridge, *Phys. Fluids 15*, 1790 (1972).

4 F. Slujter and D. Montgomery, *Phys. Fluids 8*, 551 (1965).

5 J. Arons and C. Max, *Phys. Fluids 3*, 1983 (1974).

of possible pairs of reacting particles. So it is given by the following six-fold integral over all space for both species:

$$r = \iiiiiint f_1(\mathbf{v}_1) f_2(\mathbf{v}_2) |\mathbf{v}_1 - \mathbf{v}_2| \, \sigma(|\mathbf{v}_1 - \mathbf{v}_2|) \, d\mathbf{v}_1 \, d\mathbf{v}_2 \qquad (9.12.2)$$

If the velocity distributions $f_1(\mathbf{v}_1)$ and $f_2(\mathbf{v}_2)$ are Maxwellian at temperature T and with respective masses m_1 and m_2 per particle, then

$$f_1(\mathbf{v}_1) = N_1 \left(\frac{m_1}{2\pi KT}\right)^{3/2} \exp\left(-\frac{m_1 v^2}{2KT}\right),$$

$$f_2(\mathbf{v}_2) = N_2 \left(\frac{m_2}{2\pi KT}\right)^{3/2} \exp\left(-\frac{m_2 v^2}{2KT}\right) \qquad (9.12.3)$$

The integral (9.12.2) may be reduced by changing to the relative velocity $\mathbf{v}$ and the mean velocity or the centre of mass velocity $\mathbf{v}_g$. Obviously,

$$\mathbf{v} = \mathbf{v}_1 - \mathbf{v}_2, \; \mathbf{v}_g = \frac{(m_1 \mathbf{v}_1 + m_2 \mathbf{v}_2)}{m_1 + m_2}, \; \mathbf{v}_1 = \mathbf{v}_g + \frac{m_2 \mathbf{v}}{m_1 + m_2}, \; \mathbf{v}_2 = \mathbf{v}_g - \frac{m_1 \mathbf{v}}{m_1 + m_2} \qquad (9.12.4)$$

If the reacting mixture is in thermal equilibrium, r depends only on the temperature and can be written as a simple integral and approximated in certain cases by an explicit function. So we put

$$m = \frac{m_1 m_2}{m_1 + m_2}, \; m_s = m_1 + m_2 \qquad (9.12.5)$$

in (9.12.2) and obtain

$$r = \frac{(4\pi)^2 N_1 N_2 \sqrt{m_1 m_2}}{(2\pi KT)^3} \int_{v_g=0}^{\infty} \int_{v=0}^{\infty} v_g^2 v^3 \sigma(v) \exp\left(-\frac{m_2 v_g^2}{2KT} - \frac{m v^2}{2KG}\right) dv_g \, dv \qquad (9.12.6)$$

The integral with respect to v_g is $\int_{v_g=0}^{\infty} v_g^2 \exp\left(-\frac{m_s v_g^2}{2K_B T}\right) dv_g$. Integrating with the help of results of section 9.8, we obtain

$$r = N_1 N_2 \, . \, 4\pi \left(\frac{m_1 m_2}{2K_B T m_s}\right)^{3/2} \int_0^{\infty} v^3 \sigma(v) \exp\left(-\frac{m v^2}{2K_B T}\right) dv \qquad (9.12.7)$$

When two different interacting particle species are involved, the probability r of a reaction is given by

$$r = N_1 N_2 \langle \sigma v \rangle \qquad (9.12.8)$$

where $\langle\sigma v\rangle$ is the average of the product of the reaction cross-section σ and relative velocity v of the interacting particles. So,

$$\langle\sigma v\rangle = \iiiiiint d\mathbf{v}_1\, d\mathbf{v}_2 f(\mathbf{v}_1)\, f(\mathbf{v}_2)\, v\sigma(v) = \frac{r}{N_1 N_2} \tag{9.12.9}$$

In terms of the kinetic energy $\xi\ (= m_1 v^2/2)$ and the reduced mass m of (9.12.5), the reaction probability is given by

$$\langle\sigma v\rangle = \frac{8\pi}{m_1^2}\left(\frac{m}{2\pi K_B T}\right)^{3/2} \int \xi\sigma(\xi) \exp\left(-\frac{m\xi}{m_1 K_B T}\right) d\xi \tag{9.12.10}$$

When the two types of interacting particles are identical, in the integration, to eliminate repetition the sum over each reaction should be taken only once. If the particles of species 1 and 2 are really different then the volume elements $d\mathbf{V}_1$ and $d\mathbf{V}_2$ contain different particles and so collisions would not be counted twice. But, for identical particles, to eliminate repetition, the sum over each reaction should be taken only once; then, in place of (9.12.9), we would have

$$r = \frac{1}{2} N^2 \langle\sigma v\rangle \tag{9.12.11}$$

In the $D-D$ reaction the two reacting species are the same. Classically the energy which must be supplied to a nucleus to overcome electrostatic repulsion, for the occurrence of thermonuclear reaction, is $W = q_1 q_2/R_{12}$ where R_{12} is the distance between the centres of the charges q_1 and q_2. Taking R_{12} equal to the nuclear diameter 5×10^{-13} cm, we find that for fully ionized hydrogen gas $W = 4.6 \times 10^{-7}$ erg. Actually, detectable nuclear reactions take place even when energies are considerably below this order. So, there is no lower threshold limit for the occurrence of thermonuclear reactions. The reaction rates have been estimated for the Maxwell velocity distributions, neglecting statistical diffusion of particles, during a reaction.

Chapter 10

PLASMA KINETIC THEORY

10.1 INTRODUCTION

The system of equations for plasmas, in the six-dimensional phase space of velocity and position coordinates, is called the Vlasov equations (also called the Boltzmann-Vlasov (or $B–V$) equations) (sections 10.2 and 10.3). The collision term in the Boltzmann equation, is negligible because cumulative long range interections dominate over individual short range encounters in Vlesov plasmas. For weakly ionized plasma the distribution function is expanded in Legendre polynomials in the presence of an applied electric field in section 10.10. The collision term, in this case, has been studied in sections 10.7 and 10.8. Application of this expansion in the microscopic theory of ionospheric cross-modulation effect is considered in sectons 10.9 and 10.10.

For fluids the macroscopic behaviour is easily understood and the microscopic behaviour of these is only vaguely understood, the reason being that the microscopic intermolecular forces are only inferred from the kinetic theory calculations. However, for plasmas macroscopic properties are measured with difficulty, and often basic parameters have to be indirectly inferred, while the microscopic interparticle forces are well-known; so that the situation for plasmas is reversed.

10.2 VLASOV EQUATIONS

The self-consistent system of Vlasov equations, for fully ionized plasma, is

$$\frac{\partial f_i}{\partial t} + (\mathbf{v} \,.\, \nabla_{\mathbf{r}}) f_i + (\mathbf{F}_i \,.\, \nabla_{\mathbf{v}}) f_i = 0 \tag{10.2.1}$$

$$\nabla \times \mathbf{H} = \frac{1}{c}\frac{\partial \mathbf{E}}{\partial t} + \frac{4\pi}{c}\mathbf{j}, \tag{10.2.2}$$

$$\nabla \times \mathbf{E} = -\frac{1}{c}\frac{\partial \mathbf{H}}{\partial t}, \tag{10.2.3}$$

$$\nabla \,.\, \mathbf{E} = 4\pi q, \;\; \nabla \,.\, \mathbf{H} = 0 \tag{10.2.4}$$

where $f_i(\mathbf{r}, \mathbf{v}, t)$ is the distribution function of the ith species of charged particles, e_i is the charge and m_i is the mass of each particle of this species; $\mathbf{E}, \mathbf{H}, \mathbf{j}$ and q are, respectively, the electric field, magnetic field, current vector per unit volume and charge density, and c is the vacuum speed of light; $\mathbf{j}$ and q are given by the following integrals:

$$q = \sum_i e_i \iiint f_i \, d\mathbf{v}, \mathbf{j} = \sum_i e_i \iiint_{-\infty}^{\infty} \mathbf{v} f_i \, d\mathbf{v}, \tag{10.2.5}$$

$\mathbf{F}_i$ is the Lorentz force per unit mass for the ith species of particles; so

$$\mathbf{F}_i = \frac{e_i}{m_i} \mathbf{E} + \frac{e_i}{m_i c} (\mathbf{v} \times \mathbf{H}) \tag{10.2.6}$$

The Vlasov equations are a system of integro-differential equations in the seven coordinates $\mathbf{r}$, $\mathbf{v}$, t; in these the short range collisions are neglected. Since there are many particles in a Debye sphere, the assumption $N\lambda_D^3 >> 1$ is implied where N is the number density of particles and λ_D is the Debye shielding distance. So the equations are valid not only for not too dense plasmas, but also for sufficiently dense plasmas, with perfect randomness, which is necessary for the total neglect of the collision term. With the help of these equations, the particle motions and the electromagnetic field can be treated in a self-consistent manner. The Lorentz force, acting on any one particle, is assumed continuous and slowly varying; so it represents the effect of all other particles on it. Hence, in Vlasov plasmas, the collective effect of the distant particles is much greater than any effect due to collisions of a small number of nearby particles, whose fields are rather rapidly varying. For averaged distributions and fields, the distribution function $f(\mathbf{r}, \mathbf{v}, t)$ is regarded as the density of a continuous fluid in the six-dimensional phase space of position and velocity coordinates. Then effects of Debye shielding of charges, plasma oscillations, several types of waves and instabilities and other types of response of the plasma to applied electromagnetic field, giving the dielectric constant as a functon of the field frequency, are studied with the help of the system of Boltzmann-Vlasov equations.

10.3 LINEARIZATION OF VLASOV EQUATIONS

The fully ionized plasma is a mixture of electrons and ions only. If $f_e(\mathbf{r}, \mathbf{v}, t)$ and $f_i(\mathbf{r}, \mathbf{v}, t)$ are their distributions, then $-e$ and m are the charge and mass of an electron, e and M are the same for the ions, and the equations of Vlasov plasma are

$$\frac{\partial f_e}{\partial t} + (\mathbf{v} \cdot \nabla_{\mathbf{r}}) f_e - \frac{e}{m} (\mathbf{E} \cdot \nabla_{\mathbf{v}}) f_e - \frac{e}{mc} (\mathbf{v} \times \mathbf{H}) \cdot \nabla_{\mathbf{v}} f_e = 0, \tag{10.3.1}$$

$$\frac{\partial f_i}{\partial t} + (\mathbf{v} \cdot \nabla_{\mathbf{r}}) f_i + \frac{e}{M} (\mathbf{E} \cdot \nabla_{\mathbf{v}}) f_i + \frac{e}{Mc} (\mathbf{v} \times \mathbf{H}) \cdot \nabla_{\mathbf{v}} f_i = 0, \tag{10.3.2}$$

$$\nabla \times \mathbf{E} = -\frac{1}{c} \frac{\partial \mathbf{H}}{\partial t}, \ \nabla \cdot \mathbf{H} = 0, \tag{10.3.3}$$

$$\nabla \times \mathbf{H} = \frac{1}{c} \frac{\partial \mathbf{E}}{\partial t} + \frac{4\pi}{c} \iiint \mathbf{v} (f_i - f_e) \, d\mathbf{v}, \tag{10.3.4}$$

$$\nabla \cdot \mathbf{E} = 4\pi e \iiint (f_i - f_e) \, d\mathbf{v} \tag{10.3.5}$$

Before perturbation a field-free, charge (macroscopic) free, and current (macroscopic) free equilibrium state is assumed to exist. These conditions are ensured in the assumption that the velocity distributions of particles of both species are Maxwellian:

$$f_e^0 = N_e \left(\frac{1}{2\pi s_e}\right)^{3/2} \exp\left(-\frac{v^2}{2s_e}\right),\ f_i^0 = N_i \left(\frac{1}{2\pi s_i}\right)^{3/2} \exp\left(-\frac{v^2}{2s_i}\right) \quad (10.3.6)$$

where $s_e = \kappa T_e/m,\ s_i = \kappa T_i/M$. Then

$$\nabla_{\mathbf{v}} f_e^0 = -\frac{\mathbf{v}}{s_e} f_e^0,\ \nabla_{\mathbf{v}} f_i^0 = -\frac{\mathbf{v}}{s_i} f_i^0, \quad (10.3.7)$$

$$(\mathbf{v} \times \mathbf{H}) \cdot \nabla_{\mathbf{v}} f_e^0 = -\frac{f_e^0}{s_e}((\mathbf{v} \times \mathbf{H}) \cdot \mathbf{v}) = 0, \text{ etc.} \quad (10.3.8)$$

and equations (10.3.1) and (10.3.2) reduce to

$$\frac{e f_e^0}{m s_e}(\mathbf{E}_0 \cdot \mathbf{v}) = 0,\ -\frac{e f_i^0}{M s_i}(\mathbf{E}_0 \cdot \mathbf{v}) = 0 \quad (10.3.9)$$

where $\mathbf{E}_0$ denotes the electric field in the equilibrium state. Since f_e^0 and f_i^0 are not zero for all values of $\mathbf{v}$, we obtain

$$\mathbf{E}_0 = 0,\ N_e = N_i = N_0 \text{ (say)}, \quad (10.3.10)$$

$$\iiint_{-\infty}^{\infty} \mathbf{v}\,(f_i^0 - f_e^0)\, d\mathbf{v} = 0,\ \iiint_{-\infty}^{\infty} (f_i^0 - f_e^0)\, d\mathbf{v} = 0 \quad (10.3.11)$$

So the Maxwell distribution for electrons and ions ensure zero electric field, zero net current and zero net charge. If such a state is possible in the presence of a magnetic field $\mathbf{H}_0(\mathbf{r})$, then the above equations give

$$\nabla \times \mathbf{H}_0 = 0,\ \nabla \cdot \mathbf{H}_0 = 0,\ \frac{\partial \mathbf{H}_0}{\partial t} = 0 \quad (10.3.12)$$

These relations allow a constant magnetic field in the equilibrium state. In the presence of small perturbations the distribution functions are written as

$$f_e^0(v^2) + f_e(\mathbf{r}, \mathbf{v}, t),\ f_i^0(v^2) + f_i(\mathbf{r}, \mathbf{v}, t) \quad (10.3.13)$$

where $|f_e(\mathbf{r}, \mathbf{v}, t)| << f_e^0,\ |f_i(\mathbf{r}, \mathbf{v}, t)| << f_i^0$, the electric field is $\mathbf{E}(\mathbf{r}, t)$ and the magnetic field is $\mathbf{H}_0(\mathbf{r}) + \mathbf{H}(\mathbf{r}, t)$. Retaining only terms up to the first power of the perturbed quantities $f_e(\mathbf{r}, \mathbf{v}, t)$, $f_i(\mathbf{r}, \mathbf{v}, t)$, $\mathbf{E}(\mathbf{r}, t)$, $\mathbf{H}(\mathbf{r}, t)$, we obtain the linearized system of Vlasov equations

$$\frac{\partial f_e}{\partial t} + (\mathbf{v} \cdot \nabla_{\mathbf{r}}) f_e - \frac{e}{m}(\mathbf{E} \cdot \nabla_{\mathbf{v}}) f_e^0 - \frac{e}{mc}((\mathbf{v} \times \mathbf{H}_0)\, \nabla_{\mathbf{v}}) f_e = 0, \quad (10.3.14)$$

$$\frac{\partial f_i}{\partial t} + (\mathbf{v} \, . \, \nabla_\mathbf{r}) f_i + \frac{e}{M} (\mathbf{E} \, . \, \nabla_\mathbf{v}) f_i^0 + \frac{e}{Mc} ((\mathbf{v} \times \mathbf{H}_0) \, . \, \nabla_\mathbf{v}) f_i = 0 \quad (10.3.15)$$

$$\nabla \times \mathbf{E} = -\frac{1}{c} \frac{\partial \mathbf{H}}{\partial t}, \quad (10.3.16)$$

$$\nabla \times \mathbf{H} = \frac{1}{c} \frac{\partial \mathbf{E}}{\partial t} - \frac{4\pi e}{c} \iiint \mathbf{v} \, (f_i - f_e) \, d\mathbf{v}, \quad (10.3.17)$$

$$\nabla \, . \, \mathbf{H} = 0, \nabla \, . \, \mathbf{E} = 4\pi e \iiint (f_i - f_e) \, d\mathbf{v} \quad (10.3.18)$$

Non-Maxwellian equilibrium distributions $\psi_n(\mathbf{v}), n = 1, 2, ..., \infty$ can be employed, if these satisfy the two basic requirements of zero charge density and zero current density:

$$\iiint (\psi_n^e - \psi_n^i) \, d\mathbf{v} = 0, \iiint \mathbf{v} \, \{\psi_n^e - \psi_n^i\} \, d\mathbf{v} = 0 \quad (10.3.19)$$

In fact the anisotropic velocity distributions obtained from the product of any isotropic velocity distribution (e.g., the Maxwell distribution) and $v_x^l v_y^m v_z^n$ (where, at least, two of the numbers l, m, n are positive and odd integers, and v_x, v_y, v_z are the components of $\mathbf{v}$ along three mutually perpendicular directions) satisfy the two conditions of (10.3.11) and make the static pressure anisotropic.

10.4 DISPERSIVE SMALL AMPLITUDE OSCILLATIONS

Ions being heavy compared with the electrons, their perturbation is neglected here. Then equation (10.3.11) ensures that $N_e = N_i = N_0$. Denoting, therefore, both the equilibrium distributions $f_e^0(v^2)$ and $f_i^c(v^2)$ by $f_0(v^2)$, and removing the subscript e from $f_e(\mathbf{r}, \mathbf{v}, t)$ we find that, in the simple case of zero applied dc magnetic field, equations (10.3.14)-(10.3.18) reduce to

$$\frac{\partial f}{\partial t} + (\mathbf{v} \, . \, \nabla_\mathbf{v}) f - \frac{e}{m} (\mathbf{F} \, . \, \nabla_\mathbf{v}) f_0 = 0, \quad (10.4.1)$$

$$\nabla \times \mathbf{E} = -\frac{1}{c} \frac{\partial \mathbf{H}}{\partial t}, \quad (10.4.2)$$

$$\nabla \times \mathbf{H} = \frac{1}{c} \frac{\partial \mathbf{E}}{\partial t} - \frac{4\pi e}{c} \iiint \mathbf{v} f \, d\mathbf{v}, \quad (10.4.3)$$

$$\nabla \, . \, \mathbf{E} = -4\pi e \iiint f \, d\mathbf{v}, \; \nabla \, . \, \mathbf{H} = 0 \quad (10.4.4)$$

where $\mathbf{F}$ is the Lorentz force. From (10.3.6) and (10.3.10) we find that

$$f_0(v^2) = N_0 \left(\frac{1}{2\pi s} \right)^{3/2} \exp\left(-\frac{v^2}{2s} \right)$$

with $s = K_B T/m$. To solve these equations we assume the space-time dependence $\exp(i(\mathbf{K}\,.\,\mathbf{r}) - i\omega t)$ for the variables $f(\mathbf{r}, \mathbf{v}\,t)$, $\mathbf{E}(\mathbf{r}, t)$, $\mathbf{H}(\mathbf{r}, t)$. Particularly, putting

$$f(\mathbf{r}, \mathbf{v}, t) = g(\mathbf{v}) \exp(i(\mathbf{K}\,.\,\mathbf{r}) - i\omega t), \tag{10.4.5}$$

equations (10.4.1)-(10.4.4) reduce to

$$i((\mathbf{K}\,.\,\mathbf{v}) - \omega) f - \frac{e}{m} (\mathbf{E}\,.\,\nabla_{\mathbf{v}}) f_0 = 0, \tag{10.4.6}$$

$$(\mathbf{K} \times \mathbf{E}) = \frac{\omega}{c} \mathbf{H}, \tag{10.4.7}$$

$$(\mathbf{K} \times \mathbf{H}) = -\frac{\omega}{c} \mathbf{E} + \frac{4\pi i e}{c} \iiint \mathbf{v} f\, d\mathbf{v}, \tag{10.4.8}$$

$$(\mathbf{K}\,\mathbf{E}) = 4\pi i e \iiint f\, d\mathbf{v},\ (\mathbf{K}\,.\,\mathbf{H}) = 0 \tag{10.4.9}$$

We now define three mutually perpendicular coordinate axes OX_1, OX_2, OX_3, such that OX_3 is parallel to the wave vector $\mathbf{K}$, and OX_1 and OX_2 are perpendicular to $\mathbf{K}$. Then equations (10.4.6)-(10.4.9) reduce to

$$f(\mathbf{r}, \mathbf{v}, t) = -\frac{ie}{m} \left(E_1 \frac{\partial f_0}{\partial v_1} + E_2 \frac{\partial f_0}{\partial v_2} + E_3 \frac{\partial f_0}{\partial v_3} \right) \frac{1}{kv_3 - \omega}, \tag{10.4.10}$$

$$-KE_2 = \frac{\omega}{c} H_1,\ KE_1 = \frac{\omega}{c} H_2, \tag{10.4.11}$$

$$-KH_2 = -\frac{\omega}{c} E_1 + \frac{4\pi i e}{c} \iiint v_1 f\, d\mathbf{v}, \tag{10.4.12}$$

$$KH_1 = -\frac{\omega}{c} E_2 + \frac{4\pi i e}{c} \iiint v_2 f\, d\mathbf{v}, \tag{10.4.13}$$

$$E_3 = \frac{4\pi i e}{K} \iiint f\, d\mathbf{v} = \frac{4\pi i e}{\omega} \iiint v_3 f\, d\mathbf{v}, \tag{10.4.14}$$

$$E_j = \frac{4\pi i e\, \omega}{\omega^2 - K^2 c^2} \iiint v_j f\, d\mathbf{v},\ H_3 = 0 \tag{10.4.15}$$

for $j = 1, 2$. The two expressions for E_3 in (10.4.14) give

$$\iiint v_3 f\, d\mathbf{v} = \frac{\omega}{K} \iiint f\, d\mathbf{v} \tag{10.4.16}$$

Putting the values of E_1, E_2, E_3 in (10.4.10) and cancelling $\exp(i\theta)$ where $\theta = (\mathbf{K}\,.\,\mathbf{r}) - \omega t)$ from both sides, we get

$$g(\mathbf{v}) = -\frac{ie}{m} \left[\frac{4\pi i e\, \omega}{\omega^2 - k^2 c^2} \left\{ \frac{\partial f_0}{\partial v_1} \iiint v_1 g\, d\mathbf{v} + \frac{\partial f_0}{\partial v_2} \iiint v_2 g\, d\mathbf{v} \right\} + \frac{4\pi i e}{K} \frac{\partial f_0}{\partial v_3} \iiint g\, d\mathbf{v} \right] \frac{1}{K v_3 - \omega} \tag{10.4.17}$$

Since $g(\mathbf{v})$ occurs both within and outside of the integration, equation (10.4.17) is an integro-differential equation. Also, since $\lim_{v_j \to \pm\infty} f_0(v^2) = 0$, it is evident that

$$\iiint \frac{\partial f_0}{\partial v_j} \frac{d\mathbf{v}}{K v_3 - \omega} = 0 \text{ with } j = 1, 3 \tag{10.4.18}$$

(a) Longitudinal oscillation in electron plasma

Integrating all terms of (10.4.17) with respect to $\mathbf{v}$ we get

$$\iiint g(\mathbf{v})\, d\mathbf{v} = 0 + \frac{\omega_\rho^2}{KN_0} \left(\iiint g\, d\mathbf{v} \right) \left(\iiint \frac{\partial f_0}{\partial v_3} \frac{d\mathbf{v}}{Kv_3 - \omega} \right) \tag{10.4.19}$$

This equation can be written as

$$\left(\iiint g\, d\mathbf{v} \right) \left\{ 1 - \frac{\omega_\rho^2}{KN_0} \left(\iiint \frac{\partial f_e}{\partial v_3} \frac{d\mathbf{v}}{Kv_3 - \omega} \right) \right\} = 0 \tag{10.4.20}$$

where $\omega_\rho^2 = 4\pi N_0 e^2/m$. When $(\iiint g\, d\mathbf{v}) \neq 0$ we obtain

$$\frac{KN_0}{\omega_\rho^2} = \iiint \frac{\partial f_0}{\partial v_3} \frac{d\mathbf{v}}{Kv_3 - \omega} \tag{10.4.21}$$

In this relation between the frequency ω and wave numbers K of a wave, note that the integrand, on the right hand side, is independent of v_1 and v_2, so we write v for v_3 and put

$$\iint_{-\infty}^{\infty} f_0(v^2)\, dv_1\, dv_2 = N_0 F(v) \tag{10.4.22}$$

where

$$F(v) = \left(\frac{1}{2\pi s} \right)^{3/2} \iint_{-\infty}^{\infty} \exp\left(- \frac{v_1^2 + v_2^2 + v^2}{2s} \right) dv_1\, dv_2$$

$$= \left(\frac{1}{2\pi s} \right)^{1/2} \exp\left(- \frac{v^2}{2s} \right) \text{ with } s = \frac{\kappa T}{m} \tag{10.4.22 (a)}$$

Hence the relation (10.4.21) reduces to the singular integro-differential equation

$$\frac{K\omega}{\omega_\rho^2} = - \int_{-\infty}^{\infty} \frac{\partial F}{\partial v} \frac{dv}{1 - Kv/\omega}, \qquad (10$$

Evidently, the velocity components v_1 and v_2, perpendicu
tion of wave propagation, are not important for analysis of l

sive waves. The integral of (10.4.23) has a singularity at $v = w/k$; so, there exists no definite dependence of w on k at all, and for a given value of k arbitrary values of w are possible.

(b) Landau damping of waves in electron plasma

To solve equation (10.4.23) which has a singularity at $v = w/k$ we use the following results of complex variables,

$$\frac{1}{x+i\,.\,0} = P\left(\frac{1}{x}\right) - i\pi\delta\,(x),$$

$$\int_{-\infty}^{\infty} \frac{\varphi\,(t)\,dt}{t-t_0} = P\int_{-\infty}^{\infty} \frac{\varphi\,(t)\,dt}{t-t_0} + \pi i\varphi\,(t_0) \tag{10.4.24}$$

where P denotes the principal value, t is a point on the real axis and $\varphi\,(t)$ is a continuous function of t. Then the relation (10.4.23) becomes

$$\frac{K\omega}{\omega_\rho^2} = -P\int_{-\infty}^{\infty} \frac{\partial F}{\partial v}\,\frac{dv}{1-Kv/\omega} \pm \pi i\frac{\omega}{K}F'\left(\frac{\omega}{K}\right) \tag{10.4.25}$$

where $f'\,(v) = \partial F/\partial v$. For waves of large wavelength K is small; and for computing the principal value of the integral we expand $(1-(Kv/w)-1$ in positive integral powers of kv/ω and get

$$\frac{K\omega}{\omega_\rho^2} = -\int \frac{\partial F}{\partial v}\left\{1+\frac{Kv}{\omega}+\frac{K^2\,v^2}{\omega^2}+\ldots\right\}dv + \pi i\frac{\omega}{K}F'\left(\frac{\omega}{K}\right)$$

Retaining terms up to Kv/ω, in the series, we obtain

$$\frac{K\omega}{\omega_\rho^2} = -\int\frac{\partial F}{\partial v}\,dv - \frac{K}{\omega}\int v\,\frac{\partial F}{\partial v}\,dv + \pi i\frac{\omega}{k}F'\left(\frac{\omega}{k}\right) = \frac{K}{\omega} + \pi i\frac{\omega}{K}F'\left(\frac{\omega}{K}\right) \tag{10.4.26}$$

because $\int_{-\infty}^{\infty} F\,(v)\,dv = 1$. Hence, the relation (10.4.25) becomes

$$1 = \frac{\omega_\rho^2}{\omega^2} \pm \frac{\pi i\omega_\rho^2}{K^2}F'\left(\frac{\omega}{K}\right) \tag{10.4.27}$$

If $\lim\limits_{\to\omega/k} F'\,(v) = 0$ then $\omega^2 = \omega_\rho^2$. This is the cold plasma approximation and gives a real value for ω. When $F'\,(\omega/K) \neq 0$ we find that ω is complex for a real value of the wave number K. Since $F\,(v)$ is Maxwellian, equation (10.4.27) gives

$$1=\frac{\omega_\rho^2}{\omega^2}\pm\pi i\left(\frac{\omega_\rho^2}{K^2 s}\right)\left(\frac{\omega}{K}\right)F\left(\frac{\omega}{K}\right) \tag{10.4.28}$$

In the first approximation, put $\omega=\omega_\rho-i\gamma$ where γ is real; equating the coefficients of the imaginary quantity i from both sides, we obtain

$$\gamma=\frac{\pi\omega_\rho^4}{2k^3 s}\left(\frac{1}{2\pi s}\right)^{1/2}\exp\left(-\frac{\omega_\rho^2}{2k^2 s}\right) \tag{10.4.29}$$

This is the Landau damping coefficient for electron acoustic waves. Earlier, Vlasov took only the principal value of the integral (10.4.23), and so could not predict this damping effect. We can write

$$\gamma=\left(\frac{\pi}{8}\right)^{3/2}\frac{\omega_\rho}{(K\lambda_D)^3}\exp\left(-\frac{K^2}{2}\lambda_D^2-\frac{3}{2}\right)$$

where $\lambda_D=(=K_B\,T/4\pi\,N_0\,e^2)^{1/2})$ is the Debye length. Introducing the characteristic time scale t_0 $(=1/\omega)$ of oscillation, and the time scale t_e $(1/\gamma)$, of Landau damping, since $K^2\,\lambda_D^2<<1$, it is clear that $t_0<<t_e$. For example, if $t_0=10^{-10}$ sec. and $K\,\lambda_D=1/10$, we find that $t_e=10^{19},\ t_0=10^9$ sec (= 30 years).

This is a collisionless damping, and is due to the resonant transfer of energy from waves to the small group of particles moving with velocity near the phase velocity v_P $(=w/k)$ of the wave. In the fluid approximation and for cold plasmas this effect is negligible, because, then, thermal motions are, a priori, averaged out. The energy of a wave of a specified wave-length also diffuses into other wavelengths by collisional damping.

In the neighbourhood of the wave phase velocity v_P the particles travelling faster loose energy to the wave and the particles travelling slower gain energy from the wave. Not many particles are available at velocities near v_P, because the number of particles having high velocities decreases to zero as the velocity tends to the value C for relativistic reasons. So, there would be more number of particles having velocities less than v_P (taking energy from the wave), than having velocities more than v_P (giving energy to the wave); and the wave suffers a net loss of energy, instead of gaining from this resonant and near resonant collisionless energy exchange. This is because the number of snatchers is more than the number of donors to the wave. In other words, the condition for this damping is

$$N(v_P+\delta v)<N(v_P)\langle N(v_P-\delta v)\text{ and }N(v_P+\delta v)\rangle N(v_P)\rangle N(v_P-\delta v)$$

for amplification, where N is the electron number density, where for $v_P>0$, $\delta v>0$, $v_P+\delta u>u>v_p-\delta u$.

After the bounce time ($=\sqrt{m/ek\,E}$, where m and e are the electronic mass and charge respectively, k is the wave number and E is the amplitude of the wave electric field), due to this energy gain and loss, the number of particles having more velocities than v_{PK} increases and the number having less decreases. Thus the situation becomes reverse, causing particle trapping and other effects of instability due to field amplification, which cannot be treated mathematically in a simple manner.

(c) Transverse oscillations

Multiplying all terms of (10.4.17) with v_i $(i = 1, 2)$ and then integrating them with respect to all the components of **v**, we get the dispersion relation

$$\left(\iiint v_i\, g(\mathbf{v})\, d\mathbf{v}\right)\left\{1-\frac{\omega\,\omega_\rho^2}{N_0\,(\omega^2-K^2c^2)}\iiint\frac{\partial f_0}{\partial v_i}\,\frac{v_i\, d\mathbf{v}}{K\,v_3-\omega}\right\}=0 \quad (10.4.30)$$

for electromagnetic wave field. Integrating the right hand side of this equation by parts and using (10.4.22) gives

$$\frac{(\omega^2-K^2c^2)}{\omega_\rho^2}=\int_{-\infty}^{\infty}\frac{F(v)\,dv}{1-(Kv/\omega)} \quad (10.4.31)$$

Divergence of this integral, due to the singularity at $v = \omega/k$, has no significance, because the relativistic limit of particle velocity is c, whereas the phase velocity $\omega/K\,(=\sqrt{c^2+(\omega_\rho^2/K^2)}) > c$.

(d) Elliptically polarized waves

For elliptically polarized transverse waves we write

$$f(\mathbf{r}, \mathbf{v}, t) = g_1(\mathbf{v})\cos\theta + g_2(v)\sin\theta \quad (10.4.33)$$

where the argument $\quad \theta = (\mathbf{K}\,.\,\mathbf{r}) - \omega t \quad (10.4.34)$

The integrals $\iiint_{-\infty}^{\infty} g_1(\mathbf{v})\, d\mathbf{v}$ and $\iiint_{-\infty}^{\infty} g_2(\mathbf{v})\, d\mathbf{v}$ are the two first order amplitudes of propagation of the perturbation in number density in the form of the cosine wave and the sine wave. The components of the integrals

$$\iiint_{-\infty}^{\infty}\mathbf{v}\, g_1(\mathbf{v})\, d\mathbf{v} \text{ and } \iiint_{-\infty}^{\infty}\mathbf{v}\, g_2(\mathbf{v})\, d\mathbf{v}$$

are proportional to the amplitude of the first order current waves. Calculations yield

$$E_3=\frac{4\pi e}{K}\left\{\cos\theta\iiint g_2\, d\mathbf{v}-\sin\theta\iiint g_1\, d\mathbf{v}\right\}, \quad (10.4.35)$$

$$E_1 = \frac{4\pi e\,\omega}{\Omega^2}\left\{\sin\theta\iiint v_1\,g_1\,d\mathbf{v} - \cos\theta\iiint v_1\,g_2\,d\mathbf{v}\right\}, \tag{10.4.36}$$

$$E_2 = \frac{4\pi e\,\omega}{\Omega^2}\left\{\sin\theta\iiint v_2\,g_1\,d\mathbf{v} - \cos\theta\iiint v_2\,g_2\,d\mathbf{v}\right\} \tag{10.4.37}$$

where $\qquad \Omega^2 = \omega^2 - K^2 c^2 \qquad$ (10.4.38)

To ensure that the wave is elliptically polarized and the two principal directions of the polarization ellipse are the x-axis and the y-axis, without any loss of generality, we can assume

$$\iiint v_1\,g_2\,d\mathbf{v} = 0, \quad \iiint v_2\,g_1\,d\mathbf{v} = 0 \tag{10.4.38 (a)}$$

and find that equations (10.4.36) and (10.4.37) reduce to $E_1 = E_1^0 \sin\theta$, $E_2 = E_2^0 \cos\theta$ where E_1^0 and E_2^0 are the amplitudes of the electric field parallel to the coordinate axes OX_1 and OX_2 respectively. We call E_1 as the electric field of the $\perp_1$ polarization (or perpendicular-one polarization) and E_2 as that of the $\perp_2$ polarization (or perpendicular-two polarization). Then relations (10.4.36) and (10.4.37) give

$$(E_1^0,\ E_2^0) = \frac{4\pi e\,\omega}{\Omega^2}\left\{\iiint v_1\,g_1\,d\mathbf{v},\ -\iiint v_2\,g_2\,d\mathbf{v}\right\} \tag{10.4.39}$$

The integrals $\iiint v_1\,g_1\,d\mathbf{v}$ and $\iiint v_2\,g_2\,d\mathbf{v}$ are the amplitudes of the currents of particles along the fields of $\perp_1$ and $\perp_2$ polarizations, respectively. Also, relations (10.4.39) show that these integrals are proportional to the amplitudes E_1^0 and E_2^0. Putting $E_3 = E_{l_1}\sin\theta + E_{l_2}\cos\theta$ in (10.4.35) we get

$$(E_{l_1},\ E_{l_2}) = \frac{4\pi e}{K}\left\{-\iiint g_1\,d\mathbf{v},\ \iiint g_2\,d\mathbf{v}\right\} \tag{10.4.40}$$

where E_{l_1} and E_{l_2} are the amplitudes of longitudinal field of the sine wave and the cosine wave respectively.

Using the value of $f(\mathbf{r}, \mathbf{v}, t)$ from (10.4.33) and E_1, E_2, E_3 from (10.4.35)-(10.4.39), in equation (10.4.1), equating the coefficients of $\sin\theta$ and $\cos\theta$ from both sides of this equation, dividing by $kv_3 - \omega$ and assuming, for the time being, that $v_3 = \omega/k$, we get

$$g_i(\mathbf{v}) = \left(\frac{\omega_\rho^2}{N_0\,s}\right)\left(\frac{f_0(v^2)}{\theta}\right)\left\{\frac{v_3}{k}\iiint g_i\,d\mathbf{v} + \frac{\omega}{\Omega^2}\,v_i\iiint v_i\,g_i\,d\mathbf{v}\right\} \tag{10.4.41}$$

where $\qquad i = 1, 2 : \dot{\theta} = (\mathbf{K}\,.\,\mathbf{v}) - \omega = Kv - \omega,\ S = \frac{\kappa T}{m} \qquad$ (10.4.42) because v stands for v_3. It is clear that g_1 and g_2 consist of two parts, one of which gives the amplitude of the distribution function of the longitudinal field and the other of the transverse field. To see whether the two fields are inter related or not, we integrate g_i and $v_i g_i$ in the velocity space and get

$$\iiint g_i \, d\mathbf{v} \left\{ 1 - \frac{\omega_\rho^2}{N_0 K} \int \frac{\partial F}{\partial v} \frac{dv}{\dot{\theta}} \right\} = 0, \quad \iiint v_i \, g_i \, d\mathbf{v} \left\{ 1 - \frac{\omega_\rho^2 \, \omega}{N_0 \, \Omega^2} \int \frac{F dv}{\dot{\theta}} \right\} = 0$$

where $i = 1, 2$,and obtain the dispersion relations of (10.4.21) and (10.4.31).

With the help of (10.4.33) and (10.4.41) we can write

$$f(\mathbf{r}, \mathbf{v}, t) = -\frac{e}{m} \left[\frac{\partial f_0}{\partial v} \frac{1}{\dot{\theta}_l} (E_{l_1} \cos \theta_l - E_{l_2} \sin \theta_l) - \frac{1}{\dot{\theta}_t} \left(E_1^0 \frac{\partial f_0}{\partial v_1} \cos \theta - E_2^0 \frac{\partial f_0}{\partial v_2} \sin \theta \right) \right]$$

$$= \frac{e}{ms} \left[\frac{v}{\dot{\theta}_l} f_0 \, (E_{l_1} \cos \theta_l - E_{l_2} \sin \theta_l) + \frac{f_0}{\dot{\theta}_t} (v_1 \, E_1^0 \cos \theta_t - v_2 \, E_2^0 \sin \theta_t) \right] \quad (10.4.43)$$

where θ_l and θ_t are the arguments of the wave functoin of the longitudinal field and transverse field respectively. From (10.4.43) it is clear that particles, participating in one field, do not either participate in or interact with those of the other field; so the right hand side of (10.4.43) is merely the placement of the two independent parts having the same dimensions, with a plus sign, so long as one is confined to linearized processes only. Even perturbations of the two fields at different frequencies or at different wavelengths are possible in (10.4.43). We separate the two parts of the distribution function in the following manner :

$$f_l(\mathbf{r}, \mathbf{v}, t) = \frac{e}{m} \frac{\partial f_0}{\partial v} \cdot \frac{1}{\dot{\theta}_l} (E_{l_1} \cos \theta_l - E_{l_2} \sin \theta_l), \quad (10.4.44)$$

$$f_t(\mathbf{r}, \mathbf{v}, t) = -\frac{e}{m} \frac{1}{\dot{\theta}_t} \left\{ E_1^0 \frac{\partial f_0}{\partial v_1} \cos \theta_t - E_2^0 \frac{\partial f_0}{\partial v_2} \sin \theta_t \right\}, \quad (10.4.45)$$

$$g_{l_1} = \frac{e}{m} \frac{\partial f_0}{\partial v} \frac{E_{l_i}}{\dot{\theta}_l}, \quad g_{t_i} = -\frac{\omega_\rho^2}{N_0} \frac{\omega}{\Omega^2} \frac{1}{\dot{\theta}_t} \frac{\partial f_0}{\partial v_i} \iiint v_i \, g_t \, d\mathbf{v} \quad (10.4.46)$$

where $i = 1, 2$. We ignore the singularity in g_{t_i}, but not in g_{l_i}, due to the existence of Landau damping effect in longitudinal waves.

The pressure tensor, for the perturbed electron plasma, is

$$p_{ij} = m \iiint (v_i - u_i) \; (v_j - u_j) \; (f_0 + f) \, d\mathbf{v} \quad (10.4.47)$$

where $$u_i = \iiint v_i (f_0 + f)\, d\mathbf{v} / \iiint (f_0 + f)\, d\mathbf{v} \tag{10.4.48}$$

Hence, if δ_{ij} is the Kronecker delta of (A 3.17), we find that

$$p_{ij} = m\delta_{ij} \iiint v_i v_j f_0\, d\mathbf{v} + m \iiint v_i v_j f\, d\mathbf{v} - \frac{m \left(\iiint v_i f\, d\mathbf{v}\right) \left(\iiint v_j f\, d\mathbf{v}\right)}{N_0 + \iiint f\, d\mathbf{v}} \tag{10.4.49}$$

$$= m \left[\delta_{ij} \iiint v_i v_j f_0\, d\mathbf{v} + \iiint v_i v_j f\, d\mathbf{v} - \frac{1}{N_0} \left(\iiint v_i f\, d\mathbf{v} \right) \left(\iiint v_j f\, d\mathbf{v} \right) \right] \tag{10.4.50}$$

for the electromagnetic field.

10.5 RELAXATION FOR STEADY STATE

The collision term has the dimension of f/t_0 where $f(\mathbf{r}, \mathbf{v}, t)$ is the distribution function and t_0 has the dimension of time. The collision integral is zero for Maxwellian velocity distributions. Collisions destroy any orderly motion and establish the Maxwellian velocity distribution. For other distributions, an empirical relation, involving the concept of mean-free-time, is used. In the *BKG* model, the collision term is given by

$$\left(\frac{\partial f}{\partial t} \right)_{\text{coll}} = \frac{1}{t_0} \{ f_0(\mathbf{v}) - f(\mathbf{r}, \mathbf{v}, t) \} \tag{10.5.1}$$

where $f_0(\mathbf{v})$ is the Maxwell velocity distribution function in the equilibrium state and $t_0(\mathbf{v})$ is the velocity-dependent relaxation time. Actually T is the characteristic time of variation of f due to collisions, or the average time between two successive collisions. So from (A12.6) we write $t_0 = 1/(Nv\sigma)$ where v is the mean speed $(8K_B T/m\pi)^{1/2}$ of (9.8.8).

The evolution of the system to the equilibrium state, after a disturbance has swept through it, is given by the equation

$$\frac{\partial f}{\partial t} = \frac{1}{t_0} (f_0 - f) \tag{10.5.2}$$

because no external field of force exists after the switch-off, and the system is uniform in space. The solution of this equation is

$$f(\mathbf{v}, t) = f_0 + [f(\mathbf{v}, 0) - f_0(\mathbf{v})] \exp(-t/t_0) \tag{10.5.3}$$

where $f(\mathbf{v}, 0)$ is the distribution function at the time of the switch-off of the applied field (the initial time here) and $f_0(\mathbf{v})$ is the Maxwellian distribution.

For applied electric and magnetic fields, the electron motion can be studied with the help of the kinetic equation

$$\frac{\partial f}{\partial t}+\frac{1}{m}\left(\mathbf{F}\cdot\frac{\partial f}{\partial \mathbf{v}}\right)=\frac{1}{t_0}(f_0-f) \tag{10.5.4}$$

where the applied Lorentz force $\mathbf{F}=e\mathbf{E}+e(\mathbf{v}\times\mathbf{H})/c$. More generally, the collision term of (10.5.1) leads to the modified Boltzmann equation

$$\frac{\partial f}{\partial t}+(\mathbf{v}\,.\,\nabla_{\mathbf{r}})f+\frac{e}{m}\left(\mathbf{E}+\frac{1}{c}(\mathbf{v}\times\mathbf{B})\right)\cdot\nabla_{\mathbf{v}}f=\frac{1}{t_0}(f_0-f) \tag{10.5.5}$$

for each species of charged particles. This representation does not lead to any ambiguity for transverse waves; but it is not acceptable for longitudinal waves because the condition for charge conservation, $\iiint \frac{Df}{Dt}\,dv=0$, is not satisfied, where $D/Dt=\partial/\partial t+(\mathbf{v}\,.\,\nabla_{\mathbf{r}})+(\mathbf{F}/m)\,.\,\nabla_{\mathbf{v}}$. For longitudinal waves, evidently $\iiint \frac{Df}{Dt}\,d\mathbf{v}=\iiint \frac{1}{t_0(v)}(f_0-f)\,d\mathbf{v}\neq 0$, and integration of (10.5.5) with respect to $\mathbf{v}$ and, following the deduction of (9.10.3) from (9.10.1), gives

$$\frac{\partial N}{\partial t}+\nabla\,.\,(N\mathbf{u})=\iiint \frac{1}{t_0}(f_0-f)\,d\mathbf{v} \tag{10.5.6}$$

The right hand side of this relation is a source (or sink) of particles in a volume element, because $\nabla\,.\,(N\mathbf{u})$ gives the change in $N(\mathbf{r},t)$ due to particles entering or leaving that element. The *B K G* model gives useful results.

10.6 THE SPHERICAL HARMONIC EXPANSION OF THE DISTRIBUTION FUNCTION

When there is no drift, the distribution of electrons is approximately spherically symmetric and Maxwellian in velocity space; the equilibrium is disturbed only by spatial inhomogeneities, temperature gradients and external forces. Also, the electron distribution function should contain terms depending on deviations from sphericity. To find this deviation we assume that v,θ,φ are spherical polar coordinates in velocity space, where θ is the angle measured from the v_x axis and φ is the angle measured from a fixed plane through the v_z axis (the $v_x\,v_z$ plane). Assuming the electron distribution function to be symmetric about the direction of the electric field $\mathbf{E}$, it is expanded as a series of Legendre polynomials.

$$f(\mathbf{r},\mathbf{v}\,t)=\sum_{m=0}^{\infty} P_m(\cos\theta)f_m(\mathbf{r},\mathbf{v},t) \tag{10.6.1}$$

where θ is the angle between $\mathbf{E}$ and $\mathbf{v}$, and v_z is parallel to the direction of $\mathbf{E}$. The expansion is equivalent to writing

$$f(\mathbf{r},\mathbf{v},t)=f_0(\mathbf{r},\mathbf{v},t)+\frac{1}{v}(\mathbf{v}\,.\,\mathbf{f}_1)+\frac{\mathbf{v}}{v^2}\mathbf{v}\,.\,\mathbf{f}_2+ \tag{10.6.2}$$

where $\mathbf{f}_1$ is parallel to $\mathbf{E}$. Hence $(\mathbf{E}\,.\,\mathbf{f}_1) = |\mathbf{E}|\ |\mathbf{f}_1|$ and $(\mathbf{v}\,.\,\mathbf{f}_1) = |v|\ |f_1|$ and $\mathbf{f}_1, \mathbf{f}_2$ are functions of position coordinates, time, speed and the unperturbed distribution f_0. The distribution function $f(\mathbf{r}, \mathbf{v}, t)$ of (10.6.1) is expected to converge because the average relative loss of energy by an electron with molecules and ions is small compared to the thermal energy for small applied fields and gradients.

Electrons remain in thermal equilibrium at all points, even in the presence of very weak electric fields, no spatial gradients and no sources and sinks of them. So, the thermodynamic variables may be spatially varying. If f_0 is the local Maxwellian, then

$$f_0(\mathbf{r}, \mathbf{v}, t) = N_0(\mathbf{r}) \left(\frac{m}{2\pi\, s_e}\right)^{3/2} \exp\left[-\frac{m}{2s_e}\mathbf{v}^2\right] \tag{10.6.3}$$

where $s_e = \kappa T_e(\mathbf{r})$. Generally $\mathbf{u}$ is neglected, because the flow velocity of electrons is smaller than the electron thermal velocity.

The first two terms, in the right hand side of (10.6.2), can be written as $f_0 + \frac{v_j}{v} f_1$, where f_{1j} is the jth component of the vector $\mathbf{f}_1$, and the Einstein summation convention (repeated index means summation over that index) is implied, using j as the dummy subscript. So the first two terms of the left hand side of (10.7.1) become

$$\frac{\partial}{\partial t}\left(f_0 + \frac{v_j}{v} f_{1j}\right) + v_i \frac{\partial}{\partial x_i}\left(f_0 + \frac{v_j}{v} f_{ij}\right) \equiv \frac{\partial f_0}{\partial t} + \frac{v_j}{v}\frac{\partial f_{1j}}{\partial t} + v_i \frac{\partial f_0}{\partial x_i} + \frac{v_i}{v} v_j \frac{\partial f_{1j}}{\partial x_i}, \tag{10.6.4}$$

provided the first two terms of the expansion of (10.6.8) are retained. The third term of the left hand side of (10.7.1) is then given by

$$\frac{e}{m} E_i \frac{\partial}{\partial v_i}\left(f_0 + \frac{v_j}{v} f_{1j}\right) = -\frac{e}{m} E_i \frac{\partial f_0}{\partial v_i} - \frac{e}{m} E_i \frac{\partial}{\partial v_i}\left(\frac{v_j}{v} f_{1j}\right) \tag{10.6.5}$$

Since
$$\frac{\partial}{\partial v_i} = \frac{\partial v}{\partial v_i}\frac{\partial}{\partial v} = \frac{v_i}{v}\frac{\partial}{\partial v}, \tag{10.6.6}$$

the third term becomes

$$= \frac{e}{m} E_i \frac{\partial}{\partial v_i}\left(f_0 + \frac{v_j}{v} f_{1j}\right) = -\frac{e}{m} E_i \frac{v_i}{v}\frac{\partial f_0}{\partial v} - \frac{e}{m} E_i \left\{\frac{f_{1j}}{v} + v_j \frac{\partial}{\partial v_i}\left(\frac{f_{1j}}{v}\right)\right\}$$

$$= -\frac{e}{m} E_i \frac{v_i}{v}\frac{\partial f_0}{\partial v} - \frac{e}{m} E_i \left\{f_{1i} + v_i v_j \frac{\partial}{\partial v}\left(\frac{f_{1j}}{v}\right)\right\} \tag{10.6.7}$$

The fourth term in the left hand side of (10.7.1)

$$= -\frac{e}{me}\,\varepsilon_{ijk}\,v_j\,B_k\,\frac{\partial}{\partial v_i}\left(f_0+\frac{v_l}{v}f_{1e}\right) = -\,\varepsilon_{ijk}\,\Omega_k\frac{v_j}{v}f_{1i} - \varepsilon_{ijk}\,\Omega_k\frac{v_i}{v}\,v_i\,v_j\frac{\partial}{\partial v}\left(\frac{f_{1l}}{v}\right)$$

$$= -\,\varepsilon_{ijk}\,\Omega_k\frac{v_j}{v}f_{1l} - \Omega_k\,\varepsilon_{ijk}\,v_i\,v_j\frac{v_l}{v}\frac{\partial}{\partial v}\left(\frac{f_{1l}}{v}\right) = -\,\varepsilon_{ijk}\,\Omega_K\frac{v_j}{v}f_{1i} - 0 \qquad (10.6.8)$$

because ε_{ijk} is the alternative unit tensor of (A 4.3) and $\Omega_k = eB_k/me$. Hence (10.7.1) becomes

$$\frac{\partial f_0}{\partial t}+\frac{v_j}{v}\frac{\partial f_{1j}}{\partial t}+v_i\frac{\partial f_0}{\partial x_i}+\frac{v_i}{v}v_j\frac{\partial f_{1j}}{\partial x_i}-\frac{ev_i}{mv}E_i\frac{\partial f_0}{\partial v}-\frac{eE_i}{mv}f_{1i}$$

$$-\frac{eE_i}{mv}v_i\,v_j\frac{\partial}{\partial v}\left(\frac{f_{1j}}{v}\right)-\varepsilon_{ijk}\frac{v_k\,\Omega_k}{v}f_{1i}=\left(\frac{\partial f}{\partial t}\right)_{\text{coll}} \qquad (10.6.9)$$

Multiplying all terms of this equation by the solid angle element sin θ $d\theta$ $d\varphi$, in velocity space, and integrating with respect to θ between 0 and π, and with respect to φ between 0 and 2π, we evalaute them. The resultant equation is (10.6.22). The first term of this equation is

$$\int_0^{\pi}\int_0^{2\pi}\frac{\partial f_0}{\partial t}\sin\theta\,d\theta\,d\varphi = 4\pi\frac{\partial f_0}{\partial t} \qquad (10.6.10)$$

The second term

$$=\int_0^{\pi}\int_0^{2\pi}\frac{1}{v}\frac{\partial}{\partial t}(\mathbf{v}f_1)\sin\theta\,d\theta\,d\varphi = \frac{\partial f_1}{\partial t}\int_0^{\pi}\int_0^{2\pi}\cos\theta\sin\theta\,d\theta\,d\varphi = 0$$

$$(10.6.11)$$

In velocity space the spherical polar system of coordinates is chosen, such that

$$v_1 = v\sin\theta\cos\varphi,\ v_2 = v\sin\theta\sin\varphi,\ v_3 = v\cos\theta, \qquad (10.6.12)$$

so the third term

$$=\int_0^{\pi}\int_0^{2\pi}\left(v_1\frac{\partial f_0}{\partial x}+v_2\frac{\partial f_0}{\partial y}+v_3\frac{\partial f_0}{\partial z}\right)\sin\theta\,d\theta\,d\varphi = 0, \qquad (10.6.13)$$

and the fourth term can be written as

$$=\int_0^{\pi}\int_0^{2\pi}\frac{v_i}{v}v_j\frac{\partial f_{1j}}{\partial x_i}\sin\theta\,d\theta\,d\varphi = M_{ij}\frac{1}{v}\frac{\partial f_{1j}}{\partial x_i} \qquad (10.6.14)$$

Then we find that

$$M_{11} = \int_0^{\pi}\int_0^{2\pi} v_1 \sin\theta \, d\theta \, d\varphi = v \int_0^{\pi}\int_0^{2\pi} \cos^2\varphi \sin^3\theta \, d\theta \, d\varphi$$

$$= \frac{4\pi}{3} v = M_{22} = M_{33}, \tag{10.6.15}$$

$$M_{12} = \int_0^{\pi}\int_0^{2\pi} v_1 v_2 \sin\theta \, d\theta \, d\varphi = 0$$

Therefore, the fourth term of the left hand side of (10.6.9)

$$= \frac{4\pi v}{3} \frac{\partial f_{1i}}{\partial x_i} = \frac{4\pi}{3} v (\nabla . \mathbf{f}_1) \tag{10.6.16}$$

Similarly, the fifth term of the left hand side of (10.6.9)

$$= -\frac{eE_i}{mv} \int_0^{\pi}\int_0^{2\pi} v_i \frac{\partial f_0}{\partial v} \sin\theta \, d\theta \, d\varphi = 0 \tag{10.6.17}$$

The sixth term

$$= -\frac{eE_i}{mv} \int_0^{\pi}\int_0^{2\pi} f_{1i} \sin\theta \, d\theta \, d\varphi = -\frac{4\pi e}{mv} E_i f_{1i} \tag{10.6.18}$$

The seventh term

$$= -\frac{eE_i}{mv} \frac{\partial}{\partial v}\left(\frac{f_{1j}}{v}\right) \int_0^{\pi}\int_0^{2\pi} v_i v_j \sin\theta \, d\theta \, d\phi = -\frac{4\pi}{3} \delta_{ij} v^2 \left(\frac{eE_i}{mv}\right) \frac{\partial}{\partial v}\left(\frac{f_{ij}}{v}\right) \tag{10.6.19}$$

The eighth term

$$= -\varepsilon_{ijk} \frac{\Omega_K}{v} f_{1i} \int_0^{\pi}\int_0^{2\pi} v_j \sin\theta \, d\theta \, d\varphi = 0 \tag{10.6.20}$$

The sixth term plus the seventh term

$$= -\frac{4\pi e E_i}{3mv} \frac{\partial}{\partial v} (v f_{1i}) \tag{10.6.21}$$

After integration of (10.6.9), we obtain

$$\frac{\partial f_0}{\partial t} + \frac{v}{3} (\nabla . \mathbf{f}_1) - \frac{e}{3mv^2} \frac{\partial}{\partial v} \{v (\mathbf{E} . \mathbf{f}_1)\} = c_0 (f_0) \tag{10.6.22}$$

where $$c_0(f_0) = \frac{1}{4\pi} \int_0^{\pi} \int_0^{2\pi} \left(\frac{\partial f}{\partial t} \right)_{\text{coll}} \sin\theta \, d\theta \, d\varphi \tag{10.6.23}$$

The next step is to multiply each term of (10.6.9) by $v_i \sin\theta \, d\theta \, d\varphi$, and integrate the result as in the preceding analysis. Then, the first term vanishes, the second term

$$= \int_0^{\pi} \int_0^{2\pi} \frac{v_i}{v} v_j \frac{\partial f_{1j}}{\partial t} \sin\theta \, d\theta \, d\varphi = \frac{4\pi}{3} v \delta_{ij} \frac{\partial f_{1j}}{\partial t} = \frac{4\pi}{3} v \frac{\partial f_{1i}}{\partial t}, \tag{10.6.24}$$

the third term $$= \int_0^{\pi} \int_0^{2\pi} v_i v_j \frac{\partial f_0}{\partial x_j} \sin\theta \, d\theta \, d\varphi = \frac{4\pi}{3} v^2 \frac{\partial f_0}{\partial x_i}, \tag{10.6.25}$$

and the fourth term $$= \int_0^{\pi} \int_0^{2\pi} v_i \frac{v_j}{v} v_k \frac{\partial f_{1j}}{\partial x_k} \sin\theta \, d\theta \, d\varphi = L_{ijk} \tag{10.6.26}$$

With the help of (10.6.12), it can be easily seen that $L_{111} = L_{222} = L_{333} = L_{122} = L_{113} = L_{132} = 0$; so, in general, we obtain

$$L_{ijk} = 0, \ (i, j, k) = (1, 2, 3) \tag{10.6.27}$$

Hence, the fourth term = 0, (10.6.28)

the fifth term $$= -\frac{eE_j}{mv} \frac{\partial f_0}{\partial v} \int_0^{\pi} \int_0^{2\pi} v_i v_j \sin\theta \, d\theta \, d\varphi = -\frac{4\pi}{3} \frac{eE_i}{m} v \frac{\partial f_0}{\partial v} \tag{10.6.29}$$

the sixth term $$= -\frac{eE_j}{mv} f_{1j} \int_0^{\pi} \int_0^{2\pi} v_i \sin\theta \, d\theta \, d\varphi = 0; \tag{10.6.30}$$

the seventh term $$= -\frac{eE_j}{mv} \frac{\partial}{\partial v} \left(\frac{f_{1k}}{v} \right) \int_0^{\pi} \int_0^{2\pi} v_i v_j v_k \sin\theta \, d\theta \, d\varphi = 0 \tag{10.6.31}$$

because $L_{ijk} = 0$; the eighth term $= -\frac{1}{v} \varepsilon_{klm} \Omega_m f_{1k} \int_0^{\pi} \int_0^{2\pi} v_i v_j \sin\theta \, d\theta \, d\varphi$

$$= -\varepsilon_{ijk} \Omega_j \frac{4\pi}{3} v f_{1k} = -\frac{4\pi}{3} v M_i \text{ (say)}; \tag{10.6.32}$$

Then $$M_1 = \varepsilon_{312} \Omega_2 f_{13} + \varepsilon_{213} \Omega_3, f_{12} = (\mathbf{\Omega} \times \mathbf{f}_1) \tag{10.6.33}$$

Hence, the eight term

$$= -\frac{4\pi}{3} v (\mathbf{\Omega} \times \mathbf{f}_1)_i \tag{10.6.34}$$

Dividing all the terms by $4\pi/3$, the resulting equation is

$$\frac{\partial f_{1i}}{\partial t} + v\frac{\partial f_0}{\partial x_i} - \frac{e}{m}E_i\frac{\partial f_0}{\partial v} - (\Omega \times \mathbf{f}_1)_i = C_{1i}f_{1i} \tag{10.6.35}$$

where
$$C_{1i}f_{1i} = \frac{3}{4\pi v}\int_0^{\pi}\int_0^{2\pi} v_i \left(\frac{\partial f}{\partial t}\right)_{\text{coll}} \sin\theta \, d\theta \, d\varphi \tag{10.6.36}$$

So the equation for the vector $\mathbf{f}_1$, is

$$\frac{\partial \mathbf{f}_1}{\partial t} + v\,\nabla f_0 - \frac{e}{m}\mathbf{E}\frac{\partial f_0}{\partial v} - (\Omega \times \mathbf{f}_1) = \mathbf{c}_1(\mathbf{f}_1) \tag{10.6.37}$$

where
$$\mathbf{c}_1(\mathbf{f}_1) = \frac{3}{4\pi v}\int_0^{\pi}\int_0^{2\pi} \mathbf{v}\left(\frac{\partial f}{\partial t}\right)_{\text{coll}} \sin\theta \, d\theta \, d\varphi \tag{10.6.38}$$

From (10.6.37) it is evident that $\mathbf{f}_1$ is a function of the speed v, and not of velocity $\mathbf{v}$.

10.7 COLLISION TERMS IN THE DISTRIBUTION FUNCTION FOR WEAKLY IONIZED PLASMAS

When the collision term is not negligible, the Boltzmann equation is

$$\frac{\partial f}{\partial t} + v_i\frac{\partial f}{\partial x_i} - \frac{e}{m}E_i\frac{\partial f}{\partial v_i} - \frac{e}{mc}(\mathbf{v}\times\mathbf{H})_i\frac{\partial f}{\partial v_i} = \left(\frac{\partial f}{\partial t}\right)_{\text{coll}} \tag{10.7.1}$$

where
$$\left(\frac{\partial f}{\partial t}\right)_{\text{coll}} = c\,(\text{say}) = \iiiint\!\!\iint [f(\mathbf{v}')F(\mathbf{v}_0') - f(\mathbf{v})F(\mathbf{v}_0)]\,|g|\,p\,dp\,d\varepsilon\,d\mathbf{v}_0 \tag{10.7.2}$$

and $|\mathbf{g}|\ (= |\mathbf{v} - \mathbf{v}_0|)$ is the modulus of the relative velocity, $\mathbf{v}$ and $\mathbf{v}_0$ are the velocities of electrons and molecules before collisions, the corresponding after-collision velocities are $\mathbf{v}'$ and $\mathbf{v}_0'$, respectively. The collision term of (10.7.2) has been explained in section 9.4 (*a*).

In weakly ionized plasmas (ionosphere, for example) electrons are fewer in number than the heavy molecules; so, energy and momentum are transferred to the electrons by molecules through collisions. The distribution of molecules is Maxwellian in velocity space; so

$$F(\mathbf{v}_0) = N_n\left(\frac{M}{2\pi s}\right)^{3/2}\exp\left(-\frac{M v_0^2}{2s}\right), \tag{10.7.3}$$

where $S = \kappa T$, and M and N_n, are, respectively, the mass and number density of molecules. Since the number density N and mass m of electrons satisfy the conditions

$$N_n >> N, m << M, \tag{10.7.4}$$

in a weakly ionized plasma, the influence of collision on molecules becomes negligible, so that we can write

$$F(\mathbf{v}_0') = F(\mathbf{v}_0) \tag{10.7.5}$$

The energy loss factor G, for elastic collisions between electrons and molecules, is

$$G = \frac{2m}{M} = 3{\cdot}4 \times 10^{-5} \tag{10.7.6}$$

And, including inelastic collisions (section 7.9), the accepted value is $G = 2 \times 10^{-3}$. Since $G << 1$, the root mean square of electron velocities is greater than the mean directed velocity; so the electron distribution for the symmetric part of $\mathbf{v}$ is much greater than for the antisymmetric part. Putting (10.7.3) in (10.7.5) and (10.7.2), we find that

$$\left(\frac{\partial f}{\partial t}\right)_{\text{coll}} = N_0 \int_0^\infty \int_0^\pi |\mathbf{g}| (f' - f)\, p dp\, d\varepsilon \tag{10.7.7}$$

where $$|\mathbf{g}| = |\mathbf{v} - \mathbf{v}_0| = |\mathbf{v}|, \tag{10.7.8}$$

because velocity of molecules is negligible compared to electron velocities. Since electrons do not lose any kinetic energy, during collisions with molecules, their speeds are not altered. Hence, we can write

$$f_0' = f_0, f_1' = f_1, \ |\mathbf{v}'| = |\mathbf{v}| \tag{10.7.9}$$

Then, it is obvious that

$$f' - f = \frac{1}{v}((\mathbf{v}' - \mathbf{v}) \cdot \mathbf{f}_1), \tag{10.7.10}$$

$$\left(\frac{\partial f}{\partial t}\right)_{\text{coll}} = N_n \int_0^\pi \int_0^{2\pi} ((\mathbf{v}' - \mathbf{v}) \cdot \mathbf{f}_1)\, \sigma d\Omega = \nu(v) f_1(v), \tag{10.7.11}$$

where $$\nu(v) = N_n v \int_0^\pi \int_0^{2\pi} (1 - \cos\chi)\, \sigma\, d\Omega, \tag{10.7.12}$$

$$p dp\, d\varepsilon = \sigma \sin\chi\, d\chi\, d\varepsilon = \sigma\, d\Omega, \tag{10.7.13}$$

and χ is the angle of scattering (section 6.2). Without loss of generality, $\mathbf{f}_1$ is taken parallel to the initial relative velocity of electrons, for finding the relation (10.7.11).

The orthogonal coordinate axes OX_1, OX_2, OX_3 are such that $\mathbf{v}$ is parallel to OX_3 and OX_1 is perpendicular to OX_3 in the plane $\mathbf{v}$ and $\mathbf{v}'$, as in Fig. 10.7.1. The directions of $\mathbf{f}_1$ and $\mathbf{v}'$ are given by $(\sin\theta, 0, \cos\theta)$ and $(\sin\chi \cos\varepsilon, \sin\chi \sin\varepsilon, \cos\chi)$, respectively. Hence

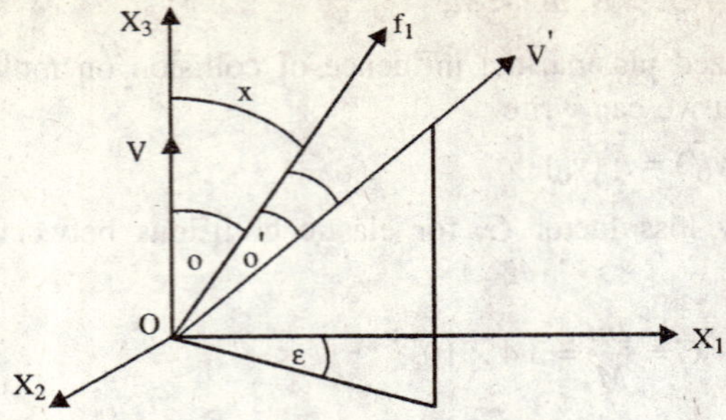

Fig. 10.7.1 Directions of $\mathbf{f}_1$, $\mathbf{v}$, $\mathbf{v}'$ and their angles, with respect to the coordinate directions.

$$\cos\theta' = \frac{(\mathbf{f}_1 \,.\, \mathbf{v}')}{|\mathbf{f}_1|\;|\mathbf{v}'|} = \cos\theta\cos\chi + \sin\theta\sin\chi\cos\varepsilon, \tag{10.7.14}$$

$$\frac{1}{v}\{(\mathbf{v}' \,.\, \mathbf{f}_1) - (\mathbf{v} \,.\, \mathbf{f}_1)\} = (\cos\theta' - \cos\theta)\, f_1$$

$$= [\sin\theta\sin\chi\cos\varepsilon - (1-\cos\chi)\cos\theta]\, f_1 \tag{10.7.15}$$

Using first (10.7.15) and then the value of σ_2 from (A11.7), in (10.7.11), we obtain

$$\frac{1}{N_n v f_1}\left(\frac{\partial f}{\partial t}\right)_{\text{coll}} = \int_{\varepsilon=0}^{2\pi}\int_{\chi=0}^{\pi} \sigma(\chi)\sin\chi\,[-(1-\cos\chi)\cos\theta + \sin\theta\sin\chi\cos\varepsilon]\, d\varepsilon\, d\chi$$

$$= -\sigma_2\cos\theta \tag{10.7.16}$$

Hence, for momentum transfer, the frequency of collisions between electrons and molecules, $\nu(v)$, is defined as

$$\nu(v) = N_n\, v\, \sigma_2; \tag{10.7.17}$$

So equation (10.7.16) reduces to

$$\left(\frac{\partial f}{\partial t}\right)_{\text{coll}} = -\frac{\nu(v)}{v}(\mathbf{f}_1 \,.\, \mathbf{v}) \tag{10.7.18}$$

Equation (10.6.23), therefore, gives

$$c_0(f_0) = \frac{1}{4\pi}\int_0^{\pi}\int_0^{2\pi}\left(\frac{\partial f}{\partial t}\right)_{\text{coll}} \sin\theta\, d\theta\, d\varphi \tag{10.7.19}$$

For the components of $\mathbf{c}_1(\mathbf{f}_1)$, equation (10.6.38), gives

$$c_{11}(\mathbf{f}_1) = \frac{3}{4\pi v}\iint v_1\left(\frac{\partial f}{\partial t}\right)_{\text{coll}} \sin\theta\, d\theta\, d\varphi = -N_n\,\sigma_2\, v f_{11}, \text{ etc.,} \tag{10.7.20}$$

So we find that

$$C_1 (\mathbf{f}_1) = -\nu (\mathbf{v})\, \mathbf{f}_1 \tag{10.7.21}$$

10.8 COLLISION TERMS

For higher order effects of collision, we consider the general relation

$$\left(\frac{\partial f}{\partial t}\right)_{\text{coll}} = \iiint_{\mathbf{v}_0} \int_{\Omega} \{f(\mathbf{v}') F(\mathbf{v}_0') - f(\mathbf{v}) F(\mathbf{v}_0)\}\, g\sigma\, d\mathbf{v}_0\, d\Omega \tag{10.8.1}$$

where $\mathbf{g} = \mathbf{v} - \mathbf{v}_0,\ \mathbf{g}' = (\mathbf{v}' - \mathbf{v}_0'),\ d\Omega = \sin\chi\, d\chi\, d\varepsilon,$

$$F(v_0) = N_0 \left(\frac{M}{2\pi s}\right)^{3/2} \exp\left(-\frac{M v_0^2}{2s}\right) \tag{10.8.2}$$

and $S = \kappa T$. We now that

$$f(v) = \varphi(v) f_M(v) \tag{10.8.3}$$

where

$$f_M(v) = N \left(\frac{m}{2\pi s}\right)^{3/2} \exp\left(-\frac{mv^2}{2s}\right) \tag{10.8.4}$$

Since $m/M << 1$, electrons lose only a small fraction of their energy in elastic collisions, though angles of deflections are large. So, the electron velocity distribution is expected to be nearly isotropic in velocity space. The condition for conservation of kinetic energy in elastic collisions is

$$mv^2 + Mv_0^2 = mv'^2 + Mv_0'^2; \tag{10.8.5}$$

so

$$f_M(v')\ F(v_0') = f_M(v)\ F(v_0), \tag{10.8.6}$$

$$\left(\frac{\partial f}{\partial t}\right)_{\text{coll}} = f_M(v) \iiint_{\mathbf{v}_0} \int_{\Omega} F(v_0) \{\varphi(v') - \varphi(v)\}\, g\sigma(g, \chi)\, d\mathbf{v}_0\, d\Omega \tag{10.8.7}$$

We multiply $(\partial f/\partial t)_{\text{coll}}$ with an arbitrary function $\psi(v)$ of the electron speed v, integrate in velocity space and obtain

$$\iiint_{v} \psi(v) \left(\frac{\partial f}{\partial t}\right)_{\text{coll}} d\mathbf{v} = \iiint_{\mathbf{v}} \iiint_{\mathbf{v}_0} \int_{\Omega} \psi(\mathbf{v}) f_M(\mathbf{v}) F(\mathbf{v}_0) \{\varphi(v') - \varphi(v)\}\, g\sigma\, d\mathbf{v}\, d\mathbf{v}_0\, d\Omega \tag{10.8.8}$$

Since all ranges of velocities are possible for direct collision, there should be an inverse collision. So the first term, within the integral on the right hand side of (10.8.8), is replaced by

$$\psi(v') f_M(\mathbf{v}') F(\mathbf{v}_0') \varphi(\mathbf{v}'')\, g'\, \sigma(g', \chi)\, d\mathbf{v}'\, d\mathbf{v}_0'\, d\Omega$$

Since $\mathbf{v}'' = \mathbf{v}$, $d\mathbf{v}'\, d\mathbf{v}_0' = d\mathbf{v}\, d\mathbf{v}_0$, this term reduces to

$$\psi(\mathbf{v}') f_M(\mathbf{v}) F(\mathbf{v}_0) \varphi(\mathbf{v})\, g\sigma(g, \chi)\, d\mathbf{v}\, d\mathbf{v}_0\, d\Omega \tag{10.8.9}$$

if the relation (10.8.6) is used. Hence (10.8.9) becomes

$$\iiint_v \psi(v)\left(\frac{\partial f}{\partial t}\right)_{coll} d\mathbf{v} = \iiint_{\mathbf{v}} \iiint_{\mathbf{v}_0} \int_{\Omega} f_M(\mathbf{v})\, F(\mathbf{v}_0)\, \varphi(\mathbf{v})\, [\psi(\mathbf{v}') - \psi(\mathbf{v})]\, g\, \sigma\, d\mathbf{v}\, d\mathbf{v}_0\, d\Omega \tag{10.8.10}$$

No energy is transferred when $v' = v$. However, to find the effects of collisional recoil of the molecules, in their elastic collisions with electrons, we set

$$v' = v + \delta v \tag{10.8.11}$$

and obtain

$$\psi(v') = \psi(v) + (\partial \psi/\partial v)\, \delta v \tag{10.8.12}$$

Next putting $d\mathbf{v} = v^2 \sin\theta\, d\theta\, d\varphi$ we integrate (10.8.10) with respect to θ between 0 to π, φ between 0 to 2π and v between 0 to ∞, and obtain

$$\int_0^\infty v^2 \psi\left(\frac{\partial f}{\partial t}\right)_{coll} dv = \int_{v=0}^{\infty} \iiint_{\mathbf{v}_0} \int_{\Omega} v^2 f_M(v)\, F(v_0)\, \varphi(v) \left\{\frac{\partial \psi}{\partial v}\delta v + \frac{1}{2}\frac{\partial^2 \psi}{\partial v^2}(\delta v)^2 + \ldots\right\} g\sigma\, dv_0\, dv\, d\Omega \tag{10.8.13}$$

Neglecting $\partial^2 \psi/\partial v^2$ we integrate the right hand side of (10.8.13) by parts, and obtain

$$\int_0^\infty v^2 \psi(v)\left(\frac{\partial f}{\partial t}\right)_{coll} dv = -\int_{v=0}^{\infty} \psi \frac{\partial}{\partial v} \iiint_{\mathbf{v}_0} \int_{\Omega} v^2 (\delta v)\, g\, f_M(v)\, \varphi(v)\, F(v_0)\, \sigma\, d\Omega\, d\mathbf{v}_0\, dv \tag{10.8.14}$$

Since $\psi(v)$ is an arbitrary function of v, we equate its coefficients from both sides and obtain

$$\left(\frac{\partial f}{\partial t}\right)_{coll} = -\frac{1}{v^2}\frac{\partial}{\partial v} \iiint_{\mathbf{v}_0} \int_{\Omega} g v^2 (\delta v)\, \sigma \varphi(v)\, f_M(v)\, F(v_0)\, d\Omega\, d\mathbf{v}_0 \tag{10.8.15}$$

Equation (A 10.43) shows that

$$\delta v = -\frac{M}{m} v\, (1 - \cos\chi) \tag{10.8.16}$$

Hence, using (10.7.8), equation (10.8.15) becomes

$$\left(\frac{\partial f}{\partial t}\right)_{coll} = \frac{m}{M v^2}\frac{\partial}{\partial v} \iiint_{\mathbf{v}_0} v^4 f_M(v)\, \varphi(v)\, F(v_0)\, \sigma_2\, d\mathbf{v}_0 \tag{10.8.17}$$

Putting the value of $\nu(v)$ from (10.8.12) and (10.7.27) in (10.8.17) we obtain

$$\left(\frac{\partial f}{\partial t}\right)_{\text{coll}} = \frac{m}{Mv^2}\frac{\partial}{\partial v}\{\nu(v)\, v^3 f_M(v)\, \varphi(v)\} \tag{10.8.18}$$

In the absence of external forces, in the steady state, collisions do not change the energy distribution for elastic and infinitely conducting molecules; so

$$c_0(f_0) = 0 \tag{10.8.19}$$

For integration over $\mathbf{v}_0'$ in (10.8.2), we assume that $\mathbf{v}_0'$ moves over a Maxwell distribution, centred about $\mathbf{v}_0' = 0$. For fixed $\mathbf{v}$ and fixed scattering angle χ, this means that $\mathbf{v}'$ moves about the value $\mathbf{v}_1'$ (say), where $\mathbf{v}' = \mathbf{v}_1'$, when $\mathbf{v}_0'$ is small. Hence

$$\mathbf{v}' = \mathbf{v}_1' + 0(\mathbf{v}_0'); \tag{10.8.20}$$

so we write $$f(\mathbf{v}') = f(\mathbf{v}_1') + 0(v_0')^2 \frac{\partial f_0}{\partial v'^2} \tag{10.8.21}$$

Incorporating this correction into $f(v)$, of (10.8.3) in (10.8.18), we obtain

$$\left(\frac{\partial f}{\partial t}\right)_{\text{coll}} = \frac{m}{Mv^2}\frac{\partial}{\partial v}\left[\nu(v)\, v^3 \left\{f_0 + 0(v_0')^2 \frac{\partial f_0}{\partial v^2}\right\}\right] \tag{10.8.22}$$

When the electrons have a Maxwell velocity distribution f_0 at the temperature of the gas, collisions do not change this distribution, and (10.8.22) vanishes. Then, writing

$$0(v_0'^2) = \frac{2\kappa T}{m}, \tag{10.8.23}$$

we replace (10.8.22) by

$$\left(\frac{\partial f}{\partial t}\right)_{\text{coll}} = \frac{m}{Mv^2}\frac{\partial}{\partial v}\left[\nu(v)\, v^3 \left(f_0 + \frac{s}{mv}\frac{\partial f_0}{\partial v}\right)\right] \tag{10.8.24}$$

It represents the expression for the rate of change of the distribution of energy of electrons due to collision with molecules. Putting (10.8.24) in (10.7.19) and using (10.8.22) we obtain the equation for f_0 :

$$\frac{\partial f_0}{\partial t} + \frac{v}{3}(\nabla \mathbf{f}_1) - \frac{e}{3Mv^2}\frac{\partial}{\partial v}\{v^2(\mathbf{E}\,.\,\mathbf{f}_1)\} = \frac{m}{Mv^2}\frac{\partial}{\partial v}\left\{\nu v^3 \left(f_0 + \frac{s}{mv}\frac{\partial f_0}{\partial v}\right)\right\} \tag{10.8.25}$$

If (10.8.21) is used, equation (10.6.37) becomes

$$\frac{\partial \mathbf{f}_1}{\partial t} + v\nabla f_0 - \frac{e}{m}\mathbf{E}\frac{\partial f_0}{\partial v} - (\boldsymbol{\Omega} \times \mathbf{f}_1) = -\nu \mathbf{f}_1 \tag{10.8.26}$$

Since $\partial/\partial \mathbf{r} = 0$, when the plasma is homogeneous, equatoins (10.8.25) and (10.8.26) reduce to

$$\frac{\partial f_0}{\partial t} - \frac{e}{3mv^2}\frac{\partial}{\partial v}\{v^2 (\mathbf{E}\,.\,\mathbf{f}_1)\} = \frac{e}{Mv^2}\frac{\partial}{\partial v}\left\{\nu v^3\left(f_0 + \frac{s}{mv}\frac{\partial f_0}{\partial v}\right)\right\}, \tag{10.8.27}$$

$$\frac{\partial \mathbf{f}_1}{\partial t} - \frac{e}{m}\mathbf{E}\frac{\partial f_0}{\partial v} - (\mathbf{\Omega} \times \mathbf{f}_1) = -\nu \mathbf{f}_1 \tag{10.8.28}$$

For an oscillating electric field,

$$\mathbf{E} = \mathbf{E}_0 \cos \omega t \tag{10.8.29}$$

Putting

$$\mathbf{f}_1 = -\mathbf{u}\,(\partial f_0/\partial v) \tag{10.8.30}$$

in (10.8.28), we find that **u** approximately satisfies the equation of motion for a single electron in presence of the applied electric field magnetic field and collisional loss,

$$\frac{\partial \mathbf{u}}{\partial t} + \nu \mathbf{u} = -\frac{e}{m}\mathbf{E} - \frac{e}{mc}(\mathbf{u} \times \mathbf{H}) \tag{10.8.31}$$

Here the collision frequency ν is a function of the random thermal velocity **v**. Its only speciality is that the collision frequency ν is a function of the thermal velocity **v**, which is not identical to the macroscopic velocity **u**.

10.9 INTERACTION OF FOUR WAVES

Cross modulation is a non-linear interaction between four waves, because, in addition to the wanted field

$$\mathbf{E_w} = \mathbf{E}_1 \cos \rho t, \tag{10.9.1}$$

there is the modulated field $\mathbf{E}_D$, given by

$$\begin{aligned}\mathbf{E}_D &= \mathbf{E}_0 \{1 + b\cos(\omega_1 t)\}\cos(\omega t)\\ &= \mathbf{E}_0 \cos(\omega t) + \frac{\mathbf{E}_0}{2} b\cos(\omega t + \omega_1 t) + \frac{\mathbf{E}_0}{2} b\cos(\omega t - \omega_1 t),\end{aligned} \tag{10.9.2}$$

where b is the constant modulation index, and the modulation frequency ω_1 is small compared with the carrier freqúencies ω. Actually, the influence of modulation is a small departure from the steady state, which is characterised by the absence of modulation. So, the interaction takes place between four fields with frequencies ω, $\omega + \omega_1$, $\omega - \omega_1$, ρ. If $\mathbf{u}_i$ is the velocity imposed by the field at frequency ω_i, then, in place of (10.8.30), we have

$$\mathbf{f}_1 = \frac{\partial f_0}{\partial v}\sum_i \mathbf{u}_i = -\frac{\partial f_0}{\partial v}\sum_i \{\mathbf{A}(\omega_i)\cos(\omega_i t) + \mathbf{B}(\omega_i)\sin(\omega_i t)\} \tag{10.9.3}$$

where now $\mathbf{A}(\omega_i) = -\dfrac{e\nu\,\mathbf{E}_i}{m(\omega^2 + \nu^2)}$, $\mathbf{B}(\omega_i) = -\dfrac{e\,\omega_i\,\mathbf{E}_i}{m(\omega^2 + \nu^2)}$ (10.9.4)

Here $\mathbf{E}_i$ is the amplitude of the wave of frequency ω_i. Hence

$$\mathbf{f}_1 = -\frac{\partial f_0}{\partial v}\{\mathbf{A}(\omega)\cos\omega t + \mathbf{A}(\omega-\omega_1)\cos(\omega t - \omega_1 t)$$
$$+ \mathbf{A}(\omega+\omega_1)\cos(\omega t + \omega_1 t) + \mathbf{A}(\rho)\cos\rho t + \mathbf{B}(\omega)\sin\omega t$$
$$+ \mathbf{B}(\omega-\omega_1)\sin(\omega t - \omega_1 t) + \mathbf{B}(\omega+\omega_1)\sin(\omega t + \omega_1 t) + \mathbf{B}(\rho)\sin\rho t\} \tag{10.9.5}$$

This relation can be wrtiten as

$$\mathbf{f}_1 = -\frac{\partial f_0}{\partial v}\{\mathbf{A}(\omega) + \{\mathbf{A}(\omega-\omega_1) + \mathbf{A}(\omega+\omega_1)\}\cos\omega_1 t$$
$$+ \{\mathbf{B}(\omega+\omega_1) - \mathbf{B}(\omega-\omega_1)\}\sin\omega_1 t]\cos\omega t - \frac{\partial f_0}{\partial v}[\mathbf{B}(\omega)$$
$$+ \{\mathbf{B}(\omega+\omega_1) + \mathbf{B}(\omega-\omega_1)\}\cos\omega_1 t + \{\mathbf{A}(\omega-\omega_1) - \mathbf{A}(\omega +$$
$$\omega_1)\}\sin\omega_1 t]\sin\omega t - \frac{\partial f_0}{\partial v}\{\mathbf{A}(\rho)\cos\rho t + \mathbf{B}(\rho)\sin\rho t\} \tag{10.9.6}$$

Since $\omega_1 << \omega$, we have $\mathbf{A}(\omega+\omega_1) \approx (b/2)\,\mathbf{A}(\omega)$,
$\mathbf{B}(\omega \pm \omega_1) \approx (b/2)\,\mathbf{B}(\omega)$. Then

$$\mathbf{f}_1 = -\frac{\partial f_0}{\partial v}[(1 + b\cos\omega_1 t)\{\mathbf{A}(\omega)\cos\omega t + \mathbf{B}(\omega)\sin\omega t] + \mathbf{A}(\rho)\cos\rho t +$$
$$+ \mathbf{B}(\rho)\sin\rho t] = -\frac{\partial f_0}{\partial v}\frac{e}{m}(\nu\cos\rho t + \rho\sin\rho t), \tag{10.9.7}$$

$$\mathbf{u} = \mathbf{u}(\omega) + \mathbf{u}(\omega+\omega_1) + \mathbf{u}(\omega-\omega_1) + \mathbf{u}(\rho)$$
$$= \frac{e}{m}\left[\frac{\mathbf{E}_0}{\omega^2+\nu^2}(\nu\cos\omega t + \omega\sin\omega t)\,(1 + b\cos\omega_1 t)\right.$$
$$\left. + \frac{\mathbf{E}_1}{\rho^2+\nu^2}(\nu\cos\rho + \rho\sin\rho t)\right] \tag{10.9.8}$$

10.10 MICROSCOPIC THEORY OF CROSS-MODULATION IN PLASMA

The mean collision frequency ν of electrons should be related to their mean thermal energy, for all ranges of values of modulation frequency. Energy balance relation that is used is suitable for modulation frequencies much less than 150 cycles per second. For higher modulation frequencies, such an energy balance relation should not be used. Electrons do not make too many collisions with molecules before recombination takes place. So the mean kinetic energy of electrons present in the atmosphere is likely to be larger than the thermal energy.

The microscopic theory ascertains the dependence of the distribution function on modulation frequency. The geomagnetic field is ignored for simplicity in this theory, which is permissible when the carrier frequencies are much greater than the gyrofrequency of electrons.

(a) This theory for unmagnetized plasma

Assuming that carrier frequencies are much greater than the gyrofrequency, we neglect **H** in (10.8.31) and obtain

$$\frac{\partial \mathbf{u}}{\partial t} + \nu \mathbf{u} = -\frac{e}{m} \mathbf{E} \tag{10.10.1}$$

Writing $\mathbf{E} = \mathbf{E}_0 \cos \omega t$, this equation gives

$$\mathbf{u} = -\frac{e \mathbf{E}_0}{m (\omega^2 + \nu^2)} (\nu \cos \omega t + \omega \sin \omega t) \tag{10.10.2}$$

Putting this value of **u** and f_1 from (10.8.30) in (10.8.27) we obtain the equation

$$\frac{\partial f_0}{\partial t} + \frac{e (\mathbf{E} . \mathbf{u})}{3mv^2} \frac{\partial}{\partial v} \left(v^2 \frac{\partial f_0}{\partial v} \right) = \frac{G}{2v^2} \frac{\partial}{\partial v} \left\{ \nu v^2 \left(f_0 + \frac{S}{mv} f_0' \right) \right\} \tag{10.10.3}$$

where $G = 2m/M$, $f_0' = \partial f_0 / \partial v$, $S = K_B T$. For finding the influence of the audio frequency ω_1 of modulation on f_0, this equation is averaged over the faster time period $2\pi/\omega_0$ of the carrier frequency ω_0 of **E**. Then, since

$$\langle (\mathbf{E} . \mathbf{u}) \rangle = -\frac{e E_0^2 \nu}{2m (\omega^2 + \nu^2)}, \tag{10.10.4}$$

equation (10.10.3) becomes

$$\frac{\partial f_0}{\partial t} - \frac{G}{2v^2} \frac{\partial}{\partial v} \left(\nu v^3 f_0 + \frac{S_0}{v} f_0' \right) = 0 \tag{10.10.5}$$

where

$$S_0 = \frac{S}{m} + \frac{F^2}{3G (\omega^2 + \nu^2)}, \quad \mathbf{E} = \frac{e \mathbf{E}_0}{m} \tag{10.10.6}$$

and $\langle x \rangle$ is the average of x over the time period $2\pi/\omega$. Writing, therefore, $S_0 = K_B T_e / m$, the modified electron temperature T_e is given by

$$T_e = T + \frac{mF^2}{3GK_B (\omega^2 + \nu^2)} \tag{10.10.7}$$

In the absence of modulational changes $\partial / \partial t = 0$; so (10.10.5) reduces to

$$f_0 + \frac{S_1}{v} \left(\frac{\partial f_0}{\partial v} \right) = \frac{\text{constant}}{\nu v^3}, \tag{10.10.8}$$

where S_1 corresponds to S_0 for fields without any modulation. Since, there should be no singularity in the static distribution, the right hand side of (10.10.8) must be zero. So

$$f_0 = C \exp\left(-\int_0^v \frac{v}{S_1}\,dv\right) \tag{10.10.9}$$

When the collision frequency ν is independent of the velocity **v**, the quantity S_0 is constant. Hence

$$f_0 = C \exp\left(-\frac{mv^2}{2K_B T_{e0}}\right) \tag{10.10.10}$$

where T_{e0} is obtained from (10.10.7) for the unmodulated field. So $S_1 = K_B T_{e0}/m$. The constant of integration, C, is determined from the normalization condition

$$N = 4\pi \int_0^\infty v^2 f_0\,dv,$$

giving
$$f_0(v) = N\left(\frac{m}{2\pi K_B T_{e0}}\right)^{1/2} \exp\left(-\frac{mv^2}{2K_B T_{e0}}\right) \tag{10.10.11}$$

(b) Solution when the field is modulated

As in the relation (10.9.2), the modulated electric field is

$$\mathbf{E} = \mathbf{E}_0 (1 + b\cos\omega_1 t)\cos\omega t \tag{10.10.12}$$

where b is the modulation index, ω_1 is the modulation frequency, and ω is the carrier frequency. Then, since $\mathbf{F} = -e\,\mathbf{E}_0 (1 + b\cos\omega_1 t)/m$, we put

$$S_2 = \frac{KT}{m} + \frac{F^2 (1 + b\cos\omega_1 t)^2}{3G(\omega^2 + \nu^2)} \tag{10.10.13}$$

For $\omega_1 << \nu$, the first three terms of (10.10.3) are, respectively, of the order of the ratios $f_0/t_\omega, f_0/t_{\omega_1}$ and f_0/t_r and t_r.

$$t_{\omega_1} = \frac{2\pi}{\omega_1},\quad t_\omega = \frac{2\pi}{\omega},\quad t_r = \frac{f_0}{c_0(f_0)} \tag{10.10.14}$$

Then, following (10.10.8)-(10.10.11), the solution is

$$f_0(\mathbf{v}, t) = N\left(\frac{m}{2\pi KT}\right)^{1/2} \exp\left[-\int_0^v \frac{v\,dv}{S_2}\right] \tag{10.10.15}$$

For weak fields, the electron distribution function does not deviate appreciably from its stationary Maxwellian form; so

$$f_0(v,t) = N\left(\frac{m}{2\pi KT}\right)^{1/2} \exp\left[-\int_0^v \frac{m}{KT} v\,dv\left\{1-\frac{F^2}{3G}\frac{m}{KT}\frac{(1+b\cos\omega_1 t)}{(\omega^2+\nu^2)}\right\}\right]$$

$$= N\left(\frac{m}{2\pi K_B T}\right)^{1/2} \exp\left(-\frac{mv^2}{K_B T}\right)$$

$$\left[1+\left(\frac{m}{K_B T}\right)^2 \int_0^v \frac{F^2}{3G}(1+b\cos\omega_1 t)^2 \cdot \frac{v\,dv}{(\omega^2+\nu^2)}\right]$$

$$\approx N\left(\frac{m}{2\pi K_B T}\right)^{1/2}\left[1+\left(\frac{m}{K_B T}\right)^2 \cdot \frac{F^2(1+b\cos\omega_1 t)^2 v^2}{6G(\omega^2+\nu^2)}\right]\exp\left(-\frac{mv^2}{2K_B T}\right) \tag{10.10.16}$$

Here the collision frequency ν has been assumed to be independent of v. The mean energy ξ of an electron is given by

$$\xi = \frac{m}{2}\frac{\int v^2 f\,dv}{\int f\,dv} = \frac{K_B T}{2}\left\{1+\frac{K_B T}{m}\frac{F^2(1+b\cos\omega_1 t)}{2G(\omega^2+\nu^2)}\right\} \tag{10.10.17}$$

Chapter 11

SMALL AMPLITUDE WAVES IN PLASMAS

11.1 INTRODUCTION

This chapter is complementary to the portion, in Chapter 3, on waves in plasmas. Regarding the plasma as a mixture of fluids of positive charges, negative charges and neutral particles, here we consider the propagation of waves in the linearized approximation (that is, propagation of infinitesimal amplitude waves) in presence of a homogeneous magnetic field, and some kinds of inhomogeneities.

11.2 LINEARIZED EQUATIONS IN A PLASMA CONTAINING NEUTRAL PARTICLES

Subscripts e, i, n denote the variables associated with electrons, ions and molecules. The unperturbed state is homogeneous; so the densities of the net charge and electric current are zero; the charge per particle, for ions is e, and for electrons, $-e$. Momentum between these fluids is exchanged through effects of resonance and collisional forces, which are proportional to the relative velocities of the elements of the fluids. Stresses at all points reduce to a scalar partial pressure for each species of particles of the plasma; the total pressure p is equal to the sum of the partial pressures p_e, p_i, p_n.

The applied magnetic field $\mathbf{H}_0$ is constant. Each species has thermal motion and perturbation in its partial pressure follows the adiabatic law of thermodynamics (Section 3.4).

The field equations for such a plasma, in the linearized approximation, are

$$\dot{\mathbf{u}}_e = c_e^2\, \nabla N_e - \frac{e}{m}\mathbf{E} - (\mathbf{u}_e \times \boldsymbol{\Omega}_e) - \nu_e\,(\mathbf{u}_e - \mathbf{u}_i) - \nu - (\mathbf{u}_e - \mathbf{u}_n), \quad (11.2.1)$$

$$\mathbf{u}_i = -\frac{c_i^2}{N_0}\nabla N_i + \frac{e}{M}\mathbf{E} + (\mathbf{u}_i \times \boldsymbol{\Omega}_i) - \nu_i\,(\mathbf{u}_i - \mathbf{u}_e) - \nu_+\,(\mathbf{u}_i - \mathbf{u}_n), \quad (11.2.2)$$

$$\dot{\mathbf{u}}_n = -\frac{c_n^2}{N_n^0}\nabla N_n - \nu_n^e\,(\mathbf{u}_n - \mathbf{u}_e) - \nu_n^i\,(\mathbf{u}_n - \mathbf{u}_i), \quad (11.2.3)$$

$$N_0\,(\nabla \cdot \mathbf{u}_e) + \dot{N}_e = 0,\ N_0\,(\nabla \mathbf{u}_i) + \dot{N}_i = 0,\ N_n^0\,(\nabla \cdot \mathbf{u}_n) + \dot{N}_m = 0, \quad (11.2.4)$$

$$\nabla \times \mathbf{E} = -\frac{1}{c}\dot{\mathbf{H}},\ \nabla \times \mathbf{H} = \frac{1}{c}\dot{\mathbf{E}} + \frac{4\pi N_0\, e}{c}(\mathbf{u}_i - \mathbf{u}_e), \quad (11.2.5)$$

$$(\nabla . \mathbf{E}) = 4\pi e\,(N_i - N_e),\ \ (\nabla . \mathbf{H}) = 0 \tag{11.2.6}$$

For neutral particles the sound speed is c_n, mass per particle is M_n, velocity perturbation is N_n, the unperturbed number density is N_n^0 and its perturbed part is N_n; the various collision frequencies are ν_e, ν_i, ν_-, ν_+, ν_n^e, ν_n^i; and $\mathbf{\Omega}_e = e\mathbf{H}/mc$, $\mathbf{\Omega}_i = e\mathbf{H}_i/Mc$.

Not all these equations are linearly independent. For instance, the conditions of (11.2.4), for electrons and ions, are equivalent to (11.2.6), together with the divergence of the second equation of (11.2.5). Eliminating N_e, N_i and N_n from equations (11.2.1)-(11.2.6), the following set of twelve linearly independent linear differential equations, in the twelve field variables, which are the components of **E**, $\mathbf{u}_e$, $\mathbf{u}_i$, $\mathbf{u}_n$, are obtained:

$$c^2\,\nabla \times (\nabla \times \mathbf{E}) + \ddot{\mathbf{E}} + 4\pi e\,N_0\,(\dot{\mathbf{u}}_i - \dot{\mathbf{u}}_e) = 0, \tag{11.2.7}$$

$$\ddot{\mathbf{u}}_e = c_e^2\,\nabla\,(\nabla . \mathbf{u}_e) - \frac{e}{m}\,\dot{\mathbf{E}} - (\dot{\mathbf{u}}_e \times \mathbf{\Omega}_e) - \nu_e\,(\dot{\mathbf{u}}_e - \dot{\mathbf{u}}_i) - \nu - (\dot{\mathbf{u}}_e - \dot{\mathbf{u}}_n), \tag{11.2.8}$$

$$\ddot{\mathbf{u}}_i = c_1^2\,\nabla\,(\nabla . \mathbf{u}_i) + \frac{e}{M}\,\dot{\mathbf{E}} + (\dot{\mathbf{u}}_i \times \mathbf{\Omega}_i) - \nu_i\,(\dot{\mathbf{u}}_i - \dot{\mathbf{u}}_e) - \nu + (\dot{\mathbf{u}}_i - \dot{\mathbf{u}}_n), \tag{11.2.9}$$

$$\ddot{\mathbf{u}}_n = c_n^2\,\nabla\,(\nabla . \mathbf{u}_n) - \nu_n^e\,(\dot{\mathbf{u}}_n - \dot{\mathbf{u}}_e) - \nu_n^i\,(\dot{\mathbf{u}}_n - \dot{\mathbf{u}}_i) \tag{11.2.10}$$

Equation (11.2.7) is deduced by taking curl of the second equation (11.2.5), and then eliminating **H** with the help of its first equation. To obtain (11.2.8), equation (11.2.1) is differentiated with respect to time and from that N_e is eliminated by using the first equation of (11.2.4). Equations (11.2.9) and (11.2.10) are also similarly obtained.

Assuming the wave solution, proportional to exp $(ikz - i\omega t)$, these relations reduce to

$$K^2\,c^2\,E_z\,\hat{\mathbf{X}}_z + (\omega^2 - K^2\,c^2)\,\mathbf{E} + 4\pi N_0\,ie\omega\,(\mathbf{u}_i - \mathbf{u}_e) = 0, \tag{11.2.11}$$

$$\omega^2\,\mathbf{u}_e - K^2\,c_e^2\,u_{ez}\,\hat{\mathbf{X}}_z + \frac{ie}{m}\,\omega\mathbf{E} + i\omega\,(\mathbf{u}_e \times \mathbf{\Omega}_e) + i\omega\,\nu_e\,(\mathbf{u}_e - \mathbf{u}_i)$$
$$+\, i\omega\nu_-\,(\mathbf{u}_e - \mathbf{u}_n) = 0, \tag{11.2.12}$$

$$\omega^2\,\mathbf{u}_i - K^2\,c_i^2\,u_{iz} - \hat{\mathbf{X}}_z - \frac{ie}{M}\,\mathbf{E} - i\omega\,(\mathbf{u}_i \times \mathbf{\Omega}_i)$$
$$+\, i\omega\,\nu_i\,(\mathbf{u}_i - \mathbf{u}_e) + i\omega\nu_+\,(\mathbf{u}_i - \mathbf{u}_n) = 0, \tag{11.2.13}$$

$$\omega^2\,\mathbf{u}_n - K^2\,c_n^2\,u_{nz}\,\hat{\mathbf{X}}_z + i\omega\,\nu_n^e\,(\mathbf{u}_n - \mathbf{u}_e) + i\omega\nu_n^i\,(\mathbf{u}_n - \mathbf{u}_i) = 0 \tag{11.2.14}$$

where $\hat{\mathbf{X}}_z$ is the unit vector along *OZ*. These yield the twelve algebraic equations, (11.2.15)-(11.2.26):

$$a_1^1 u_{ex} + a_2^1 u_{ey} + a_e^1 u_{e}, + a_4' u_{ix} + 0 + 0 + a_n^1 u_{nx} + 0$$
$$+ 0 + a_{10}^1 E_x + 0 + 0 = 0, \quad (11.2.15)$$

$$a_1^2 u_{ex} + a_2^2 u_{ey} + a_3^2 u_{ez} + 0 + a_5^2 u_{iy} + 0 + 0 + a_8^2 u_{ny}$$
$$+ 0 + 0 + a_{11}^2 E_y + 0 = 0, \quad (11.2.16)$$

$$a_1^3 u_{ex} + a_2^3 u_{ey} + a_3^3 u_{ez} + 0 + 0 + a_6^3 u_{iz} + 0 + 0$$
$$+ a_9^3 u_{nz} + 0 + 0 + a_{12}^3 E_z = 0, \quad (11.2.17)$$

$$a_1^4 u_{ex} + 0 + a_3^4 u_{ez} + a_4^4 u_{ix} + a_5^4 u_{iy} + a_6^4 u_{iz} + a_7^4 u_n$$
$$+ 0 + 0 + a_{10}^4 E_x + 0 + 0 = 0, \quad (11.2.18)$$

$$0 + a_2^5 u_{ey} + 0 + a_4^5 u_{ix} + a_5^5 u_{iy} + a_6^5 u_{iz} + a_7^5 u_{nx}$$
$$+ a_8^5 u_{ny} + 0 + 0 + a_{11}^5 E_y + 0 + 0, \quad (11.2.19)$$

$$0 + 0\, a_3^6 u_{ez} + a_4^6 u_{ix} + a_5^6 u_{iy} + a_6^6 u_{iz} + 0 + 0$$
$$+ a_9^6 u_{ny} + 0 + 0 + a_{12}^6 E_z = 0, \quad (11.2.20)$$

$$a_1^7 u_{ex} + 0 + 0 + a_4^7 u_{ix} + 0 + 0 + a_7^7 u_{nx} + 0$$
$$+ 0 + 0 + 0 + 0 = 0, \quad (11.2.21)$$

$$0 + a_2^8 u_{ey} + 0 + 0 + a_5^8 u_{iy} + 0 + 0 + a_8^8 u_{ny}$$
$$+ 0 + 0 + 0 + 0 = 0, \quad (11.2.22)$$

$$0 + 0 + a_3^9 u_{ez} + 0 + 0 + a_6^9 u_{iz} + 0 + 0$$
$$+ a_9^9 u_{nz} + 0 + 0 + 0 = 0, \quad (11.2.23)$$

$$a_1^{10} u_{ex} + 0 + 0 + a_4^{10} u_{ix} + 0 + 0 + 0 + 0$$
$$+ 0 + a_{10}^{10} E_x + 0 + 0 = 0, \quad (11.2.24)$$

$$0 + a_2^{11} u_{ey} + 0 + 0 + a_5^{11} u_{iy} + 0 + 0 + 0$$
$$+ 0 + 0 + a_{11}^{11} E_y + 0 = 0, \quad (11.2.25)$$

$$0 + 0 + a_3^{12} u_{ez} + 0 + 0 + a_6^{12} u_{iz} + 0 + 0$$
$$+ 0 + 0 + 0 + a_{12}^{12} E_z = 0 \quad (11.2.26)$$

where $a_1^1 = \omega\,(\omega + i\nu_e + i\nu_-),\ a_1^2 = i\omega\Omega_{ez},$

$a_2^3 = i\Omega\,\Omega_{ey},\ a_1^4 = -\,i\omega\nu_i,\ a_1^7 = -\,i\omega\nu_n^e,\ a_1^{10} = -\,4\pi N_0\, ie\omega,$

$a_2^1 = i\omega\,\Omega_{ez},\ a_2^2 = \omega\,(\omega + i\nu_e + i\nu_-),\ a_2^3 = -\,i\omega\,\Omega_{ex},$

$a_2^5 = -\,i\omega\nu_i,\ a_2^8 = -\,i\omega\nu_n^e,\ a_2^{11} = -\,4\,\pi\,N_0\, ie\omega,$

$$a_3^1 = -\,i\omega\,\Omega_{ey},\ a_3^2 = i\omega\,\Omega_{ex},\ a_3^3 = \omega\,(\omega + i\nu_e + i\nu_-) - K^2\,v_e^2,$$

$$a_3^4 = i\omega\,\Omega_{iy},\ a_3^6 = -\,i\omega\nu_i,\ a_3^9 = -\,i\omega\,\nu_n^e,\ a_3^{12} = -\,4\pi N_0\,ie\omega,$$

$$a_4^1 = -\,i\omega\nu_e,\ a_4^4 = \omega\,(\omega + i\nu_i + i\nu_+),\ a_4^5 = i\omega\,\Omega_{iz},$$

$$a_4^6 = -\,i\omega\,\Omega_{iy},\ a_4^7 = -\,i\omega\nu_n^i,\ a_4^{10} = 4\pi\,N_0\,ie\omega,$$

$$a_5^2 = -\,i\omega\nu_e,\ a_5^4 = -\,i\omega\,\Omega_{iz},\ a_5^5 = \omega\,(\omega + i\nu_i + i\nu_+),$$

$$a_5^6 = i\omega\,\Omega_{ix},\ a_5^8 = -\,i\omega\nu_n^i,\ a_5^{11} = 4\pi\,N_0\,ie\omega,$$

$$a_6^3 = -\,i\omega\nu_-,\ a_6^4 = i\omega\,\Omega_{iy},\ a_6^5 = -\,i\omega\Omega_{ix},$$

$$a_6^6 = \omega\,(\omega + i\nu_i + i\nu_+) - k^2\,v_e^2,\ a_6^9 = -\,i\omega\nu_n^i,\ a_6^{12} = 4\pi N_0\,ie\omega,$$

$$a_7^1 = -\,i\omega\nu_n^e,\ a_7^4 = -\,i\omega\nu_n^i,\ a_7^7 = \omega\,(\omega + i\nu_n^e + i\nu_n^i),$$

$$a_8^2 = -\,i\omega\nu_n^e,\ a_8^5 = -\,i\omega\nu_n^i,\ a_8^8 = \omega\,(\omega + i\nu_n^e + i\nu_n^i) - K^2\,v_e^2,$$

$$a_9^3 = -\,i\omega\nu_n^e,\ a_9^6 = -\,i\omega\nu_n^i,\ a_9^9 = \omega\,(\omega + i\nu_n^e + i\nu_n^i) - K^2\,v_e^2,$$

$$a_{10}^1 = -\,4\pi N_0\,ie\omega,\ a_{10}^4 = 4\,\pi\,N_0\,ie\omega,\ a_{10}^{10} = \omega^2 - K^2\,c^2,$$

$$a_{11}^2 = -\,4\pi N_0\,ie\omega,\ a_{11}^5 = 4\pi N_0\,ie\omega,\ a_{11}^{11} = \omega^2 - K^2\,c^2,$$

$$a_{12}^3 = -\,4\pi\,N_0\,ie\omega,\ a_{12}^6 = 4\pi N_0\,ie\omega,\ a_{12}^{12} = \omega^2$$

The eliminant of these equations, when equated to zero, gives the dispersion relation

$$\begin{vmatrix}
a_1^1 & a_2^1 & a_3^1 & a_4^1 & 0 & 0 & a_7^1 & 0 & 0 & a_{10}^1 & 0 & 0 \\
a_1^2 & a_2^2 & a_3^2 & 0 & a_5^2 & 0 & 0 & a_8^2 & 0 & 0 & a_{11}^2 & 0 \\
a_1^3 & a_2^3 & a_3^3 & 0 & 0 & a_6^3 & 0 & 0 & a_9^3 & 0 & 0 & a_{12}^3 \\
a_1^4 & 0 & a_3^4 & a_4^4 & a_5^4 & a_6^4 & a_7^4 & 0 & 0 & a_{10}^4 & 0 & 0 \\
0 & a_2^5 & 0 & a_4^5 & a_5^5 & a_6^5 & 0 & a_8^5 & 0 & 0 & a_{11}^5 & 0 \\
0 & 0 & a_3^6 & a_4^6 & a_5^6 & a_6^6 & 0 & 0 & a_9^6 & 0 & 0 & a_{12}^6 \\
a_1^7 & 0 & 0 & a_4^7 & 0 & 0 & a_7^7 & 0 & 0 & 0 & 0 & 0 \\
0 & a_2^8 & 0 & 0 & a_5^8 & 0 & 0 & a_8^8 & 0 & 0 & 0 & 0 \\
0 & 0 & a_3^9 & 0 & 0 & a_6^9 & 0 & 0 & a_9^9 & 0 & 0 & 0 \\
a_1^{10} & 0 & 0 & a_4^{10} & 0 & 0 & 0 & 0 & 0 & a_{10}^{10} & 0 & 0 \\
0 & a_2^{11} & 0 & 0 & a_5^{11} & 0 & 0 & 0 & 0 & 0 & a_{11}^{11} & 0 \\
0 & 0 & a_3^{12} & 0 & 0 & a_6^{12} & 0 & 0 & 0 & 0 & 0 & a_{12}^2
\end{vmatrix} = 0 \qquad (11.2.27)$$

The first six of the equations (11.2.15)-(11.2.26) show the coupling between transverse and longitudinal waves due to the field $\mathbf{H}_0$. For example, in equation (11.2.15), the transverse components u_{ex}, u_{ey}, u_{ix}, u_{iy}, E_x, E_y are linearly connected to the longitudinal component u_{ez} of $\mathbf{u}_e$ by Hoy. The single dispersion relation for all the waves, which can be sustained by the plasma model considered here, is the twelfth order determinant of (11.2.27), equated to zero. It yields the familiar dispersion relations of Chapter 3 as special cases.

In undamped plasmas, equations (11.2.8) and (11.2.9) reduce to

$$\ddot{\mathbf{u}}_e = c_e^2 \nabla (\nabla . \mathbf{u}_e) - \frac{e}{m} \dot{\mathbf{E}} - (\dot{\mathbf{u}}_e \times \mathbf{\Omega}_e), \tag{11.2.28}$$

$$\ddot{\mathbf{u}}_i = C_i^2 \nabla (\nabla . \mathbf{u}_i) + \frac{e}{M} \dot{\mathbf{E}} + (\dot{\mathbf{u}}_i \times \mathbf{\Omega}_i) \tag{11.2.29}$$

For an electromagnetic field we put $(\nabla . \mathbf{E}) = 0$, and obtain

$$\ddot{\mathbf{E}} - c^2 \nabla^2 \mathbf{E} + 4\pi N_0 e (\dot{\mathbf{u}}_i - \dot{\mathbf{u}}_e) = 0,$$

$$\ddot{\mathbf{u}}_e = -\frac{e}{M} \dot{\mathbf{E}} - (\dot{\mathbf{u}}_e \times \mathbf{\Omega}_e), \quad \ddot{\mathbf{u}}_i = \frac{e}{M} \dot{\mathbf{E}} + (\dot{\mathbf{u}}_i \times \Omega_i) \tag{11.2.31}$$

Integrating the two equations of (11.2.31), with respect to time, and putting the values of $\dot{\mathbf{u}}_e$ and $\dot{\mathbf{u}}_i$ thus obtained on their right hand side, we obtain

$$\ddot{\mathbf{u}}_e + \Omega_e^2 \mathbf{u}_e = \frac{e}{m} (\mathbf{E} \times \Omega_e) - \frac{e}{m} \dot{\mathbf{E}},$$

$$\ddot{\mathbf{u}}_i + \Omega_i^2 \mathbf{u}_i = \frac{e}{M} (\mathbf{E} \times \mathbf{\Omega}_i) + \frac{e}{M} \dot{\mathbf{E}} \tag{11.2.32}$$

The gyro frequencies Ω_e and Ω_i are some characteristic (or natural) frequencies of the system. If collisional dissipation is ignored, an applied electric field, at any of these two frequencies, would give secular solutions of instability for the velocity field (see Section 2.18).

11.3 ANISOTROPY OF MAGNETIZED PLASMA

For high frequency waves, for simplicity mobility of ions and uncharged fluids is neglected. Then, dropping the subscript e from $\mathbf{u}_e$, N_e and $\mathbf{\Omega}_e$, the field equations, for electron motion are obtained:

$$(\omega + i\nu)\, u_x + i\Omega_z\, u_y - i\Omega_y\, u_z = -\frac{ie}{m\omega} E_x, \tag{11.3.1}$$

$$- i\Omega_z\, u_x + (\omega + i\nu)\, u_y + i\Omega_x\, u_z = -\frac{ie}{m\omega} E_y, \tag{11.3.2}$$

$$i\Omega_y\, u_x - i\Omega_x\, u_y + \left(\omega + i\nu - \frac{K^2 c_e^2}{\omega} \right) u_z = -\frac{ie}{m\omega} E_z \tag{11.3.3}$$

The coefficients of u_x, u_y, u_z, in the left hand sides of these equations, form the determinant

$$D = \omega^2 \nabla = \omega^2 \begin{vmatrix} \omega + i\nu & i\Omega_z & -i\Omega_y \\ -i\Omega_z & \omega + i\nu & i\Omega_x \\ i\Omega_y & -i\Omega_x & \omega + i\nu - \dfrac{K^2 c_e^2}{\omega} \end{vmatrix} \tag{11.3.4}$$

Expanding we obtain

$$D = \omega^2 [(\omega^2 + i\omega\nu - K^2 c_e^2) \{(\omega + i\nu)^2 - \Omega_z^2\} - \omega (\omega + i\nu) (\Omega_x^2 + \Omega_y^2)] \tag{11.3.5}$$

The solution of (11.3.1)-(11.3.3) is written as

$$u_i = \frac{A_i}{D}, \; j_i = -N_0 e u_i = -\frac{N_0}{D} eA_i \tag{11.3.6}$$

where A_i is obtained by replacing the elements of the ith column of D by E_x, E_y, E_z. Hence

$$A_1 = -\frac{ie}{m}\omega \begin{vmatrix} E_x & i\omega\Omega_z & -i\omega\Omega_y \\ E_y & \omega^2 + i\nu\omega & i\omega\Omega_x \\ E_z & -i\omega\Omega_x & \omega^2 + i\omega\nu - K^2 c_e^2 \end{vmatrix}$$

$$= \frac{ie}{m}\omega^2 \; [E_x \{-(\omega + i\nu)(\omega^2 + i\omega\nu - K^2 c_e^2) + \omega\,\Omega_x^2\}$$
$$+ E_y \{\omega\Omega_x\Omega_y + i\Omega_z(\omega^2 + i\omega\nu - K^2 c_e^2)\}$$
$$+ E_z \{\omega\Omega_z\Omega_x + i\omega\Omega_y(\omega + i\nu)\}], \tag{11.3.7}$$

$$A_2 = \frac{ie}{m}\omega \begin{vmatrix} \omega^2 + i\nu\omega & E_x & -\omega\Omega_y \\ -i\omega\Omega_z & E_y & i\omega\Omega_x \\ i\omega\Omega_y & E_z & \omega^2 + i\omega\nu - K^2 c_e^2 \end{vmatrix}$$

$$= \frac{ie}{m}\omega^2 \left[E_x\omega \left\{ -i\Omega_z \left(\omega + i\nu - \frac{K^2 c_e^2}{\omega} \right) + \Omega_x\Omega_y \right\} \right.$$
$$\left. + E_y\omega \left\{ \Omega_y^2 - (\omega + i\nu)\left(\omega + i\nu - \frac{K^2 c_e^2}{\omega} \right) \right\} + E_z \{ i(\omega + i\nu)\Omega_x + \Omega_y\Omega_z \} \right], \tag{11.3.8}$$

$$A_3 = -\frac{ie}{m}\,\omega^2\,[E_x\,\{-\Omega_z\,\Omega_x - i\Omega_y\,(\omega + iv)\} + E_y\,\{-\Omega_y\,\Omega_z$$

$$+ \{i\Omega_x\,(\omega\ \ iv)\} + E_z\,\{(\omega + iv)^2 - \Omega_z^2\}] \qquad (11.3.9)$$

The components of the current **j** and the elements of the conductivity tensor σ_{ij} are

$$j_x = -\frac{i\omega_\rho^2\,\omega}{4\pi\,\Delta}\left[E_x\left\{-(\omega + iv)\left(\omega + iv - \frac{K^2\,c_e^2}{\omega}\right) + \Omega_x^2\right\}\right.$$

$$+ E_y\left\{\Omega_x\,\Omega_y + i\Omega_z\left(\omega + iv - \frac{K^2\,c_e^2}{\omega}\right)\right\}$$

$$\left. + E_z\,\{\Omega_z\,\Omega_x + i\Omega_y\,(\omega + iv)\}\right], \qquad (11.3.10)$$

$$j_y = -\frac{i\omega_\rho^2\,\omega}{4\pi\,\Delta}\left[E_x\left\{-i\Omega_z\left(\omega + iv - \frac{K^2\,c_e^2}{\omega}\right) + \Omega_x\,\Omega_y\right]\right.$$

$$+ E_y\left\{\Omega_y^2 - (\omega + iv)\left(\omega + iv - \frac{K^2\,c_e^2}{\omega}\right)\right\}$$

$$\left. + E_z\,\{i\,(\omega + iv)\,\Omega_x + \Omega_y\,\Omega_z\}\right], \qquad (11.3.11)$$

$$j_z = -\frac{i\omega_\rho^2\,\omega}{4\pi\Delta}\,[E_x\,\{-\Omega_z\,\Omega_x - i\Omega_y\,(\omega + i\nu)\} + E_y\,\{-\Omega_y\,\Omega_z + i\Omega_x\,(\omega + iv)\}$$

$$+ E_z\,\{(\omega + iv)^2 - \Omega_z^2\}], \qquad (11.3.12)$$

$$\sigma_{xx} = -\frac{i\omega_\rho^2\,\omega}{4\pi\Delta}\left\{-(\omega + iv)\left(\omega + iv - \frac{K^2\,c_e^2}{\omega}\right) + \Omega_x^2\right\}, \qquad (11.3.13)$$

$$\sigma_{xy} = -\frac{i\omega_\rho^2\,\omega}{4\pi\Delta}\left\{\Omega_x\,\Omega_y + i\Omega_z\left(\omega + iv - \frac{K^2\,c_e^2}{\omega}\right)\right\}, \qquad (11.3.14)$$

$$\sigma_{xz} = \frac{i\omega_\rho^2\,\omega}{4\pi\Delta}\,\{\Omega_z\,\Omega_x + i\Omega_y\,(\omega + iv)\}, \qquad (11.3.15)$$

$$\sigma_{yx} = -\frac{i\omega_\rho^2\,\omega}{4\pi\Delta}\left\{-i\Omega_z\left(\omega + iv - \frac{K^2\,c_e^2}{\omega}\right) + \Omega_x\,\Omega_y\right\}, \qquad (11.3.16)$$

$$\sigma_{yy} = -\frac{i\omega_\rho^2\,\omega}{4\pi\Delta}\left\{\Omega_y^2 - (\omega + iv)\ (\omega + iv) - \frac{K^2\,c_e^2}{\omega}\right\}, \qquad (11.3.17)$$

$$\sigma_{yz} = -\frac{i\omega_\rho^2 \omega}{4\pi\Delta} \{i(\omega + i\nu)\Omega_x + \Omega_y \Omega_z\}, \tag{11.3.18}$$

$$\sigma_{zx} = -\frac{i\omega_\rho^2 \omega}{4\pi\Delta} \{\Omega_z \Omega_x - i\Omega_y(\omega + i\nu)\}, \tag{11.3.19}$$

$$\sigma_{zy} = -\frac{i\omega_\rho^2 \omega}{4\pi\Delta} \{-\Omega_y \Omega_z + i\Omega_x(\omega + i\nu)\}, \tag{11.3.20}$$

$$\sigma_{zz} = -\frac{i\omega_\rho^2 \omega}{4\pi\Delta} \{(\omega + i\nu)^2 - \Omega_z^2\}, \tag{11.3.21}$$

The relations (11.3.10)-(11.3.21) give the components of the alternating current field and the components of the electrical conductivity tensor excited by an electromagnetic wave, which is an A.C. field. The direct current and the D.C. conductivity tensor are also excited by collision dominated processes; see chapter 6. We observe that $\sigma_{ij} \neq \sigma_{ji}$ when $i \neq j$, but $|\sigma_{ij}| = |\sigma_{ji}|$, $\sigma_{xy} = -\overline{\sigma}_{yx}$, $\sigma_{xx} \neq \sigma_{yy} \neq \sigma_{zz}$ where $\overline{\sigma}_{yx}$ is the complex conjugate of σ_{yx}. When $H_0 = 0$ we find that $\sigma_{ij} = 0$ for $i \neq j$ and $\sigma_{xx} = \sigma_{yy}$. The elements of the dielectric tensor are determined with the help of equation (3.12.9):

$$\varepsilon_{ij} = \delta_{ij} + \frac{4\pi i}{\omega} \sigma_{ij} \tag{11.3.22}$$

11.4 APPLETON-HARTREE EQUATION

The Appleton-Hartree equation of magneto-ionic theory is deduced from the equation for electron motion in dissipative cold plasmas in the presence of an applied homogeneous magnetic field $\mathbf{H}_0$. These equations are

$$\dot{\mathbf{u}} + \nu\mathbf{u} = -\frac{e}{m}\mathbf{E} - (\mathbf{u} \times \Omega), \ (\nabla \cdot \mathbf{H}) = 0,$$

$$\nabla \times \mathbf{H} = \frac{1}{c}\frac{\partial \mathbf{E}}{\partial t} - \frac{4\pi}{c} Ne\mathbf{u}, \ \nabla \times \mathbf{E} = -\frac{1}{c}\frac{\partial \mathbf{H}}{\partial t} \tag{11.4.1}$$

where $\mathbf{\Omega} = e\mathbf{H}_0/mc$ and ν is the mean frequency of collision, and the electronic charge is $-e$. If the time dependence $\exp(-i\omega t)$ is assumed, these equations become

$$(\nu - i\omega)\, u_i = -\frac{e}{m} E_i - \varepsilon_{ijk}\Omega_k u_j, \ \frac{\partial H_i}{\partial x_i} = 0,$$

$$\varepsilon_{ijk}\frac{\partial E_k}{\partial x_j} = \frac{i\omega}{c} H_i, \ \varepsilon_{ijk}\frac{\partial H_k}{\partial x_j} = -\frac{i\omega}{c} D_i \tag{11.4.2}$$

where $$D_i = \varepsilon_{ij} E_j = \left(\delta_{ij} + \frac{4\pi i}{\omega}\sigma_{ij}\right) E_j, \ j_i = -Ne\, u_i = \sigma_{ij} E_j, \tag{11.4.3}$$

ε_{ijk} is the alternating unit tensor of (A 9.3), and the Einstein summation convention is adopted. If the space dependence is proportional to $\exp(ik_j x_j)$, then

$$H_i = \frac{c}{\omega}\varepsilon_{ijk} E_k,\ D_i = -\frac{c}{\omega} k_j \varepsilon_{ijk} H_k \tag{11.4.4}$$

Elimination of H_i gives

$$-\frac{\omega^2}{c^2} D_i = k_l k_m \delta_{il} E_m - k^2 E_i \tag{11.4.5}$$

Replacing D_i by $\varepsilon_{ij} E_j$, in this equation, we get

$$\left\{\varepsilon_{ij} + n^2\left(\frac{k_i k_j}{k^2} - \delta_{ij}\right)\right\}, E_j = 0 \tag{11.4.6}$$

where $n\ (= kc/\omega)$ is the refractive index. The matrix form of this equation is

$$\begin{bmatrix} \varepsilon_{11} - n^2\left(1 + \frac{k_1^2}{k^2}\right) & \varepsilon_{12} + n^2\frac{k_1 k_2}{k^2} & \varepsilon_{13} + n^2\frac{k_1 k_3}{k^2} \\ \varepsilon_{21} + n^2\frac{k_2 k_1}{k^2} & \varepsilon_{22} - n^2\left(1 - \frac{k_2^2}{k^2}\right) & \varepsilon_{23} + n^2\frac{k_2 k_3}{k^2} \\ \varepsilon_{31} + n^2\frac{k_3 k_1}{k^2} & \varepsilon_{32} + n^2\frac{k_3 k_2}{k^2} & \varepsilon_{33} - n^2\left(1 - \frac{k_3^2}{k^2}\right) \end{bmatrix} \begin{bmatrix} E_1 \\ E_2 \\ E_3 \end{bmatrix} = 0 \tag{11.4.7}$$

The eliminant equation is the determinantal equation

$$\begin{vmatrix} \varepsilon_{11} - n^2\left(1 + \frac{k_1^2}{k^2}\right) & \varepsilon_{12} + n^2\frac{k_1 k_2}{k^2} & \varepsilon_{13} + n^2\frac{k_1 k_3}{k^2} \\ \varepsilon_{21} + n^2\frac{k_2 k_1}{k^2} & \varepsilon_{22} - n^2\left(1 - \frac{k_2^2}{k^2}\right) & \varepsilon_{23} + n^2\frac{k_2 k_3}{k^2} \\ \varepsilon_{31} + n^2\frac{k_3 k_1}{k^2} & \varepsilon_{32} + n^2\frac{k_3 k_2}{k^2} & \varepsilon_{33} - n^2\left(1 - \frac{k_3^2}{k^2}\right) \end{vmatrix} = 0 \tag{11.4.8}$$

This cubic in n^2 is the Appleton-Hartree equation of ionospheric physics. Here the values of ε_{ij} are obtained from the momentum transfer equation. If OX_3 is parallel to $\mathbf{H}_0$, and the direction of propagation is parallel to the plane of

OX_3 and OX_2, then $k_1 = 0$, $k_2 = k \sin\theta$, $k_2 = k \sin\theta$, $k_3 = k \cos\theta$ and the Appleton-Hartree dispersion equation reads

$$\begin{vmatrix} \varepsilon_{11} - n^2 & \varepsilon_{12} & \varepsilon_{13} \\ \varepsilon_{21} & \varepsilon_{22} - n^2 \cos^2\theta & \varepsilon_{23} + n^2 \sin\theta \cos\theta \\ \varepsilon_{31} & \varepsilon_{32} + n^2 \sin\theta \cos\theta & \varepsilon_{33} - n^2 \sin^2\theta \end{vmatrix} = 0 \quad (11.4.9)$$

11.5 DIELECTRIC AND CONDUCTIVITY TENSORS

Some standard symbols of magneto-ionic theory are

$$Z = \frac{v}{\omega}, \; u = 1 + iZ, \; \mathbf{y} = \frac{\mathbf{\Omega}}{\omega}, \; x = \frac{\omega_\rho^2}{\omega^2} \quad (11.5.1)$$

Using these and assuming that the applied magnetic field $\mathbf{H}_0$ is parallel to the OX_3 axis, the momentum transfer equation reads

$$(u\delta_{ij} + iy\,\varepsilon_{ij3})\,u_j = -\frac{ie}{m\omega} E_i \quad (11.5.2)$$

where $y = y_3 = \Omega/\omega$. The matrix form of (11.5.2) is

$$\begin{bmatrix} u & iy & 0 \\ -iy & u & 0 \\ 0 & 0 & u \end{bmatrix} \begin{bmatrix} u_1 \\ u_2 \\ u_3 \end{bmatrix} = -\frac{ie}{m\omega} \begin{Bmatrix} E_1 \\ E_2 \\ E_3 \end{Bmatrix} \quad (11.5.3)$$

Defining the new matrix $\mathbf{P}$ by

$$\mathbf{P} = \begin{bmatrix} u & iy & 0 \\ -iy & u & 0 \\ 0 & 0 & u \end{bmatrix}, \quad (11.5.4)$$

this equation becomes

$$\begin{bmatrix} u_1 \\ u_2 \\ u_3 \end{bmatrix} = -\frac{ie}{m\omega} \mathbf{P}^{-1} \begin{bmatrix} E_1 \\ E_2 \\ E_3 \end{bmatrix} \quad (11.5.5)$$

where $\mathbf{P}^{-1}$ is the inverse matrix of $\mathbf{P}$. Since

$$\det \mathbf{P} = u\,(u^2 - y^2), \quad (11.5.6)$$

this equation becomes

$$\begin{bmatrix} u_1 \\ u_2 \\ u_3 \end{bmatrix} = -\frac{ie}{m\omega u\,(u^2 - y^2)} \begin{bmatrix} u^2 & -iyu & 0 \\ iyu & u^2 & 0 \\ 0 & 0 & u^2 - y^2 \end{bmatrix} \begin{bmatrix} E_1 \\ E_2 \\ E_3 \end{bmatrix} \quad (11.5.7)$$

These give the following expression for elements of the dielectric and conductivity tensors:

$$\sigma_{ij} = \frac{i\omega\rho^2}{4\pi\,\omega u\,(u^2 - y^2)} \begin{bmatrix} u^2 & -iyu & 0 \\ iyu & u^2 & 0 \\ 0 & 0 & u^2 - y^2 \end{bmatrix} \tag{11.5.8}$$

$$\varepsilon_{ij} = \delta_{ij} + \frac{4\pi i}{\omega}\sigma_{ij} = \begin{bmatrix} 1 - \dfrac{xu}{u^2 - y^2} & -\dfrac{ixy}{u^2 - y^2} & 0 \\ \dfrac{ixy}{u^2 - y^2} & 1 - \dfrac{xu}{u^2 - y^2} & 0 \\ 0 & 0 & 1 - \dfrac{x}{u} \end{bmatrix}, \tag{11.5.9}$$

where (3.12.9), has been used. Putting $S = 1 - xu/(u^2 - y^2)$, $D = xy/(u^2 - y^2)$, $T = 1 - X/u$ in ε_{ij} of (11.5.9) we obtain

$$\varepsilon_{ij} = \begin{bmatrix} S & -iD & 0 \\ iD & S & 0 \\ 0 & 0 & T \end{bmatrix} \tag{11.5.10}$$

The real parts of the elements ε_{ij} are

$$\varepsilon_{12} = \varepsilon_{22} = 1 - \frac{X}{2}\left\{\frac{(1+y)}{(1+y)^2 + z^2} + \frac{(1-y)}{(1+y)^2 + z^2}\right\}, \tag{11.5.11}$$

$$\varepsilon_{12} = -\varepsilon_{21} = -\frac{X}{2}\left\{\frac{z}{(1+y)^2 + z^2} + \frac{z}{(1+y)^2 + z^2}\right\},$$

$$\varepsilon_{13} = \varepsilon_{31} = 0,\ \varepsilon_{32} = \varepsilon_{23} = 0,\ \varepsilon_{33} = 1 - x/(1 + z^2);$$

So the dispersion relation becomes

$$\begin{vmatrix} S - n^2 & -iD & 0 \\ iD & S - n^2\cos^2\theta & n^2\cos\theta\ \sin\theta \\ 0 & n^2\sin\theta\ \cos\theta & T - n^2\sin^2\theta \end{vmatrix} = 0 \tag{11.5.12}$$

Expansion yields, for n, the equation

$$A_0 n^2 + c_0 = 0 \tag{11.5.13}$$

where
$$A_0 = S\sin^2\theta + T\cos^2\theta,$$
$$B_0 = RL\sin^2\theta + TS\,(1 + \cos^2\theta),$$
$$C_0 = TRL,\ R = S + D,\ L = S - D \tag{11.5.14}$$

The solutions for n^2 are

$$n^2 = 1 - \frac{X}{\left\{ u - \frac{y_t^2}{2(U-X)} \pm \left[y_l^2 + \frac{y_t^4}{4(U-X)^2} \right]^{1/2} \right\}} \tag{11.5.15}$$

where $$y_t = y \sin \theta, \; y_l = y \cos \theta \tag{11.5.16}$$

For $\theta = \pi/2$ the two roots, n_1^2 and n_2^2 of (11.5.15), are given by

$$n_1^2 = 1 - \frac{X}{U}, \; n_2^2 = 1 - \frac{X}{U - \frac{Y^2}{U-X}} \tag{11.5.17}$$

The refractive index of the ordinary wave is n_1, and n_2 is that of the extraordinary wave. The ordinary wave is linearly polarized with **E** parallel to $\mathbf{H}_0$; its dispersion relation is not affected by $\mathbf{H}_0$. The extraordinary wave, excited by a linearly polarized wave, has electric field in the plane perpendicular to $\mathbf{H}_0$. A square horn and wave guide, separating the waves polarized at right angles to each other, are used together with a polarization diplexer to observe both the ordinary and extraordinary waves simultaneously on a dual beam oscilloscope for propagation perpendicular to the applied magnetic field $\mathbf{H}_0$.

11.6 WAVES IN THE PRINCIPAL COORDINATE SYSTEM

In the principal coordinate system the non-diagonal elements of the dielectric tensor vanish; only the diagonal elements are non-zero. The three principal values of ε_{ij} are the three values of λ of the cubic equation

$$\begin{vmatrix} S-\lambda & -iD & 0 \\ iD & S-\lambda & 0 \\ 0 & 0 & T-\lambda \end{vmatrix} = 0 \tag{11.6.1}$$

These values are $\lambda_1 = S + D$, $\lambda_2 = S - D$, $\lambda_3 = T$. If l_i, m_i, n_i are the direction consines of the principal axis, corresponding to $\lambda = \lambda_i$, then

$$\begin{bmatrix} -D & -iD & 0 \\ iD & -D & 0 \\ 0 & 0 & T-S-D \end{bmatrix} \begin{bmatrix} l_1 \\ m_1 \\ n_1 \end{bmatrix} = 0 \tag{11.6.2}$$

The corresponding linear equations are

$$Dl_1 + iDm_1 = 0, \; iDl_1 - Dm_1 = 0, \; (T-S-D)\, n_1 = 0$$

Since $T - S - D \neq 0$, we have $n_1 = 0$. Also $il_1 = m_1$. Hence $(l_1, m_1, n_1) = (1, i, 0)/\sqrt{2}$. Similarly for $\lambda = \lambda_2 = S - D$ we get $(l_2, m_2, n_2) \equiv (i, -i, 0)/\sqrt{2}$ and

$(l_3, m_3, n_3) \equiv (0, 0, 1)$ for $\lambda = \lambda_3 = T$. If the elements (l_1, m_1, n_1) and (l_2, m_2, n_2) are multiplied by $\exp(-i\omega t)$, the corresponding complex directions represent unit vectors rotating in the right and left hand rotational senses in the plane normal to the third direction (l_3, m_3, n_3).

Denoting the principal dielectric constants by primes over ε_{ij}, we obtain

$$\varepsilon_{11}' = S + D,\ \varepsilon_{22}' = S - D,\ \varepsilon_{33}' = T,\ \varepsilon_{ij}' = 0 \text{ for } i \neq j \tag{11.6.3}$$

Since T is independent of Ω, the component of **D**, parallel to $\mathbf{H}_0$, is not connected to the other components. The corresponding elements of the conductivity tensor σ_{ij}' are

$$\sigma_{11}' = \frac{i\omega}{4\pi}(1 - S - D),\ \sigma_{22}' = \frac{i\omega}{4\pi}(1 - S - D),\ \sigma_{33}' = \frac{i\omega}{4\pi}(1 - T) \tag{11.6.4}$$

If the real parts of ε_{ij}' and σ_{ij}' are also denoted by the same symbols, then

$$\varepsilon_{11}' = 1 - \frac{\omega_\rho^2(\omega + \Omega)}{\omega\{(\omega + \Omega)^2 + v^2\}},\ \varepsilon_{22}' = 1 - \frac{\omega_\rho^2(\omega - \Omega)}{\omega\{(\omega - \Omega)^2 + v^2\}},\ \varepsilon_{33}' = 1 - \frac{\omega_\rho^2}{\omega^2 + v^2}, \tag{11.6.5}$$

$$\sigma_{11}' = \frac{\omega_\rho^2 v}{4\pi\{(\omega + \Omega)^2 + v^2\}},\ \sigma_{22}' = \frac{\omega_\rho^2 v}{4\pi\{(\omega - \Omega)^2 + v^2\}},\ \sigma_{33}' = \frac{\omega_\rho^2 v}{4\pi(\omega^2 + v^2)} \tag{11.6.6}$$

The power P, absorbed by the plasma from the wave, is given by

$$\rho = \sigma_{11}'(\text{Re } E_x')^2 + \sigma_{22}'(\text{Re } E_y')^2 + \sigma_{33}'(\text{Re } E_z')^2 \tag{11.6.7}$$

where ReX is the real part of X, and

$$E_x' = \frac{1}{\sqrt{2}}(E_x + iE_y),\ E_y' = \frac{1}{\sqrt{2}}(E_x - iE_y),\ E_z' = E_z \tag{11.6.8}$$

Averaging equation (11.6.7), over the time period $2\pi/\omega$, we get

$$\langle P \rangle = \sigma_{11}'\langle(\text{Re } E_x')^2\rangle + \sigma_{22}'\langle(\text{Re } E_y')^2\rangle + \sigma_{33}'\langle(\text{Re } E_z')^2\rangle \tag{11.6.9}$$

So heating will be maximum for maximum value of σ_{11}', σ_{22}', σ_{33}'. For $v << \omega$ the value of σ_{11}' is largest at $\omega = -\Omega$ which is the condition of resonance for the extraordinary wave at the Larmor frequency.

11.7 APPLETON-HARTREE EQUATION IN THERMAL PLASMAS

In warm dissipative plasmas the momentum equation for electron motion is

$$\dot{\mathbf{u}} + v\mathbf{u} = -\frac{c_e^2}{N_0}\nabla N - \frac{e}{m}\mathbf{E} - (\mathbf{u} \times \mathbf{\Omega}) \tag{11.7.1}$$

where $c_e^2 = 3K_B T/m$, K_B is the Boltzmann constant and T is the kinetic temperature in degrees Kelvin for electrons; the continuity condition is $\dot{N} + N_0 (\nabla . \mathbf{u}) = 0$ or the flux relation is $(\nabla . \mathbf{E}) = -4\pi eN$. Then $\mathbf{u}$ and $\mathbf{E}$ are connected by the relation

$$(u\delta_{ij} - i\varepsilon_{ijk} y_k) u_j = -\frac{ie}{m\omega} E_i + \frac{c_e^2 k_1 N}{\omega N_0} \tag{11.7.2}$$

where the mass conservation equation, in the form $\omega N = N_0 K_j U_j$, has been used and, as in (11.5.1), $U = 1 + iZ$. For convenience of calculations the axis OX_3 is now chosen parallel to the wave vector $\mathbf{K}$, and $\mathbf{H}_0$ is in the plane of OX_2 and OX_3. Then (10.7.1) reduces to

$$\left.\begin{aligned} &Uy_1 + iy_3 u_2 - iy_2 u_3 = -\frac{ie}{m\omega} E_1, \; -iy_3 u_1 + Uu_2 + iY_1 u_3 = -\frac{ie}{m\omega} E_2, \\ &iy_2 u_1 - iy_1 u_2 + \left(U - \frac{K^2 C_e^2}{\omega^2} \right) u_3 = -\frac{ie}{m\omega} E_3 \end{aligned}\right\} \tag{11.7.3}$$

The matrix form of these equations is

$$\mathbf{P} \begin{bmatrix} u_1 \\ u_2 \\ u_3 \end{bmatrix} = -\frac{ie}{m\omega} \begin{bmatrix} E_1 \\ E_2 \\ E_3 \end{bmatrix} \tag{11.7.4}$$

where
$$\mathbf{P} = \begin{bmatrix} U & iy_l & -iy_t \\ -iy_l & U & -iy_1 \\ iy_t & iy_1 & U_1 \end{bmatrix}, \tag{11.7.5}$$

$$y_l = y_3 = y\cos\theta, \; y_t = y_2 = y\sin\theta, \; U_1 = U - \frac{K^2 c_e^2}{\omega^2}$$

Since $\det \mathbf{P} = U_1 (U^2 - y^2) - y_t^2 \beta^2 n^2$ where $n^2 \beta^2 = k^2 c_e^2/\omega^2$, we obtain

$$\mathbf{P}^{-1} = \frac{1}{U_1 (U^2 - y^2) - y_t^2 \beta^2 n^2} \begin{bmatrix} UU_1 & iy_l U_1 & -iy_t U \\ -iy_1 U_1 & UU_1 - y_t^2 & -iy_l y_t \\ iy_t U & y_l y_t & U^2 - yl^2 \end{bmatrix} \tag{11.7.6}$$

$\mathbf{P}^{-1}$ is the inverse matrix of $\mathbf{P}$. The dielectric and conductivity tensors are obtained from the relations

$$\sigma_{ij} = \frac{i\omega_\rho^2}{4\pi}\mathbf{P}^{-1}, \quad \varepsilon_{ij} = \delta_{ij} - X\mathbf{P}^{-1} \tag{11.7.7}$$

where $X = \omega_\rho^2/\omega^2$. Using now the equation

$$\left\{\varepsilon_{ij} + n^2\left(\frac{k_i - k_j}{k^2} - \delta_{ij}\right)\right\} E_j = 0 \tag{11.7.8}$$

the dispersion relation, for warm plasma, is obtained:

$$\begin{vmatrix} \varepsilon_{11} - n^2 & \varepsilon_{12} & \varepsilon_{13} \\ \varepsilon_{21} & \varepsilon_{22} - n^2 & \varepsilon_{23} \\ \varepsilon_{31} & \varepsilon_{32} & \varepsilon_{33} \end{vmatrix} = 0 \tag{11.7.9}$$

Simplifying we obtain

$$n^2 = 1 - \frac{X}{\left[U - \frac{y_t^2}{2(U - X - \beta^2 n^2)} \pm \left\{y_l^2 + \frac{y_t^4}{4(U - X - \beta^2 n^2)}\right\}^{1/2}\right.} \tag{11.7.10}$$

Other forms of this equation are

$$\left[\frac{X^2}{1-n^2} - U + \frac{y_t^2}{2(U - X - \beta^2 n^2)}\right]^2 = y_l^2 + \frac{y_t^4}{4(U - X - \beta^2 n^2)^2}, \tag{11.7.11}$$

$$(1-n^2)^2\ (U - X - \beta^2 n^2)\ (U^2 - y_l^2) - Uy_t^2(1-n^2) - 2U_x(1-n^2)$$

$$(u - X - \beta^2 n^2) + x^2 (U - X - \beta^2 n^2) + xy_t^2 (1-n^2) = 0 \tag{11.7.12}$$

For $\nu = 0$, it reduces to

$$\frac{(1-n^2)}{n} = \frac{1}{\left[1 - \frac{y_t^2}{2(1 - X - \beta^2 n^2)} \pm \left\{y_l^2 + \frac{y_t^4}{4(1 - X - \beta^2 n^2)}\right\}^{1/2}\right]} \tag{11.7.13}$$

When $|\, y_l^2 |\, y_t^4/4\,(1 - x - \beta^2 n^2)^2 |$, (11.7.14)

Equation (11.7.13) becomes

$$\frac{\omega^2 - k^2 c^2}{\omega_\rho^2} = \frac{1}{1 - \frac{y_t^2}{1 - x - \beta^2 n^2}}, \quad \frac{1}{1 + \frac{y_l^2 (1 - x - \beta^2 n^2)}{y_t^2}}, \tag{11.7.15}$$

and the dispersion relations is

$$k^2 - \frac{(\omega^2 - \omega_\rho^2)}{c^2} = \frac{y_t^2\,\omega_\rho^2\,(1 + X + \beta^2 n^2)}{c^2},\ \frac{y_l^2\,\omega_\rho^2\,(1 - X - \beta^2 n^2)}{c^2\,y_t^2} \tag{11.7.16}$$

When $$|\,y_l^2\,| >> \left|\frac{y_t^4}{4\,(1 - X - \beta^2 n^2)^2}\right| \tag{11.7.17}$$

these relations read

$$\frac{\omega^2 - k^2 c^2}{\omega_\rho^2} = \frac{1}{1 \pm y_l} = 1 \mp y_l \tag{11.7.18}$$

And the solution of the Appleton-Hartree equation (11.7.10) for $\beta^2 n^2$ is

$$\beta^2 n^2 = U - X - y_t^2\left(\frac{X}{1 - n^2} - 1\right)\Big/\left[y_l^2 - \left(\frac{X}{1 - n^2} - 1\right)^2\right] \tag{11.7.19}$$

For $v = 0$ it reduces to

$$k^2 - \frac{\omega^2 - \omega_\rho^2}{c_e^2} = -\frac{\Omega^2}{c_e^2}\sin^2\theta\left(\frac{X}{1 - n^2} - 1\right)\Big/\left[\frac{\Omega^2}{\omega}\cos^2\theta - \left(\frac{X}{1 - n^2} - 1\right)^2\right] \tag{11.7.20}$$

Eliminating, from (11.4.4), (11.7.1) and (11.7.2), the variables N and the components of **u** and **H**, we obtain

$$\begin{aligned}
&(X + Un^2 - U)\,E_1 = i\,(n^2 - 1)\,y_l\,E_2,\\
&(X + Un^2 - U)\,E_2 = -\,iy_l\,(n^2 - 1)\,E_1 + iy_t\,(n^2 - 1)\,E_3,\\
&(X + U_1\,n^2 - U_1)\,E_3 = iy_t^2\,(n^2 - 1)\,E_2
\end{aligned} \tag{11.7.21}$$

The second and third of these relations give

$$\frac{E_2}{E_3} = \frac{(X + U_1\,n^2 - U_1)}{-\,iy_t\,(n^2 - 1)}, \tag{11.7.22}$$

$$\frac{E_1}{E_3} = \frac{y_t}{y_l} - (X + Un^2 - U)\,\frac{(X + U_1\,n^2 - U_1)}{y_t\,y_l\,(n^2 - 1)^2} \tag{11.7.23}$$

Defining now the electrical energy densities

$$W_1 = \frac{E_1^2}{8\pi},\ W_2 = \frac{E_2^2}{8\pi},\ W_l = \frac{E_3^2}{8\pi},\ W_t = \frac{1}{8\pi}\,(E_1^2 + E_2^2), \tag{11.7.24}$$

we find that

$$\frac{W_t}{W_1} = 1 - \frac{(X + Un^2 - U)^2}{y_l^2 (n^2 - 1)^2} \tag{11.7.25}$$

where W_t is the energy of the transverse part of the electric field and W_l is that of the longitudinal part of this field. The ratios W_t/W_1 and W_e/W_l show the distribution of electrical energy in different directions, relative to the directions of **K** and $\mathbf{H}_0$. For $\mathbf{\Omega}_t = 0$ the longitudinal field E_l and the transverse components E_1 and E_2 are not interdependent.

11.8 WHISTLERS

Whistlers are natural bursts, at very low frequencies (below $15Kc/S$), of highly dispersive electromagnetic type, produced by lightening discharges at heights of about 85 km above the earth's surface. These are low frequency sounds, like whistles, produced by the electromagnetic energy of the discharge light, detected with simple audio frequency equipments, and are heard through a loud speaker if it is connected through amplifier to an antenna. The whistlers generally start at a high frequency, in about one second they drop in frequency to a lower limit of about one thousand cycles per second and last for about 2 seconds as a sound of descending pitch.

Due to interaction with free electrons, the wave becomes highly dispersive and, like guided waves, follows the lines of $\mathbf{H}_0$. These lines provide the field with sufficient focussing influence and thus prevent the scattering of field energy. Their paths reach into the outer space as far as three or four times the earth radius (earth radius is about 6370 km), in the plane of the equator, and bring energy from lightening discharge to the earth at a point in the opposite hemisphere which is the magnetic conjugate of the position of radio emission for whistlers. From there the whistler wave is reflected back to the hemisphere from where it started. So the length of one way path can easily be as long as 300000 or more of kilometers. The energy is almost perfectly reflected from earth's surface, four or five times, with increased dispersion and diminishing amplitude. Along such long paths the speed of propagation of energy is between $C/10$ and $C/100$, and the local value depends strongly on frequency.

Whistlers give information on the nature of the region they travel through, particularly on the structure of the earth's exosphere which directly determines their characteristics. In substances like sodium, a similar phenomenon is observed at low temperature, at low wave phase velocity and high electron density. It is called helicon mode in solid state physics because the induced magnetic field vector traces a helical path.

The pulse of electromagnetic energy of a lightening discharge, producing whistlers, contains a wide range of frequencies below the electron cyclotron frequency ($|e| H_0/mc$), but near it. The rate of collision is negligibly small and the plasma refractive index is high, especially if the density is high enough for the cutting-off of the ordinary left-circular polarization component of the wave.

The wave phase velocity is small, the wave number k is large and the frequency range is about 15000 to 1000 cycles per second. For the whistler dispersion the angle between the geomagnetic field $\mathbf{H}_0$ and the wave vector $\mathbf{K}$ of the extraordinary wave is small, Y is large and $\frac{\mathbf{X}}{\mathbf{Y}}$ is also large. Therefore, since $\mathbf{Y}_l^2 >> \mathbf{Y}_t^4/4\ (U-X)$, equation (11.5.15) gives $n^2 = 1 - X/(1 - \mathbf{Y} \cos \theta)$. Moreover, when $\mathbf{Y} \cos \theta >> 1$, this equation reduces to

$$n^2 = \frac{\omega_\rho^2}{\omega \Omega \cos \theta}, \tag{11.8.1}$$

giving
$$\frac{\partial n}{\partial \theta} = \left(\frac{\mathbf{X}}{\mathbf{Y}}\right)^{1/2} \sin\left(\frac{\theta}{2}\right) \cos^{3/2} \theta \tag{11.8.2}$$

Defining the group refractive index n_g as $(\partial/\partial\omega)\ (n\omega)$ and the group velocity v_g as $c/(n_g \cos \gamma)$, we find that

$$n_g = \frac{\partial}{\partial \omega}\left\{\omega\left(\frac{\omega_\rho^2}{\omega \Omega \cos \theta}\right)^{1/2}\right\} = \frac{n}{2},\quad v_g = \frac{2c}{\cos \gamma}\left(\frac{\omega \Omega \cos \theta}{\omega_\rho^2}\right)^{1/2} \tag{11.8.3}$$

where γ is the angle the group velocity vector makes with the geomagnetic field. It can be shown that

$$\tan \gamma = \frac{1}{n}\frac{\partial n}{\partial \theta} = \frac{1}{2} \tan \theta$$

A wave packet, consisting of waves with frequencies near ω, along an element dl of length, propagates in time

$$dt = \frac{dl}{v_g} = \frac{dl}{2c}\,\frac{\omega_\rho \cos \gamma}{(\omega \Omega \cos \theta)^{1/2}} \tag{11.8.4}$$

Hence the time of travel is

$$t = \int_{\text{path}} \frac{\omega_\rho \cos \gamma\, dl}{(\omega \Omega \cos \theta)^{1/2}} = \frac{D}{\sqrt{\omega}} \tag{11.8.5}$$

where
$$D = \frac{1}{2c}\int_{\text{path}} \frac{\omega_\rho \cos \gamma\, dl}{(\Omega \cos \theta)^{1/2}} \tag{11.8.6}$$

For sufficiently small frequencies the time delay between the instant of lightening discharge and the arrival of energy, at the given frequency ω, for a one-way path, is also given by (11.8.5). The dispersion constant D of the whistler mode is of the order of 50 to 100. When the wave normal and $\mathbf{H}_0$ are parallel,

$$D = \frac{1}{2c}\int_{\text{path}} \frac{\omega_\rho\, dl}{\sqrt{\Omega}} = \frac{5.38}{2c}\int_{\text{path}} \left(\frac{N_0}{H_0}\right)^{1/2} dl \tag{11.8.7}$$

where N_0 is given per *cc* and H_0 is the local geomagnetic field in gauss.

Since in (11.8.5), the travel time t is inversely proportional to $\sqrt{\omega}$, the waves of higher frequencies travel faster and arrive ahead of the lower frequency sounds, giving the whistlers a descending tone. The angle ψ $(= \theta - \gamma)$, between the group velocity and the magnetic field $\mathbf{H}_0$, is given by

$$\tan \psi = \tan (\theta - \gamma) = \frac{\tan \theta}{2 + \tan^2 \theta} \tag{11.8.8}$$

For maximum of the values of ψ, the derivative, with respect to θ of $\tan \psi$, is equated to zero. Then solving for θ and γ from that equation and (11.8.8), and denoting the maximum of θ, thus obtained, by θ_m, we find that

$$\tan \theta_m = \sqrt{2},\ \tan \gamma = \frac{1}{\sqrt{2}},\ \theta_m = 54^{\circ}.44',\ \theta_m - \gamma = 19^{\circ}29' \tag{11.8.9}$$

The wave packet is therefore, confined to a narrow beam of angles less than 20° around H_0. Since $\theta - \gamma$ is small, whistlers are guided by the geomagnetic field, and they travel the entire path of the lines of force to which they were attached initially.

11.9 ELECTROMAGNETIC FIELD IN DISSIPATIVE PLASMAS

For an electromagnetic wave, propagating parallel to the applied homogeneous magnetic field $\mathbf{H}_0$, equations (11.2.7)-(11.2.10) simplify to

$$\ddot{\mathbf{E}} - c^2 \nabla^2 \mathbf{E} + 4\pi N_0\, e\, (\dot{\mathbf{u}}_i - \dot{\mathbf{u}}_e) = 0, \tag{11.9.1}$$

$$\ddot{\mathbf{u}}_e + \frac{e}{m} \dot{\mathbf{E}} + (\dot{\mathbf{u}}_e \times \mathbf{\Omega}_e) + \nu_e\, (\dot{\mathbf{u}}_e - \dot{\mathbf{u}}_i) + \nu_-\, (\dot{\mathbf{u}}_e - \dot{\mathbf{u}}_n) = 0, \tag{11.9.2}$$

$$\ddot{\mathbf{u}}_i - \frac{e}{m} \dot{\mathbf{E}} - (\dot{\mathbf{u}}_i \times \mathbf{\Omega}_i) + \nu_i\, (\dot{\mathbf{u}}_i - \dot{\mathbf{u}}_e) + \nu_+\, (\dot{\mathbf{u}}_i - \dot{\mathbf{u}}_n) = 0 \tag{11.9.3}$$

$$\ddot{\mathbf{u}}_n + \nu_n^e\, (\dot{\mathbf{u}}_n - \dot{\mathbf{u}}_e) + \nu_n^i\, (\dot{\mathbf{u}}_n - \dot{\mathbf{u}}_i) = 0 \tag{11.9.4}$$

The dispersion in space is entirely determined by (11.9.1), and the remaining equations are free from space derivatives. The space derivative term appears in (11.2.8)-(11.2.10) only when the longitudinal wave is included. Ignoring ion motion, and considering the coupling between motions of electrons and molecules, we get

$$(D^2 - c^2 \nabla^2)\, \mathbf{E} = 4\pi N_0\, eD\mathbf{u}, \tag{11.9.5}$$

$$(D + \nu_e)\, \mathbf{u} + (\mathbf{u} \times \mathbf{\Omega}) + \frac{e}{m} \mathbf{E} = \nu_e\, \mathbf{u}_n,\ (D + \nu_n)\, \mathbf{u}_n = \nu_n\, \mathbf{u} \tag{11.9.6}$$

where now $D \equiv \partial/\partial t$, $\mathbf{u}$ and $\mathbf{u}_n$ are the velocities of these two species and ν_- and ν_n^0 have been replaced by ν_e and ν_n respectively. Eliminating $\mathbf{u}_n$ between the two equations of (11.9.6), we get

$$(D+\nu_n)\left\{(D+\nu_e)\,\mathbf{u}+(\mathbf{u}\times\mathbf{\Omega})+\frac{e}{m}\mathbf{E}\right\}=\nu_e\,\nu_n\,\mathbf{u} \tag{11.9.7}$$

This equation can be written as

$$D\,(D+\nu_e+\nu_n)\,\mathbf{u}+(D+\nu_n)\left\{(\mathbf{u}\times\mathbf{\Omega})+\frac{e}{m}\mathbf{E}\right\}=0 \tag{11.9.8}$$

Taking vector product of all the terms of (11.9.8) with $\mathbf{\Omega}$ then eliminating $(\mathbf{u}\times\mathbf{\Omega})$ between the resulting equation and equation (11.9.8) itself we obtain

$$\{D^2\,(D+\nu_e+\nu_n)^2+\Omega^2\,(D+\nu_n)^2\}\,\mathbf{u}_e+\frac{e}{m}D\,(D+\nu_n)\,(D+\nu_e+\nu_n)\,\mathbf{E}$$

$$-\frac{e}{m}(D+\nu_n)^2\,(\mathbf{E}\times\mathbf{\Omega})=0 \tag{11.9.9}$$

When D is replaced by $-i\omega$, equation (11.9.8) gives

$$u_x=-\frac{ie\,(\omega^2-k^2c^2)}{m\omega\,\omega_{\rho e}^2}E_x=\frac{e}{m}\{i\omega\,(\nu_n-i\omega)\,(\nu_e+\nu_n-i\omega)\,E_x$$

$$+\Omega\,(\nu_n-i\omega)^2\,E_y\}/\{\Omega^2\,(\nu_n-i\omega)^2-\omega^2\,(\nu_e+\nu_n-i\omega)^2\}, \tag{11.9.10}$$

$$u_y=-\frac{ie\,(\omega^2-k^2c^2)}{m\omega\,\omega_{\rho e}^2}E_y=\frac{e}{m}\{i\omega\,(\nu_n-i\omega)\,(\nu_e+\nu_n-i\omega)\,E_y$$

$$\{-\Omega\,(\nu_n-i\omega)^2\,E_x\}/\{\Omega^2\,(\nu_n-i\omega)^2-\omega^2\,(\nu_e+\nu_n-i\omega)^2\} \tag{11.9.11}$$

where $\omega_{\rho e}^2=4\pi N_0\,e^2/m$. These yield the following two linear equations:

$$E_x\left[\frac{\omega\,(\nu_n-i\omega)\,(\nu_e+\nu_n-i\omega)}{\Omega^2\,(\nu_n-i\omega)^2-\omega^2\,(\nu_e+\nu_n-i\omega)^2}+\frac{\omega^2-k^2c^2}{\omega\,\omega_{\rho e}^2}\right]$$

$$=\frac{i\Omega\,(\nu_n-i\omega)^2\,E_y}{\Omega^2\,(\nu_n-i\omega)^2-\omega^2\,(\nu_e+\nu_n-i\omega)^2} \tag{11.9.12}$$

$$E_y\left[\frac{\omega\,(\nu_n-i\omega)\,(\nu_e+\nu_n-i\omega)}{\Omega^2\,(\nu_n-i\omega)^2-\omega^2\,(\nu_e+\nu_n-i\omega)^2}+\frac{\omega^2-k^2c^2}{\omega\omega_{\rho e}^2}\right]$$

$$=-\frac{i\Omega\,(\nu_n-i\omega)^2\,E_x}{\Omega^2\,(\nu_n-i\omega)^2-\omega^2\,(\nu_e+\nu_n-i\omega)^2} \tag{11.9.13}$$

Eliminating E_x and E_y between (11.9.12) and (11.9.13) we get the relation

$$\left[\frac{\omega\,(\nu_n-i\omega)\,(\nu_e+\nu_n-i\omega)}{\Omega^2\,(\nu_n-i\omega)^2-\omega^2\,(\nu_e+\nu_n-i\omega)^2}+\frac{\omega^2-k^2c^2}{\omega\omega_{\rho e}^2}\right]$$

$$= \frac{\Omega^2 (\nu_n - i\omega)^4}{[\Omega^2 (\nu_n - i\omega)^2 - \omega^2 (\nu_e + \nu_n - i\omega)^2]^2} \tag{11.9.14}$$

The two roots of this equation are

$$\frac{\omega (\nu_n - i\omega)\ (\nu_e + \nu_n - i\omega)}{\Omega^2 (\nu_n - i\omega)^2 - \omega^2 (\nu_e + \nu_n - i\omega)^2} + \frac{\omega^2 - k^2 c^2}{\omega\omega_{\rho e}^2}$$

$$= \frac{\pm \Omega (\nu_n - i\omega)^2}{\Omega^2 (\nu_n - i\omega)^2 - \omega^2 (\nu_e + \nu_n - i\omega)^2} \tag{11.9.15}$$

The real part of these relations gives the waves dispersion relation. Putting $K = 0$ in (11.9.14) and solving for ω we obtain the value of the cut-off frequency. At frequencies below the cut-off, the wave will not propagate beyond the skin depth. In (11.9.14) we write $\omega = \omega_r + i\omega_i$, where ω_r is the real part and $i\omega_i$ is the imaginary part; putting $K = 0$, the real and imaginary parts of the cut-off frequency are determined.

The ion acoustic component of the longitudinal waves, in undamped plasma, has the stop band (frequencies for which it cannot propagate) at $\omega = \omega_{\rho i}$ (section 3.6). The permissible ranges of frequencies of propagation of waves are the pass bands or windows. The blackout of radio communication, during the reentry of space vehicles into earth's atmosphere, is due to the stop band effects. The different cut-off ranges, the stop bands and pass bands of waves are used in the probing of the ionosphere by signals from laboratories and in radar communication.

These relations reduce (11.9.12) and (11.9.13) to $E_x = \pm iE_y$. So the two waves are circularly polarized. If ω_1 and ω_2 are the two values of ω of (11.9.15), then (11.9.10) and (11.9.11) give

$$(u_{1x}, u_{1y}) = -\frac{ie}{m} \frac{(\omega_1^2 - k^2 c^2)}{\omega_1 \omega_{\rho e}^2} (E_{1x}, E_{1y})$$

$$= -\frac{e}{m} \frac{(\omega_1 + i\nu_n)\ (E_{1x}, E_{1y})}{\Omega \nu_n - (\omega_1 (\nu_e + \nu_n) + i\omega_1 (\omega_1 - \Omega)}, \tag{11.9.16}$$

$$(u_{2x}, u_{2y}) = -\frac{ie}{m\omega_2 \omega_{\rho e}^2} (\omega_2^2 - k^2 c^2)\ (E_{2x}, E_{2y})$$

$$= -\frac{e}{m} \frac{(\omega_2 + i\nu_n)\ (E_{2x}, E_{2y})}{\Omega \nu_n + \omega_2 (\nu_e + \nu_n) - i\omega_2 (\omega_2 + \Omega)} \tag{11.9.17}$$

where u_{1x}, E_{1x} are the values of u_x, E_x for the wave at ω_1, etc.

Expressions for the power density, averaged over the time period $2\pi/\omega_1$, for the wave at ω_1, and over $2\pi/\omega_2$, for the wave at ω_2, are

$$\langle(\mathbf{E}_1 \cdot \mathbf{j}_1)\rangle = \langle -N_0\, e\,(\mathbf{E}_1 \cdot \mathbf{u}_1) \tag{11.9.18}$$

$$= R_e \left[\frac{e\,|E_0|^2/8\pi)\ (\omega_1 + i v_n)}{\Omega\, v_n - \omega_1\, v_e - \omega_1\, v_n + i\omega_1\,(\omega_1 - \Omega)} \right],$$

$$\langle(\mathbf{E}_2 \cdot \mathbf{j}_2)\rangle = R_e \left[\frac{(e\,|E_0|^2/8\pi m)\ (\omega_2 + i v_n)}{\Omega\, v_n - \omega_2\, v_e - \omega_2\, v_n + i\omega_2\,(\omega_2 - \Omega)} \right] \tag{11.9.19}$$

where $\mathbf{E}_0$ is the wave amplitude and Re X means real part of X. When $\omega_1 = \Omega$, equation (11.9.18) reduces to

$$\langle(\mathbf{E}_1 \cdot \mathbf{j}_1)\rangle = -\frac{e\,|E_0|^2}{8\pi m v_e} \tag{11.9.20}$$

Hence motion of uncharged particles does not affect directly the resonant power absorption because (11.9.20) does not depend on v_n; moreover, (11.9.20) diverges to infinity as $v_e \to 0$.

11.10 WAVE COUPLING IN HOMOGENEOUS MAGNETIZED PLASMA

We consider electron plasma which is thermally active, uniform, compressible, in the bed of constant magnetic field $\mathbf{H}_0$. The guiding equations are

$$\dot{\mathbf{u}}_e = -\frac{e}{m}\,\mathbf{E} - (\mathbf{u} \times \mathbf{\Omega}) - \frac{c_e^2}{\rho_0}\,\nabla \rho^m + \boldsymbol{\varepsilon}_1 \tag{11.10.1}$$

$$\dot{\rho}^m + \rho_0^m\,(\nabla \cdot \mathbf{u}) = \varepsilon_2, \tag{11.10.2}$$

$$\nabla \cdot \mathbf{E} = -\frac{4\pi e}{m}\,\rho^m, \tag{11.10.3}$$

$$\nabla \times \mathbf{H} = \frac{1}{c}\,\dot{\mathbf{E}} - \frac{4\pi e}{mc}\,\rho_0^m\,\mathbf{u} + \boldsymbol{\varepsilon}_3, \tag{11.10.4}$$

$$\nabla \times \mathbf{E} = -\frac{1}{c}\,\dot{\mathbf{H}} \tag{11.10.5}$$

where c_e is the sound velocity in the electron fluid, the well-known symbols have their usual meaning and $\boldsymbol{\varepsilon}_1$, ε_2, $\boldsymbol{\varepsilon}_3$ represent terms of collisional loss and non-linearity. For instance $\boldsymbol{\varepsilon}_1$ contains the dissipative collisional loss force $-\nu\mathbf{u}$ per unit mass (where γ is the collisional frequency) and non-linear terms $-(\mathbf{u} \cdot \nabla)\,\mathbf{u}$ and $-e\,(\mathbf{u} \times \mathbf{H})/mc$; ε_2 and $\boldsymbol{\varepsilon}_3$ contain respectively the non-linear terms $-\rho_0^m\,(\nabla \cdot \mathbf{u})$ and $-4\pi e\,\rho_0\,\mathbf{u}^m\ mc$.

We would now find two differential equations of wave coupling which for zeros of $\boldsymbol{\varepsilon}_1, \varepsilon_2, \boldsymbol{\varepsilon}_3$, connect the transverse field vector $\mathbf{H}$ with ρ^m–the perturbation in the density of electronic mass. We find that $(\nabla \cdot \boldsymbol{\varepsilon}_3) = (4\pi e\,\varepsilon_2)/mc$. Moreover, equation (11.10.2) shows that the mass density ρ^m and the non-solenoidal part of **u** is in phase. Similarly, **H** and the solenoidal part of **E** are in phase because of equation (11.10.5). Equation (11.10.1) indicates that not all the components of **u** are in the same phase, specifically a component of **u** is out of the phase with the components normal to that.

The vector **E** will be eliminated from equations (11.10.1) and (11.10.3) to (11.10.5). From the resultant set of equations in **u**, **H** and ρ^m eliminating **u** the desired equations of coupling relating **H** and ρ^m will be obtained. Thus proceeding, the **E** eliminant equations are

$$\nabla^2 \mathbf{H} = \frac{1}{c^2}\ddot{\mathbf{H}} + \frac{4\pi e N_0}{c}\nabla \times \mathbf{u} - \nabla \times \boldsymbol{\varepsilon}_3, \tag{11.10.6}$$

$$(\nabla \cdot \dot{\mathbf{u}}) = \frac{\omega_0^2}{\rho_0^m}\rho^m - \nabla \cdot (\mathbf{u} \times \boldsymbol{\Omega}) - \frac{c_e^2}{\rho_0^m} + (\nabla \cdot \boldsymbol{\varepsilon}), \tag{11.10.7}$$

$$(\nabla \times \dot{\mathbf{u}}) = \frac{e\dot{\mathbf{H}}}{mc} - \nabla \times (\mathbf{u} \times \boldsymbol{\Omega}) + \nabla \times \boldsymbol{\varepsilon}, \tag{11.10.8}$$

where $\omega_0^2 = 4\pi N_0 e^2/m$. Evidently, the first equation is obtained by taking curl of (11.10.4) and using in it (11.10.5). For the second equation divergence of (11.10.1) is used together with (11.10.3) to eliminate **E**. Equation (11.10.8) is obtained by taking curl of (11.10.1) and using (11.10.5). Using the formulas of expansion of $\nabla \cdot (\mathbf{u} \times \boldsymbol{\Omega})$ and $\nabla \times (\mathbf{u} \times \boldsymbol{\Omega})$ we get

$$\frac{1}{c^2}(\ddot{\mathbf{H}} - c^2\nabla^2 \mathbf{u}) = -\frac{4\pi}{c} eN_0 (\nabla \times \mathbf{u}) + (\nabla \times \boldsymbol{\varepsilon}_3), \tag{11.10.9}$$

$$(\nabla \cdot \dot{\mathbf{u}}) = \frac{\omega_0^2 \rho^m}{\rho_0 m} - (\boldsymbol{\Omega} \cdot (\nabla \times \mathbf{u})) - \frac{c_e^2}{\rho_0^m}\nabla^2 \rho^m + (\nabla \cdot \boldsymbol{\varepsilon}_3), \tag{11.10.10}$$

$$(\nabla \times \dot{\mathbf{u}}) = \frac{e\dot{H}}{mc} - (\boldsymbol{\Omega} \cdot \nabla)\,\mathbf{u} + \boldsymbol{\Omega}\,(\nabla \cdot \mathbf{u}) + (\nabla \times \boldsymbol{\varepsilon}) \tag{11.10.11}$$

From (11.10.9) the solution for curl **U** is

$$\nabla \times \mathbf{u} = -\frac{c^2\nabla^2 \mathbf{H} - \ddot{\mathbf{H}} + c^2(\nabla \times \boldsymbol{\varepsilon}_3)}{4\pi e N_0 c} \tag{11.10.12}$$

Now elimination of curl **U** between (11.10.10) and (11.10.11) gives

$$c^2\nabla^2 \dot{\mathbf{H}} - \ddot{\mathbf{H}} - \omega_0^2 \mathbf{H}$$
$$= 4\pi e N_0 c\,[\boldsymbol{\Omega}\,(\nabla \cdot \mathbf{u}) - (\boldsymbol{\Omega} \cdot \nabla)\,\mathbf{u} + (\nabla \times \boldsymbol{\varepsilon}_1)] - c^2\,\nabla \times \boldsymbol{\varepsilon}_3 \tag{11.10.13}$$

Similarly elimination of curl **U** between (11.10.10) and (11.10.12) yields

$$(\nabla \cdot \dot{\mathbf{u}}) = \frac{1}{\rho_0 m}(\omega_0^2 \rho^m - c_e^2\nabla^2 \rho^m) + \frac{c}{4\pi e N_0}\left[\frac{1}{c^2}(\boldsymbol{\Omega} \cdot \ddot{\mathbf{H}}) - (\boldsymbol{\Omega} \cdot \nabla^2 \mathbf{H})\right]$$

$$-\frac{c}{4\pi e N_0}(\mathbf{\Omega} \,.\, (\nabla \times \boldsymbol{\varepsilon}_3)) + (\nabla \,.\, \boldsymbol{\varepsilon}_1) \quad (11.10.14)$$

Taking divergence of (11.10.14), we get

$$(\nabla \,.\, \mathbf{u}) = -\frac{\dot{P}^m}{P_0^m} + \frac{c}{4\pi e N_0}(\nabla \,.\, \boldsymbol{\varepsilon}_3) \quad (11.10.15)$$

Putting this value of div **U** in (11.10.13) and (11.10.14) we have

$$c^2 \nabla^2 \dot{\mathbf{H}} - \dddot{\mathbf{H}} - \omega_0^2 \mathbf{H} = 4\pi e N_0 c \left[-\frac{\mathbf{\Omega}}{P_0^m} P^m - (\mathbf{\Omega} \,.\, \nabla)\,\mathbf{u} + \frac{c\mathbf{\Omega}(\nabla \,.\, \boldsymbol{\varepsilon}_e)}{4\pi e N_0} + \right.$$

$$\left. + \frac{c(\nabla \,.\, \boldsymbol{\varepsilon}_3)}{4\pi e N_0}\mathbf{\Omega} + (\nabla \times \boldsymbol{\varepsilon}_1) - c^2 (\nabla \times \boldsymbol{\varepsilon}_3) \right. \quad (11.10.16)$$

$$c_e^2 \nabla^2 \rho^m - \ddot{\rho}^m - \omega_0^2 \rho^m = \frac{mc}{4\pi e}\left[\frac{1}{c^2}(\mathbf{\Omega} \,.\, \ddot{\mathbf{H}}) - (\mathbf{\Omega} \,.\, \nabla^2 \mathbf{H}) \right] -$$

$$-\frac{mc}{4\pi e}(\mathbf{\Omega} \,.\, \nabla \times \boldsymbol{\varepsilon}_3) + P_0^m (\nabla \,.\, \boldsymbol{\varepsilon}_1) - \frac{mc}{4\pi e}(\nabla \,.\, \dot{\boldsymbol{\varepsilon}}_3) \quad (11.10.17)$$

Since equation (11.10.17) relates P^m with **H** and the residual terms $\boldsymbol{\varepsilon}_1$ and $\boldsymbol{\varepsilon}_3$, it becomes one of the coupling equations when $\boldsymbol{\varepsilon}_1$ and $\boldsymbol{\varepsilon}_3$ are ignored. Now take curl of (11.10.16) and since $\nabla \times (\mathbf{\Omega} \,.\, \nabla)\,\mathbf{u} = (\mathbf{\Omega} \,.\, \nabla)\ (\nabla \times \mathbf{u})$, the relation (11.10.12) will be used to remove **U**. The other possibility may seem to be the elimination of **U** between (11.10.15) and the divergence of (11.10.16); but this results in zero of both sides because $\nabla \,.\, \mathbf{H} = 0$. We therefore get

$$(\nabla \times \dot{\mathbf{H}}) + \omega_0^2 (\nabla \times \dot{\mathbf{H}}) - c^2 \nabla^2 (\nabla \times \dot{\mathbf{H}}) - c^2 (\mathbf{\Omega} \,.\, \nabla)\, \nabla^2 \mathbf{H} + (\mathbf{\Omega} \,.\, \nabla)\, \ddot{\mathbf{H}}$$

$$= \frac{1}{m} 4\pi e c\,(\nabla \dot{\rho}^m \times \mathbf{\Omega}) - c^2 [\nabla (\nabla \,.\, \varepsilon_3) \times \mathbf{\Omega}] + c^2 \nabla \times (\nabla \times \dot{\boldsymbol{\varepsilon}}_3) +$$

$$+ c^2 (\mathbf{\Omega} \,.\, \nabla)\ (\nabla \times \boldsymbol{\varepsilon}_3) - 4\pi e N_0 c\, \nabla \times (\nabla \times \boldsymbol{\varepsilon}_1) \quad (11.10.18)$$

If the non-linear terms are included then for $e = 0$ only the non-linear part of $-\rho^c (\nabla \,.\, \boldsymbol{\varepsilon}_3)$ for P^c in (11.10.17) and $-4\pi e N_0 C \nabla \times (\nabla \times \boldsymbol{\varepsilon}_1)$ and $c^2 \nabla \times (\nabla \times \boldsymbol{\varepsilon}_3)$ for **H** in (11.10.18) exist.

For $\boldsymbol{\varepsilon}_1 = \boldsymbol{\varepsilon}_3 = 0,\ \boldsymbol{\varepsilon}_2 = 0$, the linearized coupling equations are

$$\ddot{P}^m + \omega_0^2 P^m - c_e^2 \nabla^2 P^m = \frac{m}{4\pi e c}\,[e^2 c\,(\mathbf{\Omega} \,.\, \nabla^2 \mathbf{H}) - (\mathbf{\Omega} \,.\, \ddot{\mathbf{H}})], \quad (11.10.19)$$

$$(\nabla \times \dddot{\mathbf{H}}) + \omega_0^2 (\nabla \times \mathbf{H}) - c^2 \nabla^2 (\nabla \times \mathbf{H}) - c^2 \nabla^2 (\mathbf{\Omega} \,.\, \nabla)\, \mathbf{H} + (\mathbf{\Omega} \,.\, \nabla)\, \mathbf{H}$$

$$= \frac{4\pi e c}{m}(\nabla \dot{P}^m \times \mathbf{\Omega}) \quad (11.10.20)$$

To understand the behaviour of a field vector which has both non-zero solenoidal and curl (rotational) parts, we now find the differential equation for **U**. So we eliminate **H** between this equation and (11.10.1) and ρ^m from the eliminant equation is removed by using (11.10.2). We thus find that

$$\ddot{\mathbf{u}} + \omega_0^2 \dot{\mathbf{u}} - c^2 \nabla^2 \dot{\mathbf{u}} + (c^2 - c_e^2) \nabla (\nabla \cdot \dot{\mathbf{u}}) + (\ddot{\mathbf{u}} + \mathbf{\Omega})$$
$$= - c^2 (\mathbf{\Omega}\ \nabla)\ (\nabla \times \mathbf{u}) + c^2 (\nabla (\nabla \cdot \mathbf{u}) \times \mathbf{\Omega}) \qquad (11.10.21)$$

Taking in turn divergence and curl of all the terms of this equation we find that

$$\nabla \cdot (\ddot{\mathbf{u}} + \omega_0^2 \mathbf{u} - c_e^2 \nabla^2 \mathbf{u}) = - (\mathbf{\Omega} \cdot (\nabla \times \dot{\mathbf{u}})),$$
$$\nabla \times [\dddot{\mathbf{u}} + \omega_0^2 \dot{\mathbf{u}} - c^2 \nabla^2 \dot{\mathbf{u}} + c^2 (\mathbf{\Omega} \cdot \nabla)\ (\nabla \times \mathbf{u})] \qquad (11.10.22)$$
$$= \mathbf{\Omega} \nabla \cdot (\ddot{\mathbf{u}} - c^2 \nabla^2 \mathbf{u}) + c^2 (\mathbf{\Omega} \cdot \nabla)\ \nabla (\nabla \cdot \mathbf{u}) \qquad (11.10.23)$$

These equations relate the rotational variables with non-solenoidal variables which are curl **U** and ∇ . **U** respectively. The right side of (11.10.22) is $\pi/2$ out of phase with the left hand side, if $\Omega \neq 0$. A phase difference of $\pi/2$ exists between the two sides of (11.10.23). So, these terms cannot be equated to zero to find the dispersion of waves of ∇ . **U** and curl **U**.

One related problem is that of wave propagation in magnetized plasma. It has been considered in books and review papers. Other problems suitable for study are:

(*i*) the Poynting radiation by pressure perturbation;

(*ii*) the pressure perturbation by electromagnetic field.

The dispersion relation obtained from equations (11.10.19) and (11.10.20) should give the Appleton Hartree equation for $\gamma = 0$. Analysis of propagation of light through anisotropic crystals gives two ray directions, which are the ordinary and the extraordinary ray directions. Experiments on wave propagation through a slab of compressible, warm magnetized plasma give several directions of wave propagation; (considering only electron motion the number of directions is three; if ion motion is also included the number of directions would be more than three. Electro-magnetic theory and geometrical optics of anisotropic crystals are considerably developed. Further study will find: (*i*) three ray directions; (*ii*) two directions of energy propagation: (*a*) the Poynting flow $\mathbf{S}^t$ and (*b*) the energy flow $\mathbf{S}^1$ due to pressure perturbation; (*iii*) four forms of energy densities W_E, W_H, W_P, W_K.

11.11 PONDERMOTIVE FORCE OF INHOMOGENEOUS ELECTRIC FIELD

In high power laser heated plasmas gradients of the electric field (the ponderomotive force) have very high value, because the radiation pressure becomes as large as several hundred atmosphere. This pressure force pushes away electrons from regions of high field intensity, and pile up in regions of low field, because the wave amplitudes vary.

For finding this force the actual position of a particle, **r** say, is distinguished from its initial position $\mathbf{r}_0$. We write

$$\mathbf{r} = \mathbf{r}_0 + \boldsymbol{\xi}\,(\mathbf{r}_0,\ t) \tag{11.11.1}$$

where $\boldsymbol{\xi}$ is the field induced displacement. The electric field with a weakly varying amplitude is taken as

$$\mathbf{E}\,(\mathbf{r},\ t) = \mathbf{A}\,(\mathbf{r})\cos\theta \tag{11.11.2}$$

where $\theta = wt$. Then the first order electric field is

$$\mathbf{E}_1\,(r,\ t) = \mathbf{A}\,(r_0)\cos\theta \tag{11.11.3}$$

And, correct up to the second order, the electric field may be written as

$$\begin{aligned}\mathbf{E}^{(2)} &= \mathbf{E}_1\,(\mathbf{r},\ t) + \mathbf{E}_2\,(\mathbf{r},\ t)\\ &= \mathbf{A}\,(\mathbf{r}_0)\cos\theta + (\boldsymbol{\xi}\,.\,\boldsymbol{\nabla}_{\mathbf{r}_0})\,\mathbf{A}\,(\mathbf{r}_0)\cos\theta\end{aligned} \tag{11.11.4}$$

The first order equation of motion is

$$\frac{\partial \mathbf{v}_1}{\partial t} = -\frac{e}{m}\,\mathbf{A}\,(\mathbf{r}_0)\cos\theta \tag{11.11.5}$$

Since $\mathbf{v}_1 = \boldsymbol{\xi}\,(\mathbf{r}_0,\ t)$ integration of (11.11.5) gives

$$\mathbf{v}_1 = \dot{\boldsymbol{\xi}} = -\frac{e}{m\omega}\,\mathbf{A}\,(\mathbf{r}_0)\sin\theta \tag{11.11.6}$$

$$\boldsymbol{\xi} = \frac{e}{mw^2}\,\mathbf{A}\,(\mathbf{r}_0)\cos\theta \tag{11.11.7}$$

where dot denotes time derivative. Then the equation of motion, correct up to the second order, is

$$\dot{\mathbf{v}}_2 = -\frac{e}{m}\cos\theta\left[\mathbf{A}\,(\mathbf{r}_0) + \frac{e}{m\omega^2}(\mathbf{A}\,(\mathbf{r}_0)\,.\,\boldsymbol{\nabla}_{r_0})\ (\mathbf{A}\,(\mathbf{r}_0)\cos\theta) + \ldots\right] \tag{11.11.8}$$

Faraday's law of *EM* induction gives the induction field $\mathbf{B}_1$ in terms of $\mathbf{E}_1$:

$$\mathbf{B}_1 = -\frac{c}{w}\,\boldsymbol{\nabla}_r \times A\,(\mathbf{r}_0)\sin\theta \tag{11.11.9}$$

$$\text{Hence } \left\langle\frac{(\mathbf{v}_1\times\mathbf{B}_1)}{c}\right\rangle = \frac{1}{c}\left\langle\left(-\frac{e}{mw}\,\mathbf{A}\,(\mathbf{r}_0)\sin\theta\right)\times\left(-\frac{c}{w}\,\boldsymbol{\nabla}_r\times\mathbf{A}\,(\mathbf{r}_0)\sin\theta\right)\right\rangle$$

$$= \frac{e}{2mw^2}\,\mathbf{A}\,(\mathbf{r}_0)\times(\boldsymbol{\nabla}_{r_0}\times\mathbf{A}\,(\mathbf{r}_0) \tag{11.11.10}$$

because, for a field variable the exit time average is determined by the relation

$$\langle x\rangle = \frac{1}{T}\int_0^T x\,dt \tag{11.11.11}$$

where $T\,(= 2\pi/\omega)$ is the wave time period. In the fluid approximation the other non-linear force term is the substantial derivative part of the left hand side of the equation of motion of the second order. The electron equation of motion, correct up to the second order is

$$\langle \dot{\mathbf{v}}_2\rangle = \langle -\frac{e}{m}\mathbf{E}_2\rangle - \frac{e}{me}\langle(\mathbf{v}_1 \times \mathbf{B}_1)\rangle \tag{11.11.12}$$

$$= -\frac{e}{m}\langle(\boldsymbol{\xi}\,.\,\nabla_{r_0})\cos\theta\rangle - \frac{e}{2m^2 w^2}\mathbf{A}\,(\mathbf{r}_0)\times(\nabla_{r_0}\times(\nabla_{r_0}\times\mathbf{A}\,(\mathbf{r}_0)) \tag{11.11.13}$$

Hence the ponderomotive force per unit mass is given by the relation

$$\langle \dot{\mathbf{v}}_2\rangle_{\text{Pon}} = -\frac{e^2}{2m^2 w^2}(\mathbf{A}\,.\,\nabla)\,\mathbf{A} - \frac{e^2}{2m^2 w^2}\mathbf{A}\times(\nabla\times\mathbf{A}) = -\nabla\left(\frac{e^2 a^2}{4m^2 w^2}\right) \tag{11.11.14}$$

Since $\langle A^2\rangle = 2\,\langle E^2\rangle$, (11.11.15)

the relation (11.14.15) reads

$$\langle \dot{\mathbf{v}}_2\rangle = \langle\dot{\xi}\rangle = -\nabla\,(e^2 E^2/2m^2 w^2) \tag{11.11.16}$$

where we have dropped the ⟨notation⟩ ⟨ ⟩ from E^2. If the electron number density is N, the ponderomotive force per unit volume is

$$\mathbf{f}_{\text{Pon}} = -\nabla\,\langle\frac{w_\rho^2 E^2}{2\pi\omega^2}\rangle = -\nabla\varphi_{\text{Pon}}, \tag{11.11.17}$$

so that the conservative ponderomotive force of potential

$$\varphi_{\text{Pon}} = \langle\frac{\omega_\rho^2 E^2}{2\pi\omega^2}\rangle \tag{11.11.18}$$

acts in the direction of decreasing electric field. So, electrons are pushed away from regions of increasing field. Since $-\nabla\varphi$ depends on e^2, this force acts in the same driection on the charges of both sign, but the magnitudes will be different because mass per particle is different. The ponderomotive force will give rise to an electric drift type drift

$$\mathbf{v}_{\text{Pon}} = \frac{ce}{2m\,H^2\,\omega^2}\,[(\nabla E^2)\times\mathbf{H}] \tag{11.11.19}$$

This drift velocity retards motion in the direction of increasing electrical energy density $E^2/8\pi$. Since it is proportional to the first power of the charge, for charges of opposite sign, the direction of $\mathbf{v}_{\text{Pon}}$ will be different. And being dependent on the mass, the magnitudes will be different when the masses are different. Due to charge separation the drift will generate the ponderomotive charge current density $\mathbf{j}_{\text{Pon}}$ of charges:

$$\mathbf{j}_{\text{Pon}} = Ne\,\mathbf{v}^i_{\text{Pon}} - Ne\,\mathbf{v}^e_{\text{Pon}} = -\frac{Ne^2 c}{2H^2 w^2}\left(\frac{1}{M}+\frac{1}{m}\right)[(\nabla E^2)\times \mathbf{H}]$$

$$= -\frac{Ne^2 c}{2H^2 w^2 m}[(\nabla E^2)\times \mathbf{H}] \tag{11.11.20}$$

If the acceleration $\langle\ddot{\boldsymbol{\xi}}\rangle$ of (11.11.16) is of the order of the gravitational acceleration $\mathbf{g}$, then

$$|g| = \left|\frac{e^2}{2m^2 w^2}\right| |\nabla E^2| \tag{11.11.21}$$

So $$E^2 = \frac{2lg\, m^2 w^2}{e^2} \tag{11.11.22}$$

where l is the characteristic length of variation of E^2 so that $\Delta E^2 = E^2/l$, and

$$W_E = \frac{\langle A^2\rangle}{8\pi} = \frac{\langle E^2\rangle}{16\pi},\quad \nabla W_E = \nabla\left(\frac{A^2}{8\pi}\right) = \frac{\nabla E^2}{16\pi} \tag{11.11.23}$$

Numerical calculations show that

$$|\nabla W_E| = \frac{2g\, m^2 w^2}{16\pi e^2} = 3\times 10^{-34}\times w^2 = \frac{E^2}{16\pi e}$$

The electric field E, obtained from this relation, can be sustained by gravity in the upper atmosphere.

Exercise 11.11.1. Assuming that the electric field is $\mathbf{E} = \mathbf{a}\cos\theta + \mathbf{b}\sin\theta$, find φ_{Pon}.

Answer. We have

$$\mathbf{E}_1 = \mathbf{a}(\mathbf{r}_0)\cos\theta + \mathbf{b}(\mathbf{r}_0)\sin\theta,\quad \mathbf{v}_1 = -\frac{e}{mw}(-\mathbf{a}\sin\theta + \mathbf{b}\cos\theta)$$

$$\boldsymbol{\xi} = \frac{e}{mw^2}(\mathbf{b}\cos\theta + \mathbf{b}\sin\theta),\quad \mathbf{B} = -\frac{e}{w}[(\nabla\times\mathbf{a})\sin\theta - (\nabla\times\mathbf{b})\cos\theta],$$

$$\langle\dot{\mathbf{v}}_2\rangle = -\frac{e}{m}\langle(\boldsymbol{\xi}\,.\,\nabla)\,\mathbf{E}_1\rangle - \frac{e}{mc}\left(-\frac{c}{w}\right)\frac{e}{mw}\langle(-\mathbf{a}\sin\theta + \mathbf{b}\cos\theta)\times(\nabla\times\mathbf{a}\sin\theta - \nabla\times\mathbf{b}\cos\theta)\rangle$$

$$= -\frac{e^2}{2m^2 w^2}[(\mathbf{a} \cdot \nabla)\,\mathbf{a} + (\mathbf{b} \cdot \nabla)\,\mathbf{b}] - \frac{e^2}{2m^2 w^2}[\mathbf{a} \times (\nabla \times \mathbf{a}) + \mathbf{b} \times (\nabla \times \mathbf{b})]$$

$$= -\frac{e^2}{2m^2 w^2}[(\mathbf{a} \cdot \nabla)\,\mathbf{a} + \mathbf{a} \times (\nabla \times \mathbf{a}) + (\mathbf{b} \cdot \nabla)\,\mathbf{b} + \mathbf{b} \times (\nabla \times \mathbf{b})]$$

$$= -\frac{e^2}{2m^2 w^2}\left[\nabla \frac{a^2}{2} + \nabla \frac{b^2}{2}\right] = -\nabla \frac{e^2 (a^2 + b^2)}{4m^2 w^2}$$

Hence $\varphi_{\text{Pon}} = \dfrac{e^2 (a^2 + b^2)}{4m^2 w^2}$

11.12 EFFECTS OF FINITE LARMOR RADIUS AND ION MASS

Consider a fine scale effect of instability in rarefied and collisionless plasmas, depending on the finiteness of the Larmor radius and ion mass. The applied magnetic field $\mathbf{H}_0$ is constant and along the z-axis and the constant gravitational acceleration $\mathbf{g}$ is along OX. The gravitational electron drift $\mathbf{v}_e$ and ion drift $\mathbf{v}_i$ are

$$\mathbf{v}_e = -\frac{mc}{eH_0^2}(\mathbf{g} \times \mathbf{H}) = \frac{g}{|\Omega_e|}\hat{\mathbf{y}},\quad \mathbf{v}_i = \frac{MC}{eH_0^2}(\mathbf{g} \times \mathbf{H}_0) = -\frac{g}{\Omega_i}\hat{\mathbf{y}} \qquad (11.12.1)$$

We note that $|\mathbf{v}_i| > |\mathbf{v}_e|$. When the time rate of change of the electric field is slow, compared with the gyration frequencies, the drift depending on the electric field, is

$$\mathbf{v} = \frac{c}{H_0^2}(\mathbf{E} \times \mathbf{H}_0) + \frac{c}{\Omega H_0}\frac{d\mathbf{E}}{dt} \qquad (11.12.2)$$

The electric field, acting on a gyrating particle, is $\mathbf{E}(\mathbf{r} + \boldsymbol{\rho})$ where $\mathbf{r}$ is the position vector of the guiding centre, and $\boldsymbol{\rho}$ is the vectorial radius of gyration of relation (2.3.11). For an electron

$$\boldsymbol{\rho}_e = \frac{(\mathbf{v} \times \mathbf{H}_0)}{\Omega H_0} = \frac{v}{\Omega}(\hat{\mathbf{X}} \cos \Omega gt\ \hat{\mathbf{Y}} \sin \Omega t) \qquad (11.12.3)$$

Now expanding $\mathbf{E}$ about its value at $\mathbf{r}$, we obtain

$$\mathbf{E}(\mathbf{r} + \boldsymbol{\rho}) = \mathbf{E}(\mathbf{r}) + (\boldsymbol{\rho} \cdot \nabla)\,\mathbf{E}(\mathbf{r}) + \frac{1}{2}(\boldsymbol{\rho} \cdot \nabla)^2\,\mathbf{E}(\mathbf{r}) + \ldots \qquad (11.12.4)$$

Average of all the terms of (11.12.4), over the Larmor period $2\pi/\Omega$, gives

$$\langle \mathbf{E} \rangle = \left(1 + \frac{\rho^2}{2}\nabla_\perp^2\right)\mathbf{E}(\mathbf{r}) \qquad (11.12.5)$$

Hence (11.12.2) may be replaced by

$$\langle \mathbf{v} \rangle = \frac{e}{H_0^2}(\langle \mathbf{E} \rangle \times \mathbf{H}_0) + \frac{c}{\Omega H_0}\left\langle \frac{d\mathbf{E}}{dt} \right\rangle$$

$$= \frac{e}{H_0^2}\left(1 + \frac{\rho^2}{2}\nabla_\perp^2\right)(\mathbf{E}\times\mathbf{H}_0) + \frac{c}{\Omega H_0}\langle\frac{d\mathbf{E}}{dt}\rangle \tag{11.12.6}$$

If the direction of propagation is perpendicular to $\mathbf{H}_0$ then, since $\langle \mathbf{v}_e\rangle \ll |\mathbf{v}_i|$, we can write

$$\frac{d}{dt} = \frac{\partial}{\partial t} + (\mathbf{v}\,.\,\nabla) = -i\omega + (\mathbf{v}_i\,\nabla) = -i\omega + ikv_i = -i\left(\omega + \frac{Kg}{\Omega_i}\right), \tag{11.12.7}$$

$$\langle \mathbf{v}\rangle = \frac{c}{H_0^2}\left(1 + \frac{\rho^2}{2}\nabla_\perp^2\right)(\mathbf{E}\times\mathbf{H}) - \frac{ic}{\Omega H_0}\left(\omega + \frac{Kg}{\Omega_i}\right)\mathbf{E} \tag{11.12.8}$$

The first term, on the right hand side, is the guiding centre drift and the second and third terms are small corrections, arising from the finite orbit size and finite ion mass. The second term is the finite Larmor radius effect. Since $\langle \mathbf{v}\rangle$ is complex, the plasma becomes unstable.

Chapter 12

NON-LINEAR EFFECTS OF WAVES IN PLASMA

12.1 INTRODUCTION

The unscreened, dispersive, self-magnetic field, which is a relativistic process (section 1.17), is responsible for the propagation of transverse waves in plasmas. Moreover, relativistic considerations are necessary for the study of electric field in moving media and for other high energy effects. Plasma becomes moderately relativistic at 10^8 °K for ions. For these and other reasons, relativistic and other nonlinear effects of waves are important in theory of physical plasma.

The relativistic Vlasov-equations, together with the Maxwell equations for self-consistent fields, are studied for understanding processes in plasmas due to application of strong electromagnetic fields. Validity and applicability of this representation are some what restricted, because correlations are ignored in the Vlasov model. A thorough discussion of these restrictions is beyond the scope of this book. So, these aspects of the formalism are not considered by us.

12.2 REMARKS ON NON-LINEAR EFFECTS IN PLASMAS

(a) Intensity dependent non-linear effects

A simple filter can isolate the periodic weak disturbances passing through a circuit. Non-linear interactions of waves with a material medium cannot be so separated, because the elementary solutions interact mutually and give rise to mixing of the waves. So, this separation is not possible for large amplitude waves. Hence, a non-linear device can be employed to induce artificial mixing. Intensity-dependent non-linear effects appear when a material, irrespective of its state (solid, liquid or gas), is subjected to strong fields. Generalization of the laws of physics is often necessary for understanding these effects, because, originally, these physical laws were developed for weak fields acting in media having simple structure.

For validity of the linearized theory, kinetic energy imparted by an applied electric field **E**, on electrons, must be small compared with the kinetic energy of random thermal motion. So

$$\frac{e^2 E^2}{m^2 \omega^2} << \frac{K_B T}{m} \qquad (12.2.1)$$

For laser beams of frequency $\omega = 10^{10}$ cycles per sec, and electric field $E = 10^6$ esu, we find that

$$\frac{e^2 E^2}{m^2 \omega^2} = \frac{25}{81} \times 10^{28}, \quad \frac{K_B T}{m} = \frac{1{\cdot}4}{9} \times 10^{12} \times T \tag{12.2.2}$$

Hence, at about million degrees and higher, of kinetic temperature, the inequality (12.2.1) is violated, and the system becomes non-linear. For the linearized approximation to be valid, the ratio (K_t/K_i) must be a small fraction, where K_t is the amplitude of the external field, and K_i is the amplitude of the internal field ($\approx e/d^2$ where d ($= 5 \times 10^{-8}$ cm) is the atomic radius). So

$$K_i = \frac{5 \times 10^{-10}}{5 \times 5 \times 10^{-16}} = 2 \times 10^5 \text{ esu} = 6 \times 10^8 \text{ volts/cm}$$

Intensity of a powerful external field ($= 10^6$ volts/cm), currently available in lasers, is not negligible compared to this K_i. Such fields effectively disrupt the structure of matter. The consequences are classically studied as non-linear effects of waves in media. One such effect is the detection of ultraviolet light at twice the frequency of a ruby laser beam traversing a quartz crystal. The quantum mechanical explanation of the doubling of the frequency is that two quanta are taken away from the incident beam, and one quantum at twice the frequency is radiated.

12.3 ELECTRON MOTION IN THE FIELD OF AN ELECTROMAGNETIC WAVE

Consider the elliptically polarized electric and magnetic fields

$$\mathbf{E} = (E_x, E_y, 0), \mathbf{H} \equiv (H_x, H_y, 0) \tag{12.3.1}$$

propagating parallel to the z-axis, where

$$E_x = A \cos \theta, \; E_y = B \sin \theta, \; \theta = \omega t - KZ \tag{12.3.2}$$

A and B are the amplitudes, and θ is the phase, ω is the wave frequency and K is the wave number.

(a) Transformation to the space-independent frame

The Lorentz transformation of

$$t = \gamma \left(t' + \frac{v}{c^2} Z' \right), \; Z = \gamma (Z' + vt') \tag{12.3.3}$$

where γ is the relativistic factor $(1 - \beta^2)^{-1/2}$ with $\beta = v/c$, and the primed system moves relative to the unprimed system with velocity $\mathbf{v}$ parallel to the z-axis. For convenience, in (12.3.2) we put

$$\beta = \tanh \psi \tag{12.3.4}$$

and find that the two relations of (12.3.2) reduce to

$$t = t' \cosh \psi + \frac{Z'}{c} \sinh \psi, \; Z = Z' \cosh \psi + ct' \sinh \psi \qquad (12.3.5)$$

where the hyperbolic functions are $\cosh \psi = (e^{\psi} + e^{-\psi})/2$, $\sinh \psi = (e^{\psi} - e^{-\psi})/2$, etc. Putting the values of t and z from (12.3.5) in (12.3.2) we obtain

$$\omega t - KZ = (\omega \cosh \psi - Kc \sinh \psi)\, t' - \left(K \cosh \psi - \frac{\omega}{c} \sinh \psi \right) Z', \qquad (12.3.6)$$

$$KC = \omega \tanh \psi, \qquad (12.3.7)$$

Then the coefficient of Z' vanishes and

$$\omega t - KZ = \omega \sqrt{1 - \beta^2}\, t' \qquad (12.3.8)$$

$$\omega' T = \omega t - KZ \qquad (12.3.9)$$

and so we have $v_z = z' = V - (V/V_0)\ (\partial T/\partial t)$ where the symbol t' is replaced by T, and

$$\omega' = \omega \sqrt{1 - \beta^2} = \frac{\omega}{\gamma} \qquad (12.3.10)$$

Hence if $V = \omega/K$ (V is the phase velocity of the wave),

$$\omega t - KZ = \omega \sqrt{1 - (c^2/v^2)}\, T, \qquad (12.3.11)$$

$$\omega' = \frac{\omega}{\gamma_0} \text{ where } \gamma_0 = \left(1 - \frac{c^2}{V^2} \right)^{-1/2} \qquad (12.3.12)$$

and the differential operators transform as

$$\frac{\partial}{\partial t} = \gamma_0 \frac{\partial}{\partial T}, \; \frac{\partial}{\partial z} = -\frac{\gamma_0}{V} \frac{\partial}{\partial T} \qquad (12.3.13)$$

Putting the results of (12.3.13) in the relations

$$\mathbf{E} = -\frac{1}{c} \frac{\partial \mathbf{A}}{\partial t}, \; \mathbf{H} = \nabla \times \mathbf{A} \qquad (12.3.14)$$

which are valid for electromagnetic (transverse) wave fields, we obtain

$$\mathbf{E} = -\frac{\gamma_0}{c} \frac{\partial \mathbf{A}}{\partial T}, \; \mathbf{H} = -\frac{\gamma_0}{V} \hat{\mathbf{Z}} \times \frac{\partial \mathbf{A}}{\partial T} \qquad (12.3.15)$$

Writing

$$\mathbf{A} = \hat{\mathbf{X}} A_x + \hat{\mathbf{Y}} A_y + \hat{\mathbf{Z}} A_z, \qquad (12.3.16)$$

we find that

$$\mathbf{E} = -\frac{\gamma_0}{c} \left(\hat{\mathbf{X}} \frac{\partial A_x}{\partial T} + \hat{\mathbf{Y}} \frac{\partial A_y}{\partial T} + \hat{\mathbf{Z}} \frac{\partial A_z}{\partial T} \right), \; \mathbf{H} = \frac{\gamma_0}{V} \left(\hat{\mathbf{X}} \frac{\partial A_y}{\partial T} - \hat{\mathbf{Y}} \frac{\partial A_x}{\partial T} \right), \qquad (12.3.17)$$

$$E_x = \frac{V}{c} H_y, \; E_y = -\frac{V}{c} H_x, \qquad (12.3.18)$$

E is found to be perpendicular to **H**.

(b) Wave induced motion of electrons

The relativistically correct equation of motion of an electron is

$$\frac{d\mathbf{p}}{dt} = -e\,\mathbf{E} - \frac{e}{c}(\mathbf{v} \times \mathbf{B}), \tag{12.3.19}$$

where

$$\mathbf{p} = m\gamma\mathbf{v},\ \gamma^2 = (1-\beta^2)^{-1},\ \boldsymbol{\beta} = \frac{\mathbf{v}}{\mathbf{c}} \tag{12.3.20}$$

m is the electronic rest mass. Putting the values of **E** and **H** from (12.3.17) in (12.3.19) we obtain

$$\frac{d}{dt}(\gamma\mathbf{v}) = \frac{e\gamma_0}{mc}\left(\hat{\mathbf{X}}\frac{\partial A_x}{\partial T} + \hat{\mathbf{Y}}\frac{\partial A_y}{\partial T} + \hat{\mathbf{Z}}\frac{\partial A_z}{\partial T}\right) - \frac{e\gamma_0}{mcV}v \times \left(\hat{\mathbf{X}}\frac{\partial A_y}{\partial T} - \hat{\mathbf{Y}}\frac{\partial A_x}{\partial T}\right)$$

$$= \frac{e\gamma_0(V-c)}{mcV}\left(\hat{\mathbf{X}}\frac{\partial A_x}{\partial T} + \hat{\mathbf{Y}}\frac{\partial A_y}{\partial T}\right) + \frac{e\gamma_0\hat{\mathbf{Z}}}{mcV}\left(v_x\frac{\partial A_x}{\partial T} + v_y\frac{\partial A_y}{\partial T} + V\frac{\partial A_z}{\partial T}\right) \tag{12.3.21}$$

Since

$$\frac{d}{dt}(\gamma\mathbf{v}) = \hat{\mathbf{X}}\frac{d}{dt}(\gamma v_x) + \hat{\mathbf{Y}}\frac{d}{dt}(\gamma v_y) + \hat{\mathbf{Z}}\frac{d}{dt}(\gamma v_z)$$

we obtain

$$\frac{d}{dt}(\gamma v_x) = \frac{e\gamma_0(V-v_z)}{mcV}\frac{\partial A_x}{\partial T},\quad \frac{d}{dt}(\gamma v_y) = \frac{e\gamma_0(V-v_z)}{mcV}\frac{\partial A_y}{\partial T} \tag{12.3.22}$$

$$\frac{d}{dt}(\gamma v_z) = \frac{e\gamma_0}{mcV}\left(v_x\frac{\partial A_x}{\partial T} + v_y\frac{\partial A_y}{\partial T} + V\frac{\partial A_z}{\partial T}\right) \tag{12.3.23}$$

Scalar product of (12.3.19) with **v** gives

$$\frac{d}{dt}(\gamma c) = -\frac{e}{mc}(\mathbf{E}\,.\,\mathbf{v}) = \frac{e\gamma_0}{mc^2}\left(\mathbf{v}\cdot\frac{\partial \mathbf{A}}{\partial T}\right) \tag{12.3.24}$$

Putting the value of $V - v_z$ from (12.3.9) in (12.3.22) we obtain

$$\frac{d}{dt}(\gamma v_x) = \frac{e}{mc}\frac{\partial T}{\partial t}\frac{\partial A_x}{\partial t},\quad \frac{d}{dt}(\gamma v_y) = \frac{e}{mc}\frac{\partial T}{\partial t}\frac{\partial A_y}{\partial T}, \tag{12.3.25}$$

$$\frac{d}{dt}\left[\gamma\left(V - \frac{V}{\gamma_0}\frac{\partial T}{\partial t}\right)\right] = \frac{e}{mcV}\left(v_x\frac{\partial A_x}{\partial T} + v_y\frac{\partial A_y}{\partial T} + V\frac{\partial A_z}{\partial T}\right) \tag{12.3.26}$$

When $A_z = 0$, eliminating the common factor $v_x(\partial A_x/\partial T) + v_y(\partial A_y/\partial T)$ between (12.3.24) and (12.3.26), we obtain

$$\frac{d}{dt}\left[\gamma c^2 - \gamma\gamma_0\left(V - \frac{V}{\gamma_0}\frac{\partial T}{\partial t}\right)\right] = 0 \qquad (12.3.27)$$

Hence we can write

$$\gamma c^2 - \gamma V^2 (\gamma_0 - (\partial T/\partial t)) = \text{constant} \qquad (12.3.28)$$

Solving for $\partial T/\partial t$ from this equation we get

$$\frac{\partial T}{\partial t} = \frac{a}{\gamma} + b \qquad (12.3.29)$$

where a is arbitrary, and

$$b = \frac{1}{V^2}(V^2\gamma_0 - c^2) \qquad (12.3.30)$$

If the electron is initially at rest in the coordinate system used here, then (12.3.9) initially gives

$$t = 0,\ \mathbf{v} = 0,\ \frac{\partial T}{\partial t} = \gamma_0 \qquad (12.3.31)$$

Putting these values in (12.3.29) we obtain

$$a = \gamma_0 - b = \frac{c^2}{V^2} \qquad (12.3.32)$$

Hence, if (12.3.9) is used, we have

$$\frac{\partial T}{\partial t} = \frac{c^2}{V^2\gamma} + \gamma_0 - \frac{c^2}{V^2} = \gamma_0\left(1 - \frac{v^2}{V^2}\right) \qquad (12.3.33)$$

Solving for v_z from this equation, we find that

$$v_z = \frac{c^2}{V\gamma_0}\left(1 - \frac{1}{\gamma}\right) \qquad (12.3.34)$$

The relations of (12.3.22) can be written as

$$\frac{d}{dt}(\gamma v_x) = \frac{e}{m}\frac{\partial A_x}{\partial t},\ \frac{d}{dt}(\gamma v_y) = \frac{e}{mc}\frac{\partial A_y}{\partial t} \qquad (12.3.35)$$

Replacing, approximately, the total time derivatives in the left hand side of these two relations by the corresponding partial time derivatives, we obtain

$$\frac{\partial}{\partial t}\left(\gamma v_x - \frac{eA_x}{mc}\right) = 0,\ \frac{\partial}{\partial t}\left(\gamma v_\gamma - \frac{eA_y}{mc}\right) = 0 \qquad (12.3.36)$$

Integration gives

$$\gamma v_x = \frac{eA_x}{mc},\ \gamma v_y = \frac{eA_y}{mc} \qquad (12.3.37)$$

The relations of (12.3.34) can be written as

$$\frac{V\gamma_0}{c}\frac{v_z}{c} = 1-\sqrt{1-\beta^2} = \frac{1}{2c^2}(v_x^2+v_y^2) = \frac{1}{2}\left(\frac{e}{mc^2}\right)^2 (A_x^2+A_y^2) \qquad (12.3.38)$$

where

$$v_x = \frac{eA_x}{mc},\quad v_y = \frac{eA_y}{mc}, \qquad (12.3.39)$$

and terms beyond the first power of v^2/c^2, in the expansion of γ, have been neglected. In presence of an elliptically polarized wave

$$A_x = a(T)\cos\omega T,\quad A_y = b(T)\sin\omega T, \qquad (12.3.40)$$

keeping terms up to the order of non-linear corrections used here we find that

$$\frac{\partial a}{\partial T} = 0,\quad \frac{\partial b}{\partial T} = 0, \qquad (12.3.41)$$

$$\frac{v_z}{c} = \frac{c}{2V\gamma_0}\left(\frac{e}{mc^2}\right)^2 (a^2\cos^2\omega T + b^2\sin^2\omega T) \qquad (12.3.42)$$

Denoting the average of v_z, in one time period $2\pi/\omega$, by $\langle v_z\rangle$, we obtain

$$\frac{\langle v_z\rangle}{c} = \frac{\int_{t=0}^{2\pi/\omega} v_z\,dt}{c\int_0^{2\pi/\omega} dt} = \frac{c}{4V\gamma_0}\,\frac{\left(\frac{e}{mc^2}\right)^2 (a^2+b^2)\cdot\frac{2\pi}{\omega}}{\int_0^{2\pi}\left(\frac{dT}{\gamma_0}+\frac{1}{V}\frac{\partial Z}{\partial T}\,dT\right)} \qquad (12.3.43)$$

Since the denominator of this relation is

$$= \frac{2\pi}{\gamma_0\,\omega} + \frac{1}{V}\int_0^{2\pi/\omega} v_z\frac{dt}{dT}\,dT = \frac{2\pi}{\gamma_0\,\omega} + \frac{1}{V}\int_0^{2\pi/\omega}\frac{v_z\,dT}{\gamma_0\left(1-\frac{v_z}{c}\right)}$$

$$= \frac{2\pi}{\gamma_0\,\omega} + \frac{1}{\gamma_0 V}\int_0^{2\pi/\omega} v_z\,dT + \ldots = \frac{2\pi}{\gamma_0\,\omega} + \frac{c^2}{2V^2\gamma_0^2}\left(\frac{e}{mc^2}\right)^2\cdot\frac{(a^2+b^2)}{2}\cdot\frac{2\pi}{\omega}$$

equation (12.3.43) reduces to

$$\frac{\langle v_z\rangle}{c} = \frac{c}{4\pi\gamma_0}\,\frac{\left(\frac{e}{mc^2}\right)^2 (a^2+b^2)}{\frac{1}{\gamma_0}+\frac{c^2}{4V^2\gamma_0^2}\left(\frac{e}{mc^2}\right)^2 (a^2+b^2)} = \frac{\left(\frac{e}{mc^2}\right)^2\cdot\frac{(a^2+b^2)\,c}{4V}}{1+\left(\frac{e}{mc^2}\right)^2\cdot\frac{c^2(a^2+b^2)}{4V^2}}$$

(12.3.44)

Averaging the Poynting flux $\mathbf{S}\,(= c\,(\mathbf{E} \times \mathbf{H})/4\pi)$ over the time period $2\pi/\omega$, we obtain

$$|\langle \mathbf{S} \rangle| = \frac{c}{4\pi} \langle K_x H_y - K_y H_x \rangle = \frac{c}{4\pi} \langle \frac{\gamma_0^2}{cV} \left\{ \left(\frac{\partial A_x}{\partial T} \right)^2 + \left(\frac{\partial A_y}{\partial T} \right)^2 \right\} \rangle$$

$$= \frac{\gamma_0^2 \omega^2}{8\pi V} (a^2 + b^2) \tag{12.3.45}$$

Eliminating $a^2 + b^2$ from (12.3.44), with the help of (12.3.45), we obtain

$$\frac{\langle v_z \rangle}{c} = \frac{\dfrac{c^2}{2\omega^2 \gamma_0^2} \left(\dfrac{e}{mc^2} \right)^2 |\langle \mathbf{E} \times \mathbf{H} \rangle|}{1 + \dfrac{c^3}{2\omega^2 V \gamma_0^2} \left(\dfrac{e}{mc^2} \right)^2 |\langle \mathbf{E} \times \mathbf{H} \rangle|} \tag{12.3.46}$$

(c) Dispersion relation of Circularly Polarized waves in Cold Plasmas

Consider the relativistically correct field equations for the two component, cold, magnetized plasmas:

$$\frac{d\mathbf{p}_e}{dt} = -e\mathbf{E} - \frac{e}{c} (\mathbf{u}_e \times (\mathbf{H} + \mathbf{H}_0)) \tag{12.3.47}$$

$$\frac{d\mathbf{p}_i}{dt} = e\mathbf{E} + \frac{e}{c} (\mathbf{u}_i \times (\mathbf{H} + \mathbf{H}_0)) \tag{12.3.48}$$

$$\nabla \times \mathbf{E} = -\frac{1}{c} \frac{\partial \mathbf{H}}{\partial t},\ (\nabla . \mathbf{H}) = 0,\ (\nabla . \mathbf{E}) = 0 \tag{12.3.49}$$

$$\nabla \times \mathbf{H} = \frac{1}{c} \frac{\partial \mathbf{E}}{\partial t} + \frac{4\pi e N}{c} (\mathbf{u}_i - \mathbf{u}_e) \tag{12.3.50}$$

with $\mathbf{H}_0 = (0, 0, H_0)$. The circularly polarized electric field is

$$\mathbf{E} = \alpha\,(\cos\theta, \pm \sin\theta, 0),\quad \theta = KZ - \omega t \tag{12.3.51}$$

Hence $\quad \nabla \times \mathbf{E} = \alpha K\,(\mp \cos\theta,\ -\sin\theta,\ 0),$

$$\frac{\partial \mathbf{H}}{\partial t} = \alpha\, KC\,(\pm \cos\theta,\ \sin\theta,\ 0)$$

Integration gives $\mathbf{H} = \alpha\,(KC/\omega)\,(\mp \sin\theta, \cos\theta, 0)$.

$$\frac{\partial \mathbf{E}}{\partial t} = \alpha\omega\,(\sin\theta, \mp \cos\theta, 0),\ \nabla \times \mathbf{H} = \alpha \frac{K^2 c}{\omega} (\sin\theta, \mp \cos\theta, 0),$$

$$\frac{4\pi Ne}{c} (\mathbf{u}_i - \mathbf{u}_e) = \nabla \times \mathbf{H} - \frac{1}{c} \frac{\partial \mathbf{E}}{\partial t} = \frac{\alpha}{\omega c} (K^2 c^2 - \omega^2)\ (\sin\theta, \mp \cos\theta, 0),$$

$$\mathbf{u}_i - \mathbf{u}_e = \frac{(K^2 c^2 - \omega^2)\,\alpha}{4\pi\, Ne\, \omega} (\sin\theta, \mp \cos\theta, 0) \tag{12.3.52}$$

For convenience we next write

$$\mathbf{u}_e = v_e\,(\sin\theta, \mp\cos\theta, 0),\ \mathbf{u}_i = v_i\,(\sin\theta, \mp\cos\theta, 0) \tag{12.3.53}$$

and obtain

$$v_i - v_e = \frac{(K^2c^2 - \omega^2)\,\alpha}{4\pi Ne\,\omega} = \frac{(K^2c^2-\omega^2)\,\alpha\, e}{m\omega\,\omega_{\rho e}^2} = \frac{(K^2c^2-\omega^2)\,\alpha e}{M\omega\,\omega_{\rho i}^2}, \tag{12.3.54}$$

$$\frac{d\mathbf{p}_e}{dt} = -e\alpha\,(\cos\theta,\ \pm\sin\theta,\ 0) - 0 - \frac{e\,v_e}{c}H_0\,(\mp\cos\theta, -\sin\theta, 0)$$

$$= -e\left(\alpha \mp \frac{v_e}{c}H_0\right)(\cos\theta,\ \pm\sin\theta,\ 0)$$

$$\mathbf{p}_e = -\frac{e}{\omega}\left(\alpha \mp \frac{v_e}{c}H_0\right)(-\sin\theta,\ \pm\cos\theta,\ 0), \tag{12.3.55}$$

$$\frac{d\mathbf{p}_i}{dt} = e\,\alpha\,(\cos\theta, \pm\sin\theta, 0) + 0 + \frac{e\,v_i}{c}H_0\,(\mp\cos\theta, -\sin\theta, 0)$$

$$= e\left(\alpha \mp \frac{v_i}{c}H_0\right)(\cos\theta, \pm\sin\theta, 0),$$

$$\mathbf{p}_i = \frac{e}{\omega}\left(\alpha \mp \frac{v_i}{c}H_0\right)(-\sin\theta, \pm\cos\theta, 0) \tag{12.3.56}$$

Hence, squaring, we obtain

$$p_e^2 = \frac{e^2}{\omega^2}\left(\alpha \mp \frac{v_e}{c}H_0\right)^2 = m^2\gamma_e^2 v_e^2, \tag{12.3.57}$$

$$p_i^2 = \frac{e^2}{\omega^2}\left(\alpha \mp \frac{v_i}{c}H_0\right)^2 = M^2\gamma_i^2 v_i^2 \tag{12.3.58}$$

with

$$\gamma_e^2 = \frac{1}{1-u_e^2/c^2} = \frac{1}{1-v_e^2/c^2} \tag{12.3.59}$$

$$\gamma_i^2 = \frac{1}{1-u_i^2/c^2} = \frac{1}{1-v_i^2/c^2} \tag{12.3.60}$$

Then, with the help of (12.3.55) and (12.3.56) we obtain

$$\frac{e}{\omega}\left(\alpha \mp \frac{v_e}{c}H_0\right) = m\gamma_e\,v_e,\ -\frac{e}{\omega}\left(\alpha \mp \frac{v_i}{c}H_0\right) = M\gamma_i\,v_i \tag{12.3.61}$$

$$\frac{e\alpha}{\omega} = \frac{m}{\omega}\gamma_e\,v_e\,(\omega \pm \Omega_e^r),\ -\frac{e\alpha}{\omega} = \frac{M}{\omega}\gamma_i\,v_i\,(\omega \mp \Omega_i^r) \tag{12.3.62}$$

where $$\Omega_e^r = \frac{eH_0}{\gamma_e mc}, \ \Omega_i^r = \frac{eH_0}{\gamma_i Mc} \tag{12.3.63}$$

Ω_e^r and Ω_i^r are the relativistic electron and ion gyrofrequencies, respectively. Equations (12.3.62) can be wrriten as

$$v_e = \frac{e\alpha}{m\gamma_e(\omega \pm \Omega_e^r)}, \ v_i = -\frac{e\alpha}{M\gamma_i(\omega \pm \Omega_i^r)} \tag{12.3.64}$$

Then equation (12.3.56) becomes

$$v_i - v_e = -\frac{e\alpha}{M\gamma_i(\omega \mp \Omega_i r)} - \frac{e\alpha}{m\gamma_e(\omega \pm \Omega_e^r)}$$

$$= \frac{e\alpha}{m\omega}\frac{(K^2c^2 - \omega^2)}{\omega_{pe}^2} \tag{12.3.65}$$

This gives the exact dispersion relation

$$\omega^2 - K^2c^2 = \frac{m\omega\,\omega_{\rho e}^2}{M\gamma_i(\omega \mp \Omega_i^r)} + \frac{m\omega\,\omega_{\rho i}^2}{m\gamma_e(\omega \pm \Omega_e^r)}$$

$$= \frac{\omega\omega_{\rho i}^2}{\gamma_i(\omega \mp \Omega_i^r)} + \frac{\omega\omega_{\rho e}^2}{\gamma_e(\omega \pm \Omega_e^r)} \tag{12.3.65 (a)}$$

It is simplified to

$$\omega^2 - K^2c^2 = \frac{\omega^2(\omega_{\rho i}^2\gamma_e + \omega_{\rho e}^2\gamma_i)}{\gamma_e\gamma_i(\omega \mp \Omega_i^r)(\omega \pm \Omega_e^r)} \tag{12.3.66}$$

(d) Induced magnetization of a circularly polarized wave in a two component magnetized plasma

Magnetization $\mathbf{H}_i$ from the magnetic moment is given by

$$\mathbf{H}_i = \frac{4\pi}{2c}\sum_p(\mathbf{r}_p \times \mathbf{j}_p) \tag{12.3.67}$$

Where $\mathbf{j}_p = N_p q_p \mathbf{u}_p$, q_p being the charge per particle, of the pth species, N_p, is the number density and $\mathbf{u}_p$ is the particle velocity of that species. So

$$H_i = \frac{4\pi Ne}{2c}[(x_i\dot{y}_i - \dot{x}_i y_i) - (x_e\dot{y}_e - \dot{x}_e y_e)] \tag{12.3.68}$$

For a single, circularly polarized wave we use equation (12.3.53) and write

$$\mathbf{u}_e = \frac{e\alpha(\sin\theta, \mp\cos\theta, 0)}{m\gamma_e(\omega \pm \Omega_e^r)}, \ \mathbf{u}_i = -\frac{e\alpha(\sin\theta, \mp\cos\theta, 0)}{M\gamma_i(\omega \mp \Omega_i^r)}, \tag{12.3.69}$$

$$\mathbf{r}_e = \frac{e\alpha\,(\cos\theta, \pm\sin\theta, 0)}{\omega m\gamma_e\,(\omega \pm \Omega_e^r)}, \quad \mathbf{r}_i = -\frac{e\alpha\,(\cos\theta, \pm\sin\theta, 0)}{\omega M\gamma_i\,(\omega \mp \Omega_i^r)} \tag{12.3.70}$$

where $\mathbf{r}_e$ and $\mathbf{r}_i$ are, respectively, the field induced displacements of the electrons and ions. Hence

$$x_i\dot{y}_i - \dot{x}_i y_i = \frac{\mp e^2\alpha^2}{\omega M^2\gamma_i^2\,(\omega \mp \Omega_i^r)^2},$$

$$x_e\dot{y}_e - \dot{x}_e y_e = \frac{\mp e^2\alpha^2}{\omega m^2\gamma_e^2\,(\omega \pm \Omega_e^r)^2},$$

Using these results in (12.3.68) we find that

$$H_i = \mp\frac{4\pi N e^3\alpha^2}{2c\omega}\left[\frac{1}{M^2\gamma_i^2\,(\omega \mp \Omega_i^r)^2} - \frac{1}{m^2\gamma_e^2\,(\omega \pm \Omega_e^r)^2}\right] \tag{12.3.71}$$

(e) Exact dispersion relation and induced magnetization for circularly polarized Alfven waves

For Alfven waves, since $\omega << |\Omega_i|$, the exact dispersion relation, equation (12.3.66), reduces to

$$\omega^2 - K^2c^2 = -\frac{\omega^2\,(\omega_{\rho i}^2\gamma_e + \omega_{\rho e}^2\gamma_i)}{\dfrac{eH_0}{mc}\,\dfrac{eH_0}{Mc}} \tag{12.3.72}$$

where $$\gamma_e = \frac{1}{\sqrt{1 - \dfrac{e^2\alpha^2}{m^2c^2}\,\dfrac{m^2c^2}{e^2H_0^2}}} = \frac{1}{\sqrt{1-\alpha^2/H_0^2}} = \gamma_i = \gamma \text{ (say)} \tag{12.3.73}$$

Hence equation (12.3.72) becomes

$$\frac{K^2c^2}{\omega^2} = 1 + \frac{\gamma M m c^2}{e^2H_0^2}\,(\omega_{\rho e}^2 + \omega_{\rho i}^2) \tag{12.3.74}$$

Since the second term in the right hand side is much larger than unity, we ignore the first term and obtain the dispersion relation

$$K^2C_A^2 = \omega^2 \tag{12.3.75}$$

where $$C_A^2 = \frac{H_0^2}{4\pi\rho\gamma} \tag{12.3.76}$$

$\rho\ (= MN)$ is the plasma mass density. Then the induced magnetization formula of (12.3.71) gives

$$H_i = \mp \frac{4\pi N e^3 \alpha^2 \,[m^2 \gamma_e^2 (\omega \pm \Omega_e')^2 - M^2 \gamma_i^2 (\omega \mp \Omega_i')^2]}{2c\,\omega M^2 m^2 \gamma_i^2 \gamma_e^2 \Omega_e'^2 \Omega_i'^2}$$

$$= \mp \frac{4\pi N e^3 \alpha^2}{2c\omega} \frac{\left[\omega^2 (m^2 \gamma_e^2 - M^2 \gamma_i^2) \pm 2\omega \left(m^2 \gamma_e \frac{eH_0}{mc} + M^2 \gamma_i \frac{eH_0}{MC}\right)\right]}{M^2 m^2 \frac{e^2 H_0^2}{M^2 c^2} \frac{e^2 H_0^2}{m^2 c^2}}$$

$$= \mp \frac{\omega_{\rho i}^2 \alpha^2 \gamma}{H_0 \Omega_i^2} \tag{12.3.77}$$

Exercise 12.3.1. Show that, in a multicomponent, unmagnetized cold, plasmas, the exact dispersion relation for circular polarization can be written as

$$\omega^2 - K^2 c^2 = 4\pi c\omega \sum_i \left(\frac{N_i q_i^2}{\sqrt{m_i^2 c^2 \omega^2 + q_i^2 \alpha^2}} \right)$$

Answer. Putting $H_0 = 0$ in (12.3.55) and (12.3.56) we obtain

$$\mathbf{u}_e = \frac{e\,\alpha\, c\,(\sin\theta, \mp\cos\theta, 0)}{\sqrt{m^2 c^2 \omega^2 + e^2 \alpha^2}}, \quad \mathbf{u}_i = -\frac{e\alpha c\,(\sin\theta, \mp\cos\theta, 0)}{\sqrt{M^2 c^2 \omega^2 + e^2 \alpha^2}}$$

Hence the relation (12.3.52) is replaced by

$$-\frac{4\pi N e^2 \alpha c}{\sqrt{M^2 c^2 \omega^2 + e^2 \alpha^2}} - \frac{4\pi N e^2 \alpha^2 c}{\sqrt{m^2 c^2 \omega^2 + e^2 \alpha^2}} = \frac{\alpha}{\omega} (K^2 c^2 - \omega^2)$$

This leads to the exact dispersion relation

$$\omega^2 - K^2 c^2 = \frac{M \omega_{\rho i}^2 c\omega}{\sqrt{M^2 c^2 \omega^2 + e^2 \alpha^2}} + \frac{m \omega_{\rho e}^2 c\omega}{\sqrt{m^2 c^2 \omega^2 + e^2 \alpha^2}}$$

So, more generally, for a mixture of charged particles of several different species, the exact dispersion relation can be expressed as

$$\omega^2 - K^2 c^2 = \sum_i \left(\frac{m_i \omega_{\rho i}^2 c\omega}{\sqrt{m_i^2 c^2 \omega^2 + q_i^2 \alpha^2}} \right) = 4\pi c\omega \sum_i \left(\frac{N_i q_i^2}{\sqrt{m_i^2 c^2 \omega^2 + q_i^2 \alpha^2}} \right)$$

where q_i is the charge and m_i is the mass per particle of the ith species, N_i is the number density of particles of that species and $\omega_{\rho\, i}^2 = 4\pi N_i q_i^2 / m_i$.

Exercise 12.3.2. Find the dispersion relation and the formula of 1*FE* magnetization for a circularly polarized wave in a multi component, cold plasma having an ambient magnetic field along the direction of wave propagation.

Answer. The exact dispersion relation is simply a generalization of the relation (12.3.65 (*a*))

$$\omega^2 - K^2 c^2 = \sum_i \frac{\omega\omega_{\rho i}^2}{\gamma_i (\omega \mp \Omega_i^r)},$$

and the induced magnetization is a generalization of the formula of (12.3.71):

$$H_i = \mp \frac{4\pi\alpha}{2c\omega} \sum_i \frac{N_i q_i^3}{m_i^2 \gamma_i^2 (\omega \mp \Omega_i^r)^2}$$

where now $\Omega_i^r = \dfrac{q_i H_0}{\gamma_i m_i c}$. The signs $\mp$ stand for the waves of the right and left circular polarizations.

12.4 BREAKING OF LARGE-AMPLITUDE OSCILLATIONS

(a) Overlapping of oscillations

In cold plasmas, electrons oscillate at their characteristic frequency ω_ρ. When the amplitude is large, ordering of the electrons, necessary for maintaining regular oscillatory patterns, is destroyed, and the crest of a wave falls into the trough of another. This wave breaking process is like the breaking of large amplitude water waves. It dissipates oscillatory motion, and transforms wave energy into kinetic energy.

Consider identical sheets of electrons, embedded in a fixed background of positive charges. The sheets in the unperturbed state are at rest and the intersheet distance is the constant length δ. The positive charge density is $n_0 e$ where e is charge of a positive particle. The charge and mass per unit area of a sheet are $-en_0\delta$ and $mn_0\delta$.

In between two sheets the electric field varies linearly with distance, because of the existence of the background positive charges, but at each sheet the field jumps to $-4\pi en_0\delta$. When the displacement of a sheet from the equilibrium state is small, it performs simple harmonic motion about its mean position. This oscillation is independent of the behaviour of the other sheets, and resembles an oscillator. For a small displacement the sheets do not cross each other, like a system of oscillators oscillating without mutual interference. In the displaced state, one sheet can cross another, and energy is exchanged between sheets. If the average energy per sheet remains constant and the neighbouring sheets execute nearly identical motion, then, motion of a very large number of sheets, without crossing, is possible.

The force between sheets is of long range; so the plasma exhibits collective motion as well as individual particle motion (which is also the motion of sheets). Each sheet surrounds itself with a Debye sheilding cloud and the sheet waves exhibit Landau damping.

(b) Random motion of statistical ensembles and electron oscillation

When large amplitude oscillations cross each other, energy is exchanged between the sheets, as between particles in a statistical ensemble, due to random thermal motion. When there is no crossing of the sheets; the system does not attain thermal equilibrium; so this state cannot be reached from the state of random thermal motion. The volume of the phase space in which crossing does not take place is small, compared with the total phase space, consistent with the energy.

Consider a number N of identical pendulums which are constrained to oscillate along the line of centres which is parallel to a fixed direction. When the first pendulum is moved along the line of the fixed direction and then released, it begins to oscillate. If the initial displacement is sufficiently large, the first pendulum gives its momentum to the second, the second in turn to the third and the process continues. But, in this process, the displacement is not transferred and the system begins to oscillate about the respective mean positions as in N-dimensional oscillators in N-dimensional space; the displacement of the ith pendulum in this space is represented by the value of the ith coordinate.

One pendulum does not cross its neighbour, if

$$X_i \le X_{i+1} + \delta,\ X_1 \ge -\delta,\ X_N \le \delta \tag{12.4.1}$$

where now δ denotes the distance between successive pendulums in the unperturbed state. The second and third conditions of (12.4.1) are the end conditions. The equalities

$$X_i = X_{i+1} + \delta,\ X_1 = -\delta,\ X_N = \delta \tag{12.4.2}$$

give a set of $N - 1$ dimensional hyperplanes which the system cannot surpass. The N-dimensional oscillator is forced to oscillate inside the N-dimensional polyhedron with the hyperplanes of (12.4.2) as the walls.

A pendulum strikes its neighbour like the striking of the N-dimensional oscillator on one of the walls, that is, on one of the hyperplanes, given by (12.4.2) of the polyhedron. The system is bounded, because the component of the velocity perpendicular to the walls can be reversed. The walls are identified with the positions of the zeroth and the $(N + 1)$th particle; so they are parts of the system and these two particles are supposed held fixed. The quantity $i\delta$ is the position of the ith pendulum in the unperturbed state.

The linear equations for orthogonal transformation from the $X_i\,S$ to the $A_j\,S$ and vice versa are

$$A_j = \left(\frac{2}{N+1}\right)^{1/2} \sum_{i=0}^{N+1} X_i \sin\left(\frac{\pi\, ij\,\delta}{L}\right) \tag{12.4.3}$$

$$X_i = \left(\frac{2}{N+1}\right)^{1/2} \sum_{i=0}^{N+1} A_j \sin\left(\frac{\pi\, ij\,\delta}{L}\right) \tag{12.4.4}$$

They represent a rotation of the coordinate axes. Regarding the system as N identical mass points, connected by identical massless springs, the quantities $A_1, A_2 ...., A_N$ are normal coordinates of the system. For displacements small, compared with the wave length $\lambda\ (= 2L/j)$ of the mode, the quantities $A_1, A_2, ..., A_N$ are the Eulerian coordinates for the actual positions (Fourier modes).

(c) Oscillations of Planes

Electrons vibrate back and forth in the direction of the X-axis. So the electron sheets are perpendicular to OX; Y and Z coordinates do not appear in the equations. The initial equilibrium position of the sheet is x_0. Its displacement at time t is $\xi\,(X_0, t)$. The length $\xi\,(X_0, t)$ is finite, and is infinitesimal in the linearized theory. If X is the displaced position of the sheet, at time t, then

$$X = X_0 + \xi \tag{12.4.5}$$

The displaced electrons have left a volume of positive charges behind them; their number per unit area is $e n_0\, \xi$. If no charge is added or removed at the planes $X = \pm\infty$, then by Gauss's divergence theorem, the electric field experienced by the electrons is

$$E = 4\pi\, e n_0\, \xi \tag{12.4.6}$$

Equation for electron motion, in the displaced sheet, is

$$m\ddot{\xi} = -eE; \tag{12.4.7}$$

So we find that

$$\ddot{\xi} = -\omega_\rho^2\, \xi \ \text{ with } \ \omega_\rho^2 = 4\pi n_0\, e^2/m \tag{12.4.8}$$

The solution is $\xi = a_1\,(X_0) \sin \omega_\rho\, t + b_1\,(X_0) \cos \omega_\rho\, t$ (12.4.9)

where $a_1\,(X_0)$ and $b_1\,(X_0)$ are arbitrary functions of X_0. Each electron, therefore executes simple harmonic motion with period $2\pi/\omega$ about its equilibrium position. If n_0 slightly varies in sheets initially in the neighbourhood of the sheet $X = X_0$ (in the perturbed state), then some sheets behind $X = X_0$ may cross those in front in a time of the order of $2\pi/\omega_\rho$.

If ordering of the electrons in the sheets is not changed, then all those electrons which were originally on the positive side of $X = X_0$ remain on that side, and all those of the negative side remain there. If $\xi > 0$, at time t, then there is an excess of positive charges per unit area on the negative side of this sheet and the excess negative charge $-en_0\, \xi$ per unit area on its positive side. When ordering of electrons along the X-axis is maintained, the simple harmonic motion of this sheet is independent of what the rest are doing; so, motion of the neighbouring sheets is not affected by motion of this sheet.

When an oscillation of two electron sheets maintains order, we fix one of them initially in the plane $X = X_0$, and the other in the plane $X = X_0 + \delta X_0$. The displaced positions, at time t of these planes are denoted by $X = X_0 + \xi$ and $X = X_0 + \xi + \delta X_0 + \delta\xi$. Then, the order is maintained if

$$\delta X_0 + \delta\xi > 0 \text{ or } \frac{\partial \xi}{\partial X_0} > -1 \tag{12.4.10}$$

Since the sheets are conserved, there will be no crossing if

$$n_0\, dX_0 = n dX \tag{12.4.11}$$

where the evolution of dx from dx_0 is determined by the relation

$$n(x,\ t) = \frac{n_0}{1 + (\partial \xi / \partial x_0)} \tag{12.4.12}$$

and n is the electron number density in the displaced position $X = X_0 + \xi$. If ω_ρ is independent of X_0, then differentiation of (12.4.8) with respect to X_0 gives

$$\frac{\partial^2}{\partial t^2}\left(\frac{\partial \xi}{\partial X_0}\right) = -\omega_\rho^2 \left(\frac{\partial \xi}{\partial X_0}\right) \tag{12.4.13}$$

So $\partial\xi/\partial X_0$, like ξ, is simple harmonic. The total energy of the harmonic oscillator is

$$W = \omega_\rho^2 \left(\frac{\partial \xi}{\partial x_0}\right)^2 + \left(\frac{\partial^2 \xi}{\partial x_0\, \partial t}\right)^2 = \omega_\rho^2 \left\{\left(\frac{\partial a_1}{\partial x_0}\right)^2 + \left(\frac{\partial b_1}{\partial x_0}\right)^2\right\} \tag{12.4.14}$$

Since the right hand side is independent of time the conclusion is that for all time $W < \omega_\rho^2$ provided, initially, this inequality is satisfied.

If K is the wave number of the displacement then, for $a_1 = 0$, $b_1(x_0) = b \sin Kx_0$, equations (12.4.5) and (12.4.9) give

$$X = X_0 + b \cos \omega_\rho t \sin Kx_0 \tag{12.4.15}$$

Equation (12.4.6) shows that $E = 4\pi e n_0 b \sin Kx_0 \cos \omega_\rho t$. At $t = 0$ the solutions are

$$X = X_0 + b \sin KX_0,\ E = 4\pi e n_0 b \sin KX_0 \tag{12.4.16}$$

Eliminating X_0 between these two relations, E is determined as a function of x: The graphs of Fig. 12.4.1 are plots of the ratio of E to its maximum value against the initial value of X. For a very small value of b the figure is a sine curve. As b increases, the points, giving the extreme value of E/K_{max}, move towards the plane $X = \pi/2$. When $Kb > 1$ the curve is multiple valued; this is absurd because E is always single valued. Hence the inequality (12.4.10) is not valid for $Kb >> 1$. Some of the consequences of this effect are : (1) order is lost, (2) fine scale mixing of the various parts destroys the oscillation, and (3) viscous effects become prominent, where electron density and density gradient are finite.

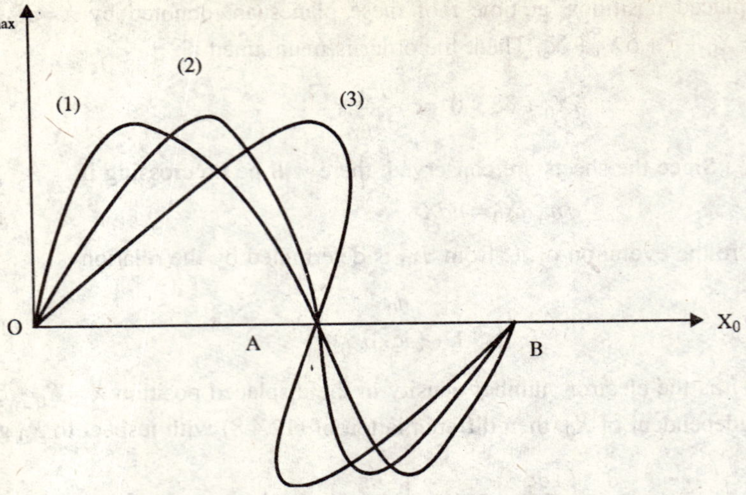

Fig. 12.4.1 Plots of E/E_{max} against X_0; curve (1) for $Kb << 1$, (2) for $Kb < 1$, (3) for $Kb > 1$. At A and B the values of X_0 are π K and $2\pi/K$ respectively.

(d) Cylindrical oscillations

Here we consider radial oscillations of electrons in a plasma which is uniformly distributed in a cylinder. Those electrons, which are initially distributed on the surface of radius r_0, are displaced to the surface of radius $r_0 + \xi$ at time t. A radially perturbed electric field is generated by the positive charges in the shell, between the surfaces $r = r_0$ and $r = r_0 + \xi$. Since n_0 is the constant number density of electrons or ions at equilibrium at any point inside this shell, $\nabla \cdot \mathbf{E} = 4\pi\, en_0$. Integrating both sides of this relation over the space in unit length of this shell, and applying Gauss's divergence theorem, we get

$$\iint E\, d\Sigma_r = 4\pi\, en_0 \iiint d\tau \tag{12.4.17}$$

where E is the radial field, $d\Sigma_r$ is a surface element on the surface $r = r_0 + \xi$ and $d\tau$ is a volume in the space between $r = r_0$ and $r = r_0 + \xi$. Since the cylinder is of unit length, $d\Sigma_r = (r_0 + \xi)\, d\theta$, we write $E \cdot 2\pi\,(r_0 + \xi) = 4\pi\, en_0$ $\{n\,(r_0 + \xi)^2 - \pi\, r_0^2\}$, and obtain

$$E = \frac{2\pi\, en_0}{r_0 + \xi}\, \{(r_0 + \xi)^2 - r_0^2\} \tag{12.4.18}$$

The equation of motion $m\ddot{\xi} = -eE$ then yields for ξ the equation

$$\ddot{\xi} + \omega_\rho^2\, \frac{(2r_0\, \xi + \xi^2)}{2\,(r_0 + \xi)} = 0 \tag{12.4.19}$$

As $r_0 \to \infty$ the motion becomes simple harmonic. Putting

$$\xi = r_0\, y \tag{12.4.20}$$

in (12.4.19) we obtain the anharmonic oscillator type of nonlinear equation

$$\ddot{y} + \omega_\rho^2 \frac{\left(y + \frac{y^2}{2}\right)}{y+1} = 0 \tag{12.4.21}$$

Solving, for the time period T, correct up to squares of the amplitudes Y_0 of the linearized approximation, we find that

$$T = \frac{2\pi}{\omega_\rho}\left(1 - \frac{y^2}{12} + \ldots\right) \tag{12.4.22}$$

The maximum value ξ_0 of ξ depends both on r_0 and the maximum value y_0 of y. Only those particles, which have the same y_0, oscillate with the same time period, and particles lying on different equilibrium surfaces will have, in general, different y_0 and so different time periods of oscillation. This clearly shows that overlapping is possible of one oscillating surface on its neighbour.

(e) Measurement of the cross-over time

Two closed cylindrical shells cross one another after the cross-over time, if their time periods differ even slightly. At one stage the inner cylinder goes out while the outer one draws inwards, there by giving rise to crossing of one by the other, fine scale mixing and to the possibility of destruction of oscillations. Time required to start the fine scale mixing is roughly the time for two concentric cylinders, separated initially by twice the amplitude of oscillation, to become half a period out of phase. If mixing of the cylinders oscillating with time periods T_1 and T_2 occurs after n_1 and n_2 oscillations, and after time t starting from $t = 0$, then

$$n_1 = n_2 \pm \frac{1}{2},\quad t = n_1 T_1 = n_2 T_2 \tag{12.4.23}$$

Expanding T_2 we can approximately write

$$T_2 = T_1 + \left(\frac{dT}{dr_0}\right) 2\xi_0 \tag{12.4.24}$$

Multiplying both sides of (12.4.24) by t/T_1^2 and then eliminating t/T_1 and T_2/t, with the help of (12.4.23), we get

$$\frac{n_1^2}{n_2} = n_1 + \frac{1}{T_1^2}\left(\frac{dT}{dr_0}\right) \cdot 2\xi_0 \tag{12.4.25}$$

After further simplification, and using the first relation of (12.4.23), we obtain

$$\pm\frac{n_1}{2n_2}=\frac{1}{T_1^2}\left(\frac{dT}{dr_0}\right)2\xi_0 \tag{12.4.26}$$

Since n is large, we find that $n_1 \approx n_2$; and

$$\frac{1}{T_1^2}\left(\frac{dT}{dr_0}\right)2\xi_0=\pm\frac{1}{2} \tag{12.4.27}$$

Equation (12.4.22) gives

$$dT=\frac{\pi}{3r_0^2\,\omega_\rho}\left(\xi_0^2\,\frac{dr_0}{r_0}\right), \tag{12.4.28}$$

and

$$t=3\pi r_0^3/(\xi_0^3\,\omega_\rho) \tag{12.4.29}$$

for the cross-over time, where, approximately $T_1 = 2\pi/\omega_\rho$.

Exercise 12.4.1. If electrons in the spherical surface $r = r_0$ are displaced to the spherical surface $r = r_0 + \xi$, then show that the cross over time

$$t = 12\pi\, r_0^3/7\omega_\rho\, \xi_0^3.$$

Solution. The radial field E and the equation for ξ are given by

$$E\,.\,4\pi\,(r_0+\xi)^2=4\pi\, en_0\cdot\frac{4\pi}{3}\{(r_0+\xi)^3-r_0^3\},\quad m\ddot{\xi}=-eE$$

Hence the equation for ξ is

$$\ddot{\xi}=\frac{\omega_\rho^2\,\{(r_0+\xi)^3-r_0^3\}}{3\,(r_0+\xi)^2}$$

Putting now $y=\xi/r_0$ in this equation, we find that

$$\ddot{y}+\frac{\omega_\rho^2}{3}\cdot\frac{(y+1)^3-1}{(y+1)^2}=0,\quad T=\frac{2\pi}{\omega}\left(1-\frac{7y_0^2}{48}\right)$$

The maximum displacement y_0 of y follows from solution of the linearized equation, giving

$$dT=\frac{7\pi\xi_0}{12\omega_\rho\, r_0^2}\cdot\frac{\xi_0\, dr_0}{r_0},$$

and the cross-over time is

$$t=\frac{12\pi\, r_0^3}{7\omega_\rho\,\xi_0^3}$$

(f) Some remarks

Non-linear effects steepen the leading edge of a wave, and portions of the wave profile, which are characterized by high velocities, tend to overtake portions which have low velocities. So, there is a competition between linear and non-linear dispersions. The increasing steepness of the leading edge generatates waves of shorter wavelength (higher harmonics), and dissipation sets a limit on the steepeness of the front. In a collisionless plasma dissipation is due to dispersion, and wave breaking from increase of steepness of the wave front. The generated higher harmonics steepen the wave front, and are not normal modes, These cause charge separation unless the steepening is balanced by dispersive effects.

12.5 NON-LINEAR FIELD EQUATIONS IN COLD RELATIVISTIC PLASMAS

The relativistic field equations in cold and collisionless plasmas consist of the equation of motion of continuum mechanics for charges of both plus and minus signs neglecting the pressure effect and the system of Lorentz equations of electrodynamics for the charge and current densities. Since ions are much heavier than electrons, their motion is neglected except for low frequency waves. Then the plasma field equations for electron motion are

$$\left\{\frac{\partial}{\partial t}+(\mathbf{v}\,.\,\nabla)\right\}\frac{\mathbf{v}}{\sqrt{1-v^2/c^2}}=-\frac{e}{m}\mathbf{E}-\frac{e}{mc}(\mathbf{v}\times\mathbf{H}), \tag{12.5.1}$$

$$\nabla\times\mathbf{E}=-\frac{1}{c}\frac{\partial\mathbf{H}}{\partial t},\ \nabla\times\mathbf{H}=\frac{1}{c}\frac{\partial\mathbf{E}}{\partial t}-\frac{4\pi c}{c}(N_0\,\mathbf{v}+N\,\mathbf{v}), \tag{12.5.2}$$

$$(\nabla\,.\,\mathbf{H})=0,\ (\nabla\,.\,\mathbf{E})=-4\pi\, eN. \tag{12.5.3}$$

Here $-e$, m and $\mathbf{v}$ are the electronic charge, mass and velocity respectively, N_0 is the equilibrium number density of both electrons and ions, N is the perturbation in the electron number density, $\mathbf{E}$ and $\mathbf{H}$ are the electric and magnetic fields respectively, and c is the vacuum speed of light. Since $v^2 << c^2$, we replace $(1-v^2/c^2)^{-1/2}$ by $1+v^2/2c^2$, and find that (12.5.1) simplifies to

$$\frac{\partial\mathbf{v}}{\partial t}+(\mathbf{v}\,.\,\nabla)\,\mathbf{v}+\frac{\partial}{\partial t}\left(\frac{v^2\,\mathbf{v}}{2c^2}\right)=-\frac{e}{m}\mathbf{E}-\frac{e}{mc}(\mathbf{v}\times\mathbf{H}) \tag{12.5.4}$$

Equations of (12.5.2) are now reduced to

$$\ddot{\mathbf{E}}-c^2\,\nabla^2\,\mathbf{E}+c^2\,\nabla\,(\nabla\,.\,\mathbf{E})+\omega_p^2\,\mathbf{E}=-4\pi e\,N_0\,\{(\mathbf{v}\,.\,\nabla)\,\mathbf{v}$$
$$+\frac{\partial}{\partial t}\left(\frac{v^2\,\mathbf{v}}{2c^2}\right)+\frac{e}{mc}(\mathbf{v}\times\mathbf{H})\}+4\pi e\frac{\partial}{\partial t}(N\,\mathbf{v}), \tag{12.5.5}$$

$$\ddot{\mathbf{H}}-c^2\,\nabla^2\,\mathbf{H}=-4\pi ec\,\{N_0\,(\nabla\times\mathbf{v})+\nabla\times(N\mathbf{v})\} \tag{12.5.6}$$

The unperturbed state is characterized by no average motion, no static magnetic field and no static electric field. Denoting the field variables of this state by the addition of the subscript o we obtain.

$$\mathbf{v}_0 = 0, \ \mathbf{E}_0 = 0, \ \mathbf{H}_0 = 0, \ N_0 \neq 0 \tag{12.5.7}$$

These equations are solved by successive approximation, assuming plane wave solution for the perturbed field variables. The linearized solution is the first approximation, and bears the subscript 1. The excited fields of second approximation bear the subscript 2; the non-linear terms of second order are linear combinations of products, taken two at a time, of the fields of linear approximation. Thus the excited fields of the nth order approximation bear the subscript n, and the non-linear terms of the nth order are linear combinations of products, taken n at a time, of the fields of first order. But, according to the British practice, the unperturbed state is the ground state; the linearized approximation is the zeroth approximation, bearing the subscript O; accordingly $\mathbf{E}_n$ is their electric field of the $(n-1)$th order of approximation.

The relativistic non-linear rate of change of momentum per unit rest mass, $(\partial (v^2 \mathbf{v})/2c^2)/\partial t$ is one non-linear term. The other nonlinear terms are the force $-(e/mc)(\mathbf{v} \times \mathbf{H})$, the current $-eN\mathbf{v}$ and the term $(\mathbf{v} \cdot \nabla)\mathbf{v}$ in the momentum transfer equation (12.5.4). The contribution of the relativistic term is numerically even more than from the sum of the other non-linear terms. The nature of the nth order excitation fields depends on the nature of the lower order non-linear terms responsible for their excitation.

12.6 SECULAR-FREE SOLUTION OF NON-LINEAR EQUATIONS

(a) Linearized solution

For transverse waves, the divergence free relations

$$(\nabla \cdot \mathbf{E}_1) = 0, \ (\nabla \cdot \mathbf{v}_1) = 0, \ N_1 = 0 \tag{12.6.1}$$

and equations (12.5.4)-(12.5.6) reduce to

$$\dot{\mathbf{v}}_1 = -\frac{e}{m}\mathbf{E}_1, \ \ddot{\mathbf{E}}_1 - c^2 \nabla^2 \mathbf{E}_1 = \frac{m}{e}\omega_\rho^2 \dot{\mathbf{v}}_1,$$

$$\ddot{\mathbf{H}}_1 - c^2 \nabla^2 \mathbf{H}_1 = -\frac{m}{e} c\omega_\rho^2 (\nabla \times \mathbf{v}_1) \tag{12.6.2}$$

Elliptically polarized form of $\mathbf{v}_1$ is

$$\mathbf{v}_1 = \hat{\mathbf{X}} v_1 \cos\theta + \hat{\mathbf{Y}} v_2 \sin\theta \tag{12.6.3}$$

where $$\theta = KZ - \omega t, \ (\mathbf{K} \cdot \hat{\mathbf{X}}) = 0, \ (\mathbf{K} \cdot \hat{\mathbf{Y}}) = 0 \tag{12.6.4}$$

$\hat{\mathbf{X}}, \hat{\mathbf{Y}}, \hat{\mathbf{Z}}$ are the unit vectors parallel to OX, OY, OZ respectively. With the help of (12.6.1) and (12.6.2) we find that

$$\mathbf{E}_1 = \hat{\mathbf{X}} K_1 \sin\theta + \hat{\mathbf{Y}} K_2 \cos\theta = \frac{m}{e}\omega(-\hat{\mathbf{X}} v_1 \sin\theta + \hat{\mathbf{Y}} v_2 \cos\theta) \tag{12.6.5}$$

$$\mathbf{H}_1 = \hat{\mathbf{X}} H_1 \cos\theta + \hat{\mathbf{Y}} H_2 \sin\theta, \ N_1 = 0, \ (\mathbf{v} \cdot \nabla) = 0, \ K^2 c^2 = \omega^2 - \omega_\rho^2 \tag{12.6.6}$$

(b) Second harmonic fields

For these the relevant equations are

$$\dot{\mathbf{v}}_2 + \frac{e}{m}\mathbf{E}_2 = -\frac{e}{mc}(\mathbf{v}_1 \times \mathbf{H}_1), \tag{12.6.7}$$

$$\ddot{\mathbf{E}}_2 - c^2\nabla^2\mathbf{E}_2 + c^2\nabla(\nabla\,.\,\mathbf{E}_1) - \frac{m}{e}\omega_\rho^2\dot{\mathbf{v}}_2 = 0,$$

$$\ddot{\mathbf{H}}_2 - c^2\nabla^2\mathbf{H}_2 + \frac{m}{e}c\,\omega_\rho^2(\nabla\times\mathbf{v}_2) = 0 \tag{12.6.8}$$

Differentiating all terms of the second equation of (12.6.8) with respect to time and using (12.6.7) and the vector formula (A3.9) we obtain

$$\ddot{\mathbf{H}}_2 - c^2\nabla^2\mathbf{H}_2 = -\omega_\rho^2\dot{\mathbf{H}}_2 \tag{12.6.9}$$

To avoid a trivial repetition, we put $\mathbf{H}_2 = 0$. Then the second equation of (12.6.2) and (12.6.6) gives

$$\dot{\mathbf{E}}_2 = 4\pi e N_0\mathbf{v}_2; \tag{12.6.10}$$

and the right hand side of (12.6.7)

$$= -\frac{Ke^2\hat{\mathbf{Z}}}{2m^3\omega^3}(K_1^2 - K_2^2)\sin 2\theta \tag{12.6.11}$$

Therefore, the second harmonic field of non-linear excitations, being parallel to $\hat{\mathbf{Z}}$, is longitudinal, and (12.6.7) and (12.6.10) become

$$\ddot{\mathbf{v}}_2 + \frac{e}{m}\dot{\mathbf{E}}_2 = \ddot{\mathbf{v}}_2 + \omega_\rho^2\mathbf{v}_2 = \frac{Ke^2\hat{\mathbf{Z}}}{2m^2\omega^2}(E_1^2 - E_2^2)\frac{\partial}{\partial t}(\sin 2\theta), \tag{12.6.12}$$

$$\mathbf{v}_2 = \frac{Ke^2\hat{\mathbf{Z}}(E_1^2 - E_2^2)\sin 2\theta}{m^2\omega(4\omega^2 - \omega_\rho^2)}, \tag{12.6.13}$$

$$\mathbf{E}_2 = -\frac{\hat{\mathbf{Z}}Ke^2\omega_\rho^2(E_1^2 - E_2^2)}{2m\omega^2(4\omega^2 - \omega_\rho^2)},\quad N_2 = \frac{K^2N_0(v_1^2 - v_2^2)\cos 2\theta}{4\omega^2 - \omega_\rho^2} \tag{12.6.14}$$

(c) Third order terms and equations

The sum of four terms of third order in the right hand side of (12.6.5) is

$$\frac{m}{e}\omega_\rho^2\left[-\frac{\partial}{\partial t}\left(\frac{v_1^2\mathbf{v}_1}{2c^2}\right) - (\mathbf{v}_2\,.\,\nabla)\mathbf{v}_1 - \frac{e}{mc}(\mathbf{v}_2\times\mathbf{H}_1) + \frac{1}{N_0}\frac{\partial}{\partial t}(N_2\mathbf{v}_1)\right] \tag{12.6.15}$$

Since all these terms give rise to only transverse waves, in the left hand side, for $\mathbf{E}_3$ also, we write

$$(\nabla\,.\,\mathbf{E}_3) = 0 \tag{12.6.16}$$

$$\nabla (\nabla \cdot (\mathbf{E}_1 + \mathbf{E}_3)) = 0 \tag{12.6.17}$$

Then, equation (12.5.5) gives

$$\left(\frac{\partial^2}{\partial t^2} + \omega_\rho^2 - c^2 \nabla^2 \right) (\mathbf{E}_1 + \mathbf{E}_3) = \frac{m}{e} \omega_\rho^2 \left\{ -\frac{\partial}{\partial t} \left(\frac{v_1^2 \mathbf{v}_1}{2c^2} \right) \right.$$

$$\left. + \frac{1}{N_0} \frac{\partial}{\partial t} (N_2 \mathbf{v}_1) - (\mathbf{v}_2 \cdot \nabla) \mathbf{v}_1 - \frac{e}{mc} (\mathbf{v}_2 \times \mathbf{H}_1) \right\} \tag{12.6.18}$$

Putting the values of $\mathbf{E}_1$, $\mathbf{H}_1$, $\mathbf{v}_1$, θ from (12.6.3)-(12.6.6) in all the terms of the right hand side of (12.6.18), we evaluate them separately and obtain

$$-\frac{e}{mc} (\mathbf{v}_2 \times \mathbf{H}_1) = -\left\{ \frac{K^2 \omega (v_1^2 v_2^2)}{2 (4\omega^2 - \omega_\rho^2)} \right\} \{ \hat{\mathbf{X}} v_1 (\sin 3\theta - \sin \theta) -$$

$$\hat{\mathbf{Y}} v_2 (\cos 3\theta + \cos \theta) \}, \tag{12.6.19}$$

$$-\frac{\partial}{\partial t} \left(\frac{v_1^2 \mathbf{v}_1}{2c^2} \right) = -\frac{\omega}{8c^2} \{ \hat{\mathbf{X}} v_1 (3v_1^2 + v_2^2) \sin \theta - \hat{\mathbf{Y}} v_2 (v_1^2 + 3v_2^2) \cos \theta$$

$$+ 3 (\hat{\mathbf{X}} v_1 \sin 3\theta - \hat{\mathbf{Y}} v_2 \cos 3\theta) (v_1^2 - v_2^2) \}, \tag{12.6.20}$$

$$-(\mathbf{v}_2 \cdot \nabla) \mathbf{v}_1 = -\left\{ \frac{K^2 \omega (v_1^2 - v_2^2)}{2 (4\omega^2 - \omega_\rho^2)} \right\} \{ \hat{\mathbf{X}} v_1 (\sin \theta - \sin 3\theta)$$

$$+ \hat{\mathbf{Y}} v_2 (\cos \theta - 3 \cos 3\theta) \}, \tag{12.6.21}$$

$$\frac{1}{N_0} \frac{\partial}{\partial t} (N_2 \mathbf{v}_1) = \left\{ \frac{K^2 \omega (v_1^2 - v_2^2)}{2 (4\omega^2 - \omega_\rho^2)} \right\} \{ \hat{\mathbf{X}} v_1 (\sin \theta + 3 \sin 3\theta) -$$

$$+ \hat{\mathbf{Y}} v_2 (\cos \theta - 3 \cos 3\theta) \} \tag{12.6.22}$$

These calculations show that

$$\frac{e}{mc} (\mathbf{v}_2 \times \mathbf{H}_1) + (\mathbf{v}_2 \cdot \nabla) \mathbf{v}_1 = 0 \tag{12.6.23}$$

Hence (12.6.17) reduces to

$$\left(\frac{\partial^2}{\partial t^2} - c^2 \nabla^2 + \omega_\rho^2 \right) (\mathbf{E}_1 + \mathbf{E}_3) = \frac{m}{e} \omega_\rho^2 \left[-\frac{\omega}{8c^2} \{ \hat{\mathbf{X}} v_1 (3v_1^2 + v_2^2) \sin \theta \right.$$

$$- \hat{\mathbf{Y}} v_2 (v_1^2 + 3v_2^2) \cos \theta + 3 (\hat{\mathbf{X}} v_1 \sin 3\theta - \hat{\mathbf{Y}} v_2 \cos 3\theta) (v_1^2 - v_2^2) \}$$

$$+\left\{\frac{K^2}{2}\,\omega\,(v_1^2 - v_2^2)\right\}\;\{\hat{\mathbf{X}}\,v_1\,(\sin\theta + 3\sin 3\theta) + \hat{\mathbf{Y}}\,v_2\,(\cos\theta - 3\cos 3\theta)\}\Bigg]$$

(12.6.24)

The left hand sides of (12.6.19)-(12.6.24) have been reduced to linear combinations of sines and cosines of θ and 3θ and also to linear combination of the unit vectors $\hat{\mathbf{X}}$ and $\hat{\mathbf{Y}}$. Since there is no term of $\hat{\mathbf{Z}}$, only transverse waves are excited for $\mathbf{E}_1$ and $\mathbf{E}_3$, hence the validity of equations (12.6.16) and (12.6.1).

Equating the first harmonic fields of the right hand side of (12.6.24) with the $\mathbf{E}_1$ part of the left hand side, and the third harmonic part with the $\mathbf{E}_3$ part, we obtain

$$\ddot{\mathbf{E}}_1 - c^2\nabla^2\mathbf{E}_1 + \omega_\rho^2\,\mathbf{E}_1 = \frac{m\omega}{2e}\,\omega_\rho^2\left[\frac{K^2\,(v_1^2 - v_2^2)}{(4\omega^2 - \omega_\rho^2)}(\hat{\mathbf{X}}\,v_1\sin\theta + \hat{\mathbf{Y}}\,v_2\cos\theta)\right.$$

$$\left. - \left\{\frac{(3v_1^2 + v_2^2)}{4c^2}\,\hat{\mathbf{X}}\,v_1\sin\theta - \frac{(v_1^2 + 3v_2^2)}{4c^2}\,\hat{\mathbf{Y}}\,v_2\cos\theta\right\}\right], \qquad (12.6.25)$$

$$\ddot{\mathbf{E}}_3 - c^2\nabla^2\mathbf{E}_3 + \omega_\rho^2\,\mathbf{E}_3 = qm\omega\,\omega_\rho^4\left\{\frac{(v_1^2 - v_2^2)\;(\hat{\mathbf{X}}\,v_1\sin 3\theta - \hat{\mathbf{Y}}\,v_2\cos 3\theta)}{8ec^2\,(4\omega^2 - \omega_\rho^2)}\right\}$$

(12.6.26)

Using (12.6.5), equation (12.6.25) gives

$$(K^2c^2 - \omega^2 + \omega_\rho^2)\;(\hat{\mathbf{X}}\,K_1\sin\theta + \hat{\mathbf{Y}}\,K_2\cos\theta)$$

$$= \frac{e^2\,\omega_\rho^2}{2m^2\,\omega^2\,c^2}\,[(K_1^2 - K_2^2)\;(\hat{\mathbf{X}}\,K_1\sin\theta - \hat{\mathbf{Y}}\,K_2\cos\theta)\,g$$

$$+ \frac{1}{2}\,(K_1^2 + K_2^2)\;(\hat{\mathbf{X}}\,K_1\sin\theta + \hat{\mathbf{Y}}\,K_2\cos\theta)] \qquad (12.6.27)$$

where
$$g = \frac{1}{4} - \frac{K^2c^2}{4\omega^2 - \omega_\rho^2} \qquad (12.6.28)$$

If the field is plane polarized in the *ZX* plane, then putting $K_2 = 0$, we obtain from (12.6.27) the non-linearly correct dispersion relation,

$$K^2c^2 - \omega^2 + \omega_\rho^2 = \left(g + \frac{1}{2}\right)K_1^2\cdot\frac{e^2\,\omega_\rho^2}{2m^2\,\omega^2\,c^2} \qquad (12.6.29)$$

Hence the frequency shift, in the temporal evolution problems, is

$$\frac{\delta\omega}{\omega} = -\frac{e^2\,\omega_\rho^2\,K_1^2}{4m^2\,\omega^4\,c^2}\left(g+\frac{1}{2}\right) \tag{12.6.30}$$

In the spatial evolution problem, the corresponding wave number shift is

$$\frac{\delta K}{K} = \frac{e^2\,\omega_\rho^2\,K_1^2}{4m^2\,\omega^2\,K^2\,c^4}\left(g+\frac{1}{2}\right) \tag{12.6.31}$$

(d) Properties of strong waves in plasmas

A wave will propagate through a plasma only if its displacement current $\omega E/c$ is larger than the plasma current $(4\pi N_0\, e/c)\,(eE/m\omega)$, or if the condition for free transmission, obtained from the dispersion relation (3.4.4), namely $\omega^2 > \omega_\rho^2$, is fulfilled, where E is the amplitude of the electric field, ω is its frequency and $\omega_\rho^2 = 4\pi N_0\, e^2/m$. This condition is written as

$$\frac{\omega}{c}E > 4\pi N_0\, e\alpha \tag{12.6.32}$$

where $\alpha = eE/m\,\omega\,c = u/c$, with u denoting the directed electron velocity. The inequality (12.6.32) can be replaced by

$$\frac{\omega}{c}E > 4\pi e\, N_0 \;\text{ if }\; \alpha > 1, \tag{12.6.33}$$

and so, by
$$\omega^2 > \frac{\omega_\rho^2}{\alpha} \tag{12.6.34}$$

This means that relativistically strong waves can propagate through a normally overdense plasma at a frequency which is less then ω_ρ and greater than $\omega_\rho/\sqrt{\alpha}$, though the linear theory predicts $\omega > \omega_\rho$ for the same. For strong fields this fact (the cut-off frequency is not ω_ρ but a quantity less than that) is evident from the non-linearly correct dispersion relation (12.6.29), because, putting $K = 0$ in that equation, we obtain.

$$\omega^2 = \omega_\rho^2\left\{1-\left(g+\frac{1}{2}\right)\frac{e^2\,K_1^2}{2m^2\,\omega^2\,c^2}\right\} \tag{12.6.35}$$

The condition $\alpha > 1$ of (12.6.33) is equivalent to the condition $S^2 > m^2\,c^3\,\omega^2/8\pi e^2$ where S is the time averaged Poynting flux $(= cE^2/8\pi)$, in ergs/cm^2 sec. Using the values of e, m, c, we find that power for which

$$S^2 > 3{\cdot}4 \times 10^{-6} \times \omega$$

is obtained from the focussing of very intense laser beams.

12.7 INTENSITY-INDUCED PRECESSIONAL ROTATION OF STRONG ELECTROMAGNETIC WAVES

(a) Some general remarks

A strong elliptically polarized electromagnetic wave develops an intensity-dependent precessional rotation about its direction of propagation in a non-linear plasma. This is the rotation of the polarization ellipse of vibration of the wave due to non-linearities of the medium, as the wave advances in it in time (at a fixed point) or in space (at a given instant). In fact, the shift in a wave parameter (frequency or wave number) and wave precession are complementary non-linear effects, so that when one occurs the other must also occur if the wave is elliptically polarized. This precession also generates the rotation of the direction of the maximum of the electric field. The photons cannot change the magnitude of their own angular momentum; hence the ellipticity of the polarization ellipse of a wave remains unchanged in its non-linear evolution, and propagation in isotropic media involves no inelastic scattering. There is the rotation of the field vectors of the electromagnetic field, the rate of which is the wave frequency. The other is the non-linear rotation of the principal polarization directions, the rate of which is the precession frequency (Fig. 12.7.1). It is dependent on the former and is evolved by that due to the interaction of the wave with non-linearities of the medium.

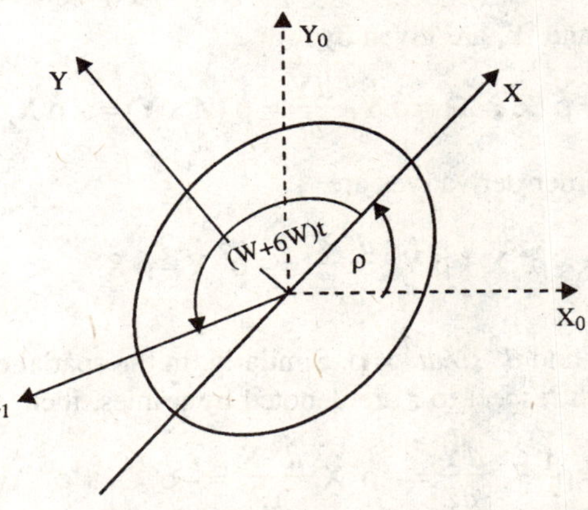

Fig. 12.7.1 Two rotations of an elliptically polarized field; of these the processional rotation is that of OX about OX_0 through the angle ρ.

The large length scales available in astrophysical plasmas allow the wave precession to become well developed. So, wave precession results may be applied to the emitting synchrotron radiation in crab nebule (dimensions 10^{18} cm). A strong wave, if emitted by the pulsar, would thus undergo large precession as it propagates across the crab nebula.

(b) Experimental background

Focussed laser beam is used to measure changes in the real part of intensity induced refractive index n for monochromatic light in isotropic centro-symmetric materials, they found that (1) n is proportional to the light intensity and (2) in addition to a polarization-independent change in velocity, the axes of the vibration ellipse rotate and that this rotation is a function of the distance traversed by the wave.

(c) Rotating Unit Vectors

For a wave propagating parallel to the z-axis, the principal axes of the polarization ellipse rotate in the xy plane, keeping the ellipticity constant. The angle ρ of rotation is a function of time t or distance z. In the coordinate system OX, OY, OZ of Fig. 12.7.1, OX and OY are fixed with the principal polarization directions of the polarization ellipse, and OZ is perpendicular to this plane. The field equations of section 12.5 for transverse waves, and their non-linear evolution, are described in this system. This coordinate frame rotates about the direction of propagation, and may be called the principal polarization reference frame (or PPRF, in short). Compared to this frame, the inertial laboratory frame (LF) of axes OX_0, OY_0, OZ_0 (= OZ) is non-rotating.

If the unit vectors of the PPRF are $\hat{\mathbf{X}}, \hat{\mathbf{Y}}, \hat{\mathbf{Z}}$, and those of the LF are $\hat{\mathbf{X}}_0, \hat{\mathbf{Y}}_0, \hat{\mathbf{Z}}_0$ then $\hat{\mathbf{X}}$ and $\hat{\mathbf{Y}}$ rotate about OZ; then these are functions of time in the temporal evolution problems, and of z in the spatial evolutions. The time derivative of $\hat{\mathbf{X}}$ and $\hat{\mathbf{Y}}$, are given by

$$\frac{d\hat{\mathbf{X}}}{dt} = \dot{\rho}\,(\hat{\mathbf{Z}} \times \hat{\mathbf{X}}) = \dot{\rho}\,\hat{\mathbf{Y}}, \quad \frac{d\hat{\mathbf{Y}}}{dt} = \dot{\rho}\,(\hat{\mathbf{Z}} \times \hat{\mathbf{Y}}) = -\dot{\rho}\,\hat{\mathbf{X}}, \tag{12.7.1}$$

and the second order derivatives are

$$\frac{d^2\hat{\mathbf{X}}}{dt^2} = -\dot{\rho}^2\,\hat{\mathbf{X}} + \ddot{\rho}\,\hat{\mathbf{Y}}, \quad \frac{d^2\hat{\mathbf{Y}}}{dt^2} = -\dot{\rho}^2\,\hat{\mathbf{Y}} - \ddot{\rho}\,\hat{\mathbf{X}} \tag{12.7.2}$$

where $\dot{\rho} = d\rho/dt$ and $d^2\rho/dt^2 = \ddot{\rho}$. Similarly, in the spatial evolution problems, if derivatives with respect to z are denoted by primes, then

$$\frac{d\hat{\mathbf{X}}}{dZ} = \rho^1\,\hat{\mathbf{Y}}, \quad \frac{d\hat{\mathbf{Y}}}{dz} = -\rho'\,\hat{\mathbf{X}}, \quad \frac{d^2\hat{\mathbf{X}}}{dz^2} = -\rho'^2\,\hat{\mathbf{X}} + \rho''\,\hat{\mathbf{Y}}, \text{ etc.} \tag{12.7.3}$$

The quantity $\dot{\rho}$ (having the dimension of the wave frequency ω) is called the precession frequency and ρ' (having the dimension of the wave number K) is called the precession number. The unit vectors $\hat{\mathbf{X}}_0$, $\hat{\mathbf{Y}}_0$ of the LF are free from the influence of the precessional rotation, the non-linearly generated angle ρ evolves like other parameters of the wave (frequency or wave number), so, it should have the series expansion

$$\rho = \rho_0 + \varepsilon\rho_1 + \varepsilon^2\rho_2 + \ldots, \tag{12.7.4}$$

where ρ_0 is the angle generated in the first (linearized) approximation, ρ_1 in the second harmonic generation, ρ_2 in the third harmonics, and like that. However, the first approximation takes account of only the rotation of the field vectors of the elliptically polarized wave, at the rate of the wave frequency, so we write $\rho_0 = 0$. Since longitudinal waves do not have any precessional rotation, and the second order fields are purely longitudinal, we put $\rho_1 = 0$ and replace the series of (12.6.4) by

$$\rho = \varepsilon^2 \rho_2 + \ldots \tag{12.7.5}$$

We, therefore, replace ρ by $\varepsilon^2 \rho_2$, and drop the subscript 2 from ρ_2. Also, $\dot{\rho}$ being complementary to the frequency shift $\delta\,\omega$ (or ρ' to the wave number shift δK), the higher order quantities $\ddot{\rho}$ and $\dot{\rho}^2$ must be neglected; so $\ddot{\hat{\mathbf{X}}}$ and $\ddot{\hat{\mathbf{Y}}}$ are ignored here.

(d) Expansion of fields and time

Following the expansion in (A16.4) and using (12.5.7) we write

$$\mathbf{E} = \varepsilon \mathbf{E}_1 + \varepsilon^2 \mathbf{E}_2 + \ldots,\ N = \varepsilon^2 N_2 + \varepsilon^4 N_4 + \ldots,$$

$$\mathbf{v} = \varepsilon \mathbf{v}_1 + \varepsilon^2 \mathbf{v}_2 + \ldots,\ \mathbf{H} = \varepsilon \mathbf{H}_1 + \varepsilon^3 H_3 + \ldots, \tag{12.7.6}$$

and evaluate the third order terms from the coefficients of ε^3 in the right hand side of (12.5.5) as in (12.6.15). These expansions include the intensity-dependent higher harmonics as a series of stationary wave patterns; for example, both the third and first harmonic waves are excited in the third order, because the combination $\pm\,\omega \pm \omega \pm \omega$ gives $\pm\,3\,\omega$ and $\pm\,\omega$. The vector differentiation results of (12.7.1) and (12.7.2) give some more terms containing $\dot{\rho}$.

Applying the unit vector differentiation results of (12.7.1) and (12.7.2) in the left hand side of (12.5.5), we obtain

$$\hat{\mathbf{X}} \frac{\partial^2}{\partial t^2}(K_1 \sin\theta) + \hat{\mathbf{Y}} \frac{\partial^2}{\partial t^2}(K_2 \cos\theta) + 2\dot{\rho}\,\varepsilon^2 \left\{ \hat{\mathbf{Y}} \frac{\partial}{\partial t}(K_1 \sin\theta) - \hat{\mathbf{X}} \frac{\partial}{\partial t}(K_2 \cos\theta) \right\} \tag{12.7.7}$$

Next, putting the expansion of t from (A16.9) and (12.7.7), in the right hand side of (12.5.5), we obtain

$$\left\{ \frac{\partial^2}{\partial T^2} + (1 + 2\varepsilon^2 b)\,(\omega_\rho^2 - c^2 \nabla^2) \right\} (\varepsilon \mathbf{E}_1 + \varepsilon^2 \mathbf{E}_2 + \ldots) + 2\dot{\rho}\,\varepsilon^2 \left\{ \hat{\mathbf{Z}}\, X \frac{\partial}{\partial T}(\varepsilon\, \mathbf{E}_1) \right\} \tag{12.7.8}$$

Then, in place of (12.6.17) the appropriate equation is

$$\left\{ \frac{\partial^2}{\partial T^2} + (1 + 2\,\varepsilon^2 b)\ (\omega_\rho^2 - c^2 \nabla^2) \right\} (\varepsilon \mathbf{E}_1 + \varepsilon^3 \mathbf{E}_3 + \ldots)$$

$$+ 2\dot{\rho}\,\varepsilon^3 \left(\hat{\mathbf{X}} \times \frac{\partial \mathbf{E}_1}{\partial T} \right) = \varepsilon^3 \frac{m}{e}\, \omega_\rho^2 \left\{ -\frac{\partial}{\partial T} \left(\frac{v_1^2 \mathbf{v}'}{2c^2} \right) \right.$$

$$\left. + \frac{1}{N_0} \frac{\partial}{\partial T} (N_2 \mathbf{v}_1) - (\mathbf{v}_2 \,.\, \nabla)\, \mathbf{v}_1 - \frac{e}{mc} (\mathbf{v}_2 \times \mathbf{H}_1) \right\} \quad (12.7.9)$$

Equating, from both sides, the coefficients of ε^3 we obtain

$$\ddot{\mathbf{E}}_3 + (\omega_\rho^2 - c^2 \nabla^2)\, \mathbf{E}_3 + b\,\ddot{\mathbf{E}}_1 + 3b\,(\omega_\rho^2 - c^2 \nabla^2)\, \mathbf{E}_1 + 2\dot{\rho}\,(\hat{\mathbf{Z}} \times \dot{\mathbf{E}})$$

$$= \frac{m}{e}\, \omega_\rho^2 \left\{ -\frac{\partial}{\partial T} \left(\frac{v_1^2 \mathbf{v}_1}{2c^2} \right) + \frac{1}{N_0} \frac{\partial}{\partial T} (N_2 \mathbf{v}_1) - (\mathbf{v}_2 \,.\, \nabla)\, \mathbf{v}_1 - \frac{e}{mc} (\mathbf{v}_2 \times \mathbf{H}_1) \right\}$$

$$(12.7.10)$$

where dots denote differentiation with respect to time T and not t here. It can be seen from (12.6.2) that $\ddot{\mathbf{E}}_1 - c^2 \nabla^2 \mathbf{E}_1 + \omega_\rho^2 \mathbf{E}_1 = 0$. So equation (12.7.10), with the help of (12.6.16), gives

$$\ddot{\mathbf{E}}_3 - c^2 \nabla^2 \mathbf{E}_3 + \omega_\rho^2 \mathbf{E}_3 - 2b\, \dot{\mathbf{E}}_1 + 2\dot{\zeta}\, (\hat{\mathbf{Z}} \times \dot{\mathbf{E}})$$

$$= \frac{m}{e}\, \omega_\rho^2 \left\{ \frac{1}{N_0} \frac{\partial}{\partial T} (N_2 \mathbf{v}_1) - \frac{\partial}{\partial T} \left(\frac{v_1^2 \mathbf{v}_1}{2c^2} \right) - (\mathbf{v}_2 \,.\, \nabla)\, \mathbf{v}_1 - \frac{e}{mc} (\mathbf{v}_2 \times \mathbf{H}_1) \right\}$$

$$(12.7.11)$$

In the first and second order equations of section 12.16 we replace t by T, because for these, effectively, $b = 0$; so $t = T$. In place of θ of (12.6.4) we henceforth take, in the temporal evolution problems, $\theta = KZ - \omega\, t$ and stick to T and not t.

(e) Secular-free behaviour of the first harmonic fields

To avoid the secularities we equate the left hand side of (12.7.11) with that part of the right hand side which contains the linear combination of purely third harmonic fields sin 3 θ and cos 3 θ, and equate the remaining part of the right hand side to the remaining terms of the left hand side. Thus, we obtain

$$\ddot{\mathbf{E}}_3 - c^2 \nabla^2 \mathbf{E}_3 + \omega_\rho^2 \mathbf{E}_3 = \frac{9m\omega\, \omega_\rho^4\, (v_1^2 - v_2^2)}{8e\, c^2\, (4\omega^2 - \omega_\rho^2)} (\hat{\mathbf{X}}\, v_1 \sin 3\theta - \hat{\mathbf{Y}}\, v_2 \cos 3\theta),$$

$$(12.7.13)$$

$$- 2b\, \ddot{\mathbf{E}}_1 + 2\dot{\rho}\, (\hat{\mathbf{Z}} \times \dot{\mathbf{E}}_1) = \left(\frac{m\omega}{2e}\, \omega_\rho^2 \right) \left[\frac{K^2\, (v_1^2 - v_2^2)}{(4\omega^2 - \omega_\rho^2)} (\hat{\mathbf{X}}\, v_1 \sin \theta + \hat{\mathbf{Y}}\, v_2 \cos \theta) \right.$$

$$\left. - \frac{1}{4c^2} \{ \hat{\mathbf{X}}\, v_1 \sin \theta\, (3v_1^2 + v_2^2) - \hat{\mathbf{Y}}\, v_2 \cos \theta\, (v_1^2 + 3v_2^2) \} \right] \quad (12.7.14)$$

Equation (12.7.14) is written as

$$-2b\,\ddot{\mathbf{E}}_1 + 2\dot{\rho}\,(\hat{\mathbf{Z}} \times \dot{\mathbf{E}}_1) = \frac{e^2\,\omega_\rho^2}{2m^2\,\omega^2\,e^2}\{(K_1^2 - K_2^2)\,(\hat{\mathbf{X}}\,K_1 \sin\theta - \hat{\mathbf{Y}}\,K_2 \cos\theta)$$

$$+\frac{1}{2}(K_1^2 + K_2^2)\,(\hat{\mathbf{X}}\,K_1 \sin\theta + \hat{\mathbf{Y}}\,K_2 \cos\theta)\} \qquad (12.7.15)$$

We now define the dimensionless amplitude parameters α and β as

$$(\alpha, \beta) = \frac{e}{m\,\omega\,c}(E_1, K_2) \qquad (12.7.16)$$

Then the right hand side of (12.7.14) reads

$$\frac{\omega_\rho^2}{2}\left\{\frac{1}{2}(\alpha^2 + \beta^2)\,(\hat{\mathbf{X}}\,K_1 \sin\theta + \hat{\mathbf{Y}}\,K_2 \cos\theta)\right.$$

$$\left. + g\,(\alpha^2 - \beta^2)\,(\hat{\mathbf{X}}\,K_1 \sin\theta - \hat{\mathbf{Y}}\,K_2 \cos\theta)\right\} \qquad (12.7.17)$$

where g is defined in (12.6.28). Putting the value of $\mathbf{E}_1$ from (12.6.5) in the left hand side of (12.7.15) and using (12.7.17) in its right hand side we obtain

$$2b\omega^2\,(\hat{\mathbf{X}}\,K_1 \sin\theta + \hat{\mathbf{Y}}\,K_2 \cos\theta) - 2\dot{\rho}\omega\,(\hat{\mathbf{X}}\,K_2 \sin\theta + \hat{\mathbf{Y}}\,K_1 \cos\theta)$$

$$= \frac{\omega_\rho^2}{2}\left\{\frac{1}{2}(\alpha^2 + \beta^2)\,(\hat{\mathbf{X}}\,K_1 \sin\theta + \hat{\mathbf{Y}}\,K_2 \cos\theta)\right.$$

$$\left. + g\,(\alpha^2 - \beta^2)\,(\hat{\mathbf{X}}\,K_1 \sin\theta - \hat{\mathbf{Y}}\,K_2 \cos\theta)\right\} \qquad (12.7.18)$$

The coefficients of $\hat{\mathbf{X}}$ and $\hat{\mathbf{Y}}$, when equated from both sides of this equation, give

$$\frac{\omega_\rho^2}{2}\left\{\frac{1}{2}(\alpha^2 + \beta^2) + g\,(\alpha^2 - \beta^2)\right\}\alpha = 2b\omega^2\,\alpha - 2\dot{\rho}\,\beta\,\omega \qquad (12.7.19)$$

$$\frac{\omega_\rho^2}{2}\left\{\frac{1}{2}(\alpha^2 + \beta^2) - g\,(\alpha^2 - \beta^2)\right\}\beta = 2b\omega^2\,\beta - 2\dot{\rho}\,\alpha\omega \qquad (12.7.20)$$

Subtracting α times (12.7.19) from β times (12.7.20), we eliminate ρ and obtain

$$b = \frac{\omega_\rho^2}{4\omega^2}\left(\frac{1}{2} + g\right)(\alpha^2 + \beta^2) \qquad (12.7.21)$$

Similarly, substracting β times (12.7.19) from α times (12.7.20), we eliminate b and obtain

$$\dot{\rho} = \alpha \beta \frac{\omega_\rho^2 g}{2\omega} \tag{12.7.22}$$

Since b is a small quantity, less than unity, equation (A16.8) gives

$$T = \frac{t}{1 + \varepsilon^2 b} \approx t (1 - \varepsilon^2 b), \ \omega T = \omega t - \varepsilon^2 b \omega t; \tag{12.7.23}$$

so putting $\varepsilon = 1$ we define the frequency shift $\delta\omega$ by $\delta\omega = - b\omega$, and obtain

$$\frac{\delta \omega}{\omega} = - b = - \frac{\omega_\rho^2}{4\omega^2} \left(\frac{1}{2} + g \right) (\alpha^2 + \beta^2) \tag{12.7.24}$$

For $\beta = 0$ the wave precession vanishes and equation (12.6.30) which is valid for a plane polarized wave, is recovered. The non-linearly correct dispersion relation is

$$K^2 c^2 - \omega^2 + \omega_\rho^2 = (\alpha^2 + \beta^2) \frac{\omega_\rho^2}{2} \left(\frac{3}{4} - \frac{K^2 c^2}{4\omega^2 - \omega_\rho^2} \right) \tag{12.7.25}$$

Exercise 12.7.1. For a two-component, cold plasma in which both electrons and ions are mobile, show that the precession frequency $\dot{\rho}$ and the non-linearly correct dispersion relation, respectively, are

$$\dot{\rho} = \frac{1}{2\omega} (\alpha_e \beta_e \omega_{\rho e}^2 + \alpha_i \beta_i \omega_{\rho i}^2) \left[\frac{K^2 c^2}{4\omega^2 - \omega_{\rho e}^2 - \omega_{\rho i}^2} - \frac{1}{4} \right],$$

$$K^2 c^2 - (\omega^2 - \omega_{\rho e}^2 - \omega_{\rho i}^2)$$
$$= \frac{1}{2} \ [\omega_{\rho e}^2 (\alpha_e^2 + \beta_e^2) + \omega_{\rho i}^2 (\alpha_i^2 + \beta_i^2)] \left[\frac{3}{4} - \frac{K^2 c^2}{4\omega^2 - \omega_{\rho e}^2 - \omega_{\rho i}^2} \right]$$

where

$$(\alpha_e, \ \beta_e) = \frac{e \, (\alpha, \ \beta)}{m \omega c}, \ (\alpha_i, \ \beta_i) = \frac{e \, (\alpha, \ \beta)}{M \omega c}, \ \omega_{\rho e}^2 = \frac{4\pi N e^2}{m}, \ \omega_{\rho i}^2 = \frac{4\pi N e^2}{M}$$

(f) Remarks on precession of waves in plasmas

Electron orbits of atoms and molecules are much distorted by the impact of a strong radiation (laser beam, for example). Thus, the neutral atoms develop non-linear polarization. One consequence is the precessional rotation (PR) of an elliptically polarized wave. It appears in the laboratory inertial frame (LF). The *LF* and the PPRF coincide for infinitesimal amplitude waves, and their difference increases with increasing field intensity. Thus, all inertial frames of reference become non-inertial by dynamical changes occurring in a plasma. Since this non-inertial motion (rotation) is negligible, even for very powerful laser fields, the phenomena can be studied within the framework of the relations of transformation (12.7.1)-(12.7.2) of Newtonian rotation.

The *PR* is the only known source of self-generated rotation of a plasma in the presence of electromagnetic waves. This rotation generates synchrotron radiation and induced magnetization of the inverse Faraday effect (IFE, section 3.14 (*b*)). The plane of polarization of a strong plane polarized wave is thus found to be non-linearly rotated by a second wave, which must be elliptically polarized, in a plasma. This effect can be summed up for several elliptically polarized internal noise fields, having arbitrary phases. Hence, during the passage of a plane polarized wave through a volume of plasma, the plane of polarization is rotated by noise fields, like the Faraday rotation effect (Section 3.14), which is due to a magnetic field. But, unlike the Faraday rotation, the contribution from local noise is cumulative, so that considerable enhancement is possible to the *PR* effect in this manner, in hot regions of a plasma.

Radially emerging strong waves, from the inner core of sun can thus gain a large amount of this *PR*, the resulting magnetization and synchrotron radiation would reduce the energy gained from the *PR*. The induced magnetization is responsible for freezing the region as in sun spots. This dynamical process of gaining *PR* followed by losing through radiation and magnetization can continue till the disturbance reaches the outer region of the solar atmosphere.

Chapter 13

LASER-PLASMA INTERACTIONS

13.1 LASERS AND PLASMA

Laser (Light Amplification by Stimulated Emission of Radiation) is a source of light, Like fields of radio transmitters laser fields are expressed by a well-defined amplitude and phase. Radiation induced particle motion shows that light can emerge from a blackbody only in the form of packets of energy, lumped according to its frequency. Spontaneous emission of energy from a laser source is in quantas, and not continuous. Left to itself, an excited state will radiate without any observable cause and emit its energy in the form of an appropriate energy packet. The dispersive damping is due to electromagnetic shift of atomic levels.

Laser light has a very high degree of spatial and temporal coherence; so it is a source of intense monochromatic electromagnetic waves. Its radiation pressure is so strong that the optical phenomena generate powerful mechanical phenomena. Some of the lasers are (1) the ruby laser, (2) the Nd-glass laser, (3) the CO_2 laser, (4) the iodine laser, (5) chemical laser and (6) cyanide laser. The wave length λ, and the photon energy ε (= 1.24 ev/λ) are (1) for the ruby laser $\lambda = 0{\cdot}69\mu$ and $\varepsilon = 1{\cdot}73ev$; (2) for the CO_2 laser $\lambda = 10{\cdot}6\mu$ and $\varepsilon = 0{\cdot}12ev$; (3) for the Nd-glass laser $\lambda = 1{\cdot}06\mu$ and $\varepsilon = 1{\cdot}17ev$. The frequency in rad/sec and the corresponding cut-off density per cm^3 are (1) for the Nd-glass laser, $1{\cdot}8 \times 10^5$ and 10^{21}, (2) for the CO_2 laser, $1{\cdot}8 \times 10^{14}$ and 10^{19}; (3) for the cyanide laser $5{\cdot}6 \times 10^{12}$ and 10^{16}. The Nd-glass laser and the CO_2 laser have peak electric field amplitudes of about one million volts/cm, which means a power flux of 10^9 watts/cm^2. At a laser focal spot the field intensity is about 10^{15} watts/cm^2. So, the oscillating electric and magnetic fields, there have peak values of 3×10^8 volts/cm and 3×10^6 gauss respectively, the radiation pressure is 3×10^5 atmospheres, and the field induced quiver velocity is that of 16 Kev electrons.

The laser field is of the order of internal fields holdings the electrons in orbits about their inner cores. The internal field E_0 is given by $eE_0\, 1 = I$ where e is the electron charge, I is the potential of ionization and 1 is the characteristic distance in which these fields vanish. For hydrogen atoms $E_0 = 5 \times 10^9$ volts/cm^{-1}, but for condensed matter E_0 is smaller and for semiconductors in some situations $E_0 \approx 10^7$ volt/cm. These fields are comparable to the pressure of fields of focussed intense laser beams. When the wave is in resonance with the atomic transition frequency, the nature of the light is changed. The relevant non-linear state relation is

$$D_i = \varepsilon_{ij} E_j + \varepsilon_{ijk} E_j E_k + \ldots \qquad (13.1.1)$$

where ε_{ijk}, etc., are material constants. The linearized approximation being $D_i = \varepsilon_{ij} E_j$, this relation explains the appearance of higher harmonics at twice and thrice the fundamental frequency. Non-linear overtones or higher combination frequencies in non-linear crystals have been experimentally observed; for instance, the second harmonic power 8×10^{-14} watts was obtained from $1{\cdot}4 \times 10^{-3}$ watts at a fundamental.

Three-wave, non-linear parametric interaction processes are very important in laser produced plasmas. In these the incident light scatters two other waves, one of which is an *EM* wave and the other is an acoustic wave. The fields of *EM* waves are identified with corpuscles called photons. Fields of high frequency longitudinal (acoustic) waves (frequencies greater than the plasma characteristic frequency ω_ρ) arise from electron vibration, these are identified with corpuscles called plasmons. The fields of low frequency longitudinal wave (frequencies less than the characteristic frequency of oscillation of ions $\omega_{\rho i}$) cause mass vibration,which, for plasmas, is effectively the vibration of ions; so these fields are identified with corpuscles called phonons.

13.2 THE SRS AND SBS IN LASER PLASMA INTERACTIONS

When an intense laser beam is shined on liquids or solids, the results are (1) absorption of the incident laser, (2) coherent scattering and (3) excitation of molecular vibration. These are stimulated Raman scattering (SRS) or stimulated Brillouin scattering (SBS) type of decay instabilities. In three-wave non-linear parametric interaction processes (section 13.3) an intense EM wave is transformed into two growing waves in the plasma. These strong waves give their energy to random particle motion via Landau damping or collisional damping, and thus convert electromagnetic energy into plasma energy of anomalous absorption and back scattering.

In SRS a strong transverse pump wave at frequency ω_3 decays into another such wave at frequency ω_1 and an electron acoustic wave at frequency ω_2. The kinematic resonance condition $\omega_3 = \omega_1 + \omega_2$ relates the frequencies with the corresponding wave numbers, through the dispersion relations

$$\omega_1^2 = K_1^2 c^2 + \omega_\rho^2,\ \omega_2^2 = K_2^2 c_e^2 + \omega_\rho^2,\ \omega_3^2 = K_3^2 c^2 + \omega_\rho^2 \qquad (13.2.1)$$

Here c_e (= $\sqrt{\kappa T_e/m}$, where T_e is the kinetic temperature of electrons, m is the electronic mass and κ is the Boltzman constant) is the velocity of sound in the electron fluid. In the SBS process

$$\omega_2^2 = K^2 c_s^2,\ c_s^2 = \frac{\kappa T_e}{M} \qquad (13.2.2)$$

where M is the ion mass.

In SBS a photon is split into a photon and a phonon ion sound; and in SRS a photon is split into a photon and a plasmon. The SBS processes occur in a certain layer of the plasma where the frequency of the incident wave ω_3 is approximately equal to the electron plasma frequency $\omega_\rho\,(=\omega_2)$ because $\omega_1\,(=K_1\,C_s)$ is very small compared with ω_3 and ω_ρ. The SRS occurs when $\omega_3 = 2\omega_\rho$ because then $\omega_1 = \omega_\rho$ and $\omega_2 = \omega_\rho$. So, it occurs at approximately one quarter of the critical density, because ω_ρ is proportional to the square root of the number density N; so $N = N_c/4$, if N_c is the critical number density.

The kinematic conditions of resonance $\omega_3 = \omega_1 + \omega_2$, $K_3 = K_1 + K_2$ are valid in linear crystals. These conditions do not forbid the occurence of processes such as the SBS and SRS in the outer part of the corona of a laser irradiated target, where the density $N << N_c/4$. The incident electromagnetic waves get the time to scatter thus other waves, before reaching the region where $N = N_c/4$. So, these other waves are parametrically excited as they start from the initial thermal noise level and propagate in the inhomongeneous plasma corona. Thus the corona expands at supersonic velocity for the SBS with phonon and due to the inhomogeneities of the corona density for SRS with plasmon. The resulting inhomogeneous Doppler effect is responsible for frequency detuning by parametric resonance. The gain in SBS is known to be appreciably larger than in SRS. The high pressure of light causes a sharp decrease of the plasma density at $N < N_c$, inside a non-linear skin layer; its absorption means overheating and the appearance of instabilities.

It seems that SRS occurs first because of larger growth rate, but then decays as electrons get trapped and the waves break. SRS then reflects almost 50 per cent of the incident energy. SBS starts later, but reaches a large amplitude. The light eventually pushes its way through any finite slab of plasma. In SRS the strong instability splits the plasma at the point where $\omega_3 = \omega_\rho$ because then the back scattered wave has the frequency ω_ρ and cannot propagate out of it.

The dispersion relation, for a collisionally damped electromagnetic wave, equation (3.8.25), is $\omega^2 = K^2\,c^2 + \omega_\rho^2\,(1 - i\nu/\omega)$. For a small collision frequency ν absorption will be small and the wave entering into the plasma will be reflected from the region where $\omega^2 = \omega_\rho^2$. For the decay of a large amplitude pump wave into two other waves the growth rate is greater than the damping rate in the medium. So there is a threshold power, depending on the damping factor, above which the value of the pump power must be maintained for the growth of this parametric instability.

Hybrid processes also occur in which both parametric and non-parametric effects are inseparable. These include a four-wave mixing in which three frequencies $\omega_1, \omega_2, \omega_3$ are mixed to generate a fourth ω_4, given by $\omega_4 = \omega_1 - \omega_2 - \omega_3$. If the arrangement is such that ω_1 and ω_2 satisfy a two-photon resonance of the Raman type, that is if $\omega_1 - \omega_2$ is equal to the Raman transition frequency of the medium, then the effect is enhanced. The stimulated Raman scattering of the powerful pump wave ω_1 can also generate the wave ω_2 within the medium instead of being given externally, from a laser source. The process can therefore be also a combination of non-parametric generation (of ω_2 by SRS) and parametric generation of ω_4 $(= \omega_1 - \omega_2 - \omega_3)$ by four-wave mixing. The two-photon resonance of $\omega_3 + \omega_4$ $(= \omega_1 - \omega_2)$ also gives rise to the two photon absorption of the generated wave. For this reason the efficiency is less critically dependent on phase matching than in a purely parametric process.

13.3 PARAMETRIC INTERACTION OF THREE WAVES

If a parameter is made to vary at the frequency 2ω, ω being the natural frequency of the system, oscillations begin at the frequencies $2\omega \pm \omega$. The word 'parametric excitation' (or parametric interaction) was originally used to emphasize the fact that this effect is due to the variation of a parameter with a double frequency. Parametric excitation is an amplification of an oscillation due to a periodic modulation of a parameter that characterizes the oscillation. It is a non-linear instability of two waves (an idler and a signal) by a modulating wave (a pump wave field) supplied from external devices, and occurs due to a mode-coupling interaction. The idler and the signal need not have the same natural frequency. These can be very different normal modes with different natural time periods.

Parametric non-linear interactions are wave-wave interactions in media. These cause (1) Non-linear mode conversion from electromagnetic to electrostatic fields; (2) Mode conversion from high frequency to low frequency; (3) Anomalous absorption and scattering of electromagnetic waves; (4) Anomalous heating of plasmas including laser heating and lower hybrid resonance heating.

The conditions of matching of frequencies and wave numbers in the simplest example of non-linear three-wave parametric excitation are (13.3.1) and (13.3.2.). The idler and the signal waves become parametrically unstable, at new frequencies, when the pump field power exceeds a threshold value. In the absence of the pump field both these waves are only weak local noises. Due to this parametric mode conversion the external pump energy is deposited into the plasma. The local fields can thus accelerate the plasma particles, and increase their thermal energy, resulting in heating of plasma by an applied field. In some cases, one of the two noise wave frequencies ω_1 and ω_2, say ω_1, is much greater than the other ω_2, so that $\omega_1 + \omega_2 \approx \omega_1$. Properties of non-linear parametric excitation phenomena are: (1) The modulation and the excited oscillation should satisfy phase matching conditions such as (13.3.1) and (13.3.2) or (13.3.3), (2) Amplification occurs only when the modulation amplitude exceeds a certain critical value; (3) The frequencies of the amplified oscillations are determined by the modulation frequency, rather than by the natural frequency.

The most powerful coupling of waves are those between the first harmonic waves. In the temporal evolution problems a linear combination of the wave frequencies of the first harmonic of these waves is equated to zero. In the spatial evolution problems, a linear combination of the wave numbers of the first harmonics of these wave is equated to zero. So, a possible frequency matching condition is

$$\omega_3 = \omega_1 + \omega_2 \tag{13.3.1}$$

in three-wave interactions. When the dispersion relation is linear between ω and *K*, the corresponding exact wave number matching condition is

$$\mathbf{K}_3 = \mathbf{K}_1 = \mathbf{K}_2 \tag{13.3.2}$$

If the dispersion relation is not linear, and (13.3.1) is valid, then, in place of (13.3.2), we write

$$\mathbf{K}_3 - (\mathbf{K}_1 + \mathbf{K}_2) = \mathbf{\Delta} \tag{13.3.3}$$

where the vector $\mathbf{\Delta}$ is the wave number mismatch for an exact resonance between the frequencies. Similarly, for $\mathbf{\Delta} = 0$, $\delta\ (= \omega_3 - (\omega_1 \pm \omega_2))$, is the frequency detuning (or the frequency mismatch) of the parametric source.

13.4 LASER PRODUCED PLASMA

Plasma is produced by laser induced breakdown in a gas. For instance, plasmas are produced by the focussed beam from a ruby laser or a Nd-glass laser with a short focal length lens, at atmospheric pressure, flux density $= 10^{11}$ W/cm^2 ($= 10^{30}$ photon/cm^2) and light field of 10^6 to 10^7 v/cm, and by irradiation of solids or liquids with laser beams if the flux density is of the order of that for break down in gas or more. Plasma densities are of the order of 10^{21} cm^{-3}, in the inner core of the flare, and 10^{11} cm^{-3}, near the periphery. Emission spectrum indicates a temperature of the order of 10^5 to 10^6 °K. Irradiation of a small pellet of hydrogen or deuterium in vacuum, in a magnetic trap, by intense laser beams, generates heat for the occurance of thermo-nuclear fusion reaction.

Dense, energetic and highly localized expanding plasmoids, their composition depending on the solid state target materials, are created "in situ" within a system of confinement fields. These are studied in the after glow; the ambipolar electric fields thus produced probably set up convective cells in the region of confinement.

When the energy accumulated by an electron is sufficient for ionization of an atom, one such electron is replaced by two slower ones and the process continues. If *K* is the number of generation in time *t*, and $N(0)$ is the initial electron density, then the electron density $N(t)$ in time *t* is given by the relation $N(t) = N(0)\, 2^K$. So, the cascade begins with one electron within the local volume of amount 10^{-6} cm^3, and after 43 generations the electron density is 10^{19} cm^{-3}. The growth of the electron density, within the focal volume, for pulse duration of the order of 10^{-11} sec, is probably due to multiphoton excitation; for normal Q-switched pulses of the order of 10^{-8} sec duration, it may be due to cascade ionization.

Resonant absorption of laser light of the order of one joule in a small volume is possible by electrons in collisions with the atoms, in the inverse of bermsstrahlung process. The energy deposited by accelerated electrons into the plasma is generated by the non-linear conversion of electromagnetic energy into energy of electrostatic oscillations. Important consequences are: particle trapping, wave breaking and production of super thermal electrons. The enormous speeds acquired by the particles radically change the thermodynamic state of matter over a wide range of temperature. A sharp intensity gradient is created in the neighborhood of the focus, producing thereby, an acceleration which builds up a shock wave, expanding outwards, from the focus at speed 10^7 cm/sec. The laser light sees at nearly the normal incidence an expanding plasma profile of increasing density running from zero to solid densities. In a region where the laser frequency equals the characteristic plasma frequency ω_p, various parametric instabilities are expected to arise. The growth rate is maximum for oscillation wave length somewhat shorter than those of the incoming light, and when the corresponding wave vectors are oriented at the angle $\pi/4$ to that of light in the plane determined by its polarization vector.

The transition between different states shows that energy is absorbed in packets of size $\hbar\,\omega$, and in impulses of magnitude $\hbar\,\mathbf{K}$ where ω is a wave frequency, **K** is the corresponding wave vector and $\hbar$ is the Planck constant.

For breakdown at optical frequencies the quantum energy of radiation (1.78*ev* for ruby) has to be much greater than the energy of oscillation ($= 10^{-2}$ *ev*). But in the break down by microwave frequencies, the quantum of energy ($= 10^{-4}$ to 10^{-5} *ev*) is much less than the oscillation energy ($= 10^{-3}$ *ev*); so the classical analysis is then adequate.

An electron plasma wave can decay into another electron plasma wave and an ion acoustic wave. Moreover, (1) coherent scattering, molecular vibration and absorption of the incident light are possible; (2) Microwave radiation at frequencies near the electron plasma frequency can be excited and coupled with both electron plasma wave and ion acoustic wave; and (3) A laser light at frequency ω_1 can be converted to coherent radiation at $\omega_1 - \omega_2$ and ω_2. All these effects are studied classically.

13.5 THE NISHIKAWA MODEL FOR NON-LINEAR PARAMETRIC PROCESSES

The Nishikawa model, and its further developed form, are convenient for study of parametric processes in plasmas. A system of two waves (the idler and the signal or the local weak noise fields) is parametrically amplified by a third wave of large amplitude (the pump wave or the driver wave). Frequency of one of the signal waves is much higher than that of the other signal wave.

(a) The basic equations and their solutions

The three waves, denoted by X, Y, Z, are damped and non-interacting in the linear approximation. The X wave is excited at the frequency ω_1, Y at ω_2 and Z at ω_3; the corresponding damping factors are $\Gamma_1, \Gamma_2, \Gamma_3$ respectively. The Z wave is initially excitable to a much greater amplitude than X and Y. The weaker waves X and Y interact with each other non-linearly, through the action of Z in such a manner that X (or Y) is forced to oscillate at the beat frequency of Y (or X) and Z. If ω_3 is close to $\omega_1 + \omega_2$, then this forced oscillation resonates with the natural oscillation for a resonant energy transfer between the applied wave field Z and the set of X and Y. Unstable modes of X and Y are possible if the intensity of Z is above a certain threshold value, for overcoming the collisional decay. The spatially homogeneous Z-wave of large amplitude, obeys the oscillation equation

$$\ddot{Z} + 2\Gamma_3 \dot{Z} + (\omega_3^2 + \Gamma_3^2) Z = 0 \tag{13.5.1}$$

where dot denotes time derivative. Putting

$$Z(t) = \exp(-\Gamma_3 t)\, Z_1(t), \tag{13.5.2}$$

this equation gives $\ddot{Z}_1 + \omega_3^2 Z_1 = 0$, so that we can write

$$Z(t) = Z_0 \exp(-\Gamma_3 t - i\omega t),\; Z_1(t) = Z_0 \exp(-i\omega t) \tag{13.5.3}$$

$$\omega^2 + 2i\,\Gamma_3\,\omega - \omega_3^2 - \Gamma_3^2 = 0 \tag{13.5.4}$$

where Z_0 is the constant amplitude of the Z wave.

For parametric coupling of oscillations of $X(t)$ and $Y(t)$, at the natural frequencies ω_1 and ω_2, one frequency, say ω_1, should be greater than the other, ω_2. Initially $X(t)$ and $Y(t)$ are weak uncoupled internal noise fields. When Z is switched on, X and Y are nonlinearly coupled by Z, and the damped normal modes of X at ω_1 and Y at ω_2 obey the inhomogeneous oscillator equations

$$LX(t) \equiv \ddot{X} + 2\Gamma_1 \dot{X} + (\omega_1^2 + \Gamma_1^2) X = \lambda Y(t)\, Z(t), \tag{13.5.5}$$

$$LY(t) \equiv \ddot{Y} + 2\Gamma_1 \dot{Y} + (\omega_2^2 + \Gamma_2^2) Y = \mu\, X(t)\; Z(t) \tag{13.5.6}$$

The frequencies ω_1 and ω_2, the damping factors Γ_1 and Γ_2 and the parameters λ and μ are constants; moreover, $\lambda, \mu, \Gamma_1, \Gamma_2$ are assumed positive. The equations for the uncoupled oscillations of X and Y are

$$\ddot{X} + 2\Gamma_1 \dot{X} + (\omega_1^2 + \Gamma_1^2) X = 0,\; \ddot{Y} + 2\Gamma_2 \dot{Y} + (\omega_2^2 + \Gamma_2^2) Y = 0 \tag{13.5.7}$$

The solutions are the real parts of

$$X(t) = X_0 \exp(-\Gamma_1 t - i\,\omega_1 t),\; Y(t) = Y_0 \exp(-\Gamma_2 t - i\omega_2 t) \tag{13.5.8}$$

The harmonic oscillator series solutions $\sum_{\omega} X(\omega) \exp(-i\,\omega\, t)$ and $\sum_{\omega} Y(\omega) \exp(-i\,\omega\, t)$, give the relations

$$(\omega^2 - \omega_1^2 - \Gamma_1^2 + 2i\omega\,\Gamma_1)\,\mathbf{X}(\omega) + \lambda\,Z_0\,[Y(\omega - \omega_3) + Y(\omega + \omega_3)] = 0 \quad (13.5.9)$$

$$(\omega^2 - \omega_2^2 - \Gamma_2^2 + 2i\omega\,\Gamma_2)\,Y(\omega) + \mu Z_0\,[X(\omega - \omega_3) + X(\omega + \omega_3)] = 0 \quad (13.5.10)$$

So, $X(\omega)$ couples with $Y(\omega \pm \omega_3)$, $Y(\omega)$ with $X(\omega \pm \omega_3)$; then $Y(\omega \pm \omega_3)$ couples with $X(\omega \pm 2\omega_3)$. Moreover, these two relations can be written as

$$X(\omega) = -\frac{\lambda Z_0\,[Y(\omega - \omega_3) + Y(\omega + \omega_3)]}{(\omega^2 - \omega_1^2 - \Gamma_1^2 + 2i\omega\,\Gamma_1)}, \quad (13.5.11)$$

$$Y(\omega) = -\frac{\mu Z_0\,[X(\omega - \omega_3) + X(\omega + \omega_3)]}{(\omega^2 - \omega_2^2 - \Gamma_2^2 + 2i\omega\,\Gamma_2)} \quad (13.5.12)$$

Solving for $Y(\omega \pm \omega_3)$, from these we obtain

$$Y(\omega - \omega_3) = -\frac{\mu Z_0\,[X(\omega - 2\omega_3) + X(\omega)]}{(\omega - \omega_3^2)^2 - \omega_2^2 - \Gamma_2^2 + 2i\,(\omega - \omega_3)\,\Gamma_2}, \quad (13.5.13)$$

$$Y(\omega + \omega_3) = -\frac{\mu Z_0\,[X(\omega) + X(\omega + 2\,\omega_3)]}{(\omega + \omega_3)^2 - \omega_2^2 + 2i\,(\omega + \omega_3)\,\Gamma_2} \quad (13.5.14)$$

The excited modes only at the resonant frequencies are important. So, for interest in the frequency range of real ω $(= \omega_1)$, we neglect $X(\omega \pm 2\omega_3)$ as two of the off-resonant modes, retain both $Y(\omega + \omega_3)$ and $Y(\omega - \omega_3)$, and obtain

$$Y(\omega - \omega_3) = -\frac{\mu Z_0\,X(\omega)}{(\omega - \omega_3)^2 - \omega_2^2 - \Gamma_2^2 + 2i\,(\omega - \omega_3)\,\Gamma_2}, \quad (13.5.15)$$

$$Y(\omega + \omega_3) = -\frac{\mu Z_0\,X(\omega)}{(\omega + \omega_3)^2 - \omega_2^2 - \Gamma_2^2 + 2i\,(\omega + \omega_3)\,\Gamma_2} \quad (13.5.16)$$

The eliminant equation, obtained from the three linearly independent linear simultaneous equations (13.5.11), (13.5.15) and (13.5.16) for $X(\omega)$, $Y(\omega - \omega_3)$ and $Y(\omega + \omega_3)$, is

$$\begin{vmatrix} \omega^2 - \omega_1^2 - \Gamma_1^2 + 2i\omega\,\Gamma_1 & \lambda Z_0 & \lambda Z_0 \\ \mu Z_0 & (\omega - \omega_3)^2 - \omega_2^2 - \Gamma_2^2 + 2i\,(\omega - \omega_3)\,\Gamma_2 & 0 \\ \mu Z_0 & 0 & (\omega_1 + \omega_3)^2 - \omega_2^2 - 2i\,(\omega + \omega_3)\,\Gamma_2 \end{vmatrix}$$

(13.5.17)

If $Z_0 = 0$, then this dispersion equation for ω reduces to the three equations

$$\begin{aligned} &\omega^2 - \omega_1^2 - \Gamma_1^2 + 2i\omega\,\Gamma_1 = 0, \\ &(\omega - \omega_3)^2 - \omega_2^2 - \Gamma_2^2 + 2i\,(\omega - \omega_3)\,\Gamma_2 = 0, \\ &(\omega + \omega_3)^2 - \omega_2^2 - \Gamma_2^2 + 2i\,(\omega + \omega_2)\,\Gamma_2 = 0 \end{aligned} \quad (13.5.18)$$

When $Z_0 \neq 0$, equation (13.5.17) gives three non-linearly parametrically excited deviations of ω (dependent on the intensity of the pump field) from the values obtained from equations (13.5.18). A conventional form of the equations (13.5.11), (13.5.15) and (13.5.15) is

$$\begin{aligned} D_1(\omega)\,X(\omega) &= \lambda Z_0\,[Y(\omega+\omega_2)+Y(\omega-\omega_2)], \\ D_2(\omega+\omega_3)\,Y(\omega+\omega_3) &= \mu Z_0\,X(\omega), \\ D_2(\omega-\omega_3)\,Y(\omega-\omega_3) &= \mu Z_0\,X(\omega) \end{aligned} \tag{13.5.19}$$

where

$$D_j(\omega) = -\omega^2 - 2i\,\Gamma_j\,\omega + \omega_j^2 + \Gamma_j^2 \tag{13.5.20}$$

with $j = 1, 2$. Then the eliminant dispersion relation is

$$1 = \frac{\lambda\mu Z_0^2}{D_1(\omega)}\left[\frac{1}{D_2(\omega+\omega_3)} + \frac{1}{D_2(\omega-\omega_3)}\right] \tag{13.5.21}$$

In laser plasma interactions the frequency ω_3 of the pump field is sufficiently large, so that

$$\omega_3 = \omega_1 + \omega_2, \tag{13.5.22}$$

but $|K| << 1$ where $K = \lambda\mu Z_0^2$ (13.5.23)

This K is proportional to the pump power.

(b) The case of nearly free resonance

In this case $|\omega_1|$ is large compared to the nonlinear shift ω of (13.5.17) when $Z_0 \neq 0$. The response $Y(\omega + \omega_3)$ is then off-resonant, so negligible. Hence, putting

$$\omega = \omega_1, \tag{13.5.24}$$

the relations (13.5.11) and (13.5.13) are found to reduce to

$$X(\omega) = -\frac{\lambda Z_0\,Y(\omega-\omega_3)}{\omega^2 - \omega_1^2 - \Gamma_1^2 + 2i\omega\,\Gamma_1} \tag{13.5.25}$$

$$Y(\omega-\omega_3) = -\frac{\mu Z_0\,X(\omega)}{(\omega-\omega_3)^2 - \omega_2^2 - \Gamma_2^2 + 2i(\omega-\omega_3)\,\Gamma_2} \tag{13.5.26}$$

These give the intensity-dependent frequency determining dispersion relation

$$(\omega^2 - \omega_1^2 - \Gamma_1^2 + 2i\omega\,\Gamma_1)\;[(\omega-\omega_3)^2 - \omega_2^2 - \Gamma_2^2 + 2i(\omega-\omega_3)\,\Gamma_2] = K \tag{13.5.27}$$

where K is given by (13.5.23). Putting $\omega = \omega_1$, from (13.5.24) in the first factor in the left hand side of (13.5.27), we obtain

$$\omega^2 - \omega_1^2 - \Gamma_1^2 + 2i\omega\,\Gamma_1 = (\omega-\omega_1)\,(\omega+\omega_1) + 2i\omega\,\Gamma_1 - \Gamma_1^2$$

$$= 2\omega_1 (\omega - \omega_1 + i\,\Gamma_1) - \Gamma_1^2 \tag{13.5.28}$$

The second factor can be written as

$$(\omega - \omega_3)^2 - \omega_2^2 + 2i\,(\omega - \omega_3)\,\Gamma_2$$

$$= (\omega - \omega_3 - \omega_2)\;(\omega - \omega_3 + \omega_2) + 2i\,(\omega - \omega_3)\,\Gamma_2 - \Gamma_2^2$$

$$= - 2\omega_2 (\omega - \omega_3 + \omega_2 + i\,\Gamma_2) \tag{13.5.29}$$

Here we have used the relations $\omega \approx \omega_1 \approx \omega_3 - \omega_2$ of (13.5.24) and (13.5.22). Hence, equation (13.5.27) becomes

$$\left(\omega - \omega_1 + i\Gamma_1 - \frac{\Gamma_1^2}{2\omega_1}\right)\left(\omega - \omega_3 + \omega_2 + i\Gamma_2 + \frac{\Gamma_2^2}{2\omega_2^2}\right) = - \frac{K}{4\omega_1\,\omega_2} \tag{13.5.30}$$

Since $\omega_1, \omega_2, \omega_3$ are real, the frequency mismatch $\Delta\,\omega$, given by

$$\omega_3 - \omega_1 - \omega_2 = \Delta\omega, \tag{13.5.31}$$

is evaluated from the relation

$$\left(\omega - \omega_1 + i\Gamma - \frac{\Gamma_1^2}{2\omega_1}\right)\left(\omega - \omega_1 - \Delta\omega + i\Gamma_2 + \frac{\Gamma_2^2}{2\omega_2}\right) = - \frac{K}{4\omega_1\,\omega_2} \tag{13.5.32}$$

The parametrically excited complex frequency ω can be written as

$$\omega = \mathbf{X} + iY \tag{13.5.33}$$

where $\mathbf{X}$ and $\mathbf{Y}$ are real. Then, separating the real and imaginary parts of (13.5.32) we obtain

$$\left(\mathbf{X} - \omega_1 - \frac{\Gamma_1^2}{2\omega_1}\right)\left(\mathbf{X} - \omega_1 - \Delta\omega + \frac{\Gamma_2^2}{2\omega_2}\right) - (Y + \Gamma_1)\,(Y + \Gamma_2) + \frac{K}{4\omega_1\,\omega_2} = 0, \tag{13.5.34}$$

$$\left(\mathbf{X} - \omega_1 - \frac{\Gamma_1^2}{2\omega_1}\right)(Y + \Gamma_2) + \left(\mathbf{X} - \omega_1 - \Delta\omega + \frac{\Gamma_2^2}{2\omega_2}\right)(Y + \Gamma_1) = 0 \tag{13.5.35}$$

Equation (13.5.35) transforms to

$$(\mathbf{X} - \omega_1)\;(2Y + \Gamma_1 + \Gamma_2) - \frac{\Gamma_1^2\,(Y + \Gamma_2)}{2\omega_1} + \frac{\Gamma_2^2\,(Y + \Gamma_1)}{2\omega_2} = \Delta\omega\,(Y + \Gamma_1) \tag{13.5.36}$$

Eliminating now $\mathbf{X} - \omega_1$ between (13.5.34) and (13.5.35) we obtain

$$(Y + \Gamma_1)\;(Y + \Gamma_2) = \frac{K}{4\omega_1\,\omega_2} + \left[\frac{\Delta\omega\,(Y + \Gamma_1)}{2Y + \Gamma_1 + \Gamma_2} - \frac{\Gamma_1^2\,(Y + \Gamma_2)}{2\omega_1\,(2Y + \Gamma_1 + \Gamma_2)}\right.$$

$$+\frac{\Gamma_2^2 (Y+\Gamma_1)}{2\omega_2 (2Y+\Gamma_1+\Gamma_2)} - \frac{\Gamma_1^2}{2\omega_1}\Bigg]\Bigg[\frac{\Delta\omega (Y+\Gamma_1)}{2Y+\Gamma_1+\Gamma_2} - \frac{\Gamma_1^2 (Y+\Gamma_2)}{2\omega_1 (2Y+\Gamma_1+\Gamma_2)}$$

$$+\frac{\Gamma_2^2 (Y+\Gamma_1)}{2\omega_2 (2Y+\Gamma_1+\Gamma_2)} - \Delta\omega + \frac{\Gamma_2^2}{2\omega_2}\Bigg] \tag{13.5.37}$$

because equation (13.5.36) gives

$$\mathbf{X} - \omega_1 = \frac{\Delta\,\omega (Y+\Gamma_1)}{2Y+\Gamma_1+\Gamma_2} - \frac{\Gamma_1^2 (Y+\Gamma_2)}{2\omega_1 (2Y+\Gamma_1+\Gamma_2)} + \frac{\Gamma_2^2 (Y+\Gamma_1)}{2\omega_2 (2Y+\Gamma_1+\Gamma_2)} \tag{13.5.38}$$

Equation (13.5.37) is written as

$$(Y+\Gamma_1)(Y+\Gamma_2)\left[1+\frac{(\Delta\omega)^2}{(2Y+\Gamma_1+\Gamma_2)^2}\right] - \frac{K}{4\omega_1\,\omega_2}$$

$$=\frac{\Delta\,\omega (Y+\Gamma_1)}{(2Y+\Gamma_1+\Gamma_2)^2}\left[-\frac{\Gamma_1^2 (\Delta\,\omega+\Gamma_1+3\,\Gamma_2)}{2\omega_1} + \frac{\Gamma_2^2 (\Delta\,\omega+3\,\Gamma_1+\Gamma_2)}{3\omega_2}\right] \tag{13.5.39}$$

The roots of Y are positive, so, exponentially growing solutions for $X(t)$ and $Y(t)$ are obtained. If K_t is the threshold value of K (which is proportional to the pump power) then $K = K_t$ when $Y = 0$; so

$$K_t = K_0\left[1+\frac{(\Delta\omega)^2}{(\Gamma_1+\Gamma_2)^2} + \frac{\Delta\omega\,\Gamma_1^2 (\Gamma_1+3\,\Gamma_2)}{2\omega_1\,\Gamma_1 (\Gamma_1+\Gamma_2)^2} - \frac{\Delta\omega\,\Gamma_2^2 (3\,\Gamma_1+P_2)}{2\omega_2\,\Gamma_2 (\Gamma_1+\Gamma_2)^2}\right] \tag{13.5.40}$$

where $K_0 = \Delta\,\omega_1\,\omega_2\,\Gamma_1\,\Gamma_2$ (13.5.41)

For exact phase matching $\Delta\omega = 0$, then $K_t = K_0$. Also, this K_t is finite when at least one of Γ_1 and Γ_2 is non-zero. At the threshold, we put $Y = 0$ in (13.5.36), and obtain

$$\mathbf{X} = \omega_1 + \frac{\Gamma_1\,\Delta\omega}{\Gamma_1+\Gamma_2} - \frac{\Gamma_1^2\,\Gamma_2}{2\omega_1 (\Gamma_1+\Gamma_2)} + \frac{\Gamma_1\,\Gamma_2^2}{2\omega_2 (\Gamma_1+\Gamma_2)} \tag{13.5.42}$$

Then using (13.5.31) we find that

$$\mathbf{X} - \omega_3 = \omega_1 - \omega_3 + \frac{\Gamma_1\,\Delta\omega}{\Gamma_1+\Gamma_2} - \frac{\Gamma_1^2\,\Gamma_2}{2\omega_1 (\Gamma_1+\Gamma_2)} + \frac{\Gamma_1\,\Gamma_2^2}{2\omega_1 (\Gamma_1+\Gamma_2)}$$

$$= -\,\omega_2 - \frac{\Gamma_2\,\Delta\omega}{\Gamma_1+\Gamma_2} - \frac{\Gamma_2^2\,\Gamma_2}{2\omega_1 (\Gamma_1+\Gamma_2)} + \frac{\Gamma_1\,\Gamma_2^2}{2\omega_2 (\Gamma_1+\Gamma_2)} \tag{13.5.43}$$

Since the wave field $X(t)$ is excited at ω (real part **X**) and $Y(t)$ is excited at $\omega - \omega_3$ (real part $\mathbf{X} - \omega_3$), equation (13.5.42) gives the frequency of $X(t)$ at the threshold power, and (13.5.43) gives the same of $Y(t)$. Thus, for $\Delta\omega = 0$ the frequency shifts of $X(t)$ and $Y(t)$ vanish. Putting $\Delta\omega = 0$ in (13.5.39), we solve for Y and obtain

$$Y = -\frac{1}{2}(\Gamma_1 + \Gamma_2) - \frac{1}{2}\sqrt{(\Gamma_1 - \Gamma_2)^2 + \frac{K}{\omega_1\omega_2}} \tag{13.5.44}$$

In the region of power much above the threshold value, we put $\Gamma_1 = 0$, $\Gamma_2 = 0$ in (13.5.39) and (13.5.38), and obtain

$$Y = \sqrt{\frac{K}{4\omega_1\omega_2} - \frac{1}{4}(\Delta\omega)^2}\,, \quad \mathbf{X} = \omega_1 + \frac{1}{2}\Delta\omega \tag{13.5.45}$$

Hence the growth rate increases with the square root of the pump power and the frequency shift $\mathbf{X} - \omega_1$ does not change, because it does not depend on the pump power.

13.6 NON-LINEAR POLARIZATION

Denoting the electron displacement at time t, from the equilibrium position, by $\mathbf{r}$, the polarization density $\mathbf{P}$ is defined as $\mathbf{P} = -Ne\,\mathbf{r}$, where N is the number density and $-e$ is the charge. In general, if e_i is the charge of a particle of the ith species, N_i is the number density and $\mathbf{r}_i$ is the displacement, then $\mathbf{P} = \sum_i N_i e_i \mathbf{r}_i$.

The electron velocity is $\mathbf{v}\,(=\dot{\mathbf{r}}$ where dot denotes the time derivative. The equations for electron oscillation in cold plasma are

$$\dot{\mathbf{v}} = -\frac{e}{m}(\mathbf{E} + \mathbf{E}_a) - 2\gamma\mathbf{v}, \quad \dot{\mathbf{E}} = 4\pi e\,(N_0 + n)\,\mathbf{v} \tag{13.6.1}$$

where γ is the damping constant, $\mathbf{E}_a$ is the applied electric field, $\mathbf{E}$ is the self electric field of the plasma, $N = N_0 + n$, N_0 is the equilibrium number density of electrons and n is the perturbed part. Hence

$$\ddot{\mathbf{v}} = -\frac{e}{m}4\pi e\,(N_0 + n)\,\mathbf{v} - \frac{e}{m}\dot{\mathbf{E}}_a - 2\gamma\dot{\mathbf{v}} \tag{13.6.2}$$

It can be written as

$$\ddot{\mathbf{v}} + \omega_\rho^2\mathbf{v} + 2\gamma\dot{\mathbf{v}} + \omega_\rho^2\frac{n}{N_0}\mathbf{v} = -\frac{e}{m}\dot{\mathbf{E}}_a \tag{13.6.3}$$

where $\omega_\rho = \sqrt{(4\pi N_0 e^2/m)}$. Putting $\dot{\mathbf{r}} = \mathbf{v}$, we integrate (13.6.3) with respect to t and remove the vector sign from all the terms because the oscillation is unidirectional along the applied electric field $\mathbf{E}_a$.

We replace the integral $(\omega_\rho^2/N_0)\int n\mathbf{v}\,dt$ in (13.6.3) by the term $-\xi r^2$, for bi-linear (or quadric) crystals and other such dielectrics, and obtain

$$\ddot{\mathbf{r}} + \omega_\rho^2 r + 2\gamma \dot{\mathbf{r}} = -\frac{e}{m} E_a + \xi r^2 \qquad (13.6.4)$$

In linear crystals the linearized approximation of this polarization is used to study dynamics of electrons as those of harmonically bound oscillators. The resonant frequencies of these oscillators correspond to the atomic spectral lines, and the valence electrons are bound by the Coulomb field of the ion cores. For large deviations from equilibrium, due to the presence of strong applied fields the anharmonicity is included in the oscillatory motion of the electrons. The non-linear polarization expression

$$P = \alpha E\,(1 + a_1 E + a_2 E^2 + ...) \qquad (13.6.5)$$

is a linear combination of positive integral powers of E; all the quantities $\alpha, a_1, a_2, \ldots$ are constants. The anharmonic non-linear term ξr^2 is small, compared to the other terms which are linear in r and E_a. So we can write

$$r = r_1 + r_2 + r_3 + \ldots \qquad (13.6.6)$$

and treat r_2 to be of the order of r_1^2, r_3 of the order of $r_1 r_2$, and like that, for the higher order terms. Then the polarization density in series is

$$P = -\Sigma N_0\, e r_i = \Sigma P_i \qquad (13.6.7)$$

where $P_i = -N_0\, e r_i$. The equations of linearized approximation, and of the next higher order, are

$$\ddot{r}_1 + 2\gamma \dot{r}_1 + \omega_\rho^2 r_1 = -\frac{e}{m} E_a, \qquad (13.6.8)$$

$$\ddot{r}_2 + 2\gamma \dot{r}_2 + \omega_\rho^2 r_2 = \xi r_1^2 \qquad (13.6.9)$$

The applied electric field E_a can be written as

$$E_a = \sum_j \left[E(\omega_j)\, e^{-i\omega_j t} + \bar{E}(\omega_j)\, e^{i\omega_j t} \right], \qquad (13.6.10)$$

where $\bar{E}$ is the cc of E. We put this value of E_a in (13.6.8), solve for r_1, and write r_1 as the sum $r_1(\omega_j) + \bar{r}_1(\omega_j)$. Then, equating the coefficients of $\exp(-i\omega_j t)$ from both sides of the equation for r_1, we get

$$r_1(\omega_j) = -\frac{e}{m}\, \frac{E(\omega_j)\exp(-i\omega_j t)}{\omega_\rho^2 - 2i\gamma\,\omega_j - \omega_j^2} \qquad (13.6.11)$$

Summing over all ωs, we then find that

$$r_1 = \sum_j \left[r_1(\omega_j) + \bar{r}_1(\omega_j) \right] = -\frac{e}{m} \sum_j \left[\frac{E(\omega_j)\exp(-i\omega_j t)}{\omega_\rho^2 - 2\gamma i\,\omega_j - \omega_j^2} + cc \right] \qquad (13.6.12)$$

where cc stands for the complex conjugate. Squaring both sides of this equation gives

$$r_1^2 = \frac{e^2}{m^2} \sum_{j,k} \left[\frac{E(\omega_j) E(\omega_k) \exp\{-i(\omega_j + \omega_k) t\}}{(\omega_\rho^2 - 2\gamma i \omega_j - \omega_j^2)\ (\omega_\rho^2 - 2\gamma i \omega_k - \omega_k^2)} \right.$$

$$\left. + \frac{E(\omega_j) \overline{E}(\omega_k) \exp\{-i(\omega_j - \omega_k) t\}}{(\omega_\rho^2 - 2\gamma i \omega_j - \omega_j^2)\ (\omega_\rho^2 - 2\gamma i \omega_k - \omega_k^2)} \right] \quad (13.6.13)$$

Using the notation $-\omega_j = \omega_{-j}$ and taking for j (or K) both positive and negative integral values (that is, putting $j = \pm 1, \pm 2, \pm 3,$) in (13.6.13) we write $\overline{E}(\omega_j) = E(-\omega_j)$, and obtain

$$r_1^2 = \frac{e^2}{m^2} \sum_{j,k} \left[\frac{E(\omega_j) E(\omega_k) \exp\{-i(\omega_j + \omega_k) t\}}{(\omega_\rho^2 - 2\gamma i \omega_j - \omega_j^2)(\omega_\rho^2 - 2\gamma i \omega_k - \omega_k^2)} \right] \quad (13.6.14)$$

Thus the equation for r_2 becomes

$$\ddot{r}_2 + 2\gamma \dot{r}_2 + \omega_\rho^2 r_2 = \frac{e^2 \xi}{m^2} \sum_{j,k} \left[\frac{E(\omega_j) E(\omega_k) \exp\{-i(\omega_j + \omega_k) t\}}{(\omega_\rho^2 - 2\gamma i \omega_j - \omega_j^2)\ (\omega_\rho^2 - 2\gamma i \omega_k - \omega_k^2)} \right]$$

$$(13.6.15)$$

The solution of this equation is

$$r_2 = \frac{e^2 \xi}{m^2} \sum_{j,k} \left[\frac{E(\omega_j) E(\omega_k) \exp\{-i(\omega_j + \omega_k) t\}}{(\omega_\rho^2 - 2i\gamma \omega_j - \omega_j^2)\ (\omega_\rho^2 - 2\gamma i \omega_k - \omega_k^2)\ \{(\omega_\rho^2 - 2\gamma_i (\omega_j + \omega_k) - (\omega_j + \omega_k)^2} \right]$$

$$(13.6.16)$$

Then the polarization densities of first and second orders are

$$P_1 = -N_0 e r_1 = \frac{\omega_\rho^2}{4\pi} \sum_j \left[\frac{E(\omega_j) \exp(-i\omega_j t)}{(\omega_\rho^2 - 2\gamma i \omega_j - \omega_j^2)} \right] = \sum_j \chi^{(j)}(\omega_j) E(\omega_j) \exp(-i\omega_j t)$$

$$(13.6.17)$$

$$P_2 = -N_0 e r_2 = \sum_{j,k} \chi^{(2)}(\omega_j, \omega_k) E(\omega_j) E(\omega_k) \exp\{-i(\omega_j + \omega_k) t\}$$

$$(13.6.18)$$

where $\chi^{(1)}(\omega_j) = \dfrac{\dfrac{\omega_\rho^2}{4\pi}}{\omega_\rho^2 - 2\gamma_i \omega_j - \omega_j^2}$,

$$\chi^{(2)}(\omega_j, \omega_k) = -\frac{e^2 \xi}{m^2} N_0 e \left(\frac{4\pi}{\omega_\rho^2} \right)^3 \chi^{(1)}(\omega_j) \chi^{(1)}(\omega_k) \chi^{(1)}(\omega_j + \omega_k) \quad (13.6.19)$$

$$= -\frac{m\,\xi}{N_0^2\,e^3}\,\chi^{(1)}\,(\omega_j)\,\chi^{(1)}\,(\omega_k)\,\chi^{(1)}\,(\omega_j + \omega_k)$$

The second order susceptibility $\chi^{(2)}$ depends on the product of the first order susceptibilities of the three waves at frequencies ω_j, ω_k, ω_{j+k}, and is a dimensionless quantity.

APPENDICES

APPENDIX 1 : DELTA FUNCTION OF DIRAC

Some properties of the delta function

1. $\delta(x - x_0) = 0$ if $x \neq x_0$ and $\delta(x - x_0)$ is undefined at $x = x_0$.

2. $\int_{x_0-\varepsilon}^{x_0+\varepsilon} \delta(x - x_0)\, dx = 1$ if $\varepsilon > 0$.

3. $\int_a^b f(x)\, \delta(x - x_0)\, dx = f(x_0)$ when $a < x < b$

   $= 0$ when $x < a$ or $x > b$ with $a < b$.

4. $\int_a^b F(x)\, \delta(x)\, dx = F(0)$ if $a < 0$ and $b > 0$.

5. $\delta(-x) = \delta(x)$.

6. $x\, \delta\, x = 0$.

7. $\delta(ax) = \dfrac{\delta(x)}{|a|}$.

8. $f(x)\, \delta(x - a) = f(a)\, \delta(x - a)$.

9. $\int \delta(a - x)\, \delta(x - b)\, dx = \delta(a - b)$.

10. $\delta(x^2 - a^2) = \dfrac{\delta(x - a) + \delta(x + a)}{2\,|a|}$.

11. $\delta[\varphi(x)] = \sum_i \dfrac{\delta(x - x_i)}{|(d\varphi/dx)_{x = x_i}|}$, so $\int_{-\infty}^{\infty} \delta[\varphi(x)]\, f(x)\, dx =$

    $\sum_i \dfrac{f(x_i)}{|d\varphi/dx|_{x = x_i}}$ where x_i are the roots of the equation $\varphi(x) = 0$.

12. $\nabla^2 \left(\dfrac{1}{r}\right) = -4\pi\delta(\mathbf{r})$.

13. In three dimensions $\delta(\mathbf{r} - \mathbf{r}_i) = \delta(x - x_i)\, \delta(y - y_i)\, \delta(z - z_i)$ where the components of $\mathbf{r}$ and $\mathbf{r}_i$ in rectangular Cartesian coordinates are (x, y, z) and (x_1, y_1, z_1) respectively.

APPENDIX 2 : THEOREMS OF GAUSS, GREEN AND STOKES

Let **A** be a vector field, φ and ψ be scalar fields, τ be a volume bounded by the surface Σ, and **n** be the unit vector normal to the surface element $d\Sigma$. Then $\mathbf{n}\, d\Sigma = d\mathbf{\Sigma}$; Gauss's theorem is

$$\iiint_V (\nabla . \mathbf{A})\, dV = \iint_\Sigma (\mathbf{A} . d\mathbf{\Sigma}) \text{ or } \nabla . \mathbf{A} = \lim_{\tau \to 0} \left(\frac{1}{V} \iint_\Sigma (\mathbf{A} . d\mathbf{\Sigma}) \right); \quad \text{(A2.1)}$$

$$\text{Green's theorem : } \iint_\Sigma \left(\psi \frac{\partial \varphi}{\partial n} - \varphi \frac{\partial \psi}{\partial n} \right) d\Sigma = \iiint_V (\psi \nabla^2 \varphi - \varphi \nabla^2 \psi)\, dV \quad \text{(A2.2)}$$

where $$\frac{\partial \varphi}{\partial n} = (\mathbf{n} . \nabla)\, \varphi;$$

$$\text{Stoke's theorem } \iint_\Sigma ((\nabla \times \mathbf{A}) . d\mathbf{\Sigma}) = \oint_\Gamma (\mathbf{A} . d\mathbf{l}) \quad \text{(A2.3)}$$

where Γ is a closed contour described on Σ, and dl is a line element of Γ. Other relevant formulas are

$$\iiint_V \nabla \varphi\, dV = \iint_\Sigma \varphi \mathbf{n}\, d\Sigma = \iint_\Sigma \varphi d\mathbf{\Sigma}, \iint_\Sigma (\mathbf{n} \times \nabla \varphi)\, d\Sigma = \oint_\Gamma \varphi d\mathbf{l} \quad \text{(A2.4)}$$

$$\oint (d\mathbf{r} \times \mathbf{A}) = \iint_\Sigma (d\mathbf{\Sigma} \times \nabla) \times \mathbf{A} \quad \text{(A2.5)}$$

$$\iint_\Sigma (\hat{\mathbf{n}} \times \nabla \varphi)\, d\Sigma = \oint_\Gamma \psi\, \mathbf{d}l, \quad \text{(A2.6)}$$

Green's Identities

$$\iiint_v [\varphi \nabla^2 \psi - \psi \nabla^2 \varphi]\, dv = \iint_\Sigma [(\varphi \nabla \psi - \psi \nabla \varphi) . d\mathbf{\Sigma}] \quad \text{(A2.7)}$$

$$\iiint_v [\varphi \nabla^2 \psi - (\nabla \varphi . \nabla \psi)]\, dv = \iint_\Sigma \varphi\, (\nabla \psi . d\mathbf{\Sigma}) \quad \text{(A2.8)}$$

$$\iiint_v \nabla \varphi\, dv = \iint_\Sigma \varphi\, d\mathbf{\Sigma} \quad \text{(A2.9)}$$

$$\iiint_v (\nabla \times \mathbf{A})\, dv = \iint_\Sigma (\hat{\mathbf{n}} \times \mathbf{A}) \times d\mathbf{\Sigma} \quad \text{(A2.10)}$$

APPENDIX 3 : VECTOR FORMULAS

1. $\mathbf{A} . (\mathbf{B} \times \mathbf{C}) = \mathbf{B} . (\mathbf{C} \times \mathbf{A}) = \mathbf{C} . (\mathbf{A} \times \mathbf{B}) = (\mathbf{A\ B\ C})$ (A3.1)
2. $\mathbf{A} \times (\mathbf{B} \times \mathbf{C}) = (\mathbf{A} . \mathbf{C})\, \mathbf{B} - (\mathbf{A} . \mathbf{B})\, \mathbf{C}$ (A3.2)

3. $(\mathbf{A}\times\mathbf{B})\times(\mathbf{C}\times\mathbf{D}) = (\mathbf{A\,B\,D})\,\mathbf{C} - (\mathbf{A\,B\,C})\,\mathbf{D}$
$= (\mathbf{A\,C\,D})\,\mathbf{B} - (\mathbf{B\,C\,D})\,\mathbf{A}$ (A3.3)
4. $(\mathbf{A}\times\mathbf{B})\,.\,(\mathbf{C}\times\mathbf{D}) = (\mathbf{A}\,.\,\mathbf{C})\,(\mathbf{B}\,.\,\mathbf{D}) - (\mathbf{A}\,.\,\mathbf{D})\,(\mathbf{B}\,.\,\mathbf{C})$ (A3.4)
5. $\text{grad}\ (\varphi\,\psi) = \varphi\ \text{grad}\ \psi + \psi\ \text{grad}\ \varphi$ (A3.5)
6. $\text{div}\ (\varphi\,\mathbf{A}) = \varphi\,(\nabla\,.\,\mathbf{A}) + (\mathbf{A}\,.\,\text{grad}\ \varphi)$ (A3.6)
7. $\text{curl}\ (\varphi\,\mathbf{A}) = \varphi\ \text{curl}\ \mathbf{A} + (\text{grad}\ \varphi\times\mathbf{A})$ (A3.7)
8. $\text{div}\ (\mathbf{A}\times\mathbf{B}) = (\mathbf{B}\,.\,\text{curl}\ \mathbf{A}) - (\mathbf{A}\,.\,\text{curl}\ \mathbf{B})$ (A3.8)
9. $\text{curl}\ (\mathbf{A}\times\mathbf{B}) = (\mathbf{B}\,.\,\nabla)\,\mathbf{A} - (\mathbf{A}\,.\,\nabla)\,\mathbf{B} + \mathbf{A}\,(\nabla\,.\,\mathbf{B}) - \mathbf{B}\,(\nabla\,.\,\mathbf{A})$ (A3.9)
10. (*a*) $\text{grad}\ (\mathbf{A}\,.\mathbf{B}) = (\mathbf{B}\,.\,\nabla)\,\mathbf{A} + (\mathbf{A}\,.\,\nabla)\,\mathbf{B} + (\mathbf{B}\times(\nabla\times\mathbf{A})) + (\mathbf{A}\times(\nabla\times\mathbf{B}))$
(A3.10 (*a*))

(*b*) $$\mathbf{A}\ \text{div}\ \mathbf{A} = \sum_{i,j} \hat{\mathbf{X}}_j \frac{\partial}{\partial x_i}\left(A_i A_j - \frac{1}{2} A^2 \delta_{ij}\right) + \mathbf{A}\times\text{curl}\ \mathbf{A} \qquad \text{(A3.10 (b))}$$

(*c*) $(\mathbf{A}\times\nabla)\times\mathbf{B} - \mathbf{A}\times(\nabla\times\mathbf{B}) = (\mathbf{A}\,.\,\nabla)\,\mathbf{B} - \mathbf{A}\,(\nabla\,.\,\mathbf{B})$ (A3.10 (*c*))

11. $\text{curl}\ \text{curl}\ \mathbf{A} = \text{grad}\ \text{div}\ \mathbf{A} - \nabla^2\,\mathbf{A}$ (A3.11)
12. $\text{div}\ \mathbf{r} = 3$ where $\mathbf{r}$ is a position vector (A3.12)
13. $\text{curl}\ \mathbf{r} = 0$ where $\mathbf{r}$ is a position vector (A3.13)
14. $\text{curl}\ \text{grad}\ \varphi = 0$ (A3.14)
15. $\text{div}\ \text{curl}\ \mathbf{A} = 0$ (A3.15)
15. (*a*) $(\mathbf{A}\,.\,((\mathbf{B}\times\nabla)\times\mathbf{C})) - (\mathbf{B}\,.\,((\mathbf{A}\times\nabla)\times\mathbf{C}))$
$= (\mathbf{A}\,.\,(\mathbf{B}\times(\nabla\times C)))$ (A3.15 (*a*))

In the formula of (A3.10 (*b*)) $\hat{\mathbf{X}}_j$ is the unit vector parallel to the *j*th coordinate direction.

Exercise A2.1. Apply Stoke's theorem, to show that

$$\oint_\Gamma (d\mathbf{r}\times\mathbf{B}) = \iint_\Sigma (d\Sigma\times\nabla)\times\mathbf{B} \qquad (1)$$

where $\mathbf{B}$ is solenoidal, *i.e.*, $\nabla\,.\,\mathbf{B} = 0$ (2)

Answer. The *x* component of LHS $= \oint_\Gamma (dy\,B_z - dz\,B_y)$ (3)

We write it as $I = \oint_\Gamma (\mathbf{A}\,.\,d\mathbf{r})$, so that $A_x = 0,\ A_y = +\,B_z\ A_z = -\,B_y$ (4)

So, *I* is the *x* component of LHS of the section (1). Apply Stoke's theorem on *I* and obtain

$$I = \iint_\Sigma (\nabla\times\mathbf{A})\,.\,d\Sigma = \iint_\Sigma \left[\left(\frac{\partial A_z}{\partial y} - \frac{\partial A_y}{\partial z}\right) dy\,dz + \right.$$

$$\left. + \left(\frac{\partial A_x}{\partial z} - \frac{\partial A_z}{\partial x}\right) dz\,dx + \left(\frac{\partial A_y}{\partial x} - \frac{\partial A_x}{\partial y}\right) dx\,dy\right]$$

$$= \iint_{\Sigma} \left[\left(-\frac{\partial B_y}{\partial y} - \frac{\partial B_z}{\partial z} \right) dy\,dz + \left(0 + \frac{\partial B_y}{\partial x} \right) dz\,dx + \left(\frac{\partial B_z}{\partial x} - 0 \right) dx\,dy \right]$$

$$= \iint_{\Sigma} \frac{\partial B_x}{\partial x} d\Sigma_x + \frac{\partial B_y}{\partial x} d\Sigma_y + \frac{\partial B_z}{\partial x} d\Sigma_z \tag{5}$$

Now x component of the RHS of (1)

$$= \mathbf{i} \,.\, \iint_{\Sigma} \left[(\mathbf{i}\,dy\,dz + \mathbf{j}\,dz\,.\,dx + \mathbf{k}\,dx\,dy) \times \left(\mathbf{i}\frac{\partial}{\partial x} + \mathbf{j}\frac{\partial}{\partial y} + \mathbf{k}\frac{\partial}{\partial z} \right) \right]$$

$$\times (\mathbf{i}B_x + jB_y + \mathbf{k}B_z)$$

$$= \mathbf{i} \,.\, \iint_{\Sigma} \left[\mathbf{i} \left(dz\,dx \frac{\partial}{\partial z} - dx\,dy \frac{\partial}{\partial y} \right) + \ldots \right] \times (\mathbf{i}B_x + \mathbf{j}B_y + \mathbf{k}B_z)$$

$$= \mathbf{i} \,.\, \iint_{\Sigma} \left[\mathbf{i} \left(dx\,dy \frac{\partial}{\partial x} - dy\,dz \frac{\partial}{\partial z} \right) B_z - \left(dy\,dz \frac{\partial}{\partial y} - dz\,dx \frac{\partial}{\partial x} \right) B_y \right] + \ldots$$

$$= \iint_{\Sigma} dx\,dy \frac{\partial B_z}{\partial x} - dy\,dz \frac{\partial B_z}{\partial z} - dy\,dz \frac{\partial B_y}{\partial y} + dz\,dx \frac{\partial B_y}{\partial z} + \ldots$$

$$= \iint_{\Sigma} \left[dx\,dy \frac{\partial B_z}{\partial x} - dy\,dz \left(\frac{\partial B_z}{\partial z} + \frac{\partial B_y}{\partial y} \right) + dz\,dx \frac{\partial B_y}{\partial x} \right] + \ldots$$

$$= \iint_{\Sigma} \left[dx\,dy \frac{\partial B_z}{\partial x} + dy\,dz \frac{\partial B_x}{\partial x} + dz\,dx \frac{\partial B_y}{\partial z} + \ldots \right.$$

Since $\nabla \,.\, \mathbf{B} = 0$. Hence the X component of RHS of (1)

$$= \iint \left[d\Sigma_z \frac{\partial B_z}{\partial x} + d\Sigma_x \frac{\partial B_x}{\partial x} + d\Sigma_y \frac{\partial B_y}{\partial z} \right]$$

APPENDIX 4 : TENSORS

A scalar field variable φ has only one component or element, so it is a tensor of zero order. A vector **F** has three components, so **F** is a first order tensor. A second order tensor P_{ij} has nine (= 3^2) elements and is written in the 3 × 3 matrix form

$$P_{ij} = \begin{bmatrix} P_{11} & P_{12} & P_{13} \\ P_{21} & P_{22} & P_{23} \\ P_{31} & P_{32} & P_{33} \end{bmatrix} \tag{A4.1}$$

The Kronecker delta is defined as

$$\delta_{ij}=1 \text{ if } i=j \text{ or } \delta_{ij}=\begin{bmatrix}1 & 0 & 0\\ 0 & 1 & 0\\ 0 & 0 & 1\end{bmatrix} \tag{A4.2}$$

$$=0 \text{ if } i\neq j$$

The alternative unit tensor ε_{ijk} is defined as

$$\varepsilon_{ijk}=+1 \text{ if } i, j, k \text{ are in cyclic order : 123, 231, 312}$$

$$=-1 \text{ if } i, j, k \text{ are in noncyclic order : 321, 132, 213.}$$

$$=0 \text{ if any of the subscripts is repeated.} \tag{A4.3}$$

The Levi-Civita symbol of fourth order, $\varepsilon_{\alpha\beta\gamma\delta}$, is antisymmetric in all the four indices α, β, γ, δ. The non-vanishing components of $\varepsilon_{\alpha\beta\gamma\delta}$, are those for which all four indices are different, and they are equal to + 1 or – 1, according as (α, β, γ, δ) is obtained through an even or odd permutation of (1, 2, 3, 4). $\varepsilon_{\alpha\beta\gamma\delta}$ vanishes if any two or more of the indices is identical.

APPENDIX 5 : THE LAGRANGIAN AND HAMILTONIAN FORMALISMS FOR A SYSTEM OF PARTICLES

A system of n particles is specified by $3n$ position numbers q_k and $3n$ velocity coordinates $\dot{q}_k$ (or momentum coordinates P_k). The Lagrangian $\mathcal{L}$ is defined by

$$\mathcal{L}=T-V \tag{A5.1}$$

where T is the kinetic energy of the system and V is the potential energy of mutual interaction of its elements. The Hamiltonian $\mathcal{H}$, (the total energy) is defined as

$$\mathcal{H}=\sum_k p_k\dot{q}_k-\mathcal{L} \tag{A5.2}$$

where
$$p_k=\frac{\partial\mathcal{L}}{\partial\dot{q}_k} \tag{A5.3}$$

For n particles, the $3n$ equations of motion, in terms of $\mathcal{L}$ (the Euler-Lagrange equations), are

$$\frac{d}{dt}\left(\frac{\partial\mathcal{L}}{\partial\dot{q}_k}\right)-\frac{\partial\mathcal{L}}{\partial q_k}=0 \tag{A5.4}$$

The corresponding $3n$ pairs of equations, involving the Hamiltonian $\mathcal{H}$, are

$$\frac{\partial q_k}{dt}=\frac{\partial\mathcal{H}}{\partial p_k},\ \frac{dp_k}{dt}=-\frac{\partial\mathcal{H}}{\partial q_k} \tag{A5.5}$$

Since $\dot{q}_j\,s$ are independent of $q_k\,s$, for all possible values of j and K, we have

$$\frac{\partial q_i}{\partial \dot{q}_j} = 0, \quad \frac{\partial \dot{q}_i}{\partial q_j} = 0 \tag{A5.6}$$

The n particles of a closed system interact among themselves only and not with those others which do not belong to the system. If $V(\mathbf{r}_1, \mathbf{r}_2, ..., \mathbf{r}_n)$ is the potential of their mutual interaction, then

$$\mathcal{L} = \sum_i \left(\frac{1}{2} m_i v_i^2 \right) - V(\mathbf{r}_1, \mathbf{r}_2, ..., \mathbf{r}_n) \tag{A5.7}$$

The rules of transformation from the coordinates $q_k, \dot{q}_k$ to the coordinates x_k and $\dot{x}_k$, are

$$x_k = \varphi_k(q_1, q_2, ..., q_k), \dot{x}_k = \sum_i \frac{\partial \varphi_k}{\partial q_i} \dot{q}_i \tag{A5.8}$$

Then, since $v_i^2 = \dot{x}_i^2 + \dot{y}_i^2 + \dot{z}_i^2$ we can write

$$\mathcal{L} = \frac{1}{2} \sum_{i,k} p_{ik}(q)\, \dot{q}_i \dot{q}_k - V(\mathbf{q}) \tag{A5.9}$$

where the p_{ik} are functions of the coordinates only, and the kinetic energy in generalized coordinates still remains a quadratic expression of velocities, but now it is also a function of the coordinates.

Since the Lagrangian $\mathcal{L}$ is a function of $q_1, q_2, ..., q_p ; \dot{q}_1, \dot{q}_2, ..., \dot{q}_s$, and time t, an infinitesimal change in these coordinates and time yields the relation

$$\delta \mathcal{L} = \sum_i \frac{\partial \mathcal{L}}{\partial q_i} dq_i + \sum_i \frac{\partial \mathcal{L}}{\partial \dot{q}_i} d\dot{q}_i + \mathcal{L}(t + \delta t) - \mathcal{L}(t) \tag{A5.10}$$

where the last two terms are included when $\mathcal{L}$ depends explicitly on time. For a single particle, the Hamiltonian $\mathcal{H}$, (the energy of the particle), is a function of the momentum $\mathbf{p}$, position vector $\mathbf{r}$ and velocity $\mathbf{v}$; so

$$\mathcal{H} = (\mathbf{p} \cdot \mathbf{v}) - \mathcal{L}, \ \nabla \mathcal{H} = -\frac{d\mathbf{p}}{dt} = -\frac{d}{dt}\left(\frac{\partial \mathcal{L}}{\partial \mathbf{v}}\right) \tag{A5.11}$$

APPENDIX 6: ORTHOGONAL COORDINATES AND CURVATURE

When three families of surfaces u_1 = constant, u_2 = constant and u_3 = constant are orthogonal to one another, the line element

$$ds = (h_1^2\, du_1^2 + h_2^2\, du_2^2 + h_3^2\, du_3^2)^{1/2} \tag{A6.1}$$

and the unit vectors, $\hat{\mathbf{X}}_1$, $\hat{\mathbf{X}}_2$, $\hat{\mathbf{X}}_3$, are along the three directions in which u_1, u_2, u_3 increase. Then the curl and divergence of a vector, and gradient and Laplacian of a scalar quantity, in orthogonal curvilinear coordinates are

$$\nabla \times \mathbf{F} = \frac{1}{h_1 h_2 h_3} \begin{vmatrix} h_1 \hat{\mathbf{X}}_1 & h_2 \hat{\mathbf{X}}_2 & h_3 \hat{\mathbf{X}}_3 \\ \frac{\partial}{\partial u_1} & \frac{\partial}{\partial u_2} & \frac{\partial}{\partial u_3} \\ h_1 F_1 & h_2 F_2 & h_3 u_3 \end{vmatrix} \tag{A6.2}$$

$$\nabla . \mathbf{F} = \frac{1}{h_1 h_2 h_3} \left[\frac{\partial}{\partial u_1} (h_2 h_3 F_1) + \frac{\partial}{\partial u_2} (h_3 h_1 F_2) + \frac{\partial}{\partial u_3} (h_1 h_2 F_3) \right], \tag{A6.3}$$

$$\nabla^2 \varphi = \nabla . (\nabla \varphi) = \frac{1}{h_1 h_2 h_3} \left[\frac{\partial}{\partial u_1} \left(\frac{h_2 h_3}{h_1} \frac{\partial \varphi}{\partial u_1} \right) + \frac{\partial}{\partial u_2} \left(\frac{h_3 h_1}{h_2} \frac{\partial \varphi}{\partial u_2} \right) + \frac{\partial}{\partial u_3} \left(\frac{h_1 h_2}{h_3} \frac{\partial \varphi}{\partial u_3} \right) \right], \tag{A6.4}$$

$$\nabla \varphi = \frac{\hat{\mathbf{X}}_1}{h_1} \frac{\partial \varphi}{\partial u_1} + \frac{\hat{\mathbf{X}}_2}{h_2} \frac{\partial \varphi}{\partial u_2} + \frac{\hat{\mathbf{X}}_3}{h_3} \frac{\partial \varphi}{\partial u_3} \tag{A6.5}$$

The Frenet-Serret formulas of differential geometry for three-dimensional curves are

$$\frac{\partial \hat{\mathbf{X}}_1}{\partial x_1} = \frac{\hat{\mathbf{X}}_2}{R}, \quad \frac{\partial \hat{\mathbf{X}}_2}{\partial x_1} = -\frac{\hat{\mathbf{X}}_3}{R_1} - \frac{\hat{\mathbf{X}}_1}{R}, \quad \frac{\partial \hat{\mathbf{X}}_3}{\partial x_1} = -\frac{\hat{\mathbf{X}}_3}{R_1} \tag{A6.6}$$

where $\hat{\mathbf{X}}_1$ is along the tangent, $\hat{\mathbf{X}}_2$ is along the principal normal and $\hat{\mathbf{X}}_3$ along the binormal at a point of the curve; dx_1, dx_2, dx_3 are the line elements along the directions of these unit vectors; R is the principal radius of curvature along the principal normal; and R_1 is the radius of curvature of torsion, along the binormal. The orthogonality condition

$$(\hat{\mathbf{X}}_i . \hat{\mathbf{X}}_j) = \delta_{ij} \tag{A6.7}$$

where δ_{ij} is the Kronecker delta, gives for the curvatures, $R_2, R_3, T_2, T_3, \sigma_2, \sigma_3$, the relations

$$\frac{\partial \hat{\mathbf{X}}_1}{\partial x_2} = R_1 \hat{\mathbf{X}}_2 - T_2 \hat{\mathbf{X}}_3, \quad \frac{\partial \hat{\mathbf{X}}_1}{\partial x_3} = T_3 \hat{\mathbf{X}}_2 + R_3 \hat{\mathbf{X}}_3, \tag{A6.8}$$

$$\frac{\partial \hat{\mathbf{X}}_2}{\partial x_2} = \sigma_2 \hat{\mathbf{X}}_3 - R_2 \hat{\mathbf{X}}_1, \quad \frac{\partial \hat{\mathbf{X}}_2}{\partial x_3} = \sigma_3 \hat{\mathbf{X}}_3 - T_3 \hat{\mathbf{X}}_1, \tag{A6.9}$$

$$\frac{\partial \hat{\mathbf{X}}_3}{\partial x_2} = T_2 \hat{\mathbf{X}}_1 - \sigma_2 \hat{\mathbf{X}}_2, \quad \frac{\partial \hat{\mathbf{X}}_3}{\partial x_3} = -R_3 \hat{\mathbf{X}}_1 - \sigma_3 \hat{\mathbf{X}}_2 \tag{A6.10}$$

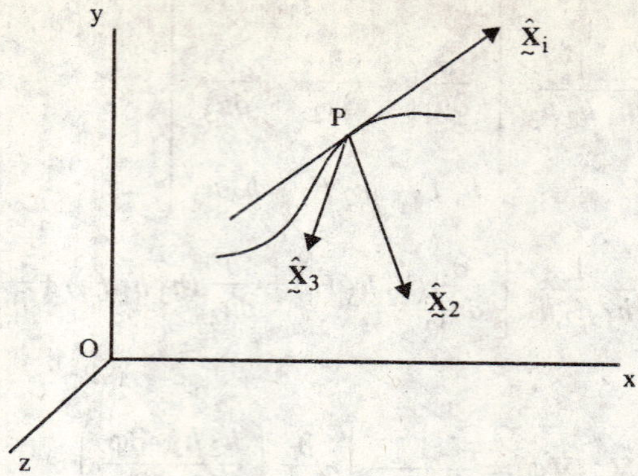

Fig. A6.1 The tangent (unit vector $\hat{\mathbf{X}}_1$), normal (unit vector $\hat{\mathbf{X}}_2$) and binormal (unit vector $\hat{\mathbf{X}}_3$) of a tortuous curve at a point P on it.

for the derivatives of the unit vectors with respect to the independent variables x_2 and x_3, which are the displacements along the principal normal and binormal respectively. Moreover, we find that

$$\text{div}\ \hat{\mathbf{X}}_1 = (\nabla \cdot \hat{\mathbf{X}}_1) = \left(\hat{\mathbf{X}}_j \cdot \frac{\partial \hat{\mathbf{X}}_1}{\partial x_j} \right) = R_2 + R_3 \tag{A6.11}$$

$$\text{curl}\ \hat{\mathbf{X}}_1 = (\nabla \times \hat{\mathbf{X}}_1) = \left[\hat{\mathbf{X}}_j \times \frac{\partial \hat{\mathbf{X}}_1}{\partial x_j} \right] = R_1 \hat{\mathbf{X}}_2 - (T_2 + T_3)\, \hat{\mathbf{X}}_1 \tag{A6.12}$$

APPENDIX 7 : PHYSICAL CONSTANTS

1. Boltzmann constant $K_B = 1{\cdot}38 \times 10^{-16}$ erg/^{o}K;

   Planck's constant $= \hbar = 6{\cdot}625 \times 10^{-34}$ joules sec.

2. **Electron :** mass = $9{\cdot}1 \times 10^{-28}$ gm; charge = $4{\cdot}8 \times 10^{-10}$ esu,

   Hydrogen atom : mass = $1{\cdot}67 \times 10^{-24}$ gm.

3. **Light :** Velocity in vacuum, c (= $2{\cdot}998 \times 10^{10}$ cm/sec), visible wavelength is between 4×10^{-5} cm and 7×10^{-5} cm; one Angstrom 1Å = 10^{-8} cm; one micron 1μ m = 10^{-4} cm, one nano meter = 10Å = 10^{-7} cm; Compton wave length $\lambda_c = \hbar/mc = 2{\cdot}43 \times 10^{-12}$ cm.

4. Avogadro number $A^o = 6{\cdot}0224 \times 10^{23}$ particles/moles, Loschmidt number $L = 2{\cdot}69 \times 10^{19}$ particles/cm^3.

5. Number density, at one m.m of mercury, and 273°K : $N_{1mm} = 3{\cdot}54 \times 10^{16}$ particles/cm^3.

6. Mean free path $1 = 10^{-5}$ cm at NTP; average distance between two charged particles $= 1/N^{1/3}$ (where N is the charge number density).

7. One Bar $= 10^8$ dynes/cm^2; one atmosphere = 1·0133 Bar, One Torr. = 1 mm of mercury = 1333·2 dynes/cm^2 = 10^6 dynes/cm^2.

8. One Calorie = 4·185 joules = 4·185 × 10^7 ergs; power of one erg/sec = 10^{-7} watt.

9. The gas kinetic cross section $\sigma = 3 \times 10^{-15}$ cm^2; Bohr radius $= (h/2\pi)(1/me^2) = 0{\cdot}53 \times 10^{-8}$ cm; cross section of the first Bohr orbit $\pi a^2 = 0{\cdot}88 \times 10^{-16}$ cm^2; classicial electron radius $r_0 = \alpha^2 a_0 = e^2/mc^2 = 0{\cdot}28 \times 10^{-12}$ cm; fine structure constant $\alpha = 2\pi e^2/\hbar c = 1/137{\cdot}03$; Thomson cross-section $\sigma_T = 8\pi r_0^2/3 = 6{\cdot}65 \times 10^{-25}$ cm^2, electric field in the first orbit of the hydrogen atom $= 5 \times 10^9$ V/cm; de Broglie wave length $\lambda_1 = (\hbar/p)$ cm where $\hbar$ is the Planck constant and p is the particle momentum.

10. One Coulomb = 2·998 × 10^9esu of charge; = 10^{-11} emu of charge; one volt per metre = 1/(2·998 × 10^4) esu of electric field intensity; one volt/cm = (10^{-3}/3) esu of electric field intensity; one electron volt = 1.6021 × 10^{-12} erg; one tesla = 10,000 gauss; velocity of a one 1 ev electron = 5.93 × 10^7 cm/sec; temperature corresponding to 1 ev energy = 11600°K.

11. Microsecond = 10^{-6} sec; nanosecond = 10^{-9} sec; pico second = 10^{-12} sec.

12. Solar radius = 7 × 10^{10} cm; distance between the photosphere and surface of the earth = 1.5 × 10^{13} cm; earth radius = 6370 km; earth mass = 6 × 10^{24} kgm; radiation pressure of sun light = 0.45 × 10^{-4} dynes/cm^2;

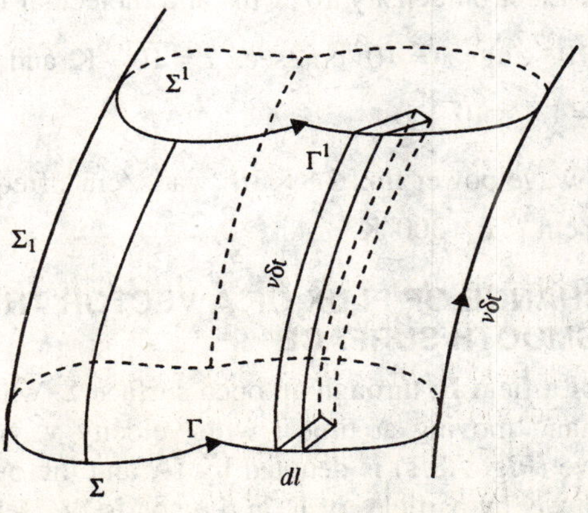

Fig. A8.1 Motion of a closed curve Γ and of the surface Σ swept away by it.

the force of this pressure for the whole earth = 5.7×10^7 kg. wt.; gravitational attraction of sun on earth 3.6×10^{21} kgm. wt/cm^3; velocity of earth round sun = 30 km/sec.

13. **Bohr magneton :** 9.27×10^{-24} J/T = 5.79×10^{-5} $2v/T$

1 radian = 57.3° = 0.159 rev.

Earth magnetic dipole moment = 8×10^{22} A . m^2

Radius of geocentries orbit = 42,000 km. Escape speed = 11.2 km/s

Earth-moon distance = 3.82×10^8 m, Earth–sun distance = 1.5×10^{11} m

Solar Corona : Linear dimensions $10^{11} - 10^{12}$ cm, $K_B T = 100$ ev

Density : $10^8 - 10^6$, $H = 20 - 0.02$ gauss. At normal temperature and density 3.6×10^{16} particles/cm^3 = gas press 1 mm Hg

When $H = 10^6$ gauss, velocity = 3×10^5 cm/sec, the excited electric field $E = 10$ esu = 3000 volt/cm

For Alfven waves $H = 10{,}000$ gauss

Solar field at surface and above is that of a dipole field with pole strength 25 gauss, the axis coinciding with the exis of rotation. Pressure of solar *EM* radiation = 4.5×10^{-5} dynes/cm^2. For the whole earth the force of this pressure = 5.7×10^7 kg wt. Compared to the gravitational attraction (3.6×10^{21} kgm wt/cm^3). Pressure free of *EM* radiation of earth is very small.

Alfven waves : $C_A = 100$ cm/sec, $P = 80$ gm/cm^3, $v = 65$ cm/sec

100 cm/sec

120 cm/sec

Ionosphere : Electron density 10^{11}/m^3 and molecular density

$= 10^{15} \approx 10^{22}/\text{m}^3$, $v = 10^3$ cm/sec, $T = 10^6$ °K and $N = 10^9$ mol/cm^3, $p = NK_B T = 1.4 \times 10^{-2}$

Flayer Radio wave power flux 2.8×10^2 watts/cm^2, frequency 10^9 cycles/sec, $N = 10^6$/cm^3, $T = 300$°K.

APPENDIX 8 : CHANGE OF FLUX OF A VECTOR THROUGH A SMOOTH SURFACE

Consider the flux of a field **B**, through an open surface Σ, which is bounded by a closed curve Γ, and moving at time t with velocity **v**. At time $t + \delta t$, the position of the curve (Fig. A8.1) is denoted by Γ', and the open surface by Σ'. The surface swept away by particles of Γ, in time δt, is Σ_1. Denoting the volume enclosed by Σ, Σ' and Σ_1 by τ, Green's theorem gives

$$\iiint_{\tau} (\nabla \cdot \mathbf{B})\, d\tau = \iint_{\Sigma+\Sigma'+\Sigma_1} (\mathbf{B} \cdot d\mathbf{\Sigma}) = \iint_{\Sigma+\Sigma'+\Sigma_1} B_n\, d\Sigma$$

where B_n is the normal component of **B** drawn outward to the volume τ. Since **B** is inward in Σ we write

$$\iiint_{\tau} (\nabla \cdot \mathbf{B})\, d\tau = \iint_{\Sigma'} (\mathbf{B} \cdot d\mathbf{\Sigma}) - \iint_{\Sigma} (\mathbf{B} \cdot d\mathbf{\Sigma}) + \iint_{\Sigma_1} (\mathbf{B} \cdot d\mathbf{\Sigma}) \quad \text{(A8.1)}$$

A line element dl of Γ sweeps away in time δt the length $\mathbf{v}\,\delta t$. The vector $d\mathbf{l}$ is such that in Σ_1 $d\mathbf{\Sigma} = (d\mathbf{l} \times \mathbf{v})\,\delta t$. Then, by Stoke's theorem, the last term on the right hand side of (A8.1) reduces to

$$\iint_{\Sigma_1} (\mathbf{B} \cdot d\mathbf{\Sigma}) = \oint_{\Gamma} (\mathbf{B} \cdot (d\mathbf{l} \times \mathbf{v}))\,\delta t = \delta t \oint_{\Gamma} ((\mathbf{v} \times \mathbf{B}) \cdot d\mathbf{l}) = \delta t$$

$$\iint_{\Sigma} (\nabla \times (\mathbf{v} \times \mathbf{B}) \cdot d\mathbf{\Sigma}) \quad \text{(A8.2)}$$

because the surface bounded by Γ is Σ and Σ_1 or Σ'. If F is the flux at time t of **B** (t) through $\bar{\Sigma}$, and $F + \delta F$ is the same at time $t + \delta t$ through Σ', then

$$\delta F = \iint_{\Sigma'} (\mathbf{B}\,(t + \delta t) \cdot d\mathbf{\Sigma}) - \iint_{\Sigma} (\mathbf{B}\,(t)\, d\mathbf{\Sigma}) \quad \text{(A8.3)}$$

Using (A8.1), at time $t + \delta\, t$, this integral becomes

$$\delta F = \iiint_{\tau\,(t+\delta t)} (\nabla \cdot \mathbf{B})\, d\tau - \iint_{\Sigma_1} (\mathbf{B}\,(t + \delta t) \cdot d\mathbf{\Sigma}) + \delta t \iint_{\Sigma} \left(\frac{\partial \mathbf{B}}{\partial t} \cdot d\mathbf{\Sigma} \right) \quad \text{(A8.4)}$$

In the first term $d\,\tau = \delta\, t\,(d\,\mathbf{\Sigma} \cdot \mathbf{v})$ (where $d\,\mathbf{\Sigma}$ is the surface element in Σ', when $\Sigma' \to \Sigma$, as $\delta\, t \to 0$). Utilizing, in the second term (A8.2), we get

$$\frac{dF}{dt} = \lim_{\delta t \to 0} \frac{\delta F}{\partial t} = \frac{d}{dt} \iint_{\Sigma} (\mathbf{B} \cdot d\mathbf{\Sigma}) = \iint_{\Sigma} \left[\left\{ \mathbf{v}\,(\nabla \cdot \mathbf{B}) + \frac{\partial \mathbf{B}}{\partial t} - \nabla \times (\mathbf{v} \times \mathbf{B}) \right\} \cdot d\mathbf{\Sigma} \right] \quad \text{(A8.5)}$$

APPENDIX 9 : DISCONTINUITY OF A FUNCTION AT A MOVING SURFACE

In Fig. A9.1, P is a point on S, the moving wave front surface, at time t, P_1 and P_2 are points near P on the opposite sides of S. The jump in the value of a field variable ψ as the current point moves from one side of S to the other, is

$$[\psi] = \lim_{\substack{P_1 \to P \\ P_2 \to P}} [\psi\,(P_2) - \psi\,(P_1)],$$

If $[\psi] \neq 0$, then ψ has a strong discontinuity across S. So, ψ has formed a shock front. If S is a surface of weak discontinuity of ψ, then, at least one of its first derivatives must be discontinuous. The flux $\mathbf{A}$ and the source Q of production of the quantity φ, satisfy the condition for conservation

$$\frac{\partial \varphi}{\partial t} + (\nabla \cdot \mathbf{A}) = Q \tag{A9.1}$$

When Q is the divergence of a vector $\mathbf{B}$, we write

$$\frac{\partial \varphi}{\partial t} + \nabla \cdot \mathbf{A}' = 0, \ \mathbf{A}' = \mathbf{A} - \mathbf{B} \tag{A9.2}$$

All terms of (A9.1) or (A9.2) are integrated in a volume τ which is cut-off by the wave front S as in Fig. A9.1. The total amount P of the function φ in τ, at time t,

$$P = \iiint_{\tau} \varphi \, d\tau \tag{A9.3}$$

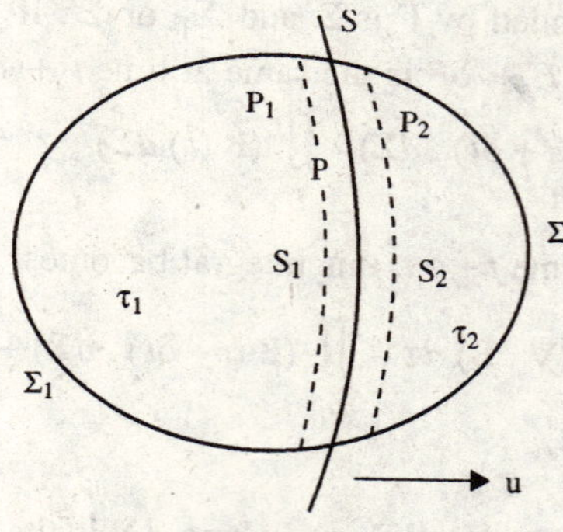

Fig. A9.1 Wave front surfaces dividing a volume τ into two parts τ_1 and τ_2.

where $d\tau$ is a volume element in τ. At time $t + \delta t$, retaining terms up to first order of δt, we get

$$\varphi(t + \delta t) = \varphi(t) + \frac{\partial \varphi}{\partial t} \delta t \tag{A9.4}$$

The surface element $d\mathbf{\Sigma}$, on τ, at the time t, moving with velocity $\mathbf{u}$, sweeps off the volume element $(\mathbf{u} \cdot d\mathbf{\Sigma})\, \delta t$ in time δt. So,

$$P + \delta P = \iiint_{\tau} \left\{ \varphi + \frac{\partial \varphi}{\partial t} \delta t \right\} d\tau + \iint_{\Sigma} (\mathbf{u} \cdot d\mathbf{\Sigma})\, \varphi \, \delta t \tag{A9.5}$$

As $\delta t \to 0$, using Gauss's divergence theorem we find that

$$\frac{dP}{dt} = \frac{d}{dt} \iiint_{\tau} \varphi \, d\tau = \iiint_{\tau} \frac{\partial \varphi}{\partial t} d\tau + \iint_{\Sigma} (\mathbf{u} \cdot d\mathbf{\Sigma})\, \varphi$$

$$= \iiint_{\tau} \left\{ \frac{\partial \varphi}{\partial t} + \nabla . (\varphi \mathbf{u}) \right\} d\tau \qquad \text{(A9.6)}$$

Hence, we obtain

$$\frac{d}{dt} \iiint_{\tau} \varphi \, d\tau = \iint_{\Sigma} \{\varphi (\mathbf{u} \, . \, d\mathbf{\Sigma}) - (\mathbf{A} \, . \, d\mathbf{\Sigma})\} + \iiint_{\tau} Q \, d\tau \qquad \text{(A9.7)}$$

The discontinuity surface S, of Fig. A9.1, within the volume τ, divides it into two parts, τ_1 and τ_2. So S divides the total surface into the surfaces $\Sigma_1 + S_1$ and $\Sigma_2 + S_2$, and the two sides of the thin shell are S_1 and S_2. Equation (A9.6) in the portion τ_1, of volume τ, is

$$\frac{d}{dt} \iiint_{\tau_1} \varphi \, d\tau = \iint_{\Sigma_1 + S_1} ((\varphi \mathbf{u} - \mathbf{A}) \, . \, d\mathbf{\Sigma}) + \iiint_{\tau_1} Q d\tau$$

Similarly, in the portion τ_2,

$$\frac{d}{dt} \iiint_{\tau_2} \varphi d\tau = \iint_{\Sigma_2 + \Sigma_2} ((\varphi \mathbf{u} - \mathbf{A}) \, . \, d\mathbf{\Sigma}) + \iiint_{\tau_2} Q d\tau$$

Since $\iiint_{\tau} \varphi d\tau = \iiint_{\tau_1} \varphi d\tau + \iiint_{\tau_2} \varphi d\tau$, these two equations and (A3.7) give

$$0 = \iint_{S_1} ((\mathbf{u} \, . \, d\mathbf{\Sigma}) \, \varphi - (\mathbf{A} \, . \, d\mathbf{\Sigma})) + \iint_{S_2} ((\mathbf{u} \, . \, d\mathbf{\Sigma}) \, \varphi - (\mathbf{A} \, . \, d\mathbf{\Sigma})) \qquad \text{(A9.8)}$$

The unit vector normally outward from τ_1 is $\mathbf{n}$. At the integration point of S_1, then, $d\mathbf{\Sigma}_1 = \mathbf{n} d\Sigma = - d\mathbf{\Sigma}_2$. Also S being an arbitrary part of the wave front surface, equation (A9.8) reduces to $[(\mathbf{u} \, . \, \mathbf{n}) \, \varphi - (\mathbf{A} \, . \, \mathbf{n})]_1^2 = 0$. It is valid at every point of S in τ, and can be written as

$$[\varphi \, u_n - A_n] = 0 \qquad \text{(A9.9)}$$

APPENDIX 10 : TWO-BODY ELASTIC COLLISIONS

(a) Dynamics of two-body Coulomb collisions

Consider two spherical masses, A_i and A_j, of given radii, Figs. A10.1 (*a*) and A10.1 (*b*). The test particle A_1, moving with velocity $\mathbf{u}_i$, collides with the field particle A_j, which is moving with velocity $\mathbf{u}_j$. The masses and charges of these two particle are m_i, m_j and e_i, e_j, and the forces acting on them are $\mathbf{F}_{ij}$ and $\mathbf{F}_{ji}$, respectively. Figure A10.1 (*a*) depicts the attractive scattering of A_i by A_j. For repulsive forces, the position of the scattering centre, A_j', is at the focus on the convex side of the trajectory. Figure A10.1 (*b*) depicts the mutual repelling of

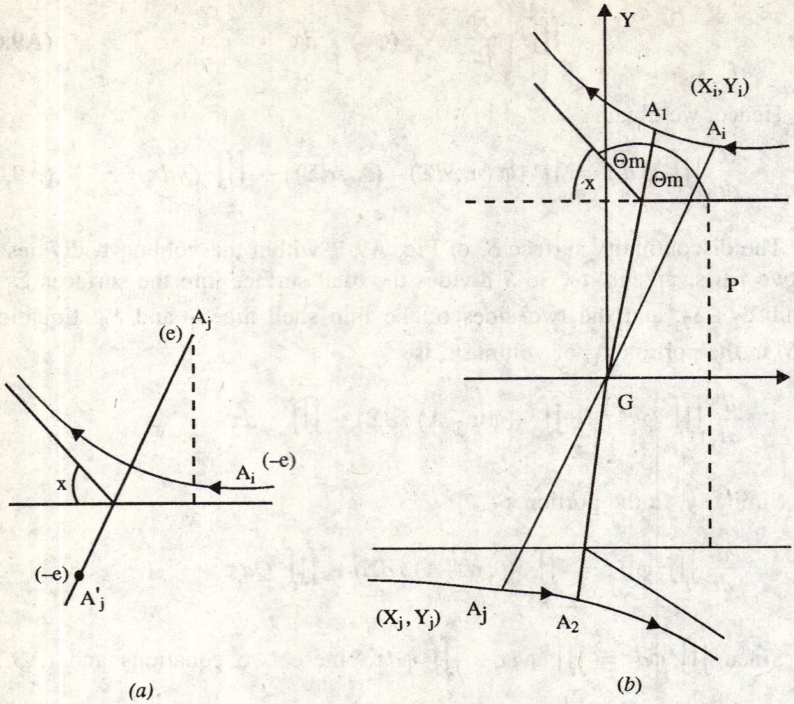

Fig. A10.1 Geometry of a Coulomb collision between two particles. In Fig. A10.1 (*b*) $A_1 A_2 = r_m$ and θ_m is the angle between $A_1 A_2$ and any of the asymptotic directions.

two identical particles. Consider the centre of mass coordinate system (or the *c* system), and the laboratory system (or the *L* system), which moves with a given velocity. The position vectors of A_i and A_j, about the origin O of the L system, are denoted by $\mathbf{r}_i$ and $\mathbf{r}_j$, respectively. The instantaneous centre of mass in this system is G, and $\mathbf{r}_G$ is its position vector, so

$$\mathbf{r}_G = \frac{m_i r_i + m_j r_j}{m_i + m_j} \tag{A10.1}$$

Equations of motion of A_i and A_j are

$$m_i \ddot{\mathbf{r}}_i = \mathbf{F}_{ij}, m_j \ddot{\mathbf{r}}_j = \mathbf{F}_{ji} \tag{A10.2}$$

where $\dot{\mathbf{r}}_i = \mathbf{u}_i = \dfrac{d\,\mathbf{r}_i}{dt}$, etc. By Newton's third law of motion $E_{ij} = -\mathbf{F}_{ji}$; so $\ddot{\mathbf{r}}_G = 0$, giving

$$\dot{\mathbf{r}}_G = \text{constant} = \mathbf{v}_G \tag{A10.3}$$

$\mathbf{F}_{ij}$ is a central conservative force, with respect to either A_i or A_j. If $\mathbf{A}_i\,\mathbf{A}_j = \mathbf{r} = \mathbf{r}_j - \mathbf{r}_i$, then

$$\ddot{\mathbf{r}} = -\left(\frac{1}{m_i} + \frac{1}{m_j}\right)\mathbf{F}_{ij}, \tag{A10.4}$$

and the equation of motion of the reduced mass m, defined by $1/m = 1/m_i + 1/m_j$, is

$$m\ddot{\mathbf{r}} = -\mathbf{F} = \frac{|e_i\, e_j|}{r^3}\mathbf{r}$$

where $\mathbf{F}$ is the $\mathbf{F}_{ij}$ of (A10.4). Since $(\mathbf{r} \times \ddot{\mathbf{r}}) = 0$, integration gives the conservation condition for angular momentum;

$$(\mathbf{r} \times \dot{\mathbf{r}}) = \text{constant} = \hat{\mathbf{K}}h \tag{A10.5}$$

where $\hat{\mathbf{K}}$ is a fixed unit vector perpendicular to the plane of $\mathbf{r}$ and $\dot{\mathbf{r}}$. Hence, motion of A_j, relative to A_i, in the collision plane (which is the plane containing the centre of mass, and the initial directions of their relative motions), is perpendicular to the fixed direction of $\mathbf{K}$. Relative to the centre of mass, the motion follows similar relations. This plane moves parallel to itself with velocity $\mathbf{v}_G$. Since $\dot{\mathbf{r}}_i = \mathbf{u}_i$ and $\dot{\mathbf{r}}_j = \mathbf{u}_j$, denoting these velocities at time t' by u_i' and u_j', we get

$$(m_i + m_j)\,\mathbf{v}_G = m_i\,\mathbf{u}_i + m_j\,\mathbf{u}_j = m_i\,\mathbf{u}_i' + m_j\,\mathbf{u}_j' \tag{A10.6}$$

If $\mathbf{G}\,A_i = \mathbf{R}_i$, $G\,\mathbf{A}_j = \mathbf{R}_j$, with the help of (A10.1) we obtain

$$(\mathbf{R}_i,\ \mathbf{R}_j) = \frac{(-m_j, m_i)\,\mathbf{r}}{m_i + m_j} \tag{A10.7}$$

If GX is parallel to the initial asymptotic directions of A_i or A_j, GY is perpendicular to GX in the collision plane and the components of $\mathbf{R}_i$ are (x_i, y_i), etc., then from (A10.7) we obtain

$$(x_i, y_i) = -\frac{rm_j\,(\cos\theta, \sin\theta)}{m_i + m_j}, \text{ etc.} \tag{A10.8}$$

The minimum distance of $A_i A_j$, Fig. A10.16, is $A_1 A_2\ (= r_m)$. θ_m is the angle of inclination of $A_1 A_2$ to GX, χ is the angle of deviation of the path of the scattered particle due to a collision, and

$$\chi = \pi - 2\theta_m \tag{A10.9}$$

$\chi = 0$ when there is no force of interaction.

(b) Some conservation conditions

Since $\mathbf{F}$ is conservative, we write

$$\mathbf{F} = -\nabla\varphi,\ (\mathbf{F}\,.\,d\mathbf{r}) = d\,(mv^2/2) = -\,d\varphi,$$

then integration gives

$$\frac{1}{2}mv^2 + \varphi(\mathbf{r}) = \frac{1}{2}mv_0^2 = \frac{1}{2}mv_0'^2 \tag{A10.10}$$

where $\mathbf{v}_0$ and $\mathbf{v}_0'$ are the velocities of the reduced mass, much before and much after collision, that is, at $r = \pm\infty$, because $\varphi(\pm\infty) = 0$. Equation (A10.5) now becomes

$$h = r^2\theta^{\circ} = v_0 p, \tag{A10.11}$$

where the collision impact parameter (the minimum impact parameter) p is the perpendicular distance between the initial directions of A_i and A_j. When A_i and A_j attract each other, $r_m < p$, when they repel each other $r_m > p$. If p is infinite, the path of A_i is not changed by A_j. Eliminating $\dot{\theta}$ between (A10.10) and (A10.11), we get

$$\dot{r} = -v_0\left\{1 - \frac{p^2}{r^2} - \frac{2\varphi(r)}{mv_0^2}\right\}^{1/2} \tag{A10.12}$$

where the minus sign is chosen in the right hand side, because $\dot{r}$ is negative at $r = \infty$. When t is eliminated between (A10.11) and (A10.12) we get

$$\frac{dr}{d\theta} = -\frac{r^2}{p}\left\{1 - \frac{p^2}{r^2} - \frac{2\varphi(r)}{mv_0^2}\right\}^{1/2} \tag{A10.13}$$

The extreme values r_m of r are the roots of the equation

$$1 - \frac{p^2}{r_m^2} - \frac{2\varphi(r_m)}{mv_0^2} = 0 \tag{A10.14}$$

The angle θ_m, for the real positive value of r_m, is given by

$$\theta_m = p\int_{r_m}^{\infty}\frac{dr}{r^2\left\{1 - \frac{p^2}{r^2} - \frac{2\varphi(r)}{mv_0^2}\right\}^{1/2}} \tag{A10.15}$$

because $\theta = 0$ when $r = \infty$. After finding θ_m in terms of p, v_0, m and other constants, x is determined with the help of (A10.9).

In the centre of mass system, with the help of (A10.10), we get

$$m_i v_i^2 + m_j v_j^2 = m_i v_i'^2 + m_j v_j'^2 \tag{A10.16}$$

where $\mathbf{v}_i$, $\mathbf{v}_j$ are the velocities much before and $\mathbf{v}_i'$, $\mathbf{v}_j'$ are those much after the collision. Similarly, in the L system, we have

$$m_i u_i^2 + m_j u_j^2 = m_i u_i'^2 + m_j u_j'^2 \tag{A10.17}$$

because $$\Sigma m_i u_i^2 = (\Sigma m_i v_i^2) + (m_i + m_j) v_G^2 \tag{A10.18}$$

Then, with the help of (A10.3) we obtain

$$\mathbf{v}_G = \mathbf{v}_G' \tag{A10.19}$$

$$m_i \mathbf{u}_i + m_j \mathbf{u}_j = m_i \mathbf{u}_i' + m_j \mathbf{u}_i' = (m_i + m_j) \mathbf{v}_G \tag{A10.20}$$

If $\mathbf{g}$ and $\mathbf{g}'$ be the relative velocities of A_i, with respect to A_j, before and after a collision, then

$$\mathbf{g} = \mathbf{u}_i - \mathbf{u}_j = \mathbf{v}_i - \mathbf{v}_j, \mathbf{g}' = \mathbf{u}_i' - \mathbf{u}_j' = \mathbf{v}_i' - \mathbf{v}_j' \tag{A10.21}$$

From the definition of $\mathbf{g}$ and $\mathbf{g}'$ it is clear that

$$(\mathbf{g} \cdot \mathbf{g}') = gg' \cos \chi \tag{A10.22}$$

(c) Relations between conditions before and after a collision

Equations (A10.19) and (A10.21) give

$$\frac{1}{2} \{(m_i + m_j) v_G^2 + mg^2\} = \frac{1}{2} (m_i u_i^2 + m_j u_j^2), \tag{A10.23}$$

$$\frac{1}{2} \{(m_i + m_j) v_G'^2 + mg'^2\} = \frac{1}{2} (m_i u_i'^2 + m_j u_j'^2) \tag{A10.24}$$

These relations and (A10.17) show that

$$g^2 = g'^2 \tag{A10.25}$$

Collisions, therefore, do not change the magnitude of the velocities, only directions are altered.

In terms of $\mathbf{v}_G$, $\mathbf{g}$, $\mathbf{g}'$ the expressions for $\mathbf{u}_i$, $\mathbf{u}_j$, $\mathbf{u}_i'$, $\mathbf{u}_j'$ are

$$\mathbf{u}_i = \mathbf{v}_G + \frac{m_i \mathbf{g}}{m_i + m_j}, \quad u_j = \mathbf{v}_G - \frac{m_i \mathbf{g}}{m_i + m_j}, \tag{A10.26}$$

$$\mathbf{u}_i' = \mathbf{v}_G + \frac{m_i \mathbf{g}'}{m_i + m_j}, \quad \mathbf{u}_j' = \mathbf{v}_G - \frac{m_i \mathbf{g}'}{m_i + m_j}$$

The kinetic energies W_i and W_i' of A_i, relative to A_j, before and after a collision, are

$$W_i = \frac{1}{2} m_i (\mathbf{u}_i - \mathbf{u}_j)^2, = \frac{1}{2} m_i g^2, \tag{A10.27}$$

$$W_i' = \frac{1}{2} m_i (\mathbf{u}_i' - \mathbf{u}_j')^2, \tag{A10.28}$$

$$= \frac{m_i g^2 (m_i^2 + m_j^2 + 2m_i m_j \cos \chi)}{2 (m_i + m_j)^2} \tag{A10.29}$$

Hence we find that

$$W_i - W_i' = \frac{2m_i m_j W_i (1-\cos\chi)}{(m_i+m_j)^2} \tag{A10.30}$$

$$= \frac{2m_i}{m_j} W_i (1-\cos\chi) \text{ when } m_j >> m_i \tag{A10.31}$$

Equations (A10.19) and (A10.25), and Fig. A10.2, show that

$$\mathbf{g}' = \mathbf{g} - \alpha\mathbf{e}, \quad \mathbf{g} = \mathbf{g}' + \alpha\mathbf{e} \tag{A10.32}$$

where **e** is the unit vector along the base AB of the isosceles triangle OAB in the sense shown in Fig. A10.2 and α is a constant. Since $O\mathbf{A} = \mathbf{g}'$ and $O\mathbf{B} = \mathbf{g}$, we find that

$$\alpha = |AB| = 2\,|g| \sin\left(\frac{\chi}{2}\right) = 2\,(\mathbf{e}\,.\,\mathbf{g}) \tag{A10.33}$$

$$(\mathbf{e}\,.\,\mathbf{g}) = |g| \sin\frac{\chi}{2}, \quad \mathbf{g} - \mathbf{g}' = 2\mathbf{e}\,(\mathbf{e}\,.\,\mathbf{g}),\ g_i - g_i' = 2e_i\,(e_j g_j) \tag{A10.34}$$

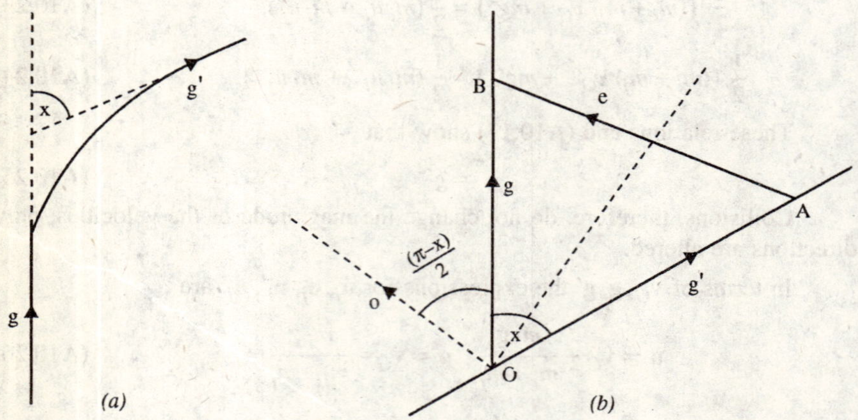

Fig. A10.2 Relative velocities of a scattered particle before and after a collision.

Putting $M_i = m_i/(m_i + m_j)$, in (A10.26) we get

$$\mathbf{v}_i' = \mathbf{v}_i + M_j\,(\mathbf{g}' - \mathbf{g}) = \mathbf{v}_i - 2M_j\,\mathbf{e}\,(\mathbf{e}\,.\,\mathbf{g})$$

Squaring this relation gives

$$v_i'^2 = v_i^2 - 4M_j\,(\mathbf{e}\,.\,\mathbf{v}_j)\,(\mathbf{e}\,.\,\mathbf{g}) - 4M_i M_j\,(\mathbf{e}\,.\,\mathbf{g})^2 \tag{A10.35}$$

Putting $\mathbf{v}_i = \mathbf{v}_j + \mathbf{g}$ in the middle term in the right hand side of (A10.35), and using (A10.21) we obtain

$$v_i'^2 = v_i^2 - 4M_j\,(\mathbf{e}\,.\,\mathbf{v}_j) - 4M_i M_j\,(\mathbf{e}\,.\,\mathbf{g})^2 \tag{A10.36}$$

If we put $v_i' = v_i + \delta v$ we can write this equation as

$$(\delta v)^2 + 2v_i\, \delta v + 4M_j\, (\mathbf{e}\, .\, \mathbf{v}_j)\, (\mathbf{e}\, .\, \mathbf{g}) + 4M_i\, M_j\, (\mathbf{e}\, .\, \mathbf{g})^2 = 0 \qquad \text{(A10.37)}$$

The heavier scattering mass being M_j, $|\mathbf{v}_j|$ is much smaller than v_i. Neglecting, therefore, v_j we get

$$\delta v \approx -\frac{2m_i}{m_j\, v_j}\, (\mathbf{e}\, .\, \mathbf{g})^2 \qquad \text{(A10.38)}$$

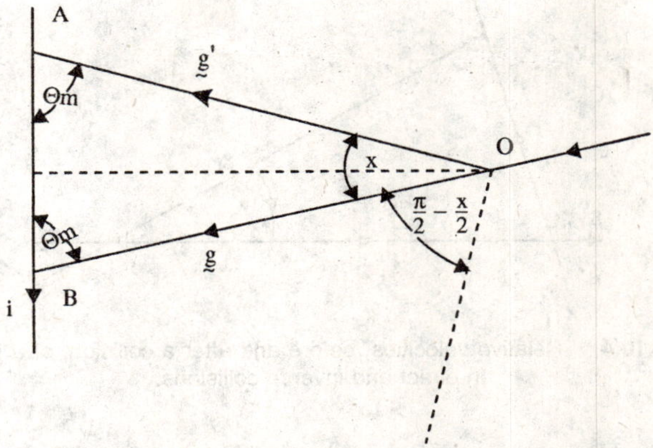

Fig. A10.3 The geometrical representation of the velocities **g**, **g′** and the unit vector **i**, as the sides of the isosceles triangle *OAB*.

Again since $\mathbf{g} = \mathbf{u}_i - \mathbf{u}_j$, for the same reason, we get

$$g \approx v_i - \frac{(\mathbf{v}_i\, .\, \mathbf{v}_j)}{v_i} = v_i - v_j \cos \theta \qquad \text{(A10.39)}$$

where θ is the angle between $\mathbf{v}_i$ and $\mathbf{v}_j$. Then since χ is a function of g, expanding it in powers of $v_j \cos \theta$, we get

$$\chi = \chi_0 + d\,\chi = \chi\, (v_i - v_j \cos \theta) = \chi_0\, (v_i) - v_j \cos \left(\frac{\partial\, \chi_0}{\partial\, v_i}\right) \qquad \text{(A10.40)}$$

Since $d\,\chi$ is small, using (A10.33) and (A10.39) we obtain

$$(\mathbf{e}\, .\, \mathbf{g}) = g \sin \frac{\chi}{2} = (v_i - v_j \cos \theta) \sin \frac{\chi}{2} \qquad \text{(A10.41)}$$

Putting this value of $(\mathbf{e}\, .\, \mathbf{g})$ in (A10.38) we find that

$$\delta v \approx -\left(\frac{2m_i}{m_j\, v_i}\right)\left(v_i \sin \frac{\chi}{2} - v_j \cos \theta \sin \frac{\chi}{2}\right)^2 = -\left(\frac{2m_i}{m_j}\right) v_i \sin^2 \frac{\chi_0}{2}$$

$$\text{(A10.42)}$$

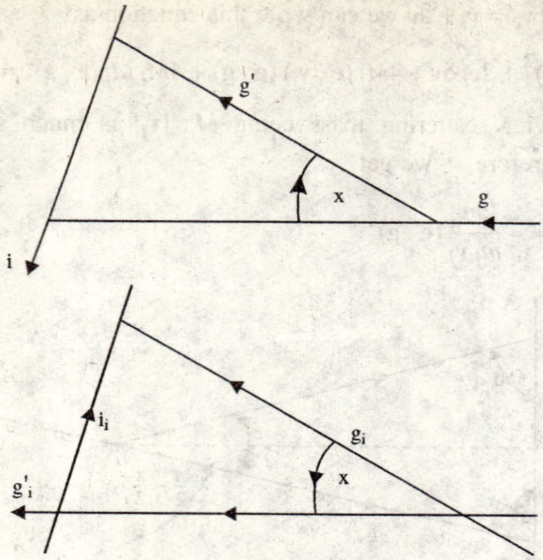

Fig. A10.4 Relative velocities, before and after a collision, of a particle in direct and inverse collisions.

where only the largest terms have been retained and m_i/m_j and $v_j \cos\theta \sin\frac{\chi}{2}$ are regarded as small quantities. Hence we find that

$$\delta v \approx -\left(\frac{m_i}{m_j}\right) v_i (1 - \cos\chi_0) \qquad \text{(A10.43)}$$

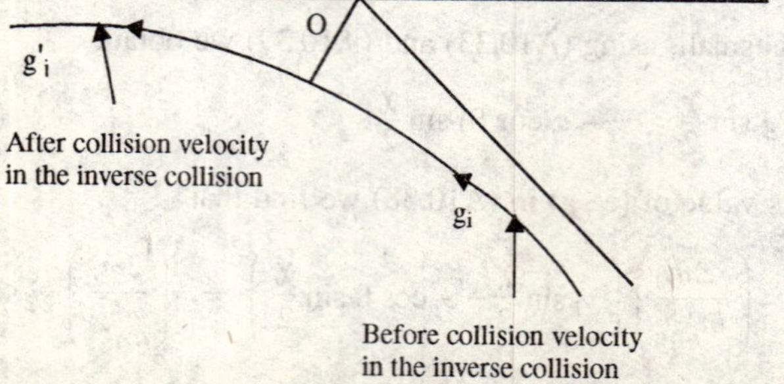

Fig. A10.5 Relative speeds of a particle in direct and inverse collisions.

(d) Inverse collisions

Both in inverse and direct collisions masses of two interacting particles M_i and M_j, the impact parameter P and the deflection angle χ remain unchanged. Then in inverse collisions the problem is to determine the after-collision velocities $\mathbf{u}_i$ and $\mathbf{u}_j$ in terms of the velocities before the collisions. The before-collision velocities are $\mathbf{u}_i'$ and $\mathbf{u}_j'$; $\mathbf{u}_i'$ and $\mathbf{u}_j'$ are identical to $\mathbf{u}_i$ and $\mathbf{u}_j$ in direct collisions. By equations (A10.19) and (A10.25), we can write

$$\mathbf{g}' = \mathbf{g} - \mathbf{i}\alpha, \quad \mathbf{g} = \mathbf{g}' - \mathbf{i}\alpha' \tag{A10.44}$$

where $\mathbf{i}$ is the unit vector along the base AB of the isoscles triangle of Fig. A10.3. The equal sides are $O\mathbf{A} = \mathbf{g}'$, $O\mathbf{B} = \mathbf{g}$ and

$$\alpha = |AB| = 2\,|g|\sin\frac{\chi}{2} \tag{A10.45}$$

Hence
$$\mathbf{i} = \frac{\mathbf{g} - \mathbf{g}'}{2\,|g|\sin\left(\dfrac{\chi}{2}\right)}, \tag{A10.46}$$

$$(\mathbf{g}\,.\,\mathbf{i}) = -(\mathbf{g}'\,.\,\mathbf{i}) = g\sin\frac{\chi}{2} = g\cos\theta_m, \tag{A10.47}$$

$$\mathbf{g}' = \mathbf{g} - 2\mathbf{i}\,|g|\cos\theta_m = \mathbf{g} - 2\mathbf{i}\,(\mathbf{g}\,.\,\mathbf{i}) \tag{A10.48}$$

Similarly
$$\mathbf{g} = \mathbf{g}' - 2\mathbf{i}\,(\mathbf{g}'\,.\,\mathbf{i}) \tag{A10.49}$$

For inverse collisions

$$g_i'\sin\left(\frac{\chi_i}{2}\right) = (\mathbf{g}_i\,.\,\mathbf{i}_i) \tag{A10.50}$$

where $\mathbf{i}_i$ is the unit vector corresponding to the unit vector $\mathbf{i}$ of direct collisions. Again, by definition of inverse collisions,

$$\chi_i = \chi, \quad \mathbf{g}_i = \mathbf{g}, \quad \mathbf{g}_i' = \mathbf{g}_i - 2\mathbf{i}_i\,(\mathbf{g}_i\,.\,\mathbf{i}_i) \tag{A10.51}$$

Using, therefore, double prime to denote the after-collision quantities, in an inverse collision, we obtain

$$\mathbf{g}_i' = \mathbf{g}'' \tag{A10.52}$$

and $\mathbf{u}_i''$, $\mathbf{u}_j''$, $\mathbf{v}_i''$, $\mathbf{v}_j''$ are the after-collision velocities in the laboratory frame and the centre-of-mass frame, respectively. The unit vector $\mathbf{i}$ is drawn from the tip of $\mathbf{g}'/|\mathbf{g}'|$, in the direction of the tip of $\mathbf{g}/|\mathbf{g}|$, about the origin O of Fig. A10.3. Similarly, the unit vector $\mathbf{i}_i$ is drawn from the tip of $\mathbf{g}_i'/|\mathbf{g}_i'|$ towards the tip of $\mathbf{g}_i/|\mathbf{g}_i|$, and in the sense of $\mathbf{g}'$. Hence, as in Fig. A10.4,

$$\mathbf{i}_i = -i; \tag{A10.53}$$

and vertices of trajectories, in direct and inverse collisions, lie on opposite sides of the centre of force, as in Fig. A10.5; so, by (A10.51) and (A10.52), we get

$$\mathbf{g}_i' = \mathbf{g}'' = \mathbf{g}_i - 2\,(\mathbf{g}_i \,.\, \mathbf{i}_i) = \mathbf{g}' - 2\mathbf{i}\,(\mathbf{g}' \,.\, \mathbf{i}) \tag{A10.54}$$

Then (A10.49) and (A10.20) give

$$\mathbf{g}_i' = \mathbf{g}'' = \mathbf{g},\ \mathbf{v}_G = \mathbf{v}_G' = \mathbf{v}_G'' \tag{A10.55}$$

Following (A10.21) we find that

$$\mathbf{u}_i'' = \mathbf{v}_G'' + \frac{m_j\,\mathbf{g}''}{m_i + m_j},\ \mathbf{u}_j'' = \mathbf{v}_G'' - \frac{m_i\,\mathbf{g}''}{m_i + m_j}, \tag{A10.56}$$

so

$$\mathbf{u}_i'' = \mathbf{u}_i,\ \mathbf{u}_j'' = \mathbf{u}_j \tag{A10.57}$$

The Jacobian J of transformation, from velocities after collision to those before, and for the corresponding inverse collision, are same. So

$$d\mathbf{u}_i'\,d\mathbf{u}_j' = |J|\,d\mathbf{u}_i\,d\mathbf{u}_j = |J|\,|J'|\,d\mathbf{u}_i'\,d\mathbf{u}_j' \tag{A10.58}$$

Hence $|J|\ |J'| = 1$. From symmetry of equations (A10.26), or since a negative Jacobian corresponds to an improper transformation, we obtain

$$|J|\ |J'| = |J|^2 = 1 \text{ or } |J| = 1 \tag{A10.59}$$

(e) Collisional invariants

Since elastic collisions do not involve any change of mass,

$$m_i + m_j = m_i' + m_j', \tag{A10.60}$$

where masses m_i and m_j before collision become m_i' and m_j' after collision. Moreover, by (A10.6) and (A10.17) we have

$$m_i\,\mathbf{u}_i + m_j\,\mathbf{u}_j = m_i\,\mathbf{u}_i' + m_j\,\mathbf{u}_j', \tag{A10.61}$$

$$\frac{1}{2}\,m_i\,u_i^2 + \frac{1}{2}\,m_j\,u_j^2 = \frac{1}{2}\,m_i\,u_i'^2 + \frac{1}{2}\,m_j\,u_j'^2 \tag{A10.62}$$

The number of quantities conserved, is five. These are mass, the three components of momentum, and kinetic energy of two colliding particles, these are called summational invariants or collisional invariants. Schematically these conservation conditions in two-body elastic collisions are represented as

$$m_i\,(\mathbf{u}_i) + m_j\,(\mathbf{u}_j) \rightarrow m_i\,(\mathbf{u}_i') + m_j\,(\mathbf{u}_j') \tag{A10.63}$$

Conservation of any other quantity for elastic collisions depends on these invariants.

APPENDIX 11 : COLLISION CROSS-SECTION

In a frame fixed with the particle A_j (Appendix 10) the target particle is at rest. The uniform flux, per unit area per unit time of particles of type i, having per particle the mass m_i and kinetic energy $m_i\,v_i^2/2$, is incident, with relative velocity $\mathbf{v}_0$, on the centre (target) A_j, of their attraction (or repulsion). So, the scattered part dN of type i, per unit time, Fig. A11.1, lying within the solid angle

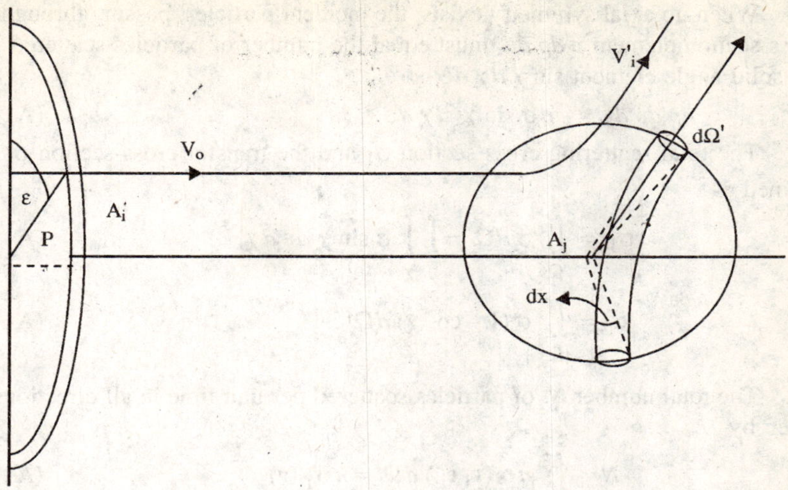

Fig. A11.1 Scattering cross-section of an incident beam by the scattering centre A_j.

element $d\Omega'$, about the direction of the unit vector $\mathbf{e}_i'$ $(= \mathbf{v}_i'/|\mathbf{v}_i'|)$, will be scattered ultimately to final velocities lying between $\mathbf{v}_i'$ and $\mathbf{v}_i' + d\mathbf{v}_i'$. Since dN is proportional to the incident flux n and $d\Omega'$, we can write

$$dN = n d\Omega' \, \sigma(\chi, v) \tag{A11.1}$$

where the constant of proportionality, σ, has the dimension of area; it is called the effective scattering cross-section or the differential scattering correction of the test particle A_i in the coordinate system fixed with the scatterer. dN depends on the magnitude of $\mathbf{v}_i'$ and the direction of $\mathbf{e}_i'$ and $\sigma d\Omega'$ is the ratio of the number of particles scattered into the solid angle $d\Omega'$ per unit time, to the incident intensity. Since the scattered directions of dN have polar angles between χ and $\chi + d\chi$, for axial symmetry, we obtain.

$$d\Omega' = 2\pi \sin \chi \, d\chi \tag{A11.2}$$

So equating the number of particles scattered into the solid angle $d\Omega'$ with the number of incident particles passing through the cross sectional elementary area lying between circles of radii ρ and $\rho + d\rho$, we obtain

$$|n \, . \, 2\pi p \, dp| = |n \sigma d\Omega'|$$

The impact parameter p is generally a single valued function of $d\Omega'$. As ρ increases the intensity of the force of interaction decreases and χ decreases. Hence $dp/d\chi$ should be negative and σ is a function of χ. So

$$n \, . \, 2\pi p \, dp = -n\sigma \, d\Omega' \tag{A11.3}$$

or

$$\sigma = -\left(\frac{p(\chi)}{\sin \chi}\right) \frac{dp}{d\chi} \tag{A11.4}$$

When no axial symmetry exists, the incident particles, passing through the cross-section element $p\,dp\,d\varepsilon$, must equal the number of particles scattered into the solid angle element $\sin\chi\,d\chi\,d\varepsilon$; so

$$np\,dp\,d\varepsilon = -n\sigma\sin\chi\;d\chi\,d\varepsilon \tag{A11.5}$$

The total scattering cross section σ_1 and the transfer cross-section σ_2 are defined as

$$\sigma_1 = \iint_\Omega \sigma\,d\Omega' = \int_\varepsilon\int_\chi \sigma\sin\chi\,d\varepsilon\,d\chi, \tag{A11.6}$$

$$\sigma_2 = \iint_\Omega \sigma(1-\cos\chi)\,d\Omega' \tag{A11.7}$$

The total number N, of particles scattered per unit time in all directions, is given by

$$N = \iint_\Omega n\sigma(\mathbf{v}, \mathbf{e}_i')\,d\Omega' = n\sigma_1(v) \tag{A11.8}$$

Since a deflection of the incident particles through the angle χ means the loss of momentum of amount, $mv(1-\cos\chi)$, σ_2 is a measure of the total momentum loss due to scattering.

Exercise A11.1. Find the cross-section for collisions between two hard spheres of radii r_1 and r_2 when the potential is

$$\varphi(r) = 0,\; r > (r_1 + r_2)$$
$$= \infty\; r < (r_1 + r_2) \tag{A11.9}$$

Answer. When the scattering cross-section is greater than $\pi(r_1 + r_2)^2$, the scattering angle $\chi = 0$. The sphere of interaction of the particles of radii r_1 and r_2, is the sphere of radius $r_1 + r_2$, Fig. A11.2. Hence

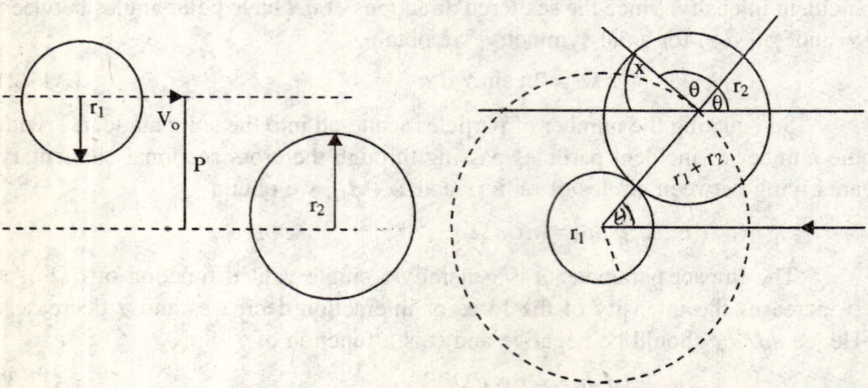

Fig. A11.2 Elastic collision between two hard spheres of radii r_1 and r_2. In (*a*) the interaction potential is zero, and in (*b*) it is infinite.

$$\left.\begin{aligned} \chi &= \pi - 2\,\theta,\ p = (r_1 + r_2)\sin\theta,\ \ \sigma(\chi) = \frac{1}{4}(r_1 + r_2)^2, \\ \frac{dp}{d\chi} &= -\frac{1}{2}(r_1 + r_2)\sin\left(\frac{\chi}{2}\right), \end{aligned}\right\} \qquad \text{(A11.10)}$$

σ is independent of χ, and the scattering is isotropic.

APPENDIX 12 : COLLISION PROBABILITY

(a) Analytical treatment

A beam of particles of number density n and velocity **v**, in presence of uniformly distributed immobile scatterers of number density N is deflected from its original path. The number dn of beam particles, scattered parallel to the X-axis, which is perpendicular to the direction of **v**, from a fixed origin, is proportional to N, n and dx. Hence.

$$dn = -\sigma Nn\,dx \qquad \text{(A12.1)}$$

where σ is the constant scattering cross-section, Appendix 11. The negative sign explains the fact that dn decreases as x increases. Integrating we get

$$n = n_0 \exp(-N\sigma x) \qquad \text{(A12.2)}$$

where n_0 is the number density of the beam at $x = 0$. Hence, the probability that a particle does not suffer any collision, in a distance x, is $\exp(-N\sigma x)$. Since the scatterers are not much disturbed by the scattered particles, the condition $n \ll N$ is implied in (A12.1). If dx in (A12.1) is replaced by $v\ dt$, and the right hand side is integrated with respect to t, then

$$n = n_0 \exp(-\nu t) \qquad \text{(A12.3)}$$

where ν (= $vN\sigma$) is the constant collision frequency.

(b) The mean-free path

The mean free path 1 is the average distance between successive collisions. So

$$l = \frac{\int_0^\infty x \exp(-N\sigma x)\,dx}{\int_0^\infty \exp(-N\sigma x)\,dx} = \frac{1}{N\sigma} \qquad \text{(A12.4)}$$

and

$$\nu = vN\sigma = \frac{v}{l} \qquad \text{(A12.5)}$$

$N\sigma$ is the macroscopic cross-section, l is the relaxation length; the relaxation time t_r is defined as

$$t_r = \frac{1}{N\sigma v} = \frac{l}{v} \qquad \text{(A12.6)}$$

So t_r is a measure of time of approach of a system to the equilibrium state after a disturbance. Putting, in this relation $\mathbf{V} = \mathbf{v}_i - \mathbf{v}_j$ we get

$$l = t_r \bar{v} = \bar{v}/(\bar{V} N \sigma_3) \tag{A12.7}$$

where $\bar{v}$ is the mean speed, σ_3 is the mean total scattering cross section at $\bar{v}$, and $\bar{V}$ is the magnitude of the average of $\mathbf{V}$. For identical particles, the average of $(\mathbf{v}_i \,.\, \mathbf{v}_j)$ over all angles is zero. So averaging $V^2 = v_i^2 - 2(\mathbf{v}_i \,.\, \mathbf{v}_j) + v_j^2$ we get $\overline{V}^2 = 2\overline{v}^2$; Consequently

$$l = \frac{1}{\sqrt{2} N \sigma_3} \tag{A12.8}$$

For collisions between identical particles of radius r and number density N, the mean free path is defined as

$$l = \frac{1}{\pi (2r)^2 N} \tag{A12.9}$$

Between a particle of negligible radius, and others of radius r and number density N, the mean free path formula is

$$l = \frac{1}{\pi r^2 N} \tag{A12.10}$$

To estimate the mean distance of separation in a volume distribution of charges, for isotropic scattering, in terms of the number density of particles, when there are N ions in a sphere of radius R, the probability that all these ions are found in a sphere of radius r is $(r^3/R^3)^N$, and the probability of finding none of them within this sphere is $\{1 - (r/l)^3\}^N$. If P is the probability that atleast one of the N particles is found within the sphere, then

$$P = 1 - \{1 - (r/R)^3\}^N \tag{A12.11}$$

Since N is large $P \approx 1 - \exp(-r^3 N/R^3)$. If N_i is the ion number density, then $N = (4\pi R^3/3) N_i$. Hence

$$P \approx 1 - \exp(-4\pi r^3 N_i/3) \tag{A12.12}$$

Differentiating this relation, we put $x = r^3/r_0^3$, and obtain

$$dP = \exp(-x)\, dx, \tag{A12.13}$$

Also, we can write $N_i = 3/4\pi r_0^3$ (A12.14)

The probability that at least one ion is found in the shell between concentric spheres of radii r and $r + dr$ is P, and r_0 is the mean distance of separation, in a volume distribution of charges, for isotropic scattering. Actually, r_0 is proportional to $1/(N^{1/3})$, so,

$$r_0 = \frac{\alpha}{N^{1/3}} \tag{A12.15}$$

where α is the constant of proportionality, and r_0 is expressed in terms of the number density of particles.

(c) Phenomenological derivation of collision probability

Avogadro stated that all perfect gases at same pressure and temperature contain same number of molecules per unit volume. The corresponding law is: Gases unite in simple proportions, both by weight and by volume. So the densities of different perfect gases, at same temperature and pressure, are proportional to their molecular weights. Thus PV is inversely proportional to molecular weight. The Avogadro number A is the number of molecules of an ideal gas in 22.4 litres at the pressure 760 mm of mercury and the temperature 0°C. So

$$A\rho = MN \tag{A12.15}$$

where M is the molecular weight of the gas, ρ is its density in gm/cm^3 and N is the number density per cm^3 of the target material. A quantity N_{1mm} is defined as the number density of a gas, at the pressure 1 mm of mercury at 1°C. The Loschmidt number L is the number density, at the standard pressure and temperature, Appendix 7.

The macroscopic cross-section σ_{1mm} and the mean free path l_{1mm}, at the pressure = 1 mm of mercury at 0°C, are given by

$$\sigma_{1mm} = \frac{1}{l_{1mm} N_{1mm}} \tag{A12.16}$$

The probability P of collisions is the average number of collisions by a projectile in 1 cm of path through the gas, at the pressure 1 mm of mercury at 0°C; so we write

$$P = \frac{1}{l_{1mm}} = \sigma_{1mm} N_{1mm}; \tag{A12.17}$$

Then the relation between the gas pressure p (mm of mercury), temperature T (degrees Kelvin) and the reduced pressure p_0 (normalized in mm of mercury, and reduced to 0°C at constant density) is

$$p_0 = \frac{273}{T} p \tag{A12.18}$$

The collision mean-free-path length l is then given by

$$l = \frac{1}{p_0 \rho}, \tag{A12.19}$$

the collision frequency ν by

$$\nu = \frac{v}{l} = vp_0 \, l \tag{A12.20}$$

and the collision cross-section σ by

$$\sigma = \frac{1}{Nl} = \frac{p_0}{N} P \tag{A12.21}$$

APPENDIX 13 : THE RANDOM WALK PROBLEM

A free particle moving parallel to the *X*-axis, is supposed to experience a series of random walks along the *X*-axis. The probability of a backward or a forward step, at any stage, is equal, and each step is independent of the nature of the other steps. Resultant displacement, after *N* steps, is determined in the problem of dispersion of particles. In simple cases all the steps are of equal length and of constant duration.

At every stage it is equally likely that a step taken is forward or backward; so the probability of a step, in one of these two given senses, is 1/2 and the probability of a sequence of steps, numbering N is $(1/2)^N$. Of these N_1 will be forward and N_2 backward. The number of ways in which these N_1 forward and N_2 backward steps can be taken is the number of combination of things (forward or backward steps) either N_1 (or N_2) at a time. This number, therefore, is $N!/N_1!\,N_2!$, and the probability of the occurrence of the resultant displacement is

$$P = \left(\frac{1}{2}\right)^N \frac{N!}{N_1!\,N_2!} \tag{A13.1}$$

Here $N = N_1 + N_2$. Moreover if $N_1 - N_2 = m$, the particle has covered the distance ml in N steps in the positive direction of the *X*-axis. When N is even, m is even and, when N is odd m is odd; so N_1 and N_2 cannot be fractions.

When N is large, by Stirling's formula,

$$\log N! \approx \left(N + \frac{1}{2}\right) \log N - N + \frac{1}{2} \log (2\pi) \tag{A13.2}$$

$$\log P \approx \left(N + \frac{1}{2}\right) \log N - \frac{1}{2} \log (2pi) - N \log 2$$

$$- \frac{1}{2} (N + m + 1) \log \left(\frac{N+m}{2}\right) - \frac{1}{2} (N - m + 1) \log \left(\frac{N-m}{2}\right)$$

Since $\log (1 + x) = x - \frac{1}{2} x^2 - \ldots$, we use the results

$$\log\left(\frac{N\pm m}{2}\right)=\log\left(\frac{N}{2}\right)\pm\frac{m}{N}-\frac{m^2}{2N^2}+\ldots \tag{A13.3}$$

and find that

$$\log P\approx-\frac{1}{2}\log\left(\frac{N\pi}{2}\right)-\frac{m^2}{2N}+\ldots \tag{A13.4}$$

Hence, for large N, the asymptotic formula is

$$P(m,n)\approx\left(\frac{2}{N\pi}\right)^{1/2}\exp\left(-\frac{m^2}{2N}\right) \tag{A13.5}$$

The probability $P(X, N)$ that the particle is in the interval between $x-x_0$ and $x-x_0+dx$, after N displacements, is

$$P(x,N)\,dx=\frac{1}{2l}P(m,N)\,dx \tag{A13.6}$$

Putting $ml=x-x_0$ in this relation we get

$$P(x,N)\approx\left\{\frac{1}{(2\pi Nl^2)^{1/2}}\right\}\exp\left\{-\frac{(x-x_0)^2}{2Nl^2}\right\} \tag{A13.7}$$

When a particle suffers ν number of displacements per unit time, we put $\nu t=N$. Then the probability $P(X, N)$ is replaced by the probability $P(X, t)$ that the particle is in the interval between $x-x_0$ and $x-x_0+dx$, after time t. We write (A13.7) as

$$P(x,t)=\frac{\exp\left\{-\dfrac{(x-x_0)^2}{(4NT\,.\,l^2/2T)}\right\}}{\{(4\pi NT)\;(l^2/2T)\}^{1/2}} \tag{A13.8}$$

where the new parameter T is given by $NT=t$. Hence, writing $l^2/2T=D$ in (A13.8) we obtain

$$P(x,t)=\frac{1}{\sqrt{4\pi Dt}}\exp\left\{-\frac{(x-x_0)^2}{4Dt}\right\} \tag{A13.9}$$

The diffusion coefficient D is given by

$$D=\frac{l^2}{2T}=\frac{\nu l^2}{2} \tag{A13.10}$$

The differential equation, satisfying (A13.9), is

$$\frac{\partial P}{\partial t}=D\,\nabla^2 P \tag{A13.11}$$

where ∇^2 is the Laplacian operator if x and x_0 in (A13.9) are replaced by the three-dimensional position vectors $\mathbf{r}$ and $\mathbf{r}_0$.

APPENDIX 14 : THE WKB METHOD

The simplest normal form, ordinary, second order, differential equation, having the slowly varying coefficient K, is

$$\frac{d^2 \psi(z)}{dz^2} + K^2(z)\,\psi(z) = 0 \tag{A14.1}$$

where K^2 is positive; here the quantity $2\pi/K$ plays the role of the local wave-length in the direction of the z-axis. In optics this equation describes the behaviour of a plane wave propagating perpendicular to the stratification of a medium with refractive index

$$n(z) = \frac{K(z)}{K_0} \tag{A14.2}$$

Here K_0 is the inverse of the wavelength in a standard medium, say vacuum. If K is independent of z, the solution is

$$\psi(z) = c_1 \exp(iKz) + c_2 \exp(-iKz) \tag{A14.3}$$

As K tends to a constant value with respect to Z, all solutions tend to (A14.3). When K depends on z the solution is sought in the form

$$\psi(z) = A \exp\{i\,\varphi(z)\} \tag{A14.4}$$

where A is a constant and $\varphi(Z)$ is the generalized phase. Putting (A14.4) into (A14.1) we obtain a nonlinear equation for $\varphi(Z)$:

$$\left(\frac{d\varphi}{dz}\right)^2 = K^2 + i\frac{d^2\varphi}{dz^2} \tag{A14.5}$$

It cannot be solved exactly; but a sequence of approximate solutions is developed, the first of which is

$$\text{(A14.6)} \qquad \frac{d\varphi}{dz} = \pm K,$$

when
$$\left|\frac{d^2\varphi}{dz^2}\right| << K^2 \tag{A14.6 (a)}$$

Differentiating (A14.6) with respect to z we put the value of $d^2\varphi/dz^2$ in (A14.5), and taking square roots, get the equation

$$\frac{d\varphi}{dz} = \pm\left\{K \pm \frac{i}{2K}\frac{dL}{dz}\right\} \tag{A14.7}$$

Integration gives

$$\varphi(z) = \pm \int K dz + i \log \sqrt{K}; \tag{A14.8}$$

and a standard WKB solution of (A14.1)

$$\psi(z) = \frac{1}{\sqrt{K}} \{A \exp(i \int Kdz) + B \exp(-i \int Kdz)\} \tag{A14.9}$$

The words WKB represent the names of the scientists Wentzel, Kramers and Brilliouin who first developed this technique of solving differential equations having slowly varying coefficients, for use in quantum mechanics. This technique was used by earlier mathematicians, in study of fluid flow problems. It is also known as the first WKB approximation, because, a more general solution, based on a successive application of this approximation scheme, up to any higher order, is possible.

The lower limit of the integrals in (A14.9) is set equal to zero, or any other value Z_0, since this alteration changes only the value of the arbitrary constants A and B. When the wave passes through a finite thickness Z, the change of phase $\int_0^z Kdz$ is cumulative; this constitutes the 'phase memory' concept. The wave is reflected in the region where $K = 0$. There, the WKB approximation fails.

APPENDIX 15 : RULES OF TRANSFORMATION FROM LABORATORY FRAME TO THE SPACE-INDEPENDENT FRAME

The equation of Lorentz transformation from the K frame to the K' frame, moving relatively with velocity v, parallel to the z axis, are

$$x = x', \; y = y', \; z = \gamma(z' + vt'), \; t = \gamma\left(t' + \frac{v}{c} z'\right) \tag{A15.1}$$

or

$$x' = x, \; y' = y, \; z' = \gamma(z - vt), \; t' = \gamma\left(t - \frac{vz}{c}\right) \tag{A15.2}$$

where $\gamma^2 = 1/(1 - \beta^2)$ with $\boldsymbol{\beta} = \mathbf{v}/c$, we put

$$\beta = \frac{v}{c} \tanh \psi \tag{A15.3}$$

and find that $\gamma = \cosh \psi$, $\beta\gamma = \sinh \psi$,

$$z = z' \cosh \psi + ct' \sinh \psi, \; t = t' \cosh \psi + \frac{z'}{c} \sinh \psi \tag{A15.4}$$

The reverse transformation from K' to K, obtained by changing the sign of v or ψ, is

$$t' = t \cosh \psi - \frac{z}{c} \sinh \psi, \; z' = z \cosh \psi - ct \sinh \psi \tag{A15.5}$$

The wave phase is

$$\omega t - Kz = (\omega \cosh \psi - Kc \sinh \psi)\, t' - (K \cosh \psi - \frac{\omega}{c} \sinh \psi)\; z' \quad \text{(A15.6)}$$

where K and ω are the wave number and frequency. We consider the transformation from the frame K to the frame K', at the velocity KC^2/ω, and find that

$$Kc = \omega \tanh \psi \quad \text{(A15.7)}$$

Then the coefficient of Z' vanishes, so that

$$\omega t - Kz = \omega \sqrt{1 - \beta^2}\; t' = \omega\, t' \operatorname{sech} \psi \quad \text{(A15.8)}$$

Hence the linear combination $\omega t - Kz$ is transformed to $\omega \operatorname{sech} \psi t'$ in the inertial frame K'. Since this expression is time like, the fields are described in terms of only one time-like independent variable t', in place of a linear combination of t and z, in the K' frame. For transverse waves, the phase velocity ω/K is greater than c, so $Kc^2/\omega < c$. Hence, the velocity of K', relative to K is less than c, and

$$\omega t - Kz = \omega'\, t', \; \omega' = \omega \sqrt{1 - \beta^2} = \omega \operatorname{sech} \psi, \quad \text{(A15.9)}$$

$$\dot{z} = \frac{dz}{dt} = V - \frac{V}{\gamma_0} \frac{\partial t'}{\partial t}, \quad \text{(A15.10)}$$

$$\frac{\partial}{\partial t} = \gamma_0 \frac{\partial}{\partial t'}, \; \frac{\partial}{\partial z} = -\frac{\gamma_0}{V} \frac{\partial}{\partial t'} \quad \text{(A15.11)}$$

where $$V = \frac{\omega}{K}, \; \gamma_0 = \frac{1}{\sqrt{1 - \beta_0^2}}, \; \beta_0 = \frac{V}{c} \quad \text{(A15.12)}$$

Then equation (A15.7) shows that

$$v = \frac{K}{c} c^2 = \frac{c^2}{V}, \; \omega' = \omega \sqrt{1 - \frac{c^2}{V^2}}, \quad \text{(A15.13)}$$

$$v_x = \frac{v_x'}{\gamma \left(1 + \beta \frac{v_z'}{c}\right)} = \frac{c\, v_x'}{c \cosh \psi + v_z' \sinh \psi}, \; v_y = \frac{c\, v_y'}{c \cosh \psi + v_z' \sinh \psi},$$

$$v_z = \frac{v_z' + \beta c}{1 + \beta \frac{v_z'}{c}} = \frac{c\,(v_z' \cosh \psi) + (\sinh \psi)}{c \cosh \psi + v_z' \sinh \psi} \quad \text{(A15.14)}$$

$$\left.\begin{aligned} E_x &= \gamma\,(E_x' + \beta\, E_y') = E_x' \cosh \psi + \beta_y' \sinh \psi, \; E_y = E_y' \cosh \psi - B_x' \sinh \psi, \; E_z = E_z', \\ B_x &= \gamma\,(B_x' - \beta\, E_y') = B_x' \cosh \psi - E_y' \sinh \psi, \; B_y = B_y' \cosh \psi + E_x' \sinh \psi, \; B_z = B_z' \end{aligned}\right\} \quad \text{(A15.16)}$$

The reverse transformation relations are

$$\left.\begin{aligned} v_x' &= \frac{c\, v_x}{c \cosh \psi - v_z \sinh \psi}, \quad v_y' = \frac{c\, v_y}{c \cosh \psi - v_z \sinh \psi}, \\ v_z' &= \frac{c\,(v_z \cosh \psi) - (\sinh \psi)}{c \cosh \psi - v_z \sinh \psi} \end{aligned}\right\} \quad \text{(A15.16 } (a))$$

$$E_x' = E_x \cosh \psi - B_y \sinh \psi, \;\; E_y' = E_y \cosh \psi + B_z \sinh \psi, \;\; E_z' = E_z \qquad \text{(A15.17)}$$

$$B_x' = B_x \cosh \psi + E_y \sinh \psi, \;\; B_y' = B_y \cosh \psi - E_x \sinh \psi, \;\; B_z' = B_z \qquad \text{(A15.18)}$$

Putting $\theta = Kz - \omega t$ in the components of the Maxwell equation $\nabla \times \mathbf{E} = -\,\partial \mathbf{B}/c\, \partial t$ we find that

$$\frac{d}{d\theta}\left(\frac{KC}{\omega} E_x - B_y\right) = 0, \;\; \frac{d}{d\theta}\left(\frac{KC}{\omega} E_y + B_x\right) = 0$$

$$E_x \tanh \psi - B_y = -\,B_y^0, \;\; E_y \tanh \psi + B_x = B_x^0 \qquad \text{(A15.19)}$$

$$B_x' = B_x^0 \sinh \psi, \; B_y' = B_y^0 \cosh \psi \qquad \text{(A15.20)}$$

So $\mathbf{B}' = \mathbf{B}^{0\prime}$ where $\mathbf{B}^{0\prime}$ is constant and

$$B_x^{0\prime} = B_x^0 \sinh \psi, \; B_y^{0\prime} = B_y^0 \cosh \psi \qquad \text{(A15.21)}$$

Similarly from (A15.17) we obtain

$$B_x' = E_x \operatorname{sech} \psi - B_y^0 \sinh \psi, \;\; E_y' = E_y \operatorname{sech} \psi + B_x^0 \sinh \psi \qquad \text{(A15.22)}$$

Denoting the rest mass by m_0 (and not by m), we find that

$$m = m' \gamma \left(1 + \beta \frac{v_z'}{c}\right), \;\; m' = m\gamma \left(1 - \beta \frac{v_z}{c}\right) \qquad \text{(A15.23)}$$

Then, the relations of transformation of momentum components are

$$p_x = m v_x = \frac{m' \gamma \left(1 + \beta \dfrac{v_z'}{c}\right) c\, v_x'}{c \cosh \psi + v_z' \sinh \psi} = m'\, v_x' = p_x', \; p_y = m v_y = m'\, v_y' = p_y',$$

$$p_z = m v_z = \frac{m' \gamma c \left(1 + \beta \dfrac{v_z'}{c}\right) (v_z' \cosh \psi + c \sinh \psi)}{c \cosh \psi + v_z' \sinh \psi} \qquad \text{(A15.24)}$$

Writing

$$q = \sqrt{m_0^2 c^2 + p^2}, \;\; q' = \sqrt{m_0^2 c^2 + p'^2} \qquad \text{(A15.25)}$$

we find that

$$q'^2 = m_0^2 c^2 + \frac{m_0^2 v'^2}{1 - \frac{v'^2}{c^2}} = \frac{m_0^2 c^2}{1 - \frac{v'^2}{c^2}} = m'^2 c^2 \tag{A15.26}$$

where $\mathbf{v}'$ is the velocity of the K' system relative to the rest system K_0. So

$$p_z = p_z' \cosh \psi + q' \sinh \psi, \tag{A15.27}$$

$$q = mc = m' \gamma c \left(1 + \beta \frac{v_z'}{c} \right) = m' c \gamma + m' v_z' \gamma \beta$$

$$= q' \gamma + p_z' \beta \gamma = q' \cosh \psi + p_z' \sinh \psi \tag{A15.28}$$

If N, N', N_0 represent the number density in the systems $K . K'$, K^0, respectively, then

$$N' = \frac{N_0}{\sqrt{1 - \frac{v'^2}{c^2}}}, \tag{A15.29}$$

$$M_0 N' = \frac{m_0 N_0}{\sqrt{1 - \frac{v'^2}{c^2}}} = m' N_0 \tag{A15.30}$$

Similarly $\quad m_0 N = m N_0$, and

$$\frac{N'}{N} = \frac{m'}{m} = \frac{q'}{q} \tag{A15.31}$$

Since the vector potential **A** and the scalar potential φ form a four-vector, we write

$$A_{\parallel} = \gamma (A_{\parallel}' + \beta \varphi'), \ \mathbf{A}_{\perp} = \mathbf{A}_{\perp}', \ \varphi = \gamma (\varphi' + \beta A_{\parallel}'), \tag{A15.32}$$

$$A_{\parallel}' = \gamma (A_{\parallel} - \beta \varphi), \ \mathbf{A}_{\perp}' = \mathbf{A}_{\perp}, \ \varphi' = \gamma (\varphi - \beta A_{\parallel}) \tag{A15.33}$$

and obtain

$$A_{\parallel} = A_{\parallel}' \cosh \psi + \varphi' \sinh \psi, \ \varphi = \varphi' \cosh \psi + A_{\parallel}' \sinh \psi, \ \mathbf{A}_{\perp} = \mathbf{A}_{\perp}' \tag{A15.34}$$

APPENDIX 16 : PERTURBATION TECHNIQUES FOR WEAKLY NON-LINEAR SYSTEMS

Perturbation techniques of several types exist for finding approximate solutions of weakly non-linear systems of field equations, in closed form, by series expansion of variables, both dependent and independent. The expansion of a field variable (of equations (12.5.2)-(12.5.4)), for example), for stationary wave solution of weakly nonlinear sets of field equations, is

$$f(\mathbf{r}, t) = f_0 + \sum_{n=1}^{\infty} [f_n(\mathbf{r})\, e^{-in\omega t} + \bar{f}_n(\mathbf{r})\, e^{in\omega t}] \tag{A16.1}$$

where $\bar{f}_n$ is the complex conjugate (cc) of f_n, the incident wave is at frequency ω, $|f_1| >> |f_2| >> |f_3| >> |f_n|$, and f_0 is a constant. Equating, from both sides, of an equation, terms of the same power of $\exp(\pm i\omega t)$, we obtain a set of equations of the type (12.11.2)-(12.11.4). Terms of power higher than that of $\exp(\pm i\omega t)$ do not appear in the equations of first order. In the second order equations terms of f_3, f_4 and other higher orders do not appear. Thus, for fields of the third order, terms of higher orders, yielding $\exp(\pm i\, 3\omega\, t)$, are neglected. A third order equation, therefore, is of the type.

$$L(f_3) = G(f_1, f_2)$$

where L is a differential operator and the right hand side is determined from solution of lower order equations. Equation (A16.1) is like the series

$$u = u_0 + u_1 + u_2 = \ldots \tag{A16.2}$$

where u_0 is the equilibrium state value of u, u_1 is the linearized (or the first harmonic) approximation, and u_n (for $n = 2, 3, \ldots$) is a linear combination of the nth harmonic fields of u; so,

$$u_n = f_n(\mathbf{r}) \exp(-in\omega\, t) + \bar{f}_n(\mathbf{r}) \exp(in\omega t) \tag{A16.3}$$

For convergence $(|u_n|/|u_{n-1}|) < 1$; and the term $u_r\, u_{n-r}$ is of the order of u_n for $n = 1, 2, 3, \ldots$, where $r \le n$, when the series expansion is

$$u = u_0 + \varepsilon u_1 + \varepsilon^2 u_2 + \ldots \tag{A16.4}$$

The dimensionless expansion parameter ε is a mathematical artifice, used for 'book keeping' purposes, in a systematic fashion, as in section 12.6. It is set equal to unity, in the final results. Terms of the same dimension and of comparable degree of approximation, are thus grouped together for convenience. But, in a large number of situations of another type, this parameter occurs naturally, and is not necessarily a dimensionless quantity. For example, in the computation of field induced distortions of particle orbits, from the guiding centre approximation, for finite Larmour radius of gyration, ε is the ratio e/m of electronic charge and mass. Other examples are Planck's constant $\hbar$, the reciprocal of the vacuum speed c of light, the amplitude of a forcing term. A set of equations, correct up to a specified order of smallness, obtains for both categories of problems, by equating terms of equal powers of ε up to that order, from both sides of a non-linear system of field equations.

In temporal evolution problems, the most saturated contribution to the solution, in time $t\,(> t_0)$, due to disturbances at the given time t_0, is evaluated.

For that purpose, the method of Lindstedt is simple and convenient. The field variables, and the independent variable, are expanded as in (A16.1) to (A16.4), to find the secular free solution, which is the most saturated solution, by a processes of accelerated convergence. In continuum mechanics, time is the independent variable; its expansion is

$$t = T + \varepsilon^2 a_2(T) + \ldots \tag{A16.5}$$

The parameter ε is the ratio between the fast time scale t, of the applied field, and the non-linearly emerging slow time scale T. The auxiliary functions $a_2, a_4, \ldots$ of T are determined by imposing the condition of secular free behaviour, for a closed system of non-linear differential equations. The time variable T is introduced to extract the most saturated evolution of the field variables in time. The expansion in series of t, in (A16.5), gives

$$\frac{\partial}{\partial t} = \frac{1}{1 + \varepsilon^2 a_2' + \varepsilon^4 a_4' + \ldots} \frac{\partial}{\partial t} \tag{A16.6}$$

$$\frac{\partial^2}{\partial t^2} = \frac{1}{(1 + \varepsilon^2 a_2' + \ldots)^2} \frac{\partial^2}{\partial T^2} - \frac{\varepsilon^2 a_2'' + \varepsilon^4 a_4'' + \ldots}{(1 + \varepsilon^2 a_2' + \ldots)^3} \frac{\partial}{\partial T} \tag{A16.7}$$

where $a_2' = \partial a_2/\partial T$, etc. The secular free solutions of the non-linear field equations of section 12.1, correct up to the third order, follow easily if we write

$$t = T + \varepsilon^2 bT + \ldots \tag{A16.8}$$

so that $a_2(T) = bT$, where b is constant and a_4, etc. are neglected. Then, the simpler relations,

$$\frac{\partial}{\partial t} = \frac{1}{1 + \varepsilon^2 b} \frac{\partial}{\partial T}, \quad \frac{\partial^2}{\partial t^2} = \frac{1}{(1 + \varepsilon^2 b)^2} \frac{\partial^2}{\partial T^2}, \tag{A16.9}$$

determine the secular free, first harmonic solutions, correct upto the third order. For practical purpose these are replaced by

$$\frac{\partial}{\partial t} = (1 - \varepsilon^2 b) \frac{\partial}{\partial T}, \quad \frac{\partial^2}{\partial t^2} = (1 - 2\varepsilon^2 b) \frac{\partial^2}{\partial T^2} \tag{A16.10}$$

For a closed system of differential equations, having weak nonlinearities, the words, secular free solution, solution for accelerated convergence, or the most saturated solution, mean the same thing, because a less rigorous treatment produces secular, non-convergent solutions, which are not the most saturated solutions. Perturbation techniques of several types exist for finding approximate solutions of weakly non-linear systems of field equations in closed form. These employ suitable expansion of variables, both dependent, and independent, in series. The asymptotic series expansion of the field variables (of section 12.5, for example) are used for stationary wave solution of such systems.

BIBLIOGRAPHY*

Abraham, M. and R. Becker; (1950). "The Classical theory of electricity and magnetism," Blackie, London.

Acheson, D.J. and Hide, R. (1973). "Hydromagnetics of rotating fluids," Rept. Progr. Phys. **36**, 159-221.

Akhiezer A. I., Akhiezer, I.A., Polovin, R.V., Sitenko, A.G. and Stepanov, K.N. (1975) "Plasma electrodynamics", Vol. I, II, Pergamon Press, Oxford.

Akhiezer A. I., Akhiezer, I.A., Polovin, R.V., Sitenko, A. G. and Stepanov, K.V. (1967), "Collective oscillations in a plasma", Massachusetts Institute of Technology, Mass.

Akhmanov, S.A. and Khokhlov R.V. (1964), Problems of nonlinear optics (In Russian), Moscow.

Alfven, H. (1950), "Cosmical Electrodynamics," Clarendon Press, Oxford.

Alfven, H. (1956). "Plasma physics," Rendi Conti, course XIII, Bologna, p.1.

Alfven, H.(1988), "Cosmic plasma". D. Reidel Publishing Co. Dordrecht (Netherlands).

Alfven, H. and C. G. Falthammer, (1963), "Cosmical electrodynamics," Oxford University Press, London.

Ali Kettani, M. and Hayaux, M.F. (1973). "Plasma engineering'', Butterworth, London.

Allis, W.P. (1956). "Motion of ions and electrons, " Handbuch der Physik, **21**, 384-410, Springer-Verlag, Berlin.

Allis, W.P., S. J. Buchsbaum and A. Bers (1963), " Waves in anisotropic plasma" M. I. T. Press, Mass.

Alpert. Ya L. (1974), "Waves and satellites in the near-earth plasma", Consultants Bureau, New York.

Ames, W.F., (1965), "Nonlinear partial differential equations in engineering," Academic Press, New York.

Arons, J. and C. E. Max (1974). "Self-precessions and Frequency shift for electromagnetic waves in Homogeneous Plasma". Fluids **17**, 1983-1994.

Artsimovich, L. A. (1978), "ABC on plasma", MIR Publishers, Moscow.

Artsimovich, L.A. and S. Yu. Lukyanov, (1980), "Motion of charged particles in electric and magnetic fields," Mir, Moscow.

Arzimovich. L.A. (1965), "Elementary plasma physics", Ginn (Blaisdell), Boston.

Bailey, V.A. and D.F. Martyn (1934). "Interaction of radio waves''. Nature. **133** 218.

Bailey, V.A. and D.F. Martyn. (1934). "The influence of electric waves on the ionosphere" Phil Mag. **18**, 369-396.

Bailey, V.A. (1937). "Interaction by resonance of radio waves", Nature **139**, 838-839.

Bailey, V.A. (1937). "The motion of electrons in a gas in the presence of variable electric field and a constant magnetic field", Phil. Mag. **23**, 774-791 (1937).

Bailey, V.A. (1938). "On some effects caused in the ionosphere by Electric Waves" Phil Mag. **26**, 425-453.

Bailey, V.A. (1956). "On the interaction of radio waves". Nuovo Cimento 4, Suppl. **4**, 1430-1449.

Bailey, VA., (1956). "On nonlinear phenomena in the ionosphere". Radio Science **690**, 9-24.

Bailey, V.A., R.A. Smith, K. Landecker, A.J. Higgs and F.H. Hibberd (1952). "Resonance in gyro-interaction of radio waves" Nature, **169**, 911-913.

---

* This list contains references mentioned as well as not mentioned in the text, but important. This list is alphabetical only upto the first letter.

Baldwin, G.C. (1969). "An introduction to nonlinear optics", Plenum Press, New York.

Balescu, R. (1964), "Statistical mechanics of charged particles" Inter Science, New York.

Basu, J., (1988). "Tokamak, The man made sun", Science Reporter, July, 366-372.

Batchelor, G.K., (1950), "On the Spontaneous magnetic field in a conducting liquid in turbulent motion", Proc. Roy. Soc. Lond. **A210**, 405-416.

Bateman, H. (1915). "Electrical and optical wave motion", Cambridge University Press, Cambridge, U.K.

Bateman G. (1978), "MHD, Instabilities," The MIT Press, Cambridge, Mass.

Bekefi, G. (1966), "Radiation processes in Plasmas," John Wiley, New York.

Bellman, R., (1966), "Perturbation techniques in mathematics, physics and engineering". Holt. Rinehart and Winston, New York.

Bera, B. and S. Sarkar (1980), "Nonlinearly induced precession frequency and dispersion relation for an elliptically polarized wave in a cold plasma in which both electrons and ions are mobile," unpublished (communicated privately).

Bergins, F.E. and Papas, C.H. (1958), "Electromagnetic wave guides and resonators'', Handbuch der Physik, Vol. **XVI**, pages 289-422 Springer-Verlag, Berlin.

Bernstein, I.B, E.A. Frieman, M.D. Kruskel and R.M. Kulsrud (1958), "An energy principle for hydromagnetic stability problems" Proc. Roy. Soc. London, **A244**, 17-40.

Bernstein, I.B. (1958), "Waves in a plasma in a magnetic field" Phys. Rev. **109**, 10-21.

Bernstein, I.B. and S.K. Trehan, (1961), "Plasma oscillation" Nucl. Fusion **1**. 1-40.

Bernstein, I.B. (1971), "The Motion of a charged particle in a strong magnetic field;" in Advances in plasma physics, Vol. 4, edited by A. Simon and W.B. Thompson, pages 311-333; Inter Science, John Wiley, New York.

Bershader, D. (Editor) (1962), "Plasma hydromagnetics,", Stanford University Press, California.

Bhatnagar, P.L., Gross, E.P. and Krook, M. (1954), "A Model for collision processes in Gases, I. Small amplitude processes in charged and neutral one-component systems." Phys. Rev. **94**, 511-525.

Bhattacharyya, B. (1980), "Nonlinear wave processes in plasmas;" Ph.D. Thesis, Jadavpur University, Calcutta.

Bickerton, R.J. (1960), "Pinch experiments," Rise Report **No. 18**, DAEC, Riso, Denmark, pp. 443-454.

Bishop, A.S. (1958), "Project sherwood": US program in controlled fusion, Addison-Wesley, Reading, Mass.

Bittencourt, J. A. (1986) "Fundamentals of plasma physics", Pergamon Press, Oxford.

Bloembergen, N. (1965), "Nonlinear optics," W.A,. Benjamin, New York.

Boenig, H.V. (1982), "Plasma science and technology," Cornell University Press, Cornel.

Boley, F. I. (1966), "Plasmas-laboratory and cosmic'', Van Nostrand, Princeton.

Bond, J.W., K.M. Watson and J.A. Welch (1965), "Atomic theory of gas dynamics," Addison-Wesley, Reading.

Born, M. and E. Wolf (1970), "Principles of optics", Pergamon Press, Oxford.

Bosticks, W.H. (1957), (i) "Experimental study of plasmoids" Phys. Rev. **106**, 404-412; (ii) "Plasmoids", Scientific American, **197**, 87-94.

Boyd, R.L. F. (1974). "Spce physics – the study of plasmas in space, Clarendon Press, Oxford.

Boyd, T.J.M. and J.J. Sanderson (1969), "Plasma dynamics," Nelson, London.

Brandstatter, J.J. (1963)." An introduction to waves, rays and radiation in plasma media'', McGraw-Hill, New York.

Brekhovskikh, L.M. (1960), "Waves in layered media". Academic Press, New York.

Bremmer, H. (1958), "Propagation of electromatnetic waves", Handbuch der physik, Band XVI, Springer-Verlag, Berlin, p. 423.

Brinkman, H.C. (1960), "The motion of a particle in a magnetic field, adiabatic invariance." Riso Report No. 18. DAEC, Riso, Denmark, pages 25-33.

Brinkman, H.C. (1960) "Magnetohydrodynamic theory of rotating plasmas." Riso Report No. 18, DAEC, Riso, Denmark, pages 55-60.

Broer, L. J. F., and L. V. Wijnagaarden, (1960), "On the motion of a charged particle in an almost homogeneous magnetic field." Appl. Sci. Res. **B8**, 159-176.

Brogan, T. R. (1968). "The plasma MHD power generator", in advance in plasma physics, edited by A. Simon and W.B. Thompson, John Wiley, New Yordk, Vol. 1, pp. 227-330.

Brown. S.C. (1959) "Basic data of plasma physics," John Wiley, New York.

Brown. S.C. (1966), "Introduction to electrical discharges in gases," John Wiley, New York.

Brueckner, K. A. (1973), "Laser driver fusion", IEEE, Trans. PS-1, 13-26.

Budden, K.G. (1961), "Radio waves in the ionosphere." Cambridge University Press. London.

Bullard, E. (1955), "A discussion of magnetohydrodynamics", Proc. Roy. Soc. London **A233**, 289.

Bulletin of American Physical Society, (1977) "Physics and astronomy classification scheme"-1977, **22**, 218-240.

Buti, B, (1962), "Plasma oscillations and landau damping in a relativistic gas", Phys. Fluids **5**, 1-5.

Butt., A.G. (1933), "World radio," April 28 (title not available).

Cabannes, H. (1970), "Theoretical Magnetofluid dynamics," Academic Press, New York.

Cairns. R. A. (1985) "Plasma physics," Blackie. London.

Caldirola, P. (1979), "A relativistic theory of the classical electron," La Rivista del Nuovo Cimento 2, **No. 13**, p 1-49.

Caldirola, P. and De Barbieri, O. (1965), "On some nonlinear phenomena in the ionospheric plasma" Radio Science 69D, 33-58.

Cambel, A. B. (1963), "Plasma physics and magnetofluid mechanics," McGraw Hill, New York.

Carleton, N.P. and Mcgill, L.R. (1962), "Electron energy distribution in slightly ionized air under the influence of electric and magnetic fields", Phys. Rev. **126**, 2089-2099.

Chakraborty, B. (1966). "On interaction between longitudinal, surface and transverse waves at the boundary of plasma''. VEST. MOSK. UNIV. SER. III FIZIKA I ASTRONOMIA (In Russian). No. **5**, 30.

Chakraborty, B. (1968), "Nonlinear deviation of the direction of a reflected wave in vacuum at a plasma boundary", Sov. Phys. JETP Letters **8**, 393-395.

Chakraborty, B. (1971), "Nonlinear effects due to transverse waves in statistical plasma", Sov. Phys. JEPT, **32**, 250-253.

Chakraborty, B. (1973), "Conversion of longitudinal waves into electromagnetic radiation by a nonuniform Plasma", Appl. Sci. Res. **28**, 467-476.

Chakraborty, B. (1971). "Coupling of waves in nonuniform plasma", J. Math. Phys. **12**, 529-533.

Chakraborty, B., M.R. Khan, and B. Bhattacharya (1980), "Impact of strong radiation on neutral atoms in a plasma background", Phys. Rev. **A22**.

Chakraborty, B. (1977), "Parametric rotation of principal polarization axes and other effects due to four transverse waves in a plasma", Phys. Rev. **A16**, 1297-1308.

Chakraborty, B. (1987) Variational classical electrodynamics physics, Reports, **156**, 1-65 (1987).

Chakraborty, B. and Chandra, S.K. (1978) "Self-precession and frequency shift for two-waves in a plasma", Phys. Fluids **21**, 2336-2338.

Chamberlain, J.W. (1964), "Motion of charged particles in the earth's magnetic field" in geophysics: The earth's environment, edited by C. Dewitt, J. Heiblot and A. Lebeau; (Gordon and Breach, New York, pages 141-174).

Chandra, S.K. (1979), "Some aspects of nonlinear propagation of waves through plasma", Ph.D. thesis Jadavpur University, Calcutta.

Chandrasekhar, S. (1950), "Radiative transfer", Clarendon Press, Oxford.

Chandrasekhar, S. (1960), "Plasma physics'', The University of Chicago, Chicago.

Chandrasekhar, S. (1958), "Adiabatic invariants in the motions of charged particles", pages 3-31 in The plasma in a magnetic field, edited by R.K.M. Landshoff, Stanford University Press, Stanford, California.

Chandrasekhar, S. (1961), "Hydrodynamic and hydromagnetic stability" (Oxford University Press, London).

Chandrasekhar, S. (1951), "Principles of stellar dynamics", University of Chicago Press, Chicago.

Chandrasekhar, S. (1943), "Stochastic problems in physics and astronomy" Rev. Mod. Phys. **15**, 1-89.

Chapman, S. and T.G. Cowling (1939), "The mathematical theory of nonuniform gases", Cambridge University, London. Paper back new edition (1960).

Chapman. S. G. and Cowling, T. G. (1970), "The mathematical theory of nonuniform gases," Cambridge University Press, New York.

Chatterjee, P. (1975), "Fundamentals of units and dimensions" (Macmillan, New Delhi).

Chawla, B.R and Unz. H., (1979), "Electromagnetic waves in moving magnetoplasma", Kansas University Press, Lawrence.

Chen, F.F. (1977), "Introduction to plasma physics", Plenum Press, New York.

Chen, F.F. (1973), "Nonlinear optics of plasma, paper presented at the International Congress of Waves and Instabilities in Plasmas". Innsbruck, Austria, April 2-7.

Chernov, L.A. (1960), "Wave propagation in a random medium", McGrawHill, New York.

Chu, T.K., H.W. Hendel and J.M. Dawson (1972), "Laboratory parametric instability experiments at the plasma frequency" Comments on Plasma Physics and Controlled Fusion **1**, 111-118.

Chiao, R.Y. and Godine, J. (1969), "Polarization-dependence of stimulated Rayleigh-wing scattering and the optical-frequency Kerr effect", Phys, Rev. **185**, 430-445.

Cheo, P.K. and Heer, C.V. (1964), Beat "frequency between two travelling waves in a Fabry-Perot Square Cavity", Applied Optics **3**, 788-789.

Clemmow, P.C. and A.J. Wilson (1956), "The dispersion equation in plasma oscillations" Proc. Roy. Soc. London **A237**, 117-131.

Clemmow, P.C. and J.P. Dougherty (1969), "Electrodynamics of particles and plasmas", Addison Wesley, Reading, Mass.

Clauser, F.H. (ed.). (1960) Symposium of plasma dynamics. Addison-Wesley Reading, pp. 75, 240-284.

Cohen, R.S., L.S. Spitzer and P.M. Routley, (1950). "The electrical conductivity of an ionized gas", Phys, Rev. **80**, 230-238.

Cohen, M.H. (1961). "Radiation in plasma I, cerenkov effect", Phys. Rev. **123**, 711-721.

Cohen, M.H. (1962), "Radiation in plasma II, equivalent sources" Phys. Rev. **125**, 389-397.

Coppi, B., M.N. Rosenbluth and R.N. Suden (1969). "Nonlinear interactions of positive and negative energy modes in rarefied plasmas". Ann. Phys. (New York), **55**, 207-247, 248-270.

Coulson, C.A. and Boyd. T.J.M., (1979), "Electricity", Longman, London, Second edition.

Cowling, T.G. (1945). "The electrical conductivity of an ionized gas in a magnetic field, with applications to solar atomosphere and the ionosphere", Proc. Roy. Soc. **A183**, 453.

Cowling, T.G. (1957) "Magnetohydrodynamics", John Wiley, New York.

Cowling, T. G. (1976), "Magnetohydrodynamics," Admas Hilgar, London.

Cramer, K. P. and Shih-I Pai (1973), "Magnetofluid dynamics for engineers and applied physicists", McGraw-Hill, New York.

Cross, R. (1988), "An introduction to alfven waves;" Adam Helger, New York.

Cutolo, M. (1950). "Effects of gyro interaction and their interpretation" Nature **166**, 58-100.

Dagos, L. (1975), "Magnetofluid dynamics", Abacus Press, Kent.

Davidov, B. (i) (1935) "Uber die Geschwindigkeits verteilung der sich in elekrishchen felde bewegenden electromen", Phys. Zeit. Sowjet Union **8**, 59-70; (ii) (1936) "On distribution of velocities of electrons moving in electric field", Zhurn, Eksp. Teor. Tyz. **6**, 453-480 (in Russian), (iii) (1937), "On theory of motion of electrons in gases and in semi conductors", Zhurn. Eksp. Teor, Fyz. **7**, 1069-1089.

Davidson, R.C. (1972). "Methods in nonlinear plasma theory", Academic Press, New York

Davidson, R.C. (1974), "Theory of non-neutral plasma", W.A. Benjamin, London.

Davies. K., (1969), "Ionospheric radio waves", Blaisdell, Mass.

Dawson, J. M. (1961) "On landau damping", Phys. Fluids **4**, 869-874.

Dawson. J.M. (1968), "Radiation from plasmas"; pages 1-66 in Advances in Plasmas Physics, edited by A. Simon and W.B. Thompson Interscience Publishers, New York.

Dawson, J.M. (1959). "Nonlinear oscillations in a cold plasma", Phys. Rev. **113**, 383-387.

Dawson, J.M. (1962). "One dimensional plasma model", Phys. Fluids, **5**, 445-459.

De, J. (1957), "The analogue of Kelvin's theorem in hydromagnetics", Naturwissen Schaften **44**, 256-257.

Decoster, A. (1978). "Nonlinear travelling waves is a homogeneous cold, collisionless plasma". Physics Reports **47**, 285-422.

Dendy, R.O. (1990), "Plasma dynamics," Oxford Science Publications, Oxford.

DeGroot, S.R. and L.G. Suttorp (1972), "Foundations of electrodynamics", North-Holland, Amsterdam.

De Jager, E.M. (1967), "The Lorentz-invariant solutions of the Klein-Gordon equation", SIAM Journ. Appl. Math. **15**, 944-963.

De Jager, E.M. (1970), "Theory of distributions"; Pages 52-110 in Mathematics Applied to Physics, edited by E. Routine, Springer-Verlag, Berlin.

Delcroix, J.L. (1960). "Introduction to the theory of ionized gases." Interscience, New York.

Delcroix, J. L. (1985), "Plasma physics," Vol. I & II, John Wiley, New York.

Denisse, J.F. and J.L. Delcroix (1963), "Plasma waves", John Wiley, New York.

Deschamps, J., M. Fitaire and M. Lagoutte. (1970). "Inverse Faraday effect in plasma", Phys. Rev. Lett. **25**, 1330-1332.

DeSloge, E.A. and Matthysse. S.W., (1960) "Collision term in the Boltzmann transport equation", Amer. J. Phys. **28**, 1-11.

Dolan, T.J. (1982), "Fusion research," Vol. 1, Pergamon Press, Oxford.

Dougherty, J. P. "Lagrangian methods in plasma dynamics" I (1970), "General theory of the method of the averaged Lagrangian", J. Plasma Phys. **4**, 761-585; II (1974) "Construction of Lagrangian for plasmas", **11**, 331-46.

Drummond, J.E. (Editor) (1961), "Plasma physics," McGraw-Hill, New York.

Drummond, W.E., (1970), "The theory of plasma turbulence as applied to experimental phenomena"; pages 303-309 in proceedings of the symposium on Turbulence of Fluids and plasmas, New York, April 16-18, 1968, Polytechnic Press, Brooklyn. N.Y.

Dungey, J.G. (1958), "Cosmical electrodynamics," Cambridge University Press, New York.

DuBois, D.F. (1973), "Linear and nonlinear theory of parametric instabilities in plasma"; Congress on Waves and Instabilities in Plasma. Innsbruck, Austria, April 2-7.

Ducuing, J. (1969), "Nonlinear optical processes"; pages 421-472 in Quantum Optics, edited by R.J. Glauber Academic Press, New York.

Dunphy, K.H., D. Kahn and D. Mintzer (1967), "Energy coupling of a two fluid density discontinuity" Phys. Fluids **10**, 162-171.

Dupree, T.H. (1970), "Introduction to basic phenomena of turbulence in plasma"; pages 3-12, in Proceedings of the Symposium on Turbulence of fluids and plasmas, New York, April 16-18, 1969; Polytechnic Press, Brooklyn, N.Y.

Dutta, A.K. (1977) "Electromagnetism, electron theory and special theory of relativity", Macmillan, New Delhi.

Dysthe, K. B. (1970), "Nonlinear interaction of positive and negative energy waves,", Int. J. Electronics, **29**, 401-418.

Dysthe, K.B. (1974). "A note on the application of Whitham's method to nonlinear waves in dispersive media", Journ. Plasma Phys. II, 63-76.

Ecker, G. (1972), "Theory of fully ionized plasmas," Academic Press, New York.

Einaudi, F. and Oxford. W. I. (i) (1967), "Higher order approximations in the theory of longitudinal plasma oscillations," J. Plasma Phys. **1**, 483-497; (ii) (1970), Higher Order approximations in the theory of Longitudinal plasma oscillations II, J. Plasma Phys. **4**, 819-824.

Engeles, E. and Verheest. F., (1975), "Wave propagation in rotating plasmas and influence of coriolis force", Astrophysics and Space Sci. **37**, 427-440.

Everhart, E., G. Stone and R.J.Carbone (1955). "Classical calculations of differential cross seciton for scattering from a Coulomb poteintial with exponential screening''. Phys. Rev. **99**, 1287-1290.

Eyges, L. (1972). "The classical electromagnetic field", Addison-Wesley, Reading, Mass.

Fain, V.M. (1955), "The velocity distribution of electrons in the presence of a varying electric field and a constant magnetic field", Sov. Phys. JETP, **1**, 205-211.

Fermi, E. (1932), "Quantum theory of radiations". Rev. Mod. Phys. **4**, 87-132.

Ferrari, L. (1975) "On the three-term aproximation of the solution of the Boltzmann equation for weakly ionized gases". Physica **81A**, 276-296.

Ferraro, V.C.A. (1937). "Nonuniform rotation of the sun and its magnetic field" Mon. Not. Roy. Astr. Soc, **47**, 458-472.

Ferraro, V.C.A. and C. Plumpton (1966). "An introduction to magnetofluid mechanics", Clarendon Press, Oxford.

Field, G.B. (1956), "Radiation by plasma oscillation", Astrophys. J. **124**, 555-579.

Flammer, C. (1962). "A relativistic theory of plasma oscillations". In Interaction of Electromagnetic waves with a Plasma. Stancord Reseasrch Institute, AFCRL-62-185, April, pp. 37-71.

Fowler, T. K., (1968), "Thermodynamics of unstable plasmas;" pages 201-225 in Advances in Plasma Physics, Vol. **1**, edited by A. Simon and W. B. Thompson. Inter Science, New York.

France, L. A. and Williams, E. R. (1976), "An improved Faraday rotation experiment for rocket measurement of lower ionosphere electron concentration." J. Atoms. and Terr. Phys. **38**, 957-964.

Francis, G. (1960), "Ionization phenomena in gases", Butterworth, London.

Franklin, R. N. (1976), "Plasma phenomena in gas discharges," Clarendon Press, Oxford.

Fried, B. D. and Conte, S. D. (1961), "The plasma dispersion function," Academic Press, New York.

Furth, H. P. (1975). " Tokamak research," Nuclear Fusion **15**, 487-534.

Furth, H.P. (1978), "Reaching ignition in the Tokamak", Phys. Today, **36**, 53-61.

Galeev, A. A., G. Laval, T. M. O'Neil, M.N. Rosenbluth and R.Z. Sagdeev (1973), "Parametric back scattering of a linear electromagnetic wave in a plasma", JETP Letters, **17**, 35-38, (1973), Original in Russian, Zh. ETF Pis Red. **17**, 48-52.

Gallet, R. (1965), "Whistlers" In C. Dewitt, J. Hieblot and A. Lebeua (editors), Goephysics – The earth's environment, Gordon & Breach, New York, pp. 551-590.

Gartenhaus, S. (1964), "Elements of plasma physics;" Holt, Rinehart and Winston, New York.

Gekker, I. R. (1982), "Interaction of Strong electromagnetic fields with plasmas", Clarendon Press, Oxford.

Ghosh, P.K. (1957). "Lectures on theory of generalized functions" Calcutta University (unpublished).

Giacaglia, G.E.O., (1972), "Perturbation methods in nonlinear systems," Springer-Verlag, Berlin.

Ginzburg, V.L. and A.V. Gurevich (1960) "Nonlinear Phenomena in a plasma located in an alternating electromagnetic field". Uspekhi Fiz. Nauk. **70**, 201-246 (in Russian), (1960) English translation: Sov. Phys. Uspekhi **3**, 115-146.

Ginzburg, V. L. (1963), "Propagation of electromagnetic waves in plasmas," Gordon and Breach, New York.

Glasstone, S. and R.H. Lovberg (1960). "Controlled thermonuclear reactions"; D. Van Nostrand, Princeton.

Globe, S. (1959). "Laminar steady state magnetohydrodynamic flow in an annular channel" Phys. Fluids **2**, 404-407.

Gold, L., (1965), "Generalized stream instabilities in cold plasma", Phys. Rev. **137A**, A1083-A1086.

Goldstein, S. (1961), "Concerning a continuum theory of the electrodynamics of gaseous plasma", Polytechnic Institute of Brooklyn, April 1961, Pages 65-80.

Goldstein, H., (1980), "Classical machanics", Addison-Wesley, Reading Mass.

Gould, R.W., (1960), "The dynamics of electron beams", Ch. 4 in Symposium of Plasma Dynamics, edited by F.H. Clauser, Pergamon Press, Oxford.

Gourdine, M.C. (1962), "Magnetohydrodynamic channel flow of a rotating fluid", pages 19-24 in Magnetohydrodynamics, edited by A.B. Cambel, T.P. Anderson and M.M. Slawsky, North Western University Press, Eyanston U.K.

Grad, H. (1958) "Principles of Kinetic theory of gases". Handbuch der Physik, Band XII, Springer-Verlag, Berlin, Chapter 1.

Gray, C.G., (1978), "Multipole expansions of electromagnetic field using Debye potentials", Amer. J. Phys. **46**, 169-179.

Green, H.S. (1952). "The molecular theory of fluids", John Wiley, New York.

Green, H.S. and Leipnik R.B., (1970), 'Sources of plasma physics", Vol. **1**, Wolters-Noordhoff, Groningen, Netherlands.

Griem, H. R. (1964), "Plasma spectroscopy," McGraw-Hill, New York.

Griem, H.R., (1966), "Optical radiation"; pages 278-297 in plasma physics in theory and application, edited by W.B. Kunkel, McGraw-Hill, New York.

Hagler, M. O., and Kristiansen, M. (1979) "An introduction to controlled thermonuclear fusion", Laxington Books, Health & Co. Laxington, Massachusetts.

Hanna, D.C., M.A. Yuratich and D. Coller (1979), "Nonlinear optics of free atoms and molecules", Springer-Verlag, Berlin.

Hartree, D.R. (i) (1929), "The propagation of electromagnetic waves in a stratified medium," Proc. Camb. Phil. Soc. **25**, 297; (ii) (1931), "The Propagation of electromagnetic waves in a reflecting medium in a magnetic field," Proc. Camb. Phil. Soc. **27**, 143.

Harwit, H., (1973), "Astrophysical concepts," John Wiley, New York.

Hasegawa, A. (1975), "Plasma instability and nonlinear effects," Springer Verlag, New York.

Hasegawa, A. and Uberoi, C. (1982). "The Alfven wave," Technical Informations Center, U.S. Department of Commerce, Springfield, Virginia.

Hauser, W. (1971), "Introduction to the principles of electromagnetism," Addison-Wesley, Reading.

Haynes, K. A. and D. Kahn (1965), "Wave propagation across a two plasma density discontinuity," Phep. Fluids **8**, 1681-1688.

Heald, M.A. and C. B. Wharton, (1965), "Plasma diagnestice with microwave," John Wiley, New York.

Heer, C. V. (1964), "Resonant frequencies of an electromagnetic cavity in an accelerated system of reference," Phys. Rew. **134**, A799-A804.

Helliwell, R. (1965), "Whistlers and related ionospheric phenomena," Standford University Press, Stanford.

Hellund, E. J. (1961) "The plasma staste," Reinhold, New York.

Hellwig, G. (1955), "Uberdie Bewegung geladner Teilchen in schwach veranderlichen Magnet feldern," z. f. Naturforsch **10a**, 508-516.

H.E., Knveptel magnetic field, Wiley, New York, 2000.

Hide, R. and P.H. Robertz (1962), "Some elementary problems in magnetohydrodynamics'' in Advances in Applied Mechanics. Vol. VII, Academic Press, New York, pp. 216-236, pp. 274-286.

Holt, E. H. and R. E. Haskell (1965), "Plasma dynamics," Macmillan, New York.

Hora, H. (1979). "Nonlinear plasma dynamics at laser irradiation" Springer-Verlag, Berlin.

Howatson, A. N. (1976). "An introduction to gas discharges,", Pergamon Press, Oxford.

Hughes, T.P., "Plasma and laser light " (1975), Adams Hilger, London

Hughes, W.F. and F. J. Young (1966). "The electrodynamics of fluids", John Wiley, New York.

Huschilt, J. and Baylis, W. E. (1978) "Rutherford scattering with radiation reaction," phys. Rev. **D17**, 985-993.

Hutchinson, I. H., (1987), "Principles of plasma diagnostics," Cambridge University Press, Cambridge.

Huxley, L. G. H. and R. W. Crompton, (1974). "The diffusion and drift of electrons in gases,", John Wiley, New York.

Huxley, L. G. H. and J. A. Ratlciffe (1949), "A survey of ionospheric cross-modulation" Proc. IEEE **96**, Part III, No. **43**, 433-449.

Ichimaru, S. (1973). "Basic principles of plasma physics," W. A. Benjamin, Reading, Mass.

Imre, K. (1962), "Oscillations in a relativistic plasma", Phys. Fluids **5**, 459-466.

Ise, J. Jr., (1957), "Relativistic electrodynamics of rotating media" UCRL-3636, pages 1-24.

Ise, J. and Uretsky, J. L., (1958), "Vaccum electrodynamics on a merry-go-round," Amer. J. Phys. **26**, 431-435.

Ivanenko, D. and Sokolov. A, (1949), "Klassicheskaya teoria polia" (In Russian), GITTL, Moscow.

Jackson, J.D. (1975), "Classical electrodynamics", Second edition, John Wiley, New York.

Jackson, E.A. (1960), "Nonlinear oscillations in a cold plasma" Phys. Fluids **3**, 831-833.

Jancel, R. and Th. Kahan (1966), "Electrodynamics of plasma", John Wiley, New York.

Jeans, J. (1929), "Astronomy and cosmogony", Cambridge University Press, Cambridge.

Jeffrey, A. (1966), "Magnetohydrodynamics", Oliver & Boyd, Edinburgh.

Jeffrey, A and T. Taniuti (1966), "Magnetohydrodynamic stability and thermonuclear containments: an introduction", Academic Press, New York.

John, F., (1970) "Partial differential equations"; pages 229-315 in Mathematics Applied to Physics, edited by E. Roubine, Springer-Verlag, Berlin.

Jones, R., (1980), "Laboratory plasma devices", Phys, Reports **61**, 298-325.

Jukes, J.D. (1967), "Magnetohydrodynamic stability theory", Report on Progress in Physics, Vol. XXX, The Institute of Phys. and Phys. Soc. London.

Kac, M. (1954), "Random walk and the theory of brownian motion", in N. Wax (editor). Selected papers on Noise and Stochastic Processes, Dver, New York, pp. 295-372.

Kadomtsev, B.B., (1965), "Plasma turbulence", Academic Press, London.

Kantrowitz, A, and H. E. Petschek (1966) "MHD characteristics and shock waves" in W.B. Kunkel (editor), Plasma Physics in Theory and Application, McGraw-Hill, New York, pp. 148-206.

Karpman, V. I. (1974), "Nonlinear waves in dispersive media," Pergamon Press, Oxford.

Katz, I, D.E. Parks and M. Rotenberg (1975), "Precession of high intensity elliptically polarized light in a plasma", Phys. Fluids **18**, 895-96.

Keen, B.E. (Editor) (1974), "Plasma physics," Culham Lectures, The Institute of Physics, London.

Kelso, J.M. (1964) "Radio ray propagation in the ionosphere", McGraw-Hill, New York.

Kendal, P.C., and Plumpton, C. (1964), "Magnetohydrodynamics", Vol. 1, Pergamon Press, Oxford.

Kennard, E.H. (1938), Kinetic theory of Gases, McGraw-Hill, New York.

Kerrebrock, J.L. (i) (1962), "Conduction of gases with elevated electron temperature", pages 327-346 in Proc. Second Symp. Eng. "Aspects of magnetohydrodynamics", edited by C. Mannel and N.W. Mather, Columbia Press, New York; (ii) (1964) "Nonequilibrium Ionization due to Electron Heating". I. Theory, AIAAJ **2**, 1072-1080; (iii) (1965), Magnetohydrodynamic Generators with non equilibrium Ionization; AIAAJ, **3**, 591-601.

Khan, M.R. and Chakraborty, B. (1979). "Precessional rotation and other nonlinear effects due to standing waves, and other two wave resonant interactions in a magnetized plasma", Journ. Applied Phys. **50**, 7977-7985.

Khan, M.R., (1980), "Some problems of wave processes in plasma"; Ph.D. thesis, Jadavpur University, Calcutta.

Khromykh, S.M. (1966), "Ring generator in a rotating reference system", Sov. Phys. J.E.T.P. **23**, 281-283.

Kibble, T.W.B. (1966), (i) "Refraction of electron beams by intense electromagnetic waves", Phys. Letters **16**, 1054-1057, (ii) "Mutual Refraction of Electrons and Photons", Phys. Rev. **150**; 1060-1069.

Kittel, C. (1979), "Introduction to solid state physics", John Wiley, New York.

Klimontovich, Yu. L, (1962), "The statistical theory of nonequilibrium processes in a plasma", Pergamon Press, New York.

Kline, M. and I.W. Kay (1965), "Electromagnetic theory and geometric optics", Inter Science. New York.

Knoepfel, H.E. ''Magnetic field'', Wiley, New York, 2000.

Kraichnan, R.H. and D. Montgomery, (1980), "Two-dimensional turbulence", Rep. Prog. Phys. **43**, 547-619.

Krall, N.A. and A.W. Trivelpiece (1973), "Principles of plasma physics", McGraw-Hill, New York.

Kristein. P.T., G.S. Kino and W.E. Waters (1967), "Space-charge flow,"McGraw-Hill, New York.

Kritz, A.H. and D. Mintzer (1960), "Propagation of plasma waves across a density discontinuity", Phys. Rev. **117**, 382-386.

Kruer, W.L. (1988), "The physics of laser plasma interactions", Addison-Wesley, New York.

Kruskal, M. (1964), "The Gyration of a charged particle'', in advanced plasma theory, M. Rosenbluth (editor), pp. 260-266, Academic Press, New York.

Kruskal, M. (1957), "The gyration of charged particles", Proc. III Int. Conf. on ionization Phenomena, Venice, pages 562-568.

Kruskal, M. (1962), "Asymptotic theory of hamiltonian and other systems with all solutions nearly periodic", J. Math. Phys. 3, 806-828.

Kulikovskiy, A.G. and G.A. Lyubimov (1965), "Magnetohydrodynamics", Addision-Wesley, Mass.

Kulsrud, R.M. (1957), "Adiabatic invariants of the harmonic oscillator", Phys. Rev. **106**, 205-207.

Kulsrud, R.M. (1964), "Introduction for papers on adiabatic invariance", in Advanced Plasma Theory, M. Rosenbluth (editor), Academic Press, New York.

Kulsrud, R. (1964), "General stability theory in plasma physics" in M. N. Rosenbluth (editor), Advanced Plasma Theory, Proc. Internat. School of Phys.," Enrico Fermi" course XXXV, Academic Press, New York, pp. 54-96.

Kunkel. W. B. (Editor) (1966), "Plasma physics in theory and applications,", McGraw-Hill, New York.

Lai, C.S. and Wonnacott, E. (1976), "Frequency shift and precession of elliptically polarized electromagnetic waves in plasma," Canad. J. Phys. **54**, 962-966.

Laing, E. W., "Plasma physics, "Sussex University Press, London, 1976.

Lamb, Jr., W.E. (1964), "Theory of optical maser." Phys. Rev. **A134**, 1429-1450.

Landau, L.D. and E. M. Lifshitz (1960), "Electrodynamics of continuous media,", Addison-Wesley, Reading, Mass.

Landau, L.D. and E. M. Lifshitz (1975), "The classical theory of fields," 4th edition, Pergamon Press, Oxford.

Landau, L. D. (1946), "On the vibrations of the electron plasma", Phys. J., U.S.S.R., **10**, 25-34.

Landau, L.D. and E.M. Lifshitz, (1978), "Mechanics," Pergamon Press, Oxford.

Landshoff, R.K.M. (Editor) (1957), "Magnetohydrodynamics," Stanford University Press, Stanford, California.

Larsson, J. and L. Stenflor (1974), "Threshold fields of parametric instabilities", Beitr. Plasma Phys. **14**, 7-14.

Lawson, J. D. (1957), "Some criteria for power producing thermonuclear reactions", Proc. Phys. Soc. **B70**, 5-10.

Lass, H. (1950), "Vector and tensor analysis," McGraw-Hill, New York.

Layzer, D. and Menzel, D.H. (1965), "Ionospheric crossmodulation : A microscopic theory," Radio Science **69D**, 59-68.

Lehnert, B. (1954), "Magnetohydrodynamic waves under the action of Coriolis force," I. Astrophys. Journ. **119**, 647.

Lehnert, B. (1964), "Dynamics of charged particles," North Holland, Amsterdam.

Lehnert, B. (1971), "Rotating plasmas," Nuclear Fusion, **11**, p. 485-533.

Leontovich. M.A. (1965-80), "Reviews of plasma physics," Vol. I-VIII Consultants Bureau, New York.

Leubner, C. and F. Ehlotzky, (1978). "The intensity dependent frequency shift in Thomson scattering from a thermal plasma," Journ. Phys. A. : Math. Gen. **11**, 759-763.

Linhart, J.G. (1961), "Plasma physics," North Holland, Amsterdam.

Lochte-Holtgreven. W. (Editor) (1968), "Plasma diagnostics," North Holland, Amsterdam.

Loeb. LB. (1964), "Basic processes of gaseous electronics," University of California, Berkeley, California.

Longmire, C.L. (1963), "Elementary plasma physics," John Wiley, New York.

Lorentz, H. A. (1915), "Theory of electrons'', Reprint 1952, Dover Book, New York.

Lundquist, S. (1951), "On the stability of magneto hydrostatic fields, " Phys. Rev. **83**, 307.

Lundquist, S. (1959), "Particle view compared with hydrodynamics" Rendi Conti, Course XIII, Academic Press, New York.

Macek, W.M. and Davis, D.T.M. (1977), "Rotation rate sensing with travelling wave ring lasers", Appl. Phys. Letters **2**, 67-68.

Macek, W.M., Schneider, J.R. and Salamon, R.M., (1964), "Measurement of Fresnel drag weith ring laser", J. Appl. Phys. **35**, 2556-2557.

Macek, W.M. and McCartney, E.J. (1966), "The ring laser"; Sperry Eng. Rev. **19**, 8.15.

Maitland, A. and Dunn, M.H. (1969), "Laser physics", North Holland, Amsterdam.

Maker, P.D. and Terhune, R.W. (1965), "Study of optical effects due to an induced polarization of third order in the electric field strength", Phys. Rev. **A137**, 801-818.

Maker, P.D., Terhune, R.W. and Savage, C.M., (1964), "Intensity-dependent changes in the refractive index of liquids", Phys, Rev. Let. **12**, 507-509.

Mandel, shtam, L.I. (1951), "Collected works" 1, Acd. Nauk, USSR (original paper in Russian, 1924).

Mathews, J. and R.L. Walker, (1970), "Mathematical methods of physics"; Benjamin, Inc., California.

Max, C.E. (1973), "Steady-state solutions for relativistically strong electromagnetic waves in plasmas", Phys. Fluids **16**, 1277-1288.

Melrose, R.H. (1980) "Plasma astrophysics" Vol. I & II, Gordon & Breach, New York.

McNamara, B. (1977), "Single particle behaviour in plasmas," Lecture notes in College in Theoretical and Computational Plasma Physics (March-April, 1977), ICTP, Trieste Italy).

McWane, P.D. and Sealer, D. A. (1966), "New measurements of intensity-dependent changes in the refractive index of liquids," Appl. Phys. Lett. **8**, 278-279.

Mendonca, J.T. (1979), "Nonlinear interaction of wave packets", J. Plasma Phys. **22**, 15-26.

Menegozzi, L.N. and Lamb. Jr., W.E., (1972), "Theory of a ring laser", Phys. Rev. A, **8**, 2103-2125.

Menzel, D.H. (1965), "Some problems of ionospheric nonlinearities", Radio Science, **69D**, 1-7.

Michelson, D. R. (1983), "Introduction to plasma theory," John Wiley, New York.

Mikhailovskii, A. B. (1974), "Theory of plasma instability," Vols. I & II, Consultants Bureau, New York.

Minorsky, N. (1962), "Nonlinear oscillations," D. Van Nostrand, Princeton.

Mintzer, D. (1964), "Transport theory of gases", Chapter 1 in H. Margenau and G. M. Murph (editors), Mathematics of Physics and Chemistry, Vol. 2, D. Van Nostrand, Princeton.

Mitchner, M. (Editor) (1961), "Radiation and waves in plasmas," Stanford University Press, Standford, California.

Mitra, S.K. (1947), "The upper atomosphere," Asiatic Society of Bengal, Calcutta.

Miyamoto. K. (1980), "Plasma physics for nuclear fusion," The MIT Press Cambridge, Mass.

Modesitt, G.E. (1970), "Maxwell equations in a rotating reference frame," Amer, J. Phys. **38**, 1487-1489.

Moffat, H.K., (1978), "Magnetic field generation in electrically conducting fluids," University Press, Cambridge.

Moller, C. (1972), "The theory of relativity," Oxford University Press, London.

Molmud, P., Altshuler, S. and Gardner, J. H. (1962), "The study of a method to reduce electron density in the ionospheric D-region by means of high powered ground based transmission," Space and Technology Labs. Inc. Final Reprots.

Montgomery, D., C. S. Liu and G. Vahala (1972), "Three dimensional plasma diffusion in a very strong magnetic field", Phys. Fluids, **15**, 815-19.

Montgomery, D.C. and Tidman, D. A. (1964), "Plasma kinetic theory," McGraw-Hill, New York.

Moreau, R. (1970), "On magnetohydrodynamic turbulence"; pages 359-372 in Proceedings of the symposium on Turbulence of Fluids, and Plasmas, New York, April 16-18, 1950; Polytechnic Press, Brooklyn, New York.

Morin, S. and D. Kahn (1968). "Wave coupling across a density discontinuity in a magnetized plasma" Phys. Fluids **11**, 2627-2634

Morozov, A.I. and L.S. Soloviev (1966), "The structure of magnetic fields", in M.A. Leontovich (editor), Reviews of Plasma Physics, Consultants Bureau, New York, Vol. 2, pp. 1-101.

Morozov, A.I. and L.S. Soloviev, (1966a), "Motion of charged particles in electromagnetic fields", Reviews of Plasma Physics, Vol, 2, pages 101-197; edited by M.A. Leontovich, Consultants Bureau, New York.

Motz, H. (1979), "The physics of laser fusion," Academic Press, London.

Mulser, P., R. Sigel and S. Witkowski (1973), "Plasma production by laser", Phys. Reports, **6**, 187-239.

Nasser. E. (1971), "Fundamentals of gaseous ionization and plasma electronics", Wiley-Interscience, New York.

Nayfeh, A.H. (1979), "Nonlinear oscillations", John Wiley, New York.

Nedospasov, A.V. (1977), "The physics of MHD generators", Sov. Phys. Uspekhi, **20**, 861-869.

Neufeld, J. (1964), "Interaction of a stationary plasma with an electron beam", Phys. Fluids, **7**, 306-318.

Neufeld, J. and Doyle, P.H. (1961), "Electromagnetic interaction of a charged particle with a plasma", Phys. Rev. **121**, 654-658.

Nicholson, D.D. (1983), "Introduction to plasma theory;" John Wiley, New York.

Nicolet, M. (1959) "Collision frequency of electrons in the terrestrial atmosphere", Phys. Fluids **2**, 95-99.

Nijboer, B.R.A. (1962); "General introduction in fundamental problems in statistical mechanics,", editor E.G. Cohen, North-Holland, Amsterdam.

Nishikawa, K and M. Wakatani, (1990), "Plasma physics," Springer-Verlag, Berlin.

Nishikawa, K, and Wakatani W (1990), "Plasma physics, basic theory, with fusion application'', Springer-Verlag, Berlin.

Nishikawa, K. (1968), "Parametric excitation of coupled waves, I. general formulation", J. Phys. Soc. Jap. **24**, 916-922.

Nishikawa, K. and Liu, C. S. (1976), "General formalism of parametric excitation"; pages 1-81, in Advances in Plasma Physics, Vol. 6, edited by A. Simon and W.B. Thompson, Interscience, Now York.

Northrop, T. G. (1963). "Adiabatic invariance of the charged particles, " Inter Science, John Wiley, New York.

Northrop, T.G. (1961). "The guiding center approximation to a charged particle motion", Ann. Phys. (New York), **15**, 70-101.

Northrop, T.G. (1966) "Adiabatic charged particle motion," Ch. 2 in Plasma Physics in Theory and Application, edited by W.B. Kunkel, McGraw-Hill, New York.

Owyoung, A., Hellworth, R.W. and George, N., (1972), "Intensity-induced changes in optical polarization in glasses'', Phys. Rev. **85**, 628-632.

Pai, Shih, I. (1962). "Magnetogasdynamics and plasma dynamic", Prentice Hall, Englewood Cliffs, N.J.

Panofsky, W.K.H. and M. Phillips (1964), "Classical electricity magnetism", Addison-Wesley, Heading, Mass.

Parker, E.N. (1975), "Newtonian development of the dynamic properties of ionized gases of low density", Phys. Rev. **107**, 924-933.

Parker, E.N. (1979) "Cosmical magnetic field", Clarendon Press, Oxford.

Paul, J.W.M., (1970), "Collisionless shock waves;" pages 302-345 in "Physics of hot plasmas", edited by B.J. Rye and J.C. Taylor, Oliver & Boyd, Edinburgh.

Paul, S.N. and B. Chakraborty (1983a) "Some aspects of transformation of nonlinear plasma equations to the space-independent frame," Physica B and C, **122C**, 139-149.

Paul, S.N. and B. Chakraborty (1983b). "Some applications of transfromation to the space-independent frame for strong radiation in plasma," Austral. Jour, Phys. **36**, 867-890.

Pershan, P.S. (1963), "Nonlinear optical properties of solids : Energy considerations," Phys. Rev. **130**, 919-929.

Pershan, P.S., van der Ziel, J.P. and Malmstrom, L.D. (1966), "Theoretical discussion of the inverse Faraday effect, Raman Scattering, and related phenomena", Phys. Rev. **142**, 574-583.

Phillips. N. J. and J.J. Sanderson (1966), "Trapping of electrons in a spatially inhomogeneous laser beam", Phys. Letters, **21**, 535.

Pomeau, Y. and E. Quemada, (1967), "Effect Faraday inverse dans un plasma, " Compt. Rend. **24**, 517-518.

Portis, A.M., (1978). "Electromagnetic fields, sources and media," John Wiley, New York.

Post, E.J. (1986) "Linking and enclosing magnetic flux." Phys. Lett **A119**, 47-49.

Post, E.J. and Bahulikar, D. D., (1971), "Note on electrodynamics of accelerated systems", J. Math. Phys. **12**, 1093-1099.

Post, E.J. and Yildiz. A. (1965), "Cavity resonance in accelerated systems", Phys. Rev Lett. **15**, 177-178.

Post, E.J., (1972a), "Non-intertial phenomena and inertial sensing," klein heubacher Ber. Ferrmeldetech. Zentralamt, Damstadt, Germany, **15**, 183-195.

Post, R.F. and Ribe, F.L. (1974), "Fusion reactors as future energy sources'' (1974), Science **186**, 397-407.

Post, E. J. (1972), "Interferometric path-length changes due to motion", Journ. Opt. Soc. Amer. **62**, 234-239.

Post, E. J., (1967), "Sagnac effect," Rev. Mod. Phys. **39**, 475-493.

Poverlain, H., (1962), "Sommerfeld-runge law in three and four dimensions,", Phys. Rev. **128**, 957-964.

Poynting, J. H. (1884). "On the transformation of energy in the electromagnetic field," Phil. Trans. CLXXV, 343.

Privalov, V.E. and Fridrikhov. E.A., (1969). "The ring gas laser", Sov. Phys. Uspekhi, **12**, 153-167 original Russian: Uspekhi **97**, New Delhi, **469** 377.

Rai Chowdhury, A., "The physics of fluid and plasmas", Cambridge University Press, Cambridge (1998).

Ramsden, S.A., (1970), "Laser produced plasmas"; pages 346-362 in Physics of hot plasmas, edited by B.J. Rye and J.C. Taylor, Oliver & Boyd, Edinburgh.

Ratcliffe, J.A., (1959). "The magneto-ionic theory", Cambridge University Press, London.

Ratcliffe, J. A. (1973), "Introduction to the ionosphere and magnetosphere", Cambridge University Press, London.

Rawar, K. and Suchy, K., (1967), "Radio-observations of the ionosphere", handbuch der Physil, Band XLIX/2, Springer-Verlag, Berlin.

Ray, B.S. (1957), "Lectures on advanced electricity and magnetism", Calcutta University (unpublished).

Rishbeth, H. and Garriott. O.K., (1969), "Introduction to ionospheric physics", Academic Press, New York.

Roberts, P.H. (1967) "An introduction to magnetohydrodynamics", Longman, London.

Robinson, F.N.H., (1973), "Macroscopic electromagnetism", Pergamon Press, Oxford.

Rosa, R.J. (1966), "Power conversion" in "Plasma physics in theory and applications", editor W.B. Kunkel, McGraw-Hill, New York.

Rosa, R.J. (1968), "Magnetohydrodynamics energy conversion", McGraw-Hill, New York.

Rose, D.J. and M. Clark (1961), "Plasma and controlled fusion", MIT Press, Mass.

Rosenbluth, M.N. and R.S. Sagdeev, (1983) (editors); "Handbook of plasma physics," Vol. I, II, North Holland, Amsterdam.

Rosenbluth, M.N. (1957), "Dynamics of a pinched gas" in magnetohydrodynamics, Editor R.K. M. Landshoff, Standford University Press, California.

Rosenbluth, M. (1960), "Introduction to plasma physics," Riso Report No. **18** DAEC, Riso, Denmark.

Rosenbluth, M. and C. L. Longmire (1957), "Stability of plasmas controlled by magnetic fields"; Ann. Phys. New York **1**, 120-140.

Rosenbluth, M.N. (1960), "Pinch dynamics", Riso Report No. **18**, DAEC, Riso Denmark.

Rosenbluth, M.N. and R.Z. Sagdeev (1972), "Laser fusion and parametric instabilities" Comments on Plasma Physics and Controlled Fusion **1**, 129-138.

Rushbridge, M.G., (1970), "Turbulence"; pages 256-275, in Physics of hot Plasmas, edited by B.J. Rye and J.C. Taylor; Oliver & Boyd, Edinburgh.

Russakoff, G., (1970) "A derivation of the macroscopic Maxwell equations," Amer. J. Phys. **38**, 1188-1195.

Sagdeev, R. Z. and A. A. Galeev (1969), "Nonlinear plasma theory," Benjamin, New York.

Saha, M. N. (1921), "On a physical theory of stellar spectra," Proc. Roy. Soc. (London) **A 99**, 135-153.

Samaras, D. G., (1971), "Theory of ion flow dynamic," Dover, New York.

Sard, R. D., (1970), "Relativistic mechanics, W. A. Benjamin, New York.

Schelkunoff, S. A., (1963), "Electromagnetic fields," Blaisdell, New York.

Scheuer, P. A. G., (1967), "Radiation"; pages 289-306 in Plasmas Astrophysics, edited by P. A. Sturrock, course XXXIX, IPS, Academic Press, New York.

Schiff, L. I. (1939), "A question in general relativity", Proc. Nat Acad. Sci, U.S. **25**, 391-395.

Schluter, A., (1960), "The constancy of the orbital magnetic moment in a time varying magnetic field", Riso Report No. **18**, Riso, Denmark.

Schmidt, G., (1979), "Physics of high temperature plasma", Academic Press, New York, (Second edition).

Schott, L. (1968), "Electrical Probes"; pages 668-731 in Plasma diagnostics, edited by W. Lochte Holtgreven, North Holland, Amsterdam.

Seely, S. (1958) "Introduction to electromagnetic field", McGraw Hill, New York.

Sen Gupta, N.D., (1966), "Thomson scattering of high intensity beam with radiation reaction", Zeitzsohrift für Physik, **196**, 385-392.

Seshadri, S. R., (1973), "Fundamentals of plasma physics", American Elsevier, New York.

Shafranov, V.D. (1958), "On magnetohydrodynamical equilibrium configuration", Sov. Phys,. JEPT, **6**, 545.

Shafranov, V.D. (1966), "Plasma equilibrium in a magnetic field" in Reviews of plasma physics, Vol. 2., editor M. A. Leontovich. pp. 103-151, Consultants Bureau, New York.

Shakarofsky, I.P., Johnston. T.W. and Bachynski. M., (1966). "The particle-Kinetics of plasmas", Addison-Wesley, Reading, Mass.

Shercliff, J.A., (1965), "A text book of magnetohydrodynamics", Pergamon Press, Oxford.

Shercliffe, J.A. (1958), "The flow of conducting fluids in circular pipes under transverse magnetic fields", J. Fluid, Mech. **1**, 644-660.

Shiozawa, T., (1975), "Phenomenological and electron theoretical study of the electrodynamics of rotating systems", Proc. IEEE, **61**, 1624-1707.

Shohet, J.L., (1978), "Advances in plasma heating with Alfven waves", Comments on Plasma Phys. and Controlled Fusion, **IV**, 37-53.

Silverstein, L. (1907), Annalender physik, **22**, 579; **24**, 784.

Simon, A. and Thompson, W. B. (1968-1975), "Advances in plasma physics," Vols. I-V, Wiley-Interscience, New York.

Simon, A. L. (1960), "An introduction to thermonuclear research", Pergamon Press, New York.

Slater, J.C., (1950), "Microwave electronics", D. van Nostrand, Princeton.

Sluijter, F.W. and D. Montgomery (1965), "Comments on secular and nonsecular behaviour for the cold plasma equations", Phys. Fluids, **8**, 551.

Smirnov, B.M., (1977), "Introduction to plasma physics", Mir Publishers, Moscow.

Sokolov, A. A., and Ternov. I.M. (1968), "Synchrotron radiation", Pergamon Press, Oxford.

Sommerfeld, A. (1952), "Electrodynamics", Academic Press, New York.

Sommerfeld, A. (1954), "Optics", Academic Press, New York.

Sommerfeld, A., (1934), "Atomic structure and spectral lines", Methuen, London.

Spitzer, L. (1956), "Physics of fully ionized gases", John Wiley, New York.

Stacey, W. M.; (1981), "Fusion plasma analysis", John wiley, New York.

Steele, M. C., and Vural, B. (1969) "Wave interactions in solid state plasmas," McGraw-Hill, New York.

Steiger, A. D. and C. H. Woods, (1972), "Intensity-dependent propagation characteristics of circularly polarized high power laser radiation in a dense electron plasma", Phys. Rev. **A5**, 1467-1474.

Stewart, J. M., (1977), "On transient relativistic thermodynamics and Kinetic theory", Proc. Roy. Soc., Lond. **A357**, 59-75.

Stix, T. H. (1962), "The theory of plasmas waves," McGraw-Hill, New York.

Stormer, C. (1955), "The polar aurora," Clarendon Press, Oxford.

Storey, L.R.O. (1953), "An investigation of whistling atmosphere," Phil. Trans. Roy. Soc.London **A245**, 113-141.

Stratton, J.A. (1941), "Electromagnetic theory," McGraw-Hill, New York.

Sturrock, P.A. (1957), "Nonlinear effects in electron plasmas", Proc. Roy. Soc. London, **A242**, 277-299.

Sturrock, P.A., (1967), "Waves in plasmas"; pages 38-43 Plasma Astrophysics, Italian Physical Society, international School of Physics, course **39**; Academic Press, New York.

Suchy, K. (1972), "Attenuation of waves in plasmas", Radio Science, **7**, 871-884.

Sutton, G. and A. Sherman (1965), "Engineering magnetohydrodynamics," McGraw-Hill, New York.

Sutton, G. W. and Gloersen. P., (1962), "Magnetohydrodynamic power and propulsion"; pages 243-268 in Magnetohydrodynamics, edited by A. B. Cambel, T.P. Anderson and M. M. Slawsky, North-Western University Press, Evanston.

Synge, J.L. (1957), "The relativistic gas," North-Holland, Amsterdam.

Talin, B., V. P. Kaftandjan and L. Klein, (1975), "Inverse Faraday effect in plasmas." Phys. Rev. **A11**, 648-657.

Tamm, I. E. (1979), "Fundamentals or the theory of electricity," Mir Publishers, Moscow.

Tanenbaum, B. S., (1967), "Plasma physics", McGraw-Hill, New York.

Tanenbaum, B.S., (1968), "Continuum theory of Thomson scattering," Phys. Rev. **171** 215-221.

Tatarskii, V. I. (1960), "Wave propagation in a turbulent medium," McGraw-Hill, New York

Taylor, R. J. (1970), "Plasma waves and oscillations" in physics or hot plasma, editors B. J. Rye and J. C. Taylor pp. 103-144, Oliver & Boyd, Edinburgh.

Tellegen, B.D.H. (1933), "Interaction between radio waves," Nature **131**, 840.

Thome, R. J. and Terrh. J. M. (1982), "MHD and fusion magnets," Wiley Interscience, New York.

Thompson, W. B. (1957), "Thermonuclear reaction rates", Proc. Roy. Soc., London, **B70**, 1-5.

Thompson, W.B. (1964), "An introduction to plasmas physics," Pergamon Press, Oxford.

Thompson, W.B., (1970), "Introduction to kinetic theory of plasma"; Pages 1-141 in Physics of hot plasmas, edited by Rye, B. J. and Taylor, J. C., Oliver & Boyd, Edinburgh.

Tidman, D. A. and Stainer, H. M. (1965). "Frequency and wave number shift for waves in a hot plasma" Phys. Fluids **8**, 345-353.

Tidman, D. A. (1960) "Radio emission by plasma oscillations in nonuniform plasma." Phys. Rev. **117**, 366-374.

Tidman, D. A. and G.H. Weiss (1961), (i) "Radio emission by plasma oscillations in nonuniform plasmas," Phys. Fluids **4**, 703-710; (ii) "Radiation by large Amplitude Oscillations", Phys. Fluids **4**, 866-868.

Tidman, D. A. and Krall. N. A., (1971), "Shock waves in collisionless plasma," John Wiley, New York.

Tonks, L. and Langmuir, I. (1929), "Oscillations in ionized gases". Phys. Rev. **33**, 195-210.

Townsend, J. S. (1947), "Electrons in gases,", Hutchinson, London.

Trocheris, M.G., (1947), "Electrodynamics in a rotating frame of reference," Phil, Mag. **40**, 1143-1154.

Trubnikov, B. A. (1955), "Particle interaction in a fully ionized plasma", in Reviews of Plasma Physics, Vol. **1**, editor M.A. Leontovich, pp. 105-204, Consultants Bureau, New York.

Tsytovich, V. N. (1972), "An introductison to the theory of plasma turbulence,'' Pergamon Press, Oxford.

Tsytovich, V. N. (1977), "Theory of turbulent plasmas", Consultants Bureau, New York.

Tsytovich, V. N., (1970), "Nonlinear effects in plasma,", Plenum Press, New York.

Uberoi, C. (1988), "Introduction to unmagnetized plasmas," Prentice Hall of India, New Delhi.

Uman, M. A. (1962), "Introduction to plasma physics," McGraw-Hill, New York.

Vahala, G. M. (1972), "Kinetic theory of a two-dimensional magnetized plasma", Ph. D. Thesis, University of Iowa.

Van der Pol, B. (1935), "Interaction of radio waves," Tijdschrift Van het Nederlandsch Radiogenootschap 7, 93.

Van Kampen, N. G. and B. U. Felderhoff (1967), "Theoretical methods in plasma physics," North-Holland, Amsterdam.

Vandenplas, P. E. (1968), "Electron waves and resonances in bounded plasmas," John Wiley, New York.

Vander Ziel, J. P., P. S. Pershan and L. D. Malmstron, (1965), "Optically-induced magnetization resulting from the inverse Faraday effect," Phys. Rev. Lett. **15**, 190-193.

Vedenov, A. A., (1968), Theory of turbulent plasma," Illife Books, London.

Vander Pol, B. and Vander Mark (1935), "Interaction of radio waves," Tijdscrift Van het Nederlandsch Radiogenootschep, **7**, 12.

Vlasov, A. A. (1938), "The vibrational properties of an electron gas", JETP, **8**, 291.

Vlasov, A. A. (1950) "Teoria Mnogikh chastits" (In Russian), State Publishing House, Moscow; English translation: (1961), "Many Particle theory and Its Application to Plasmas", Gordon & Breach, New York.

Vlasov, A. A. (1966), "Statistical functions of distribution," Nauka Publication, Moscow (In Russian).

Volk, H., (1970), "Collisionless shocks": pages 276-30 in Physics of Hot Plasmas, edited by B. J. Rye and J. C. Taylor, Oliver & Boyd, Edinburgh.

Volkov, A.M., and Kisilev. V. A. (1970), "Proper frequencies of a rotating ring resonator", Sov. Phys. JETP, **30**, 733-736.

Von Kerozek C., (1976), "Stability of modulated plane Couette flow," Phys. Fluids, **19**, 1288-1295.

Wait, J. R. (1962), "Electromagnetic waves in stratified media" Pergamon Press, Oxford.

Wang, C. C. (1966), "Nonlinear susceptibility constants and self focussing of optical beams in liquids," Phys. Rev. **152**, 149-156.

Werson, J. A. (1974), "Plasma stability theory"; Ch. 4 in Plasma Physics, edited by B. E. Keen (Culham Lectures). The Institute of Physics, London.

Wesson, J. (1987), "Tokamak," Clarendon Press, Oxford.

Wharton, C. B. and A. W. Trivelpiece, (1966), "Waves in laboratory plasmas", in plasma physics in theory and application, edited by W.B. Kunkel, Chapter 8, pages 232-273, McGraw-Hill, New York.

Whitham, G. B. (1965), "Nonlinear dispersive waves, Proc. Roy. Soc. London", **A283**, 238-261.

Whitham, G. B. (1965), "A general approach to linear and nonlinear dispersive waves using a Lagrangian, " J. Fluid Mech. **21**. 273-283.

Whitham, G. B., (1974), "Linear and nonlinear waves"; John Wiley, New York.

Wilhelmsson, H. (1973), "On the explosive instabilities of waves in plasmas with special regard to dissipation and phase effects," Physica Scripta **7**, 209-212.

Winkles, B. B. and O. Eldridge (1972), "Self consistent electromagnetic waves in relativistic Vlasov plasmas", Phys. Fluids **15**, 1790-1800.

Womack, G.J. (1969), MHD "Power generation", Chapman and Hall, London.

Wong, G. K. L. and Shen. Y. R. (1974), "Study of pretransitional behaviour of laser-induced molecular alignment in isotropic nematic substances," Phys. Rev. **A10**, 1278-1289.

Woods, L. C. (1987), "Principles of magnetoplasma dynamics," Oxford University Press, Oxford.

Woods, L. C. (1993). "An introduction to the Kinetic theory of gases and magnetoplasmas", Oxford University Press, Oxford.

Wu, T.Y. (1966), "Kinetic equations of gases and plasmas", Addison-Wesley, Reading, Mass.

Yeh, K.C. and Liu, C.H. (1972), "Theory of ionospheric waves", Academic Press, New York.

Yildiz, A and Tang, C.H. (1966), "Electromagnetic cavity resonance in an accelerated system of reference", Phys. Rev. **146**, 947-954.

Yoler, Y.A. (1961), "A review of magnetohydrodynamics", in plasma physics, editor J. E. Drummond, McGraw-Hill, New York, Chapter 6.

Zernike, F. and Midwinter, J.E., (1972), "Applied nonlinear Optics", John Wiley, New York.

Zhevakin, S. A. and V.M. Fain (1956), "The theory of nonlinear effects in the ionosphere", Sov. Phys. JETP, **3**, 417-425.

INDEX

A

B

C

D

E

F

G

H

I

Q

R

S

W